Essential Mathematics for Convex Optimization

With an emphasis on the timeless essential mathematical background for optimization, this textbook provides a comprehensive and accessible introduction to convex optimization for students in applied mathematics, computer science, and engineering. Authored by two influential researchers, the book covers both convex analysis basics and modern topics such as conic programming, conic representations of convex sets, and cone-constrained convex problems, providing readers with a solid, up-to-date understanding of the field. By excluding modeling and algorithms, the authors are able to discuss the theoretical aspects of convex optimization in greater depth. Over 170 in-depth exercises provide hands-on experience with the theory, while over 30 "Facts" and their accompanying proofs enhance the approachability of the text. Instructors will appreciate appendices that cover all the necessary background and the instructors-only solutions manual provided online. By the end of the book, readers will be well equipped to engage with state-of-the-art developments in optimization and its applications in decision-making and engineering.

FATMA KILINÇ-KARZAN is a Professor of Operations Research at Tepper School of Business, Carnegie Mellon University. She was awarded the 2015 INFORMS Optimization Society Prize for Young Researchers, the 2014 INFORMS JFIG Best Paper Award (with S. Yıldız), and an NSF CAREER Award in 2015. Her research focuses on the foundational theory and algorithms for convex optimization and structured nonconvex optimization and their applications in optimization under uncertainty, machine learning, and business analytics. She has been an elected member on the councils of the Mathematical Optimization Society and INFORMS Computing Society and has served on the editorial boards of the MOS-SIAM Book Series on Optimization, MathProgA, MathOR, OPRE, SIAM Journal on Optimization, IJOC, and OMS.

ARKADI NEMIROVSKI is the John P. Hunter, Jr., Chair and Professor of Industrial and Systems Engineering at the Georgia Institute of Technology. He has co-authored six optimization textbooks including this one, and he has received many rewards for his contributions to the field, including the 1982 MPS-SIAM Fulkerson Prize (with L. Khachiyan and D. Yudin), the 1991 MPS-SIAM Dantzig Prize (with M. Grotschel), the 2003 INFORMS John von Neumann Theory Prize (with M. Todd), the 2019 SIAM Norbert Wiener Prize (with M. Berger), and the 2023 WLA Prize (with Yu. Nesterov). His research focuses on convex optimization, algorithmic design and complexity analysis, optimization under uncertainty, engineering applications, and nonparametric statistics. He is a member of the National Academy of Engineering, the American Academy of Arts and Sciences, and the National Academy of Sciences.

"This new book by Fatma Kılınç-Karzan and Arkadi Nemirovski is an important contribution to the field of optimization, offering valuable insights for both theoretical research and practical applications. This thorough volume starts with the basics of convex analysis and extends to recent developments in cone-constrained convex problems. The authors include many interesting exercises that help expand on the topics discussed. Additionally, the appendices contain useful supplementary materials that enhance the overall value of the book."

Yurii Nesterov, Professor at Corvinus University of Budapest;
Emeritus Professor at Catholic University of Louvain, Belgium

"This is a well-structured textbook on the mathematical foundations of convex optimization. It focuses on the structure of convex sets and functions, separation theorems, subgradients, and the theory of duality. The treatment is rigorous but readable, balancing clarity with depth."

Osman Güler, University of Maryland, Baltimore County

Essential Mathematics for Convex Optimization

Fatma Kılınç-Karzan
Carnegie Mellon University

Arkadi Nemirovski
Georgia Institute of Technology

Shaftesbury Road, Cambridge CB2 8EA, United Kingdom

One Liberty Plaza, 20th Floor, New York, NY 10006, USA

477 Williamstown Road, Port Melbourne, VIC 3207, Australia

314–321, 3rd Floor, Plot 3, Splendor Forum, Jasola District Centre, New Delhi – 110025, India

103 Penang Road, #05–06/07, Visioncrest Commercial, Singapore 238467

Cambridge University Press is part of Cambridge University Press & Assessment, a department of the University of Cambridge.

We share the University's mission to contribute to society through the pursuit of education, learning and research at the highest international levels of excellence.

www.cambridge.org
Information on this title: www.cambridge.org/highereducation/isbn/9781009510523
DOI: 10.1017/9781009510561

© Fatma Kılınç-Karzan and Arkadi Nemirovski 2026

This publication is in copyright. Subject to statutory exception and to the provisions of relevant collective licensing agreements, no reproduction of any part may take place without the written permission of Cambridge University Press & Assessment.

When citing this work, please include a reference to the DOI 10.1017/9781009510561

First published 2026

A catalogue record for this publication is available from the British Library

A Cataloging-in-Publication data record for this book is available from the Library of Congress

ISBN 978-1-009-51052-3 Hardback

Additional resources for this publication at www.cambridge.org/KKN

Cambridge University Press & Assessment has no responsibility for the persistence or accuracy of URLs for external or third-party internet websites referred to in this publication and does not guarantee that any content on such websites is, or will remain, accurate or appropriate.

For EU product safety concerns, contact us at Calle de José Abascal, 56, 1°, 28003 Madrid, Spain, or email eugpsr@cambridge.org

Contents

Preface			*page* xiii
Main Notational Conventions			xvii

Part I	**Convex sets in R^n: From First Acquaintance to Linear Programming Duality**		1
1	**First Acquaintance with Convex Sets**		3
	1.1	Definition and Examples	3
		1.1.1 Affine Subspaces and Polyhedral Sets	3
		1.1.2 Unit Balls of Norms	5
		1.1.3 Ellipsoids	7
		1.1.4 Neighborhood of a Convex Set	7
	1.2	Inner Description of Convex Sets: Convex Combinations and Convex Hull	7
		1.2.1 Convex Combinations	7
		1.2.2 Convex Hull	8
		1.2.3 Simplex	9
		1.2.4 Cones	9
	1.3	Calculus of Convex Sets	11
		1.3.1 Calculus of Closed Convex Sets	12
	1.4	Topological Properties of Convex Sets	14
		1.4.1 The Closure	14
		1.4.2 The Interior	14
		1.4.3 The Relative Interior	16
		1.4.4 Nice Topological Properties of Convex Sets	17
	1.5	★ Conic and Perspective Transforms of a Convex Set	20
	1.6	Proofs of Facts	23
2	**Theorems of Carathéodory, Radon, and Helly**		28
	2.1	Carathéodory's Theorem	28
		2.1.1 Carathéodory's Theorem, Illustration	30
	2.2	Radon's Theorem	31
	2.3	Helly's Theorem	32
		2.3.1 Helly's Theorem, Illustration 1	34
		2.3.2 Helly's Theorem, Illustration 2	35

v

		2.3.3 ★ Helly's Theorem for Infinite Families of Convex Sets	36
	2.4	Proof of Fact	37
3	**Polyhedral Representations and Fourier–Motzkin Elimination**		38
	3.1	Polyhedral Representations	38
	3.2	Every Polyhedrally Representable Set is Polyhedral (Fourier–Motzkin Elimination)	40
		3.2.1 Some Applications	41
	3.3	Calculus of Polyhedral Representations	42
4	**General Theorem of the Alternative and Linear Programming Duality**		45
	4.1	Homogeneous Farkas' Lemma	45
	4.2	Certificates for Feasibility and Infeasibility	46
	4.3	General Theorem of the Alternative	48
	4.4	Corollaries of the GTA	50
	4.5	Application: Linear Programming Duality	51
		4.5.1 Preliminaries: Mathematical and Linear Programming Problems	52
		4.5.2 Dual to an LP Problem: The Origin	54
		4.5.3 Linear Programming Duality Theorem	56
5	**Exercises for Part I**		60
	5.1	Elementaries	60
	5.2	Around Ellipsoids	63
	5.3	Truss Topology Design	64
	5.4	Around Carathéodory's Theorem	69
	5.5	Around Helly's Theorem	72
	5.6	Around Polyhedral Representations	74
	5.7	Around the General Theorem of the Alternative	75
	5.8	Around Linear Programming Duality	77

Part II Separation Theorem, Extreme Points, Recessive Directions, and Geometry of Polyhedral Sets   81

6	**Separation Theorem and Geometry of Convex Sets**		83
	6.1	Separation: Definition	83
	6.2	Separation Theorem	85
	6.3	Supporting Hyperplanes	90
	6.4	Extreme Points and Krein–Milman Theorem	91
		6.4.1 Extreme Points: Definition	91
		6.4.2 Krein–Milman Theorem	92
	6.5	Recessive Directions and Recessive Cone	95
	6.6	Dual Cone	99
	6.7	★ Dubovitski–Milutin Lemma	102
	6.8	Extreme Rays and Conic Krein–Milman Theorem	105
	6.9	★ Polar of a Convex Set	108
	6.10	Proofs of Facts	110

7	**Geometry of Polyhedral Sets**		118
	7.1	Extreme Points of Polyhedral Sets	118
		7.1.1 Important Polyhedral Sets and Their Extreme Points	121
	7.2	Extreme Rays of Polyhedral Cones	125
	7.3	Geometry of Polyhedral Sets	126
		7.3.1 Application: Descriptive Theory of Linear Programming	129
		7.3.2 Proof of Theorem 7.10	130
	7.4	★ Majorization	132
8	**Exercises for Part II**		135
	8.1	Separation	135
	8.2	Extreme Points	136
	8.3	Cones and Extreme Rays	140
	8.4	Recessive Cones	141
	8.5	Around Majorization	142
	8.6	Around Polars	143
	8.7	Miscellaneous Exercises	143

Part III Convex Functions — 151

9	**First Acquaintance with Convex Functions**		153
	9.1	Definition and Examples	153
		9.1.1 More Examples of Convex Functions: Norms	154
	9.2	Jensen's Inequality	155
	9.3	Convexity of Sublevel Sets	156
	9.4	Value of a Convex Function Outside Its Domain	156
10	**How to Detect Convexity**		158
	10.1	Operations Preserving Convexity of Functions	159
	10.2	Criteria of Convexity	161
		10.2.1 Differential Criteria of Convexity	163
	10.3	Important Multivariate Convex Functions	167
	10.4	Gradient Inequality	169
	10.5	Boundedness and Lipschitz Continuity of a Convex Function	171
11	**Minima and Maxima of Convex Functions**		175
	11.1	Minima of Convex Functions	175
		11.1.1 Unimodality	175
		11.1.2 Necessary and Sufficient Optimality Conditions	176
		11.1.3 ★ Symmetry Principle	179
	11.2	Maxima of Convex Functions	180
12	**Subgradients**		184
	12.1	Proper Lower Semicontinuous Convex Functions and Their Representation	184
	12.2	Subgradients	193

viii Contents

| | | 12.3 | Subdifferentials and Directional Derivatives of Convex Functions | 199 |

13	★ Legendre Transform	203
	13.1 Legendre Transform: Definition and Examples	203
	13.2 Legendre Transform: Main Properties	205
	13.3 Young, Hölder, and Moment Inequalities	207
	13.3.1 Young's Inequality	207
	13.3.2 Hölder's Inequality	208
	13.3.3 Moment Inequality	209
	13.4 Proofs of Facts	210

14	★ Functions of Eigenvalues of Symmetric Matrices	212
	14.1 Proofs of Facts	216

15	Exercises for Part III	218
	15.1 Around Convex Functions	218
	15.2 Around Support, Characteristic, and Minkowski Functions	223
	15.3 Around Subdifferentials	226
	15.4 Around the Legendre Transform	227
	15.5 Miscellaneous Exercises	230

Part IV Convex Programming, Lagrange Duality, Saddle Points — 233

16	Convex Programming Problems and Convex Theorem of the Alternative	235
	16.1 Mathematical Programming and Convex Programming Problems	235
	16.1.1 Convex Programming Problem	236
	16.2 Convex Theorem of the Alternative	236
	16.3 ★ Convex Theorem of the Alternative – Cone-Constrained Form	239
	16.4 Proofs of Facts	247

17	Lagrange Function and Lagrange Duality	249
	17.1 Lagrange Function	249
	17.2 Convex Programming Duality Theorem	249
	17.3 Lagrange Duality and Saddle Points	250

18	★ Convex Programming in Cone-Constrained Form	252
	18.1 Convex Problem in Cone-Constrained Form	252
	18.2 Cone-Constrained Lagrange Function	252
	18.3 Convex Programming Duality Theorem in Cone-Constrained Form	253
	18.3.1 "Subgradient Interpretation" of Lagrange Multipliers	254
	18.4 Conic Programming and Conic Duality Theorem	255
	18.5 Proofs of Facts	261

19 Optimality Conditions in Convex Programming — 264
- 19.1 Saddle Point Form of Optimality Conditions — 264
- 19.2 Karush–Kuhn–Tucker Form of Optimality Conditions — 267
- 19.3 Cone-Constrained KKT Optimality Conditions — 269
- 19.4 Optimality Conditions in Conic Programming — 271
- 19.5 Duality in Linear and Convex Quadratic Programming — 272
  - 19.5.1 Linear Programming Duality — 273
  - 19.5.2 Convex Quadratic Programming Duality — 274

20 ★ Cone-Convex Functions: Elementary Calculus and Examples — 277
- 20.1 Epigraph Characterization of Cone-Convexity — 277
- 20.2 Testing Cone-Convexity and Cone-Monotonicity — 278
  - 20.2.1 Cone-Monotonicity — 278
  - 20.2.2 Differential Criteria for Cone-Convexity and Cone-Monotonicity — 278
- 20.3 Elementary Calculus of Cone-Convexity — 281

21 ★ Mathematical Programming Optimality Conditions — 284
- 21.1 Formulating Optimality Conditions — 285
  - 21.1.1 Default Assumption — 285
- 21.2 Justifying Optimality Conditions — 286
  - 21.2.1 Main Tool: Implicit Function Theorem — 287
  - 21.2.2 Strategy — 287
  - 21.2.3 Justifying Optimality Conditions for (21.4) — 288
  - 21.2.4 Justifying Propositions 21.3 and 21.4 — 291
- 21.3 Concluding Remarks — 293
  - 21.3.1 Illustration: $\mathcal{S}$-Lemma Revisited — 295

22 Saddle Points — 297
- 22.1 Definition and Game Theory Interpretation — 297
- 22.2 Existence of Saddle Points: Sion–Kakutani Theorem — 300
- 22.3 Proof of Sion–Kakutani Theorem — 301
  - 22.3.1 Proof of Minimax Lemma — 301
  - 22.3.2 From Minimax Lemma to the Proof of Sion–Kakutani Theorem — 302
- 22.4 Sion–Kakutani Theorem: A Refinement — 304
- 22.5 Proof of Fact — 306

23 Exercises for Part IV — 307
- 23.1 Around Conic Duality — 307
  - 23.1.1 ★ Geometry of Primal–Dual Pair of Conic Problems — 310
- 23.2 Around the $\mathcal{S}$-Lemma — 313
- 23.3 Miscellaneous Exercises — 316
- 23.4 Around Convex Cone-Constrained and Conic Problems — 320
- 23.5 ★ Cone-Convexity — 327
- 23.6 ★ Around Conic Representations of Sets and Functions — 330
  - 23.6.1 Conic Representations: Definitions — 330

		23.6.2	Conic Representability: Elementary Calculus	332
		23.6.3	$\mathfrak{R}/\mathfrak{L}/\mathfrak{S}$ Hierarchy	335
		23.6.4	More Calculus	336
		23.6.5	Raw Materials	336

Appendix A Prerequisites from Linear Algebra — 343

- A.1 Space $\mathbf{R}^n$: Algebraic Structure — 343
  - A.1.1 A Point in $\mathbf{R}^n$ — 343
  - A.1.2 Linear Operations — 343
- A.2 Linear Subspaces — 344
  - A.2.1 Examples of Linear Subspaces — 344
  - A.2.2 Sums and Intersections of Linear Subspaces — 345
  - A.2.3 Linear Independence, Bases, Dimensions — 345
  - A.2.4 Linear Mappings and Matrices — 347
  - A.2.5 Determinant and Rank — 348
- A.3 Space $\mathbf{R}^n$: Euclidean Structure — 350
  - A.3.1 Euclidean Structure — 351
  - A.3.2 Inner Product Representation of Linear Forms on $\mathbf{R}^n$ — 351
  - A.3.3 Orthogonal Complement — 352
  - A.3.4 Orthonormal Bases — 353
- A.4 Affine Subspaces in $\mathbf{R}^n$ — 355
  - A.4.1 Affine Subspaces and Affine Hulls — 356
  - A.4.2 Intersections of Affine Subspaces, Affine Combinations, and Affine Hulls — 356
  - A.4.3 Affinely Spanning Sets, Affinely Independent Sets, Affine Dimension — 358
  - A.4.4 Dual Description of Linear Subspaces and Affine Subspaces — 361
  - A.4.5 Affine Subspaces and Systems of Linear Equations — 361
  - A.4.6 Structure of the Simplest Affine Subspaces — 363
- A.5 Proof of Fact — 364
- A.6 Exercises — 365

Appendix B Prerequisites from Real Analysis — 367

- B.1 Space $\mathbf{R}^n$: Metric Structure and Topology — 367
  - B.1.1 Euclidean Norm and Distances — 367
  - B.1.2 Convergence — 370
  - B.1.3 Closed and Open Sets — 371
  - B.1.4 Local Compactness of $\mathbf{R}^n$ — 373
- B.2 Continuous Functions on $\mathbf{R}^n$ — 375
  - B.2.1 Continuity of a Function — 375
  - B.2.2 Elementary Continuity-Preserving Operations — 376
  - B.2.3 Basic Properties of Continuous Functions on $\mathbf{R}^n$ — 377
- B.3 Semicontinuity — 379
  - B.3.1 Functions with Values in the Extended Real Axis — 379
  - B.3.2 Epigraph of a Function — 380

		B.3.3	Lower Semicontinuity	380
		B.3.4	Hypograph and Upper Semicontinuity	381
	B.4	Proofs of Facts		382
	B.5	Exercises		383

Appendix C Prerequisites from Calculus — 385

	C.1	Differentiable Functions on $\mathbf{R}^n$		385
		C.1.1	The Derivative	385
		C.1.2	Derivative and Directional Derivatives	387
		C.1.3	Representations of the Derivative	389
		C.1.4	Existence of the Derivative	391
		C.1.5	Calculus of Derivatives	391
		C.1.6	Computing the Derivative	392
	C.2	Higher-Order Derivatives		394
		C.2.1	Calculus of C^k Mappings	398
		C.2.2	Examples of Higher-Order Derivatives	399
		C.2.3	Taylor Expansion	400

Appendix D Prerequisites: Symmetric Matrices and Positive Semidefinite Cone — 402

	D.1	Symmetric Matrices		402
		D.1.1	Main Facts on Symmetric Matrices	403
		D.1.2	Variational Characterization of Eigenvalues	405
		D.1.3	Corollaries of the VCE	407
		D.1.4	Spectral Norm and Lipschitz Continuity of Vector of Eigenvalues	409
		D.1.5	Functions of Symmetric Matrices	410
	D.2	Positive Semidefinite Matrices and Positive Semidefinite Cone		413
		D.2.1	Positive Semidefinite Matrices	413
		D.2.2	The Positive Semidefinite Cone	414
	D.3	Proofs of Facts		418
	D.4	Exercises		423

References — 425

Index — 426

Preface

Convex optimization serves as a cornerstone in various fields of science, engineering, and mathematics, offering powerful tools for solving a wide range of practical problems. With the latest advances in data sciences and engineering, convex optimization has flourished into a vibrant and rapidly evolving field.

This textbook aims to present the fundamental theory necessary to establish a robust foundation for doing research on convex optimization. In particular, we have aimed to cover both the indispensable basics suited for beginners – rooted in centuries-old research on convexity – as well as modern facets of convex optimization, e.g., cone-constrained conic programming, targeting more advanced readers.

Our emphasis is on foundations and mathematical prerequisites that underpin (primarily convex) optimization theory, and not operational aspects such as modeling and algorithms. This deliberate choice stems from our desire to emphasize "timeless and essential classics" and providing an accessible, self-contained, concise, rigorous, yet practical mathematical toolkit. Our goal is to illuminate the entrance to the field of convex optimization, offering readers the background necessary to engage with and comprehend the state-of-the-art "operational" optimization literature, for example the excellent books [BV04, Nes18]. While applications and algorithms naturally evolve with changing trends and advances, they will always rely on these timeless foundational blocks. Overall, we view the primary purpose of this book as providing learning and teaching as opposed to an extensive reference text to be kept on a shelf by experts.

Note to an expert in the field: the primary focus of this book is convex analysis and the basic theory of convex optimization. Convex analysis boasts a rich and profound theoretical framework, chronicled in classical treatises such as [Roc70, IT79, HUL93]. However, our approach and presentation style in this textbook (a textbook, not an academic monograph!) are tailored to meet the needs of those new to the subject. In particular, we deliberately present the classical results at the level of generality suited to the practical needs of optimization-related research, as we understand it, rather than in the full generality provided by their authors. These "savings in generality" allow us to make statements and proofs more accessible to beginners. While we have condensed the scope compared to the classical treatises, we have strived to maintain rigor and cover the fundamental descriptive mathematical groundwork that we believe is necessary for mastering the state-of-the-art research in convex optimization models and algorithms. With regard to convex optimization theory, once again our emphasis is on timeless building blocks such as duality and optimality conditions.

Our intended audience is students, practitioners, and researchers with backgrounds in mathematics, operations research, engineering, computer science, data sciences, statistics, or economics.

As prerequisites all we assume is a basic knowledge of linear algebra and calculus. In fact, we do not anticipate a deep, "ready-to-use" mastery of these subjects either; rather, we expect a basic mathematical culture. To clarify our expectations, consider the following: asserting that $2 \times 2 = 5$ does *not* necessarily indicate a deficiency in mathematical culture; it may simply be a miscalculation. In contrast, claiming that 2×2 is a triangle or a violin (occasionally encountered, albeit perhaps figuratively rather than literally, in our teaching experience) *does* signify a lack of mathematical culture: under any circumstances, the product of two real numbers should invariably be another real number and cannot be a triangle or a violin.

Our choice of material is driven by years of experience teaching graduate-level courses on nonlinear and convex optimization. We have organized this material into four main parts:

- basics on convex sets – instructive examples, "calculus" (convexity-preserving operations), the main theorems on convex sets such as those of Carathéodory and Helly, the topology of convex sets, and the "descriptive basics" of linear programming (the General Theorem of the Alternative, linear programming duality);
- the Separation Theorem and its applications – extreme points, extreme rays, and the (finite-dimensional) Krein–Milman Theorem, the structure of polyhedral sets;
- basics on convex functions – instructive examples, detecting convexity, "calculus," gradient inequality and preliminaries on subgradients, maxima and minima, Legendre transformation;
- the basic theory of convex optimization – Lagrange Duality and the Lagrange Duality Theorem for problems in standard form and in cone-constrained form, conic programming and the Conic Duality Theorem, optimality conditions in mathematical programming, and convex–concave saddle points and the Sion–Kakutani Theorem.

We envision that in a semester-long (14–15 weeks) course on convex optimization, this material may cover about 40% or 60% of the course, to build the foundational blocks before other parts (modeling and/or algorithms) of convex optimization are addressed. In a course focusing more on nonlinear aspects and going beyond convexity, an instructor may elect to skip Chapters 2, 3, and 7, and instead pay attention to Chapter 22. For a course more geared towards linear and/or combinatorial optimization, an instructor may opt to use specific material from Parts I and II. Finally, a course focusing on conic and semidefinite programming can benefit from the material in Chapters 14, 16, 18, 19, and 21.

For the reader's convenience, the elementary facts from linear algebra, calculus, real analysis, and matrix analysis are summarized in the appendices of the book (reproducing, courtesy of the World Scientific Publishing Co., appendices A–C in [Nem24]). In contrast with the main body of the book, these appendices usually do *not* feature any accompanying proofs, which are readily available in standard undergraduate textbooks covering the respective subjects (see, e.g., [Axl15, Edw12, Gel89, Pas22, Rud13, Str06]). A well-prepared reader may opt for a "fast-forward" approach by initially reviewing these appendices before delving into the main body of the book. Alternatively, they may commence their reading journey from Part I, referring back to the appendices as necessary.

Certain sections in our text, with titles starting with ★, delve into more advanced and specialized topics, such as conic, perspective, and Legendre transforms, majorization, cone-convexity, and cone-monotonicity, among others. Although these starred sections hold significance in their respective domains, they are designed to be optional and can be skipped over depending on the goals of the reader.

Our exposition adheres to the usual standards of rigor needed to present mathematical subjects. Accordingly, we provide complete formal proofs for all theorems, propositions, lemmas, and the like. In addition to these, we include formal statements of a similar nature labeled as "facts," scattered throughout a chapter, but without accompanying proofs. The claims made in these "facts" are also an essential part of our exposition, and a knowledge of them is as necessary for mastering the material as a knowledge of the theorems, propositions, etc. Nonetheless, the statements within the "facts" are sufficiently elementary to be easily verified by a diligent reader. In essence, the "facts" serve as embedded exercises, and we firmly believe that engaging with these exercises as they appear in the text provides valuable hands-on practice for effective learning. This active participation is an indispensable facet of mastering the presented material and honing mathematical skills. At the same time, we recognize the importance of providing access to detailed self-contained proofs of the "facts." To this end, nearly all the "facts" are repeated,[1] this time with accompanying proofs, at the end of the respective chapters.

The exercises are crafted to align with our educational objective of fostering hands-on learning and providing ample opportunities for practice at various levels of difficulty. In particular, they are categorized into three types. The traditional (and simpler in our opinion) "test yourself" exercises enable readers to evaluate their grasp of the material presented in the main body of the textbook. In addition to these, we also offer "try yourself" (marked [**TrYs**]) exercises, which typically require readers to prove something and are aimed at fostering and assessing their creative skills. Finally, our "educate yourself" (marked [**EdYs**]) exercises address topics that we deem significant, extending beyond the core material covered in the main body of the book. An illustrative example of this latter type of exercise is the investigation of conic representations of convex sets and functions, along with the calculus associated with these representations. This plays a crucial role in formulating and solving "well-structured" convex optimization problems, particularly those involving second-order conic and semidefinite programs. Some simple "test yourself" exercises (i.e., without any label such as [**TrYs**] or [**EdYs**]) originated and evolved from [BTN, Nem24]. We provide a separate solution manual for the "try yourself" and "educate yourself" exercises.

Acknowledgements. The main body of this textbook existed for about 25 years, in a more restricted form, as appendices to the graduate course on Modern Convex Optimization taught by the second author first at TU Delft (1998) and then at the Georgia Institute of Technology (since 2003). The first author was fortunate to have been a student at Georgia Tech, thoroughly enjoying this material, and later on she adopted it and has been teaching a similar course at Carnegie Mellon University since 2012. These appendices originate from the descriptive part of the graduate course Optimization II designed in the 1980s by Professor Aharon Ben-Tal and for over 20 years taught by him at the Technion – Israel

---

[1] The few exceptions are absolutely straightforward and are presented as "facts" solely for the sake of further reference.

Institute of Technology. This course was "inherited" and taught, in redesigned form, by the second author, first at the Technion and then at Georgia Institute of Technology. It is our pleasure to acknowledge here, with deepest gratitude, the instrumental role played by Professor Ben-Tal in selecting and structuring significant parts of the material in the book. In addition, we are greatly indebted to Dr. Sergei Gelfand for the idea of converting the above-mentioned appendices into a "standalone" textbook.

Main Notational Conventions

The letters **N, Z, Q, R, C** stand for the set of all, respectively, nonnegative integers, integers, rational numbers, real numbers, and complex numbers.

Vectors and matrices. By default, *all vectors are column vectors.*

- The space of all n-dimensional vectors with real entries is denoted by $\mathbf{R}^n$ and the set of all $m \times n$ matrices with real entries is denoted by $\mathbf{R}^{m \times n}$; the notation $\mathbf{N}^n, \ldots, \mathbf{C}^{m \times n}$ is interpreted similarly, where we restrict the entries to belong to the respective number domains. The set of symmetric $n \times n$ matrices is denoted by $\mathbf{S}^n$. By default, all vectors and matrices have real entries, and when speaking about $\mathbf{R}^n$ and $\mathbf{S}^n$ ($\mathbf{R}^{m \times n}$), n (m and n) are *positive* integers.

  By default, notation like x_i (or y_k) refers to the ith entry of a context-specified vector x (or the kth entry of a context-specified vector y). Similarly, notation such as x_{ij} (or $Y_{k\ell}$) refers to the (i, j)th entry of a context-specified matrix x (or the (k, ℓ)th entry of a context-specified matrix Y).

- Sometimes, MATLAB notation is used to save space: a vector with coordinates $x_1, \ldots, x_n$ is written down as

$$x = [x_1; \ldots; x_n]$$

(note the use of semicolons). For example, $\begin{bmatrix} 1 \\ 2 \\ 3 \end{bmatrix}$ is written as $[1; 2; 3]$.

More generally,
- if $A_1, \ldots, A_m$ are matrices with the same numbers of columns, we write $[A_1; \ldots; A_m]$ to denote the matrix obtained by writing A_2 beneath A_1, A_3 beneath A_2, and so on.
- if $A_1, \ldots, A_m$ are matrices with the same numbers of rows then $[A_1, \ldots, A_m]$ stands for the matrix obtained by writing A_2 to the right of A_1, A_3 to the right of A_2, and so on.

Examples:

$$A_1 = \begin{bmatrix} 1 & 2 & 3 \\ 4 & 5 & 6 \end{bmatrix}, A_2 = \begin{bmatrix} 7 & 8 & 9 \end{bmatrix} \implies [A_1; A_2] = \begin{bmatrix} 1 & 2 & 3 \\ 4 & 5 & 6 \\ 7 & 8 & 9 \end{bmatrix}.$$

$$A_1 = \begin{bmatrix} 1 & 2 \\ 3 & 4 \end{bmatrix}, A_2 = \begin{bmatrix} 7 \\ 8 \end{bmatrix} \implies [A_1, A_2] = \begin{bmatrix} 1 & 2 & 7 \\ 4 & 5 & 8 \end{bmatrix}.$$

$$[1,2,3,4] = [1;2;3;4]^\top$$
$$[[1,2;3,4],[5,6;7,8]] = \left[\begin{bmatrix} 1 & 2 \\ 3 & 4 \end{bmatrix}, \begin{bmatrix} 5 & 6 \\ 7 & 8 \end{bmatrix}\right] = \begin{bmatrix} 1 & 2 & 5 & 6 \\ 3 & 4 & 7 & 8 \end{bmatrix}$$
$$= [1,2,5,6;3,4,7,8].$$

- We follow the standard convention that the sum of vectors over an empty set of indices, i.e., $\sum_{i=1}^{0} x^i$, where $x^i \in \mathbf{R}^n$, has a value – it is the origin in $\mathbf{R}^n$.

Intervals in $\mathbf{R}^n$. Given two vectors $x, y \in \mathbf{R}^n$, we use the notation $[x, y]$ to denote the *segment* in $\mathbf{R}^n$ that connects x and y, where both endpoints are included, i.e., $[x, y] := \{\lambda x + (1-\lambda)y : 0 \leq \lambda \leq 1\}$. Similarly, we define the open segment $(x, y) := \{\lambda x + (1-\lambda)y : 0 < \lambda < 1\}$ in $\mathbf{R}^n$, i.e., the segment without the endpoints.

Semidefinite order. The relations $A \succeq B$, $B \preceq A$, $A - B \succeq 0$, $B - A \preceq 0$ all mean the same, namely, that A, B are real symmetric matrices of the same size such that the difference $A - B$ is positive semidefinite. The positive definiteness of the difference $A - B$ is expressed by every one of the relations $A \succ B$, $B \prec A$, $A - B \succ 0$, $B - A \prec 0$.

Diag and Dg. For $x \in \mathbf{R}^n$, $\mathrm{Diag}\{x\}$ stands for the diagonal $n \times n$ matrix with the entries of x on the diagonal. For a collection $X_1, \ldots, X_K$ of rectangular matrices, $\mathrm{Diag}\{X_1, \ldots, X_k\} = \begin{bmatrix} X_1 & & \\ & \ddots & \\ & & X_k \end{bmatrix}$ stands for the block-diagonal matrix with diagonal blocks $X_1, \ldots, X_K$. For an $n \times n$ matrix X, $\mathrm{Dg}\{X\}$ stands for the n-dimensional vector composed of the diagonal entries of X.

Extended real axis. We follow the standard conventions on the operations of summation, multiplication, and comparison in the extended real line $\mathbf{R} \cup \{+\infty\} \cup \{-\infty\}$. These conventions are as follows:

- Operations with real numbers are understood in their usual sense.
- The sum of $+\infty$ and a real number, and also the sum of $+\infty$ and $+\infty$, is $+\infty$. Similarly, the sum of a real number and $-\infty$, and also the sum of $-\infty$ and $-\infty$, is $-\infty$. The sum of $+\infty$ and $-\infty$ is undefined.
- The product of a real number and $+\infty$ is $+\infty$, 0 or $-\infty$, depending on whether the real number is positive, zero, or negative, and similarly for the product of a real number and $-\infty$. The product of two infinities is again infinity, with the usual rule for assigning the sign to the product.
- Finally, any real number is $< +\infty$ and $> -\infty$, and of course $-\infty < \infty$.

Abbreviations. From time to time we use the following abbreviations:

a.k.a. for "also known as"
iff for "if and only if"
w.l.o.g. for "without loss of generality"
w.r.t. for "with respect to"

[TrYs] and [EdYs]. These indicate respectively "try yourself" and "educate yourself" exercises.

Part I

Convex sets in R^n: From First Acquaintance to Linear Programming Duality

1

First Acquaintance with Convex Sets

1.1 Definition and Examples

In school geometry a figure is called *convex* if it contains, along with every pair of its points x, y, also the entire segment $[x, y]$ linking the points. This is exactly the definition of a convex set in the multidimensional case; all we need is to define "the segment $[x, y]$ linking the points $x, y \in \mathbf{R}^n$". We state this formally in the following definition.

Definition 1.1 *[Convex set]*

(i) *Let x, y be two points in $\mathbf{R}^n$. The set*

$$[x, y] := \{\lambda x + (1 - \lambda)y : 0 \leq \lambda \leq 1\}$$

is called a segment *with endpoints x, y.*

(ii) *A subset M of $\mathbf{R}^n$ is called* convex *if it contains, along with every pair of its points x, y, also the entire segment $[x, y]$:*

$$x, y \in M, \ 0 \leq \lambda \leq 1 \implies \lambda x + (1 - \lambda)y \in M.$$

The definition of a segment $[x, y]$ is in full accordance with our "real life experience" in two or three dimensions: when $\lambda \in [0, 1]$, the point $x(\lambda) = \lambda x + (1-\lambda)y = x + (1-\lambda)(y-x)$ is the point at which you arrive when traveling from x directly towards y after you have covered a fraction $(1 - \lambda)$ of the entire distance from x to y, and these points compose the "real world segment" with endpoints $x = x(1)$ and $y = x(0)$.

Note that an empty set is convex by the exact sense of the definition: for an empty set, you cannot present a counterexample to show that it is not convex.

Closed and open segments are clearly convex. A closed *ray* given by a direction $0 \neq d \in \mathbf{R}^n$ is also convex:

$$\mathbf{R}_+(d) := \{t \, d \in \mathbf{R}^n : t \geq 0\}.$$

Note also that the open ray given by $\{t \, d \in \mathbf{R}^n : t > 0\}$ is convex as well.

We next continue with a number of important examples of convex sets.

1.1.1 Affine Subspaces and Polyhedral Sets

We start with a simple and important fact.

Proposition 1.2 *The solution set of an arbitrary (possibly, infinite) system*

$$a_\alpha^\top x \leq b_\alpha, \ \alpha \in \mathcal{A} \tag{1.1}$$

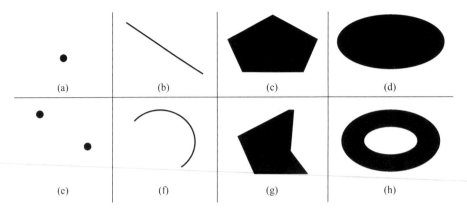

Figure 1.1 (a)–(d) Convex sets; (e)–(h) nonconvex sets

of nonstrict linear inequalities with n unknowns x, i.e., the set

$$S := \left\{ x \in \mathbf{R}^n : a_\alpha^\top x \leq b_\alpha, \alpha \in \mathcal{A} \right\}$$

is convex.

Proof Consider any $x', x'' \in S$ and any $\lambda \in [0, 1]$. As $x', x'' \in S$, we have $a_\alpha^\top x' \leq b_\alpha$ and $a_\alpha^\top x'' \leq b_\alpha$ for any $\alpha \in \mathcal{A}$. Then, for every $\alpha \in \mathcal{A}$, multiplying the inequality $a_\alpha^\top x' \leq b_\alpha$ by λ, and the inequality $a_\alpha^\top x'' \leq b_\alpha$ by $1 - \lambda$, respectively, and summing up the resulting inequalities, we get $a_\alpha^\top [\lambda x' + (1-\lambda) x''] \leq b_\alpha$. Thus, we deduce that $\lambda x' + (1-\lambda) x'' \in S$. ∎

Note that this verification of the convexity of S works also in the case when in the definition of S some nonstrict inequalities $a_\alpha^\top x \leq b_\alpha$ are replaced with their strict versions $a_\alpha^\top x < b_\alpha$.

Recall that linear and affine subspaces can be represented as the solution sets of systems of linear equations (Proposition A.47). Consequently, from Proposition 1.2 we deduce that such sets are convex.

Example 1.1 All linear subspaces and all affine subspaces of $\mathbf{R}^n$ are convex. ◇

Another important special case of Proposition 1.2 occurs when we have a finite system of nonstrict linear inequalities. Such sets have a special name as they are frequently encountered and studied.

Definition 1.3 *[Polyhedral set] A set in $\mathbf{R}^n$ is called* polyhedral *if it is the solution set of a finite system*

$$Ax \leq b$$

of m nonstrict linear inequalities with n variables (i.e., A is an $m \times n$ matrix) for some nonnegative integer m.

On the basis of this definition and as an immediate consequence of Proposition 1.2, we arrive at our second generic example of convex sets.

Example 1.2 Any polyhedral set in $\mathbf{R}^n$ is convex. ◇

1.1 Definition and Examples

Remark 1.4 Note that every set given by Proposition 1.2 is not only convex but also closed (why?). In fact, the Separation Theorem (see Theorem 6.3) implies the following:

Every closed convex set in $\mathbf{R}^n$ is the solution set of an infinite system $a_i^\top x \leq b_i$, $i = 1, 2, \ldots,$ of nonstrict linear inequalities.

$\diamond$

Remark 1.5 Replacing some of the nonstrict linear inequalities $a_\alpha^\top x \leq b_\alpha$ in the system (1.1) with their strict versions $a_\alpha^\top x < b_\alpha$ preserves, as we have already mentioned, the convexity of the solution set but can destroy its closedness.

$\diamond$

1.1.2 Unit Balls of Norms

Let $\|\cdot\|$ be a norm on $\mathbf{R}^n$ i.e., a real-valued function on $\mathbf{R}^n$ satisfying the three characteristic properties of a norm (section B.1.1), specifically:

1. *Positivity:* $\|x\| \geq 0$ for all $x \in \mathbf{R}^n$, and $\|x\| = 0$ if and only if $x = 0$.
2. *Homogeneity:* For $x \in \mathbf{R}^n$ and $\lambda \in \mathbf{R}$, we have $\|\lambda x\| = |\lambda| \|x\|$.
3. *Triangle inequality:* For all $x, y \in \mathbf{R}^n$, we have $\|x + y\| \leq \|x\| + \|y\|$.

Fact 1.6 *The* unit ball *of a norm $\|\cdot\|$, i.e., the set*

$$\{x \in \mathbf{R}^n : \|x\| \leq 1\},$$

is convex, as for every other $\|\cdot\|$-ball

$$B_r(a) := \{x \in \mathbf{R}^n : \|x - a\| \leq r\}$$

(here $a \in \mathbf{R}^n$ and $r \geq 0$ are fixed).

In particular, Euclidean balls ($\|\cdot\|$-balls associated with the standard Euclidean norm $\|x\|_2 := \sqrt{x^\top x}$) are convex.

The standard examples of norms on $\mathbf{R}^n$ are the ℓ_p-norms

$$\|x\|_p = \begin{cases} \left(\sum_{i=1}^n |x_i|^p\right)^{1/p}, & \text{if } 1 \leq p < \infty \\ \max_{1 \leq i \leq n} |x_i|, & \text{if } p = \infty. \end{cases}$$

These are indeed norms (which is not clear at the outset; for a proof, see p. 154, and for more details see p. 209). When $p = 2$, we get the usual Euclidean norm. When $p = 1$, we obtain

$$\|x\|_1 = \sum_{i=1}^n |x_i|,$$

and its unit ball is the *hyperoctahedron*

$$V = \left\{x \in \mathbf{R}^n : \sum_{i=1}^n |x_i| \leq 1\right\}.$$

When $p = \infty$, we get

$$\|x\|_\infty = \max_{1 \leq i \leq n} |x_i|,$$

and its unit ball is the *hypercube*

$$V = \{x \in \mathbf{R}^n : -1 \leq x_i \leq 1, 1 \leq i \leq n\};$$

see Figure 1.2.

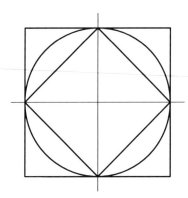

Figure 1.2 $\|\cdot\|_p$-balls in two dimensions: $p = 1$ (diamond), $p = 2$ (circle), $p = \infty$ (box).

Remark 1.7 As we have already mentioned, the fact that the ℓ_p-norms, $1 \leq p \leq \infty$, are indeed norms, is not completely trivial and will be proved in full generality later. What is evident is that $\|\cdot\|_p$ does possess the properties of positivity and homogeneity; what requires effort is the triangle inequality. There are, however, two special cases, i.e., $p = 1$ and $p = \infty$, where this inequality is easy. Indeed, for reals a, b it always holds that $|a+b| \leq |a|+|b|$. It follows that

$$\|x+y\|_1 = \sum_i |x_i + y_i| \leq \sum_i (|x_i| + |y_i|) = \sum_i |x_i| + \sum_i |y_i| = \|x\|_1 + \|y\|_1$$

and

$$\|x+y\|_\infty = \max_i |x_i + y_i| \leq \max_i \{|x_i| + |y_i|\} \leq \max_{i,j} \{|x_i| + |y_j|\}$$
$$= \max_i |x_i| + \max_j |y_j| = \|x\|_\infty + \|y\|_\infty.$$

The triangle inequality for the Euclidean norm $\|\cdot\|_2$ should be already known to the reader; this is an immediate consequence of the Cauchy–Schwarz inequality $|x^\top y| \leq \|x\|_2 \|y\|_2$; see section B.1.1. ◇

Fact 1.8 *Unit balls of norms on $\mathbf{R}^n$ are exactly the same as convex sets V in $\mathbf{R}^n$ satisfying the following three properties:*

(i) *V is symmetric with respect to the origin: $x \in V \implies -x \in V$;*
(ii) *V is bounded and closed;*
(iii) *V contains a neighborhood of the origin, i.e., there exists $r > 0$ such that the Euclidean ball of radius r centered at the origin – the set $\{x \in \mathbf{R}^n : \|x\|_2 \leq r\}$ – is contained in V.*

Any set V satisfying the above properties is indeed the unit ball of a particular norm given by

$$\|x\|_V := \inf\left\{t : t^{-1}x \in V, \ t > 0\right\}. \tag{1.2}$$

1.1.3 Ellipsoids

Fact 1.9 *Let Q be an $n \times n$ matrix which is symmetric (i.e., $Q = Q^\top$) and positive definite (i.e., $x^\top Q x > 0$ for all vectors $x \neq 0$). Then, for every nonnegative r, the Q-ellipsoid of radius r centered at a, i.e., the set*

$$\left\{x \in \mathbf{R}^n : (x - a)^\top Q (x - a) \leq r^2\right\}$$

is convex.

1.1.4 Neighborhood of a Convex Set

Example 1.3 Let M be a nonempty convex set in $\mathbf{R}^n$, and let $\epsilon > 0$. Then, for every norm $\|\cdot\|$ on $\mathbf{R}^n$, the ϵ-neighborhood of M, i.e., the set

$$M_\epsilon := \left\{y \in \mathbf{R}^n : \inf_{x \in M} \|y - x\| \leq \epsilon\right\}$$

is convex. ◇

The justification of Example 1.3 is left as an exercise at the end of this Part (see Exercise I.6).

1.2 Inner Description of Convex Sets: Convex Combinations and Convex Hull

1.2.1 Convex Combinations

Recall the notion of the *linear combination* x of vectors $x^1, \ldots, x^m$; this is a vector represented as

$$x = \sum_{i=1}^{m} \lambda_i x^i,$$

where $\lambda_i \in \mathbf{R}$ are coefficients. By including a specific restriction on which coefficients can be used in this definition, we arrive at important special types of linear combinations. For example, an *affine combination* is a linear combination where the sum of the coefficients is equal to 1. Given a nonempty set X, the smallest (w.r.t. inclusion) affine plane containing X is composed of all affine combinations of the points of X; see section A.4.2. Another beast in this genre is the *convex combination*.

Definition 1.10 *[Convex combination] A convex combination of vectors $x^1, \ldots, x^m$ is the linear combination*

$$x := \sum_{i=1}^{m} \lambda_i x^i,$$

with nonnegative coefficients summing up to 1:

$$\lambda_i \geq 0, \; \forall i = 1, \ldots, m, \quad \sum_{i=1}^{m} \lambda_i = 1.$$

Equivalently, a convex combination is an affine combination with nonnegative coefficients.

According to linear algebra, a nonempty set $X \subseteq \mathbf{R}^n$ is a linear (or an affine) subspace if and only if X is closed with respect to all linear (respectively, all affine) combinations of its elements. Convex combinations play a similar role when one is speaking about convex sets.

Fact 1.11 *A set $M \subseteq \mathbf{R}^n$ is convex if and only if it is closed with respect to taking all convex combinations of its elements. That is, M is convex if and only if every convex combination of vectors from M is again a vector from M.*

Hint: Note that, assuming $\lambda_1, \ldots, \lambda_m > 0$, one has

$$\sum_{i=1}^{m} \lambda_i x^i = \lambda_1 x^1 + (\lambda_2 + \lambda_3 + \cdots + \lambda_m) \sum_{i=2}^{m} \mu_i x^i, \quad \text{where } \mu_i := \frac{\lambda_i}{\lambda_2 + \lambda_3 + \cdots + \lambda_m}$$

(cf. Corollary A.39).

1.2.2 Convex Hull

Recall that taking the intersection of linear subspaces results in another linear subspace. The same property holds true for convex sets as well (why?).

Proposition 1.12 *Let $\{M_\alpha\}_\alpha$ be an arbitrary family of convex subsets of $\mathbf{R}^n$. Then, the intersection*

$$M = \bigcap_\alpha M_\alpha$$

is also convex.

As an immediate consequence of Proposition 1.12, we come to the notion of the *convex hull* $\mathrm{Conv}(M)$ of a subset $M \subseteq \mathbf{R}^n$ (cf. the notions of linear or affine spans):

Definition 1.13 *[Convex hull] For any $M \subseteq \mathbf{R}^n$, the convex hull of M [notation: $\mathrm{Conv}(M)$] is the intersection of all convex sets containing M (and thus, by Proposition 1.12 $\mathrm{Conv}(M)$ is the smallest (w.r.t. inclusion) convex set containing M).*

From linear algebra, the linear span of a set M – the smallest (w.r.t. inclusion) linear subspace containing M – can be described in terms of linear combinations: this is the set of all linear combinations of points from M. Analogous results hold for the affine span of a (nonempty) set and affine combinations of points from the set as well. We have an analogous description of convex hulls via convex combinations:

Fact 1.14 *[Convex hull via convex combinations] For a set $M \subseteq \mathbf{R}^n$,*

$$\mathrm{Conv}(M) = \{\text{the set of all convex combinations of vectors from } M\}.$$

1.2 Inner Description of Convex Sets: Convex Combinations and Convex Hull

We will see in section 7.3 that when M is a finite set in $\mathbf{R}^n$, $\text{Conv}(M)$ is a bounded polyhedral set. Bounded polyhedral sets are also called *polytopes*.

We next continue with a number of important families of convex sets.

1.2.3 Simplex

Definition 1.15 *[Simplex] The convex hull of $m+1$ affinely independent points $x^0, \ldots, x^m$ is called the m-dimensional simplex with vertices $x^0, \ldots, x^m$. (See section A.4.3 for affine independence.)*

Consider an m-dimensional simplex with vertices $x^0, \ldots, x^m$. Then, on the basis of section A.4.3, every point x from this simplex admits exactly one representation as a convex combination of these vertices. The coefficients λ_i, $i = 0, \ldots, m$, used in the convex-combination representation of x form the unique solution to the system of linear equations given by

$$\sum_{i=0}^m \lambda_i x^i = x, \quad \sum_{i=0}^m \lambda_i = 1.$$

This system in the variables λ_i is feasible if and only if $x \in M = \text{Aff}(\{x^0, \ldots, x^m\})$, and the components of the solution (the barycentric coordinates of x) are affine functions of $x \in \text{Aff}(M)$. The simplex itself is composed of points from M with nonnegative barycentric coordinates.

1.2.4 Cones

We next examine a very important class of convex sets.

A nonempty set $K \subseteq \mathbf{R}^n$ is called *conic* if it contains, along with every point $x \in K$, the entire ray $\mathbf{R}_+(x) = \{tx : t \geq 0\}$ spanned by the point x:

$$x \in K \implies tx \in K, \ \forall t \geq 0.$$

Note that based on our definition, any conic set is nonempty and it always contains the origin.

Definition 1.16 *[Cone] A* cone *is a nonempty, convex, and conic set.*

Fact 1.17 *A set $K \subseteq \mathbf{R}^n$ is a cone if and only if it is nonempty and it*

- *is conic, i.e., $x \in K, t \geq 0 \implies tx \in K$; and*
- *contains sums of its elements, i.e., $x, y \in K \implies x + y \in K$.*

Example 1.4 The solution set of an arbitrary (possibly, infinite) system of *homogeneous* linear inequalities with n unknowns x, i.e., the set

$$K = \left\{ x \in \mathbf{R}^n : a_\alpha^\top x \geq 0, \ \forall \alpha \in \mathcal{A} \right\},$$

is a cone.

In particular, the solution set of a *finite* system composed of m homogeneous linear inequalities

$$Ax \geq 0$$

(A is an $m \times n$ matrix) is a cone. A cone of this latter type is called *polyhedral*.[1]

Specifically, the *nonnegative orthant* $\mathbf{R}_+^m := \{x \in \mathbf{R}^m : x \geq 0\}$ is a polyhedral cone. ◇

Note that the cones given by systems of linear homogeneous nonstrict inequalities are obviously closed. From the Separation Theorem (see Theorem 6.3) we can deduce the reverse as well, i.e., every closed cone is the solution set to such a system. Thus, Example 1.4 is the generic example of a closed convex cone.

We already know that a norm $\|\cdot\|$ on $\mathbf{R}^n$ gives rise to specific convex sets in $\mathbf{R}^n$, namely, balls of this norm. In fact, a norm also gives rise to another important convex set.

Proposition 1.18 *For any norm $\|\cdot\|$ on $\mathbf{R}^n$, its* epigraph, *i.e., the set*

$$K := \left\{ [x;t] \in \mathbf{R}^{n+1} : t \geq \|x\| \right\},$$

is a closed cone in $\mathbf{R}^{n+1}$.

Proof Obviously, K is nonempty as $[x;t] = [0;0]$ is in K. Also, K is a conic set as any norm $\|\cdot\|$ is positively homogeneous. Moreover, the closedness of K with respect to summation is readily given by the triangle inequality: consider two points $[x;t] \in K$ and $[x';t'] \in K$. Then, $t \geq \|x\|$ and $t' \geq \|x'\|$, which implies that $t+t' \geq \|x\|+\|x'\| \geq \|x+x'\|$. Thus, $[x+x';t+t'] \in K$. Invoking Fact 1.17, we see that K is a cone. In order to see that K is closed, bear in mind that $\|\cdot\|$ is continuous (see Fact B.23). Thus, for any sequence of points $[x^i;t_i] \in K$ converging to a point $[x;t]$ as $i \to \infty$, we have $[\|x^i\|;t_i] \to [\|x\|;t]$ and therefore $t \geq \|x\|$. This establishes that the limit of any converging sequence from K belongs to K, proving that K is closed. ∎

A particular case of Proposition 1.18 states that the epigraph of the Euclidean norm, i.e.,

$$\mathbf{L}^m := \left\{ [x;t] \in \mathbf{R}^{n+1} : t \geq \|x\|_2 \right\},$$

is a closed cone. This is the *second-order* (or *Lorentz* or *ice cream*) cone (see Figure 1.3), and it plays a significant role in convex optimization.

To complete our first acquaintance with cones, we also mention the *semidefinite cone* $\mathbf{S}_+^m$ "living" in the space $\mathbf{S}^m$ of real symmetric $m \times m$ matrices and composed of positive semidefinite matrices from $\mathbf{S}^m$, i.e.,

$$\mathbf{S}_+^m := \left\{ X \in \mathbf{R}^{m \times m} : X = X^\top, a^\top X a \geq 0, \forall a \in \mathbf{R}^m \right\};$$

see section D.2.2.

Cones form a very important family of convex sets, and one can develop a theory of cones that is absolutely similar (and in a sense, equivalent) to that of all convex sets. For example, by introducing the notion of a *conic combination* of vectors $x^1, \ldots, x^k$ as a

---

[1] The literal interpretation of the term "polyhedral cone" is "a set of the form $\{x : Ax \leq b\}$ which is a cone;" this is not exactly the terminology just introduced. Luckily, there is no collision: *a polyhedral set $X = \{x : Ax \geq b\}$ is a cone if and only if $X = \{x : Ax \geq 0\}$*; see Exercise II.32.

1.3 Calculus of Convex Sets

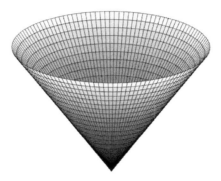

Figure 1.3 Boundary of three-dimensional Lorentz cone $\mathbf{L}^3$.

linear combination of these vectors with *nonnegative* coefficients, we can easily prove the following statements, which are completely similar to those for general convex sets but with "conic combination" substituted for convex.

- A set is a cone if and only if it is nonempty and is closed with respect to taking conic combinations of its elements.
- The intersection of a family of cones is again a cone; in particular, for every set $K \subseteq \mathbf{R}^n$ there exists the smallest (w.r.t. inclusion) cone containing K, called the *conic hull* of K:

Definition 1.19 *[Conic hull] For any $K \subseteq \mathbf{R}^n$, the* conic hull *of K [notation: Cone(K)] is the intersection of all cones containing K. Thus, Cone(K) is the smallest (w.r.t. inclusion) cone containing K.*

- We can describe the conic hull of a set $K \subseteq \mathbf{R}^n$ in terms of its conic combinations:

Fact 1.20 *[Conic hull via conic combinations] The conic hull Cone(K) of a set $K \subseteq \mathbf{R}^n$ is the set of all conic combinations (i.e., linear combinations with nonnegative coefficients) of vectors from K:*

$$\mathrm{Cone}(K) = \left\{ x \in \mathbf{R}^n : \exists N \geq 0, \lambda_i \geq 0, x^i \in K, i \leq N : x = \sum_{i=1}^{N} \lambda_i x^i \right\}.$$

Note that here we use a standard convention: the sum of vectors over an empty set of indices, such as $\sum_{i=1}^{0} z^i$, has a value – it is the origin of the space in which the vectors live. In particular, the set of conic combinations of vectors from an empty set is $\{0\}$, in full accordance with Definition 1.19.

1.3 Calculus of Convex Sets

The calculus of convex sets is, in a nutshell, the list of operations which preserve convexity.

Proposition 1.21 *The following operations preserve the convexity of sets:*

(i) Taking the intersection: If M_α, $\alpha \in \mathcal{A}$, are convex sets, so is their intersection $\bigcap_\alpha M_\alpha$.

(ii) Taking the direct product: *If $M_1 \subseteq \mathbf{R}^{n_1}$ and $M_2 \subseteq \mathbf{R}^{n_2}$ are convex sets, so is their direct product, i.e., the set*

$$M_1 \times M_2 := \left\{ x = [x^1; x^2] \in \mathbf{R}^{n_1} \times \mathbf{R}^{n_2} = \mathbf{R}^{n_1+n_2} : x^1 \in M_1, x^2 \in M_2 \right\}.$$

(iii) Arithmetic summation and multiplication by reals: *If $M_1, \ldots, M_k$ are nonempty convex sets in $\mathbf{R}^n$ and $\lambda_1, \ldots, \lambda_k$ are arbitrary reals then the set*

$$\lambda_1 M_1 + \cdots + \lambda_k M_k := \left\{ \sum_{i=1}^{k} \lambda_i x^i : x^i \in M_i, i = 1, \ldots, k \right\}$$

is convex.

Warning: *The "linear combination $\lambda_1 M_1 + \cdots + \lambda_k M_k$ of sets" as defined above is just a notation. When operating with these "linear combinations of sets," one should be careful. For example. while it is true that $M_1 + M_2 = M_2 + M_1$ and that $M_1 + (M_2 + M_3) = (M_1 + M_2) + M_3$, and even that $\lambda(M_1 + M_2) = \lambda M_1 + \lambda M_2$, it is, in general, not true that $(\lambda_1 + \lambda_2) M = \lambda_1 M + \lambda_2 M$.*

(iv) Taking the image under an affine mapping: *If $M \subseteq \mathbf{R}^n$ is a convex set and $x \mapsto \mathcal{A}(x) \equiv Ax + b$ is an affine mapping from $\mathbf{R}^n$ into $\mathbf{R}^m$ (where $A \in \mathbf{R}^{m \times n}$ and $b \in \mathbf{R}^m$) then the image of M under the mapping $\mathcal{A}(\cdot)$, i.e., the set*

$$\mathcal{A}(M) := \{\mathcal{A}(x) : x \in M\},$$

is convex.

(v) Taking the inverse image under an affine mapping: *If $M \subseteq \mathbf{R}^n$ is a convex set and $y \mapsto \mathcal{A}(y) = Ay + b$ is an affine mapping from $\mathbf{R}^m$ to $\mathbf{R}^n$ (where $A \in \mathbf{R}^{n \times m}$ and $b \in \mathbf{R}^n$) then the inverse image of M under the mapping $\mathcal{A}(\cdot)$, i.e., the set*

$$\mathcal{A}^{-1}(M) := \{y \in \mathbf{R}^m : \mathcal{A}(y) \in M\},$$

is convex.

The (completely straightforward) verification of this proposition is left to the reader.

1.3.1 Calculus of Closed Convex Sets

Numerous important convexity-related results require not just the convexity, but also the closedness, of the participating sets. Therefore, it makes sense to think to what extent the "calculus of convexity" as presented in Proposition 1.21 is preserved when passing from general convex sets to *closed convex* sets. Here are the answers:

1. *Taking the intersection*: If M_α, $\alpha \in \mathcal{A}$, are closed convex sets, so is the set $\bigcap_\alpha M_\alpha$.
2. *Taking the direct product*: If $M_1 \subseteq \mathbf{R}^{n_1}$ and $M_2 \subseteq \mathbf{R}^{n_2}$ are closed convex sets, so is the set $M_1 \times M_2$.
3. *The arithmetic summation* of nonempty closed convex sets M_i, $1 \leq i \leq k$, preserves convexity but not necessarily closedness. However, it does preserve closedness when at most one of the sets is unbounded.

1.3 Calculus of Convex Sets

An example of a pair of closed convex sets with non-closed sum is $M_1 = \{x \in \mathbf{R}^2 : x_1 > 0, x_2 \geq 1/x_1\}$, $M_2 = \{x \in \mathbf{R}^2 : x_2 = 0\}$. The sum of these two closed sets clearly is the open upper halfplane $\{x \in \mathbf{R}^2 : x_2 > 0\}$ (why?) and is not closed.

Let us verify that if at most one of the nonempty closed convex sets is unbounded then the sum of the sets is convex (this we already know from the calculus of convexity) and closed. Closedness is given by the following observation:

> *The sum of nonempty closed sets, convex or not, with at most one of the sets unbounded, is closed.*

To justify this observation, it clearly suffices to verify its validity for a pair of sets, M_1 and M_2. Assuming both sets are nonempty and closed and M_1 is bounded, we need to prove that if a sequence $\{x^i + y^i\}_i$ with $x^i \in M_1$ and $y^i \in M_2$, converges as $i \to \infty$, the limit $\lim_{i\to\infty} (x^i + y^i)$ belongs to $M_1 + M_2$. Since M_1, and thus the sequence $\{x^i\}_i$, is bounded, passing to a subsequence we may assume that the sequence $\{x^i\}_i$ converges, as $i \to \infty$, to some x. Since the sequence $\{x^i + y^i\}_i$ converges as well, the sequence $\{y^i\}_i$ also converges to some y. As M_1 and M_2 are closed, we have $x \in M_1$, $y \in M_2$, and therefore $\lim_{i\to\infty} (x^i + y^i) = x + y \in M_1 + M_2$.

4. *Multiplication by a real:* For a nonempty closed convex set M and a real λ, the set λM is closed and convex (why?).
5. *The image under an affine mapping* of a closed convex set M is convex, but not necessarily closed; it is definitely closed when M is bounded.

   As an example of closed convex set with a non-closed affine image consider the set $\{[x; y] \in \mathbf{R}^2 : x, y \geq 0, xy \geq 1\}$ (i.e., a branch of a hyperbola) and its projection onto the x-axis. This set is convex and closed, but its projection onto the x-axis is the positive ray $x \in \mathbf{R} : x > 0$, which is not closed. The closedness of the affine image of a closed and bounded set is a special case of the following general fact:

   > *The image of a closed and bounded set under a mapping that is continuous on this set is closed and bounded as well* (why?).

6. *Inverse image under affine mapping:* If $M \subseteq \mathbf{R}^n$ is convex and closed and $y \mapsto \mathcal{A}(y) = Ay + b$ is an affine mapping from $\mathbf{R}^m$ to $\mathbf{R}^n$ then the set

   $$\mathcal{A}^{-1}(M) := \{y \in \mathbf{R}^m : \mathcal{A}(y) \in M\}$$

   is a closed convex set in $\mathbf{R}^m$. Indeed, the convexity of $\mathcal{A}^{-1}(M)$ is given by the calculus of convexity, and its closedness is due to the following standard fact:

   > *The inverse image of a closed set in $\mathbf{R}^n$ under a continuous mapping from $\mathbf{R}^m$ to $\mathbf{R}^n$ is closed* (why?).

We see that the "calculus of closed convex sets" is somewhat weaker than the calculus of convexity *per se*. Nevertheless, we will see that these difficulties disappear when restricting the operands of our operations to be polyhedral, and not just closed and convex.

1.4 Topological Properties of Convex Sets

Convex sets and closely related objects – convex functions – play the central role in optimization. To play this role properly, convexity alone is not sufficient; we need convexity *and* closedness.

1.4.1 The Closure

It is clear from definition of a closed set that the intersection of a family of closed sets in $\mathbf{R}^n$ is also closed (see Fact B.13). From this fact it follows that for every subset M of $\mathbf{R}^n$ there exists the smallest (w.r.t. inclusion) closed set containing M. This leads us to the following definition.

Definition 1.22 *[Closure] Given a set $M \subseteq \mathbf{R}^n$, the closure of M [notation: cl M or cl(M)] is the smallest (w.r.t. inclusion) closed set (i.e., the intersections of all closed sets) containing M.*

From Real Analysis, we have the following inner description of the closure of a set in a metric space (and, in particular, in $\mathbf{R}^n$).

Fact 1.23 *The closure of a set $M \subseteq \mathbf{R}^n$ is exactly the set composed of the limits of all converging sequences of elements from M.*

Example 1.5 Based on Fact 1.23, it is easy to prove that, e.g., the closure of the open Euclidean ball

$$\{x \in \mathbf{R}^n : \|x - a\|_2 < r\} \quad [\text{where } r > 0]$$

is the closed Euclidean ball $\{x \in \mathbf{R}^n : \|x - a\|_2 \leq r\}$.

Another useful example is the closure of a set defined by *strict* linear inequalities, i.e.,

$$M := \left\{x \in \mathbf{R}^n : a_\alpha^\top x < b_\alpha, \alpha \in \mathcal{A}\right\}.$$

Whenever such a set M is nonempty, its closure is given by the nonstrict versions of the same inequalities:

$$\mathrm{cl}\, M = \left\{x \in \mathbf{R}^n : a_\alpha^\top x \leq b_\alpha, \alpha \in \mathcal{A}\right\}.$$

Note here that the nonemptiness of M in this last example is essential. To see this, consider the set $M = \{x \in \mathbf{R} : x < 0, -x < 0\}$. Clearly, M is empty, so that its closure is also the empty set. On the other hand, if we ignore the nonemptiness requirement on M and apply the above rule formally, we would incorrectly claim that cl $M = \{x \in \mathbf{R} : x \leq 0, -x \leq 0\} = \{0\}$. $\diamond$

1.4.2 The Interior

Consider a set $M \subseteq \mathbf{R}^n$. Referring to Definition B.9, a point $x \in M$ is an *interior* point of M if some neighborhood of the point is contained in M, i.e., if there exists a ball of positive radius centered at x which is contained in M:

$$\exists r > 0: \quad B_r(x) := \{y \in \mathbf{R}^n : \|y - x\|_2 \leq r\} \subseteq M.$$

Definition 1.24 *[Interior] The set of all interior points of a given set $M \subseteq \mathbf{R}^n$ is called the* interior *of M [notation: int M or int(M)] (see Definition B.10).*

Example 1.6 We have for the following sets their corresponding interiors:

- The interior of an open set is the set itself.
- The interior of the closed ball $\{x \in \mathbf{R}^n : \|x - a\|_2 \leq r\}$ ($r > 0$ or $n \geq 1$) is the open ball $\{x \in \mathbf{R}^n : \|x - a\|_2 < r\}$ (why?).
- The interior of the *standard full-dimensional simplex*

$$\left\{ \mu \in \mathbf{R}^n : \mu \geq 0, \ \sum_{i=1}^{n} \mu_i \leq 1 \right\}$$

  is composed of all vectors μ with $\mu_i > 0$ for all $i = 1, \ldots, n$ and with $\sum_{i=1}^{n} \mu_i < 1$ (why?).
- The interior of a polyhedral set $\{x \in \mathbf{R}^n : Ax \leq b\}$ with matrix A containing no zero rows is the set $\{x \in \mathbf{R}^n : Ax < b\}$ (why?).

  Note that here the requirement that the set is polyhedral, i.e., defined by a *finite* system of linear inequalities, is critical. In particular, this statement is *not*, generally speaking, true for solution sets of infinite systems of linear inequalities. For example, the following set, defined by an infinite system of linear inequalities

$$M := \left\{ x \in \mathbf{R} : x \leq \frac{1}{n}, \ n = 1, 2, \ldots \right\},$$

  is just the nonpositive ray $\mathbf{R}_- = \{x \in \mathbf{R} : x \leq 0\}$, i.e., $M = \mathbf{R}_-$. Thus, int $M = \{x \in \mathbf{R} : x < 0\}$, i.e., it is the negative ray. On the other hand, the following set, defined by the strict versions of the inequalities

$$M' := \left\{ x \in \mathbf{R} : x < \frac{1}{n}, \ n = 1, 2, \ldots \right\}$$

  is the same nonpositive ray, i.e., $M' = \{x \in \mathbf{R} : x \leq 0\}$. Hence, $M' \neq$ int M for this set M, defined by an infinite system of inequalities. ◇

The following observation is evident:

Fact 1.25 *For any set M in $\mathbf{R}^n$, its interior,* int M, *is always open, and* int M *is the largest (w.r.t. inclusion) open set contained in M.*

The interior of a set is, of course, contained in the set, which, in turn, is contained in its closure:

$$\text{int } M \subseteq M \subseteq \text{cl } M. \tag{1.3}$$

Definition 1.26 *[Boundary] For any $M \subseteq \mathbf{R}^n$, the* boundary *of M [notation: bd M or bd(M)] is the set*

$$\text{bd } M := \text{cl } M \setminus \text{int } M,$$

and the points on the boundary are called the boundary points *of M.*

The boundary points of M are exactly the points from $\mathbf{R}^n$ which can be approximated to whatever high accuracy both by points from M and by points from outside M (check this!).

Given a set $M \subseteq \mathbf{R}^n$, it is important to note that the boundary points do not necessarily belong to M, since $M = \operatorname{cl} M$ need not necessarily hold in general. In fact, all boundary points belong to M if and only if $M = \operatorname{cl} M$, i.e., if and only if M is closed.

The boundary of a set $M \subseteq \mathbf{R}^n$ is clearly closed as $\operatorname{bd} M = \operatorname{cl} M \cap (\mathbf{R}^n \setminus \operatorname{int} M)$ and both sets $\operatorname{cl} M$ and $\mathbf{R}^n \setminus \operatorname{int} M$ are closed (note that the set $\mathbf{R}^n \setminus \operatorname{int} M$ is closed since it is the complement of an open set). In addition, from the definition of the boundary, we have

$$M \subseteq (\operatorname{int} M \cup \operatorname{bd} M) = \operatorname{cl} M.$$

Therefore, any point from M is either an interior point or a boundary point of M.

1.4.3 The Relative Interior

Many of the constructions in optimization possess nice properties in the interior of the set to which the construction is related and may lose these nice properties at the boundary points of the set. This is why in many cases we are especially interested in interior points of sets and want the set of these interior points to be "sufficiently dense." What should we do if this is not the case, for example if there are no interior points at all (e.g., if we are looking at a line segment in the plane)? It turns out that in these cases we can use a good surrogate of the "normal" interior, namely the *relative interior*, defined as follows.

Definition 1.27 *[Relative interior] Let $M \subseteq \mathbf{R}^n$ be nonempty. We say that a point $x \in M$ is a relative interior point of M if M contains the intersection of a small enough ball centered at x with $\operatorname{Aff}(M)$, i.e., if there exists $r > 0$ such that*

$$(B_r(x) \cap \operatorname{Aff}(M)) := \left\{ y \in \mathbf{R}^n : y \in \operatorname{Aff}(M), \|y - x\|_2 \leq r \right\} \subseteq M.$$

The relative interior of M [notation: $\operatorname{rint} M$] refers to the set of all relative interior points of M.

By definition, the relative interior of an empty set is empty.

Example 1.7 We have for the following sets their corresponding relative interiors:

- The relative interior of a singleton is the singleton itself (since a point in the 0-dimensional space is the same as a ball of a positive radius).
- More generally, the relative interior of an affine subspace is the subspace itself.
- Given two distinct point $x \neq y$ in $\mathbf{R}^n$, the interior of a segment $[x, y]$ is empty whenever $n > 1$. In contrast with this, the relative interior of this set is always (independently of n) nonempty and it is precisely the interval (x, y), i.e., the segment without its endpoints. ◇

Geometrically speaking, the relative interior is the interior we get when we treat $M \subseteq \mathbf{R}^n$ as a subset of its affine hull (the latter, geometrically, is just $\mathbf{R}^k$, k being the affine dimension of $\operatorname{Aff}(M)$).

We can play with the notion of the relative interior in basically the same way as with the notion of the interior. Namely, for any $M \subseteq \mathbf{R}^n$, since $\operatorname{Aff}(M)$ is closed and contains M it contains also the smallest closed set containing M, i.e., $\operatorname{cl} M$. Therefore, we have the following analogies of inclusions, cf. (1.3):

$$\operatorname{rint} M \subseteq M \subseteq \operatorname{cl} M \quad [\subseteq \operatorname{Aff}(M)]. \tag{1.4}$$

We can also define the *relative boundary*.

Definition 1.28 *[Relative boundary] For any $M \subseteq \mathbf{R}^n$, its* relative boundary *[notation: rbd M] is defined as the set* $\operatorname{rbd} M := \operatorname{cl} M \setminus \operatorname{rint} M$.

Note that, for any $M \subseteq \mathbf{R}^n$, we naturally have that rbd M is a closed set contained in $\operatorname{Aff}(M)$, and, as for the "actual" interior and boundary, we have

$$\operatorname{rint} M \subseteq M \subseteq \operatorname{cl} M = \operatorname{rint} M \cup \operatorname{rbd} M.$$

Of course, if $\operatorname{Aff}(M) = \mathbf{R}^n$ then the relative interior becomes the usual interior, and similarly for the boundary. Note that $\operatorname{Aff}(M) = \mathbf{R}^n$ is certainly the case when $\operatorname{int} M \neq \emptyset$ (since then M contains a ball B, and therefore the affine hull of M is the entire space $\mathbf{R}^n$, which is the affine hull of B).

1.4.4 Nice Topological Properties of Convex Sets

An arbitrary set $M \subseteq \mathbf{R}^n$ may possess a very pathological topology. In particular, both inclusions in the chain

$$\operatorname{rint} M \subseteq M \subseteq \operatorname{cl} M$$

can be very "loose." For example, let M be the set of rational numbers in the segment $[0, 1] \subset \mathbf{R}$. Then, rint $M = \operatorname{int} M = \emptyset$ since every neighborhood of every rational number contains irrational numbers. On the other hand, cl $M = [0, 1]$. Thus, rint M is "incomparably smaller" than M, cl M is "incomparably larger" than M, and M is contained in its relative boundary (by the way, what is this relative boundary?).

The following theorem demonstrates that the topology of a *convex* set M is much better than what it might be for an arbitrary set.

Theorem 1.29 *Let M be a convex set in $\mathbf{R}^n$. Then:*

(i) *The interior* int M, *the closure* cl M, *and the relative interior* rint M *are convex.*
(ii) *If M is nonempty, then its relative interior* rint M *is nonempty.*
(iii) *The closure of M is the same as the closure of its relative interior, i.e.,* $\operatorname{cl} M = \operatorname{cl}(\operatorname{rint} M)$. *(In particular, every point of* cl M *is the limit of a sequence of points from* rint M.*)*
(iv) *The relative interior remains unchanged when we replace M with its closure, i.e.,* $\operatorname{rint} M = \operatorname{rint}(\operatorname{cl} M)$.

Moreover, (iii) and (iv) imply that

(v) *The relative boundary remains unchanged when we replace M with its closure.*

We will use the following basic result to prove this theorem (the proof of this lemma will be presented after the proof of the theorem).

Lemma 1.30 *Let M be a convex set in $\mathbf{R}^n$. Then, for any $x \in \operatorname{rint} M$ and $y \in \operatorname{cl} M$, we have*

$$[x, y) := \{(1 - \lambda)x + \lambda y : 0 \leq \lambda < 1\} \subseteq \operatorname{rint} M.$$

Proof of Theorem 1.29 (i): The reader is invited to prove this.

(ii): Let M be a nonempty convex set, and let us prove that $\operatorname{rint} M \neq \emptyset$. By translation, we may assume that $0 \in M$. There is nothing to prove when $M = \{0\}$, in which case $\operatorname{rint} M = \{0\}$ is nonempty. Now assume that M contains, along with the origin, a nonzero vector. Furthermore, we may assume that the linear span of M, i.e., $\operatorname{Lin}(M)$, is the whole of $\mathbf{R}^n$. Indeed, as far as linear operations and the Euclidean structure are concerned, $\operatorname{Lin}(M)$, as every other linear subspace in $\mathbf{R}^n$, is equivalent to $\mathbf{R}^k$ for a certain k. Since the notion of the relative interior deals only with linear and Euclidean structures, we lose nothing by thinking of $\operatorname{Lin}(M)$ as of $\mathbf{R}^k$ and taking it as our universe instead of the original universe $\mathbf{R}^n$. Thus, in the rest of the proof of (ii), we assume that $0 \in M$ and $\operatorname{Lin}(M) = \mathbf{R}^n$; what we need to prove is that the interior of M (which in our case is the same as the relative interior of M) is nonempty. Note that, since $0 \in M$, we have $\operatorname{Aff}(M) = \operatorname{Lin}(M) = \mathbf{R}^n$.

As $\operatorname{Lin}(M) = \mathbf{R}^n$, we can find n linearly independent vectors $a^1, \ldots, a^n$ in M. Let us also set $a^0 := 0$. The $n+1$ vectors $a^0, \ldots, a^n$ belong to M. *Since M is convex, the convex hull of these vectors, i.e.,*

$$\Delta := \left\{ x = \sum_{i=0}^{n} \lambda_i a^i : \lambda \geq 0, \sum_{i=0}^{n} \lambda_i = 1 \right\} = \left\{ x = \sum_{i=1}^{n} \mu_i a^i : \mu \geq 0, \sum_{i=1}^{n} \mu_i \leq 1 \right\}$$

also belongs to M. Note that the set Δ is the image of the *standard full-dimensional simplex*

$$\left\{ \mu \in \mathbf{R}^n : \mu \geq 0, \sum_{i=1}^{n} \mu_i \leq 1 \right\}$$

under the linear transformation $\mu \mapsto A\mu$, where A is the matrix with columns $a^1, \ldots, a^n$. Recall from Example 1.6 that the standard simplex has a nonempty interior. Since A is nonsingular (due to the linear independence of $a^1, \ldots, a^n$), multiplication by A maps open sets onto open sets, so that Δ has a nonempty interior. Since $\Delta \subseteq M$, the interior of M is nonempty.

(iii): The statement is evidently true when M is empty, so we assume that $M \neq \emptyset$. We clearly have $\operatorname{cl}(\operatorname{rint} M) \subseteq \operatorname{cl} M$ since $\operatorname{rint} M \subseteq M$. Thus, all we need to do to complete the proof of (iii) is to verify that every $y \in \operatorname{cl} M$ is the limit of a sequence of points $y^i \in \operatorname{rint} M$. Indeed, pick $x \in \operatorname{rint} M$ (recall that from part (ii) we have $\operatorname{rint} M \neq \emptyset$) and set $y^i := (1 - 1/i)y + (1/i)x$. By Lemma 1.30, we have $y^i \in \operatorname{rint} M$, and clearly $y = \lim_{i \to \infty} y^i$, completing the verification of (iii).

(iv): The statement is obviously true when M is empty, so we will assume that $M \neq \emptyset$. Since $M \subseteq \operatorname{cl} M$, we always have $\operatorname{rint} M \subseteq \operatorname{rint}(\operatorname{cl} M)$. To prove the reverse inclusion, let us consider any $z \in \operatorname{rint}(\operatorname{cl} M)$ and prove that $z \in \operatorname{rint} M$. Let $x \in \operatorname{rint} M$ (from part (ii), we already know that $\operatorname{rint} M \neq \emptyset$). As x and z are in $\operatorname{Aff}(M)$, for any $t \in \mathbf{R}$ the vectors $z^t := x + t(z - x)$ belong to $\operatorname{Aff}(M)$, and, when t approaches 1, z^t approaches z. Since $z \in \operatorname{rint}(\operatorname{cl} M)$, it follows that there exists $\epsilon > 0$ such that $y := z^{1+\epsilon} \in \operatorname{cl} M$. It remains to note that $z = (1 - \lambda)y + \lambda x$ with $\lambda = \frac{\epsilon}{1+\epsilon} \in (0, 1)$, and therefore $z = (1 - \lambda)y + \lambda x \in \operatorname{rint} M$ by Lemma 1.30 (recall that $x \in \operatorname{rint} M$, $y \in \operatorname{cl} M$). ∎

Remark 1.31 We see from the proof of Theorem 1.29(iii) that to get the closure of a (nonempty) convex set, it suffices to take its "radial" closure, i.e., to take a point $x \in \operatorname{rint} M$, take all rays in Aff($M$) starting at x, and look at the intersection of such a ray ℓ with M; such an intersection will be a convex set on the line which contains a one-sided neighborhood of x, i.e., it is either a segment $[x, y^\ell]$, or the entire ray ℓ, or a half-interval $[x, y^\ell)$. In the first two cases we do not need to do anything; in the third case, we need to add y^ℓ to M. After all rays are considered and all "missed" endpoints y^ℓ are added to M, we obtain cl M. To understand the role of convexity in this result, look at the *nonconvex* set of rational numbers from $[0, 1]$. The interior ($\equiv$ relative interior) of this "highly percolated" set is empty, the closure is $[0, 1]$, and there is no way to restore the closure in terms of the interior. ◇

Proof of Lemma 1.30 Given that $x \in M$, let us denote Aff(M) $= x + L$, where L is the linear subspace parallel to Aff(M). Then

$$M \subseteq \mathrm{Aff}(M) = x + L.$$

Let B be the unit Euclidean ball in L, i.e., $B = \{h \in L : \|h\|_2 \leq 1\}$. Since $x \in \mathrm{rint}\, M$, there exists a positive radius r such that

$$x + rB \subseteq M. \tag{1.5}$$

Now consider any $\lambda \in [0, 1)$, and let $z := (1 - \lambda)x + \lambda y$. As $y \in \mathrm{cl}\, M$, we have $y = \lim_{i \to \infty} y^i$ for a certain sequence of points y^i from M. By setting $z^i := (1 - \lambda)x + \lambda y^i$, we get $z^i \to z$ as $i \to \infty$. Then, from (1.5) and the convexity of M, it follows that the sets $Z_i := \{(1 - \lambda)x' + \lambda y^i : x' \in x + rB\}$ are contained in M. Clearly, Z_i is exactly the set $z^i + r'B$, where $r' := (1 - \lambda)r > 0$. Thus, z is the limit of the sequence z^i, and the r'-neighborhood (in Aff(M)) of every point z^i belongs to M. For every $0 < r'' < r'$ and for all i such that z^i is close enough to z, the r'-neighborhood of z^i contains the r''-neighborhood of z; thus, a neighborhood (in Aff(M)) of z belongs to M, hence $z \in \mathrm{rint}\, M$. ∎

A useful byproduct of Lemma 1.30 is as follows:

Corollary 1.32 *Let $M \subseteq \mathbf{R}^n$ be convex. Then every convex combination $\sum_{i \leq K} \lambda_i x^i$ of points $x^i \in \mathrm{cl}\, M$ such that $K \geq 2$ and at least one term with positive coefficient is associated with $x^i \in \mathrm{rint}\, M$ is in fact a point from $\mathrm{rint}\, M$.*

Another useful byproduct of Lemma 1.30 is as follows. Let $M_k, k \leq K$, be a finite collection of subsets of $\mathbf{R}^n$. The closure of the union of these sets is the union of their closures: $\mathrm{cl}(\cup_{k \leq K} M_k) = \cup_{k \leq K} \mathrm{cl}\, M_j$ (why?). Now let us ask ourselves a similar question about intersections: *what is the relation between $\mathrm{cl} \cap_{k \leq K} M_k$ and $\cap_{k \leq K} \mathrm{cl}\, M_k$?* The set $\cap_{k \leq K} \mathrm{cl}\, M_k$ is closed and clearly contains $\cap_{k \leq K} M_k$ and thus always contains the closure of the latter set:

$$\mathrm{cl}\left(\bigcap_{k \leq K} M_k\right) \subseteq \bigcap_{k \leq K} \mathrm{cl}\, M_k. \tag{1.6}$$

In general, this inclusion can be "loose" – the right-hand side set in (1.6) can be much larger than the left-hand side set, even when all M_k are convex. For example, when $K = 2$, $M_1 = \{x \in \mathbf{R}^2 : x_2 = 0\}$ is the x_1-axis and $M_2 = \{x \in \mathbf{R}^2 : x_2 > 0\} \cup \{[0; 0]\}$, both sets are convex and their intersection is the singleton $\{0\}$, so that $\mathrm{cl}(M_1 \cap M_2) = \mathrm{cl}\{0\} = \{0\}$, while the intersection of cl M_1 and cl M_2 is the entire x_1-axis, which is simply M_1. In this example the right-hand side in (1.6) is "incomparably larger" than the left-hand side. However, under suitable assumptions we can also achieve equality in (1.6).

Proposition 1.33 *Consider convex sets $M_k \subseteq \mathbf{R}^n$, $k \leq K$.*

(i) *If $\bigcap_{k \leq K} \operatorname{rint} M_k \neq \emptyset$ then $\operatorname{cl}(\bigcap_{k \leq K} M_k) = \bigcap_{k \leq K} \operatorname{cl} M_k$, i.e., (1.6) holds as an equality.*
(ii) *Moreover, if $K \geq 2$ and $M_K \cap \operatorname{int} M_1 \cap \operatorname{int} M_2 \cap \cdots \cap \operatorname{int} M_{K-1} \neq \emptyset$ then we have $\bigcap_{k \leq K} \operatorname{rint} M_k \neq \emptyset$, i.e., the premise (and thus the conclusion) in (i) holds true, so that $\operatorname{cl}(\bigcap_{k \leq K} M_k) = \bigcap_{k \leq K} \operatorname{cl} M_k$.*

Proof (i): To prove that under the premise of (i) the inclusion (1.6) becomes an equality is the same as to verify that under the same premise, given $x \in \bigcap_k \operatorname{cl} M_k$, one has $x \in \operatorname{cl}(\bigcap_k M_k)$. Indeed, under the premise of (i) there exists $\bar{x} \in \bigcap_k \operatorname{rint} M_k$. Then, for every k, we have $\bar{x} \in \operatorname{rint} M_k$ and $x \in \operatorname{cl} M_k$, implying by Lemma 1.30 that the set $\Delta := [\bar{x}, x) = \{(1 - \lambda)\bar{x} + \lambda x : 0 \leq \lambda < 1\}$ is contained in M_k. Since $\Delta \subseteq M_k$ for all k, we have $\Delta \in \bigcap_k M_k$, and thus $\operatorname{cl} \Delta \subseteq \operatorname{cl}(\bigcap_k M_k)$. It remains to note that $x \in \operatorname{cl} \Delta$.

(ii): Let $\bar{x} \in M_K \cap \operatorname{int} M_1 \cap \cdots \cap \operatorname{int} M_{K-1}$. As $\bar{x} \in \operatorname{int} M_k$ for all $k < K$, there exists an open set $U \subset \bigcap_{k<K} M_k$ such that $\bar{x} \in U$. As $\bar{x} \in M_K \subseteq \operatorname{cl} M_K$, by Theorem 1.29 $\bar{x}$ is the limit of a sequence of points from $\operatorname{rint} M_K$, so that there exists $\widehat{x} \in U \cap \operatorname{rint} M_K$. Due to the definition of U, we have $\widehat{x} \in \operatorname{rint} M_k$ for all $k \leq K$, so that the premise of (i) indeed takes place. ∎

1.5 ★ Conic and Perspective Transforms of a Convex Set

Let $X \subseteq \mathbf{R}^n$ be a nonempty convex set. We can "lift" it to $\mathbf{R}^{n+1}$ by passing to the set

$$X^+ := \{[x; 1] \in \mathbf{R}^n \times \mathbf{R} : x \in X\}.$$

Now let us look at the conic hull of X^+, given by

$$\operatorname{ConeT}(X) := \operatorname{Cone}(X^+)$$
$$= \left\{ [x; t] \in \mathbf{R}^n \times \mathbf{R}_+ : \begin{array}{l} \exists (I, \lambda_i \geq 0, x^i \in X, \forall i \leq I): \\ x = \sum_{i \leq I} \lambda_i x^i, \ t = \sum_{i \leq I} \lambda_i \end{array} \right\}.$$

We will call this the *conic transform of X*; see Figure 1.4. Note that this set is indeed a cone. Moreover, all vectors $[x; t]$ from this cone have $t \geq 0$, and, importantly, the only vector with $t = 0$ in the cone $\operatorname{ConeT}(X)$ is the origin in $\mathbf{R}^{n+1}$ (this is what you get when taking trivial – with all coefficients zero – conic combinations of vectors from X^+).

All nonzero vectors $[x; t]$ from $\operatorname{ConeT}(X)$ have $t > 0$ and form a convex set, which we call the *perspective transform* $\operatorname{Persp}(X)$ of X:

$$\operatorname{Persp}(X) := \{[x; t] \in \operatorname{ConeT}(X) : t > 0\} = \operatorname{ConeT}(X) \setminus \{0_{n+1}\}.$$

The name of this set is motivated by the following immediate observation:

Proposition 1.34 *[Perspective transform of a nonempty convex set] Let X be a nonempty convex set in $\mathbf{R}^n$. Then, its perspective transform admits the representation*

$$\operatorname{Persp}(X) = \{[x; t] \in \mathbf{R}^n \times \mathbf{R} : t > 0, \ x/t \in X\}. \tag{1.7}$$

1.5 ★ Conic and Perspective Transforms of a Convex Set

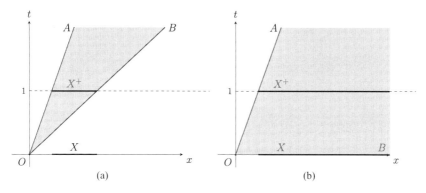

Figure 1.4 Conic transform: (a) the conic transform of segment X is the angle AOB; (b) the conic transform of ray X is the angle AOB with the relative interior of the ray OB excluded.

In other words, to get $\mathrm{Persp}(X)$, *we pass from X to X^+ (i.e., we lift X to $\mathbf{R}^{n+1}$) and then take the union of all rays $\{[sx;s] \in \mathbf{R}^n \times \mathbf{R} : s > 0, \, x \in X\}$ emanating from the origin (with the origin excluded) and passing through the points of X^+.*

Proof Let $\widehat{X} := \{[x;t] \in \mathbf{R}^n \times \mathbf{R} : t > 0, \, x/t \in X\}$, so that the claim in the proposition is $\mathrm{Persp}(X) = \widehat{X}$. Consider a point $[x;t] \in \widehat{X}$. Then $t > 0$ and $y := x/t \in X$, and thus we have $[x,t] = t[y;1]$, so that the point $[x;t]$ from $\widehat{X}$ is a single-term conic combination – just a positive multiple – of the point $[y;1] \in X^+$. As this holds for every point $[x;t] \in \widehat{X}$ we conclude that $\widehat{X} \subseteq \mathrm{Persp}(X)$. To verify the opposite inclusion, recall that every point $[x;t] \in \mathrm{Persp}(X)$ is of the form $[\sum_i \lambda_i x^i; \sum_i \lambda_i]$ with $x^i \in X$, $\lambda_i \geq 0$, and $t = \sum_i \lambda_i > 0$. Then

$$\left[\sum_i \lambda_i x^i; \sum_i \lambda_i\right] = t\left[\sum_i (\lambda_i/t) x^i; 1\right] = t[y;1],$$

where $y := \sum_i (\lambda_i/t) x^i$. Note that $y \in X$, as it is a convex combination of points from X, and X is convex. Thus, $[x;t]$ is such that $t > 0$ and $y = x/t \in X$, that is, $\widehat{X} \supseteq \mathrm{Persp}(X)$ as desired. ∎

As a byproduct of Proposition 1.34, we conclude that the right-hand side set in (1.7) is convex whenever X is convex and nonempty – a fact that is not immediately evident.

Note that X^+ is geometrically the same as X, and moreover we can view X^+ as simply the intersection of $\mathrm{ConeT}(X)$ (or $\mathrm{Persp}(X)$) with the hyperplane $t = 1$ in $\mathbf{R}^n \times \mathbf{R}$.

Example 1.8

1. $\mathrm{ConeT}(\mathbf{R}^n) = \{[x;t] \in \mathbf{R}^{n+1} : t > 0\} \cup \{0_{n+1}\}$, and $\mathrm{Persp}(\mathbf{R}^n) = \{[x;t] \in \mathbf{R}^{n+1} : t > 0\}$.
2. $\mathrm{ConeT}(\mathbf{R}^n_+) = \{[x;t] \in \mathbf{R}^{n+1}_+ : t > 0\} \cup \{0_{n+1}\}$, and $\mathrm{Persp}(\mathbf{R}^n_+) = \{[x;t] \in \mathbf{R}^{n+1}_+ : t > 0\}$.
3. Given any norm $\|\cdot\|$ on $\mathbf{R}^n$, let B be its unit ball. Then, we have $\mathrm{ConeT}(B) = \{[x;t] \in \mathbf{R}^{n+1} : t \geq \|x\|\}$ and $\mathrm{Persp}(B) = \{[t;x] \in \mathbf{R}^{n+1} : t \geq \|x\|, \, t > 0\}$. ◇

Note that, in all three cases in Example 1.8, the set X of which we are taking conic and perspective transforms is not just convex but also closed. However, in the first two cases the conic transform is a *non-closed* cone, while in the third case the conic transform is closed; however, in all three cases the intersections of $\mathrm{ConeT}(X)$ with the half-space $\{[x;t] \in \mathbf{R}^{n+1} : t \geq \alpha\}$ are closed, provided $\alpha > 0$. There is indeed a general fact underlying this phenomenon.

Proposition 1.35 *Let $X \subset \mathbf{R}^n$ be a nonempty convex set. Then, we have the following:*

(i) *For $\alpha > 0$, define $H_\alpha := \{[x;t] \in \mathbf{R}^{n+1} : t \geq \alpha\}$. When X is closed, $\mathrm{ConeT}(X) \cap H_\alpha = \mathrm{Persp}(X) \cap H_\alpha$ and this intersection is closed for any $\alpha > 0$.*

(ii) *Moreover, the cone $\mathrm{ConeT}(X)$ is closed if and only if X is closed and bounded. In fact, $\mathrm{ConeT}(X)$ is closed if and only if $\mathrm{cl}\,(\mathrm{Persp}(X)) = \mathrm{ConeT}(X)$.*

Proof (i): When $\alpha > 0$, we clearly have $\mathrm{ConeT}(X) \cap H_\alpha = \mathrm{Persp}(X) \cap H_\alpha$. To see that these intersections are closed whenever X is closed, invoking (1.7) it suffices to prove that, when $\{[x^i;t_i]\}_{i \geq 1}$ is a converging sequence such that $t_i \geq \alpha$ and $x^i/t_i \in X$, the limit $[x;t]$ of this sequence satisfies $x/t \in X$ and $t \geq \alpha$. Since $t_i \to t$ as $i \to \infty$ and $t_i \geq \alpha$ holds for all i, we clearly have $t \geq \alpha$. Moreover, we have that the converging sequence $y^i := x^i/t_i$ is in X; thus $x^i/t_i \to x/t$ as $i \to \infty$ and the point x/t is in X since X is closed.

(ii): First, we assume that the nonempty convex set X is closed and bounded and then prove that $\mathrm{ConeT}(X)$ is closed, that is, whenever a sequence $\{[x^i;t_i]\}_{i \geq 1}$ of points from $\mathrm{ConeT}(X)$ converges, the limit of the sequence belongs to $\mathrm{ConeT}(X)$. Indeed, consider such a sequence along with its limit $[x;t]$. When $t > 0$, all but finitely many terms of the sequence belong to the half-space $H_{t/2}$, and as, by part (i), $\mathrm{ConeT}(X) \cap H_{t/2}$ is closed, we have $[x;t] \in \mathrm{ConeT}(X)$. When $t = 0$, then either (a) $t_i = 0$ for infinitely many values of i, or (b) $t_i > 0$ for all but finitely many values of i. In case (a), infinitely many terms in our sequence are of the form $[0_n;0]$ (since whenever $[y;0] \in \mathrm{ConeT}(X)$ we must have $y = 0$ as well), so that $[x;t] = 0_{n+1} \in \mathrm{ConeT}(X)$. In case (b), for all large enough i we have $t_i > 0$ and $x^i/t_i \in X$, and since $t_i \to 0$ as $i \to \infty$ and X is bounded we deduce that $x^i \to 0$ as $i \to \infty$. Then this, together with $t_i \to 0$, $i \to \infty$, implies that $[x;t] = [0_{n+1};0]$, and we again have $[x;t] \in \mathrm{ConeT}(X)$. Thus, whenever X is a nonempty closed and bounded set, $\mathrm{ConeT}(X)$ is closed.

Now assume that $\mathrm{ConeT}(X)$ is closed, and let us prove that X is closed and bounded. Clearly, X is closed if and only if X^+ is closed, and since X^+ is the intersection of the closed set $\mathrm{ConeT}(X)$ with the hyperplane $t = 1$ in $\mathbf{R}^n \times \mathbf{R}$, X^+ is indeed closed. It remains to prove that X is bounded. Assume for contradiction that X is unbounded. Then we can find a sequence $x^i \in X$, $i \geq 1$, with $\|x^i\|_2 \to \infty$ as $i \to \infty$. Passing to a subsequence, we can assume that the $\|\cdot\|_2$-unit vectors $\xi^i := x^i/\|x^i\|_2$ converge to some unit vector ξ. Setting $t_i := 1/\|x^i\|_2$, we have $t_i > 0$, $t_i \to 0$ as $i \to \infty$, and $\xi^i/t_i = x^i \in X$, so that $[\xi^i;t_i] \in \mathrm{ConeT}(X)$. Since $\mathrm{ConeT}(X)$ is closed and by construction $[\xi^i;t_i] \to [\xi;0]$ as $i \to \infty$, we would have $[\xi;0] \in \mathrm{ConeT}(X)$, which is impossible as $\|\xi\|_2 = 1$.

The claim that $\mathrm{ConeT}(X)$ is closed if and only if $\mathrm{cl}\,(\mathrm{Persp}(X)) = \mathrm{ConeT}(X)$ follows immediately as well. Indeed, whenever X is nonempty and convex, we have $\mathrm{Persp}(X) = \mathrm{ConeT}(X) \setminus \{0_{n+1}\}$ and clearly $0_{n+1} \in \mathrm{cl}(\mathrm{Persp}(X))$, implying that $\mathrm{cl}(\mathrm{ConeT}(X)) = \mathrm{cl}(\mathrm{Persp}(X))$. As a result, whenever X is nonempty and convex, $\mathrm{ConeT}(X)$ is closed if and only if $\mathrm{ConeT}(X) = \mathrm{cl}(\mathrm{Persp}(X))$. ∎

For a nonempty convex set X, let us also define the *closure* of $\operatorname{ConeT}(X)$, i.e., the set

$$\overline{\operatorname{ConeT}}(X) := \operatorname{cl}\{[x;t] \in \mathbf{R}^n \times \mathbf{R} : t > 0, \ x/t \in X\}.$$

Clearly, $\overline{\operatorname{ConeT}}(X)$ is a closed cone in $\mathbf{R}^{n+1}$ containing X^+. Moreover, it can immediately be seen that $\overline{\operatorname{ConeT}}(X)$ is the smallest (w.r.t. inclusion) closed cone in $\mathbf{R}^{n+1}$ which contains X^+ and that this cone remains intact when extending X to the closure of X. We will refer to $\overline{\operatorname{ConeT}}(X)$ as the *closed conic transform of X*. In some cases, $\overline{\operatorname{ConeT}}(X)$ admits a simple characterization. An immediate illustration of this is as follows.

Fact 1.36 *Let K be a closed cone and let the set*

$$X := \{x \in \mathbf{R}^n : Ax - b \in K\}$$

be nonempty. Then $\overline{\operatorname{ConeT}}(X) = \{[x;t] \in \mathbf{R}^n \times \mathbf{R} : Ax - bt \in K, \ t \geq 0\}.$

For useful additional facts on closed conic transforms, see Exercise III.12.1–3.

1.6 Proofs of Facts

Fact 1.6 *The* unit ball *of a norm* $\|\cdot\|$, *i.e., the set*

$$\{x \in \mathbf{R}^n : \|x\| \leq 1\},$$

is convex, as for every other $\|\cdot\|$-*ball*

$$B_r(a) := \{x \in \mathbf{R}^n : \|x - a\| \leq r\}$$

(here $a \in \mathbf{R}^n$ and $r \geq 0$ are fixed).

In particular, Euclidean balls ($\|\cdot\|$-balls associated with the standard Euclidean norm $\|x\|_2 := \sqrt{x^\top x}$) are convex.

Proof Let us prove that the set $Q := \{x \in \mathbf{R}^n : \|x - a\| \leq r\}$ is convex. For any $x', x'' \in Q$ and $\lambda \in [0, 1]$, we have

$$\begin{aligned}\|\lambda x' + (1-\lambda)x'' - a\| &= \|\lambda(x' - a) + (1-\lambda)(x'' - a)\| \\ &\leq \|\lambda(x' - a)\| + \|(1-\lambda)(x'' - a)\| \\ &= \lambda\|x' - a\| + (1-\lambda)\|x'' - a\| \leq \lambda r + (1-\lambda)r = r.\end{aligned}$$

Here, the first inequality follows from the triangle inequality, the second equality follows from the homogeneity of norms, and the last inequality holds since $x', x'' \in Q$. Thus, from $\|\lambda x' + (1-\lambda)x'' - a\| \leq r$, we conclude that $\lambda x' + (1-\lambda)x'' \in Q$ as desired. ∎

Fact 1.8 *Unit balls of norms on $\mathbf{R}^n$ are exactly the same as convex sets V in $\mathbf{R}^n$ satisfying the following three properties:*

(i) *V is symmetric with respect to the origin: $x \in V \implies -x \in V$;*
(ii) *V is bounded and closed;*
(iii) *V contains a neighborhood of the origin, i.e., there exists $r > 0$ such that the Euclidean ball of radius r centered at the origin – the set $\{x \in \mathbf{R}^n : \|x\|_2 \leq r\}$ – is contained in V.*

Any set V satisfying the above properties is indeed the unit ball of a particular norm given by

$$\|x\|_V = \inf\left\{t : t^{-1}x \in V, t > 0\right\}. \tag{1.2}$$

Proof First, let V be the unit ball of a norm $\|\cdot\|$, and let us verify the three stated properties. Note that $V = -V$ since $\|x\| = \|-x\|$. The set V is bounded and contains a neighborhood of the origin owing to the equivalence between $\|\cdot\|$ and $\|\cdot\|_2$ (Proposition B.3). Moreover, V is closed. To see this, note that $\|\cdot\|$ is Lipschitz continuous with constant 1 with respect to itself since by the triangle inequality and because $\|x - y\| = \|y - x\|$ we have

$$|\|x\| - \|y\|| \leq \|x - y\|, \quad \forall x, y \in \mathbf{R}^n,$$

which implies by Proposition B.3 that there exists $L_{\|\cdot\|} < \infty$ such that

$$|\|x\| - \|y\|| \leq L_{\|\cdot\|}\|x - y\|_2, \quad \forall x, y \in \mathbf{R}^n,$$

that is, $\|\cdot\|$ is Lipschitz continuous (and thus continuous). And, of course, for any $a \in \mathbf{R}$, the sublevel set $\{x \in \mathbf{R}^n : f(x) \leq a\}$ of a continuous function is closed.

For the reverse direction, consider any V possessing properties (i)–(iii). Then, as V is bounded and contains a neighborhood of the origin, the function $\|\cdot\|_V$ is well defined, is positive outside the origin, and vanishes at the origin. Moreover, $\|\cdot\|_V$ is homogeneous – when the argument is multiplied by a real number λ, the value of the function is multiplied by $|\lambda|$ (by construction and since $V = -V$).

Now, let us show that the relation $V = \{y \in \mathbf{R}^n : \|y\|_V \leq 1\}$ holds. Indeed, the inclusion $V \subseteq \{y : \|y\|_V \leq 1\}$ is evident. So, we will verify that $\|y\|_V \leq 1$ implies $y \in V$. Consider any y such that $\|y\|_V \leq 1$ and let $\bar{t} := \|y\|_V$ (note that $\bar{t} \in [0, 1]$). There is nothing to prove when $\bar{t} = 0$, which, owing to the boundedness of V, implies that $y = 0$ and V contains the origin. When $\bar{t} > 0$, then, by the definition of $\|\cdot\|_V$, there exists a sequence of positive numbers $\{t_i\}$ that converges to $\bar{t}$ as $i \to \infty$ such that $y^i := t_i^{-1} y \in V$. Then, as V is closed, $\bar{y} := \bar{t}^{-1} y \in V$. And since $0 < \bar{t} \leq 1$, $y = \bar{t}\bar{y}$ is a convex combination of the origin and $\bar{y}$. As both $0 \in V$ and $\bar{y} \in V$ and V is convex, we conclude that $y \in V$.

Let us now check that $\|\cdot\|_V$ satisfies the triangle inequality. As $\|\cdot\|_V$ is nonnegative, all we have to check is that $\|x + y\|_V \leq \|x\|_V + \|y\|_V$ when $x \neq 0$, $y \neq 0$. Setting $\bar{x} := x/\|x\|_V$, $\bar{y} := y/\|y\|_V$, we have by homogeneity that $\|\bar{x}\|_V = \|\bar{y}\|_V = 1$. Then, from the relation $V = \{y \in \mathbf{R}^n : \|y\|_V \leq 1\}$, we deduce that $\bar{x} \in V$ and $\bar{y} \in V$. Now, as V is convex and $\bar{x}, \bar{y} \in V$, we have

$$\frac{1}{\|x\|_V + \|y\|_V}(x + y) = \frac{\|x\|_V}{\|x\|_V + \|y\|_V}\bar{x} + \frac{\|y\|_V}{\|x\|_V + \|y\|_V}\bar{y} \in V.$$

That is, $\left\|\frac{1}{\|x\|_V + \|y\|_V}(x + y)\right\|_V \leq 1$. Then, once again by the homogeneity of $\|\cdot\|_V$, we conclude that $\|x + y\|_V \leq \|x\|_V + \|y\|_V$. ∎

Fact 1.9 *Let Q be an $n \times n$ matrix which is symmetric (i.e., $Q = Q^\top$) and positive definite (i.e., $x^\top Q x > 0$ for all vectors $x \neq 0$). Then, for every nonnegative r, the Q-ellipsoid of radius r centered at a, i.e., the set*

$$\left\{x \in \mathbf{R}^n : (x - a)^\top Q(x - a) \leq r^2\right\},$$

is convex.

1.6 Proofs of Facts

Proof Note that since Q is positive definite, the matrix $P := Q^{1/2}$ (see section D.1.5 for the definition of the matrix square root) is well defined and positive definite. Then, we have that P is nonsingular and symmetric, and

$$\left\{x \in \mathbf{R}^n : (x-a)^\top Q(x-a) \leq r^2\right\} = \left\{x \in \mathbf{R}^n : (x-a)^\top P^\top P(x-a) \leq r^2\right\}$$
$$= \left\{x \in \mathbf{R}^n : \|P(x-a)\|_2 \leq r\right\}.$$

Now, note that whenever $\|\cdot\|$ is a norm on $\mathbf{R}^n$ and P is a nonsingular $n \times n$ matrix, the function $x \mapsto \|Px\|$ is a norm itself (why?). Thus, the function $\|x\|_Q := \sqrt{x^\top Q x} = \|Q^{1/2} x\|_2$ is a norm, and the ellipsoid in question clearly is just the $\|\cdot\|_Q$-ball of radius r centered at a. ∎

Fact 1.11 *A set $M \subseteq \mathbf{R}^n$ is convex if and only if it is closed with respect to taking all convex combinations of its elements. That is, M is convex if and only if every convex combination of vectors from M is again a vector from M.*
Hint: Note that assuming $\lambda_1, \ldots, \lambda_m > 0$, one has

$$\sum_{i=1}^m \lambda_i x^i = \lambda_1 x^1 + (\lambda_2 + \lambda_3 + \cdots + \lambda_m) \sum_{i=2}^m \mu_i x^i, \quad \text{where } \mu_i := \frac{\lambda_i}{\lambda_2 + \lambda_3 + \cdots + \lambda_m}.$$

Proof There is nothing to prove when M is empty, so we assume $M \neq \emptyset$. If M is closed with respect to taking arbitrary convex combinations of its points, it is closed with respect to taking two-point combinations, which is exactly the same as saying that M is convex. For the reverse direction, let M be convex. We will prove, by induction on the number of points N used in the convex combination, that a point given as a convex combination of N points from M is itself in M. The claim is clearly true when $N = 1$ (independently of M) and it is also true when $N = 2$ (since M is convex). Suppose now that the claim is true for some $N \geq 2$. Consider the $(N+1)$-term convex combination $x = \sum_{i=1}^{N+1} \lambda_i x^i$ of points x^i from M. If $\lambda_1 = 1$, we have $x = x^1 \in M$. When $\lambda_1 < 1$, we have

$$x = \lambda_1 x^1 + (1 - \lambda_1) \left(\sum_{i=2}^N \frac{\lambda_i}{1-\lambda_1} x^i \right).$$

Define $\bar{x} := \sum_{i=2}^N \frac{\lambda_i}{1-\lambda_1} x^i$. As $\sum_{i=2}^N \lambda_i = 1 - \lambda_1$, we see that $\bar{x}$ is an N-term convex combination of points from M and thus belongs to M by the inductive hypothesis. Hence, $x = \lambda_1 x^1 + (1-\lambda_1)\bar{x}$ is convex combination of $x^1, \bar{x} \in M$, and as M is convex we conclude that $x \in M$. This completes the inductive step. ∎

Fact 1.14 *[Convex hull via convex combinations]* *For a set $M \subseteq \mathbf{R}^n$,*

$$\mathrm{Conv}(M) = \{\text{the set of all convex combinations of vectors from } M\}.$$

Proof Define $\widehat{M} := \{\text{the set of all convex combinations of vectors from } M\}$. Recall that a convex set is closed with respect to taking convex combinations of its members (Fact 1.11); thus any convex set containing M also contains $\widehat{M}$. As by definition $\mathrm{Conv}(M)$ is the intersection of all convex sets containing M, we have $\mathrm{Conv}(M) \supseteq \widehat{M}$. It remains to prove that $\mathrm{Conv}(M) \subseteq \widehat{M}$. We start with the claim that $\widehat{M}$ is convex. By Fact 1.11 $\widehat{M}$ is convex if and only if every convex combination of points from $\widehat{M}$ is also in $\widehat{M}$. Indeed this criteria holds for $\widehat{M}$: let $\bar{x}^i \in \widehat{M}$ for $i = 1, \ldots, N$ and consider a convex combination of these points, i.e.,

$$\hat{x} := \sum_{i=1}^{N} \lambda_i \bar{x}^i,$$

where $\lambda_i \geq 0$ and $\sum_{i=1}^{N} \lambda_i = 1$. For each $i = 1, \ldots, N$, since $\bar{x}^i \in \widehat{M}$, by the definition of $\widehat{M}$, we have $\bar{x}^i = \sum_{j=1}^{N_i} \mu_{i,j} x^{i,j}$, where $x^{i,j} \in M$, $\mu_{i,j} \geq 0$, and $\sum_{j=1}^{N_i} \mu_{i,j} = 1$. Then, we arrive at

$$\hat{x} := \sum_{i=1}^{N} \lambda_i \bar{x}^i = \sum_{i=1}^{N} \lambda_i \left(\sum_{j=1}^{N_i} \mu_{i,j} x^{i,j} \right) = \sum_{i=1}^{N} \sum_{j=1}^{N_i} (\lambda_i \mu_{i,j}) x^{i,j}.$$

Clearly, $\gamma_{i,j} := \lambda_i \mu_{i,j}$ is nonnegative for all i, j. Moreover,

$$\sum_{i=1}^{N} \sum_{j=1}^{N_i} \gamma_{i,j} = \sum_{i=1}^{N} \sum_{j=1}^{N_i} \lambda_i \mu_{i,j} = \sum_{i=1}^{N} \lambda_i \left(\sum_{j=1}^{N_i} \mu_{i,j} \right) = \sum_{i=1}^{N} \lambda_i = 1,$$

where the third and fourth equalities follow from $\sum_{j=1}^{N_i} \mu_{i,j} = 1$ for all i and $\sum_{i=1}^{N} \lambda_i = 1$, respectively. Therefore, $\hat{x} = \sum_{i=1}^{N} \sum_{j=1}^{N_i} \gamma_{i,j} x^{i,j}$ is just a convex combination of points from M, and thus $\hat{x} \in \widehat{M}$, proving that $\widehat{M}$ is convex. Clearly, we also have $\widehat{M} \supseteq M$, and so by definition of $\text{Conv}(M)$, we deduce that $\widehat{M} \supseteq \text{Conv}(M)$, as desired. ∎

Fact 1.17 *A set $K \subseteq \mathbf{R}^n$ is a cone if and only if it is nonempty and it*

- *is conic, i.e., $x \in K, t \geq 0 \implies tx \in K$; and*
- *contains sums of its elements, i.e., $x, y \in K \implies x + y \in K$.*

Proof Suppose K is nonempty and possesses the above properties. Let us prove that K is a cone. The first property already states that K is conic, so we will show that K is convex. For any $x, y \in K$ and $\lambda \in [0, 1]$, we have

$$\lambda x + (1 - \lambda) y = \bar{x} + \bar{y},$$

where $\bar{x} := \lambda x$ and $\bar{y} := (1 - \lambda) y$. As $\lambda \in [0, 1]$, $x, y \in K$ and K is conic, we have $\bar{x}, \bar{y} \in K$. Moreover, since K contains the sum of its elements, we conclude that $\lambda x + (1 - \lambda) y = (\bar{x} + \bar{y}) \in K$, i.e., K is convex.

For the reverse direction, if K is a cone then by definition K is nonempty, conic, and convex. Then, for any $x, y \in K$, as K is convex we have $\frac{1}{2} x + \frac{1}{2} y \in K$, and as K is conic we arrive at $x + y = 2 \left(\frac{1}{2} x + \frac{1}{2} y \right) \in K$. ∎

Fact 1.20 *[Conic hull via conic combinations]* *The conic hull $\text{Cone}(K)$ of a set $K \subseteq \mathbf{R}^n$ is the set of all conic combinations (i.e., linear combinations with nonnegative coefficients) of vectors from K:*

$$\text{Cone}(K) = \left\{ x \in \mathbf{R}^n : \exists N \geq 0, \lambda_i \geq 0, x^i \in K, i \leq N : x = \sum_{i=1}^{N} \lambda_i x^i \right\}.$$

1.6 Proofs of Facts

Proof The case $K = \emptyset$ is trivial; see the comment on the value of an empty sum after the statement of Fact 1.20. When $K \neq \emptyset$, this fact is an immediate corollary of Fact 1.17. ∎

Fact 1.23 *The closure of a set $M \subseteq \mathbf{R}^n$ is exactly the set composed of the limits of all converging sequences of elements from M.*

Proof Let $\overline{M}$ be the set of the limit points of all converging sequences of elements from M. We need to show that $\mathrm{cl}\, M = \overline{M}$. Let us first prove $\mathrm{cl}\, M \supseteq \overline{M}$. Consider any $x \in \overline{M}$. Then, x is the limit of a converging sequence of points $\{x^i\} \in M \subseteq \mathrm{cl}\, M$. Since $\mathrm{cl}\, M$ is a closed set, we arrive at $x \in \mathrm{cl}\, M$.

For the reverse direction, note that by definition $\mathrm{cl}\, M$ is the smallest (w.r.t. inclusion) closed set that contains M, so it suffices to prove that $\overline{M}$ is a closed set satisfying $\overline{M} \supseteq M$. It is easy to see that $\overline{M} \supseteq M$ holds, since, for any $x \in M$, the sequence $\{x^i\}$ where $x^i = x$ is a converging sequence of points from M with the limit x, and thus by the definition of $\overline{M}$ we deduce that $x \in \overline{M}$. Now, consider a converging sequence of points $\{x^i\} \subseteq \overline{M}$, and let us prove that the limit $\bar{x}$ of this sequence belongs to $\overline{M}$. For every i, since the point $x^i \in \overline{M}$ is the limit of a sequence of points from M, we can find a point $y^i \in M$ such that $\|x^i - y^i\|_2 \leq 1/i$. The sequence $\{y^i\}$ is composed of points from M and clearly has the same limit as the sequence $\{x^i\}$, so that $\bar{x}$ is the limit of a sequence of points from M and as such belongs to $\overline{M}$. ∎

Fact 1.36 *Let K be a closed cone, and let the set*

$$X := \{x \in \mathbf{R}^n : Ax - b \in K\}$$

be nonempty. Then, $\overline{\mathrm{ConeT}}(X) = \{[x; t] \in \mathbf{R}^n \times \mathbf{R} : Ax - bt \in K, t \geq 0\}$.

Proof Define $\widehat{X} := \{[x; t] \in \mathbf{R}^n \times \mathbf{R} : Ax - bt \in K, t \geq 0\}$; so, we need to prove that $\widehat{X} = \overline{\mathrm{ConeT}}(X)$. Recall that $\overline{\mathrm{ConeT}}(M) := \mathrm{cl}\{[x; t] \in \mathbf{R}^n \times \mathbf{R} : t > 0, x/t \in M\}$ for any nonempty convex set M. Note that, for the given set X, its perspective transform is

$$\begin{aligned}
\mathrm{Persp}(X) &= \{[x; t] \in \mathbf{R}^n \times \mathbf{R} : t > 0, A(x/t) - b \in K\} \\
&= \{[x; t] \in \mathbf{R}^n \times \mathbf{R} : t > 0, Ax - bt \in K\},
\end{aligned}$$

where the last equality follows since K is conic. So, $\mathrm{Persp}(X) \subseteq \widehat{X}$, and by taking the closures of both sides we arrive at $\overline{\mathrm{ConeT}}(X) = \mathrm{cl}(\mathrm{Persp}(X)) \subseteq \mathrm{cl}(\widehat{X}) = \widehat{X}$, where the last equality follows as $\widehat{X}$ is clearly closed. Hence, $\overline{\mathrm{ConeT}}(X) \subseteq \widehat{X}$. To verify the opposite inclusion, consider $[x; t] \in \widehat{X}$ and let us prove that $[x; t] \in \overline{\mathrm{ConeT}}(X)$. Let $\bar{x} \in X$ (recall that X is nonempty). Then, $[\bar{x}; 1] \in \widehat{X}$. Moreover, as $\widehat{X}$ is a cone and the points $[x; t]$ and $[\bar{x}; 1]$ belong to $\widehat{X}$, we have $z_\epsilon := [x + \epsilon \bar{x}; t + \epsilon] \in \widehat{X}$ for all $\epsilon > 0$. Also, for $\epsilon > 0$, we have $t + \epsilon > 0$ and so $z_\epsilon \in \widehat{X}$ implies $\frac{1}{(t+\epsilon)}(x + \epsilon \bar{x}) \in X$, which is equivalent to $z_\epsilon = [x + \epsilon \bar{x}; t + \epsilon] \in \mathrm{Persp}(X)$. Finally, as $[x; t] = \lim_{\epsilon \to +0} z_\epsilon$, we have $[x; t] \in \mathrm{cl}(\mathrm{Persp}(X)) = \overline{\mathrm{ConeT}}(X)$, as desired. ∎

2

Theorems of Carathéodory, Radon, and Helly

We next examine three theorems from convex analysis that have important consequences in optimization.

2.1 Carathéodory's Theorem

Let us recall the notion of *dimension* from linear algebra. First of all, we define the *dimension* of a linear subspace. This is precisely the number of linearly independent vectors spanning the linear subspace. In the case of an affine subspace, we talk about its *affine dimension*, which is precisely the dimension of the linear subspace that underlies (is parallel to) the given affine subspace. On the basis of these notions, we are now ready to define the dimension of a nonempty set M.

Definition 2.1 *[Dimension of a nonempty set] Given a nonempty set $M \subseteq \mathbf{R}^n$, its dimension (also referred to as its affine dimension) [notation: $\dim(M)$] is defined as the affine dimension of $\mathrm{Aff}(M)$, or, which is the same, the linear dimension of the linear subspace parallel to $\mathrm{Aff}(M)$.*

Remark 2.2 Note that some subsets of $\mathbf{R}^n$ are within the scope of several definitions of dimension. Specifically, a linear subspace is also an affine subspace, and similarly, an affine subspace is a nonempty set as well. It is immediately seen that if a set is within the scope of more than one definition of dimension, all applicable definitions attribute to the set the same value of the dimension. ◇

As an informal introduction to what follows, draw several points ("red points") on the two-dimensional plane and a take a point (a "blue point") in their convex hull. You will observe that whatever your selection of red points and the blue point in their convex hull, the blue point will belong to a properly selected triangle with red vertices. The general fact is as follows.

Theorem 2.3 *[Carathéodory's Theorem] Consider a nonempty set $M \subseteq \mathbf{R}^n$, and let $m := \dim(M)$. Then, every point $x \in \mathrm{Conv}(M)$ is a convex combination of at most $m + 1$ points from M.*

Proof Let $E := \mathrm{Aff}(M)$. Then, $\dim(E) = m$. The claim we want to prove is trivially true when $m = 0$, thus let us assume that $m > 0$. Replacing, if necessary, the embedding space $\mathbf{R}^n$ of M with E (the latter is, geometrically, just $\mathbf{R}^m$), we can assume without loss of generality that $m = n > 0$.

Let $x \in \text{Conv}(M)$. By Fact 1.14 on the structure of the convex hull, there exist $x^1, \ldots, x^N$ from M and convex combination weights $\lambda_1, \ldots, \lambda_N$ such that

$$x = \sum_{i=1}^{N} \lambda_i x^i, \quad \text{where} \quad \lambda_i \geq 0, \; \forall i = 1, \ldots, N, \quad \sum_{i=1}^{N} \lambda_i = 1.$$

Among all such possible representations of x as a convex combination of points from M, let us choose one with the smallest possible N, i.e., involving the fewest points from M. Let this representation of x be the above convex combination. We claim that $N \leq m+1$; proving this claim is all we need in order to complete the proof of Carathéodory's Theorem.

Let us assume for contradiction that $N > m+1$. Now, consider the following system in N variables $\mu_1, \ldots, \mu_N$:

$$\sum_{i=1}^{N} \mu_i x^i = 0, \quad \sum_{i=1}^{N} \mu_i = 0.$$

This is a system of $m+1$ scalar homogeneous linear equations (recall that we are in the case $m = n$, that is, $x^i \in \mathbf{R}^m$). As $N > m+1$, the number of variables in this system is *strictly greater* than the number of equations. Therefore, this system has a *nontrivial* solution, say $\delta_1, \ldots, \delta_N$, i.e.,

$$\sum_{i=1}^{N} \delta_i x^i = 0, \quad \sum_{i=1}^{N} \delta_i = 0, \quad \text{and} \quad [\delta_1; \ldots; \delta_N] \neq 0.$$

Then, for every $t \in \mathbf{R}$, we have the following representation of x as a linear combination of the points $x^1, \ldots, x^N$:

$$\sum_{i=1}^{N} (\lambda_i + t\delta_i) x^i = x.$$

For ease of reference, let us define $\lambda_i(t) := \lambda_i + t\delta_i$ for all i and for all $t \in \mathbf{R}$. Note that, for any $t \in \mathbf{R}$, by the definitions of λ_i and δ_i we always have

$$\sum_{i=1}^{N} \lambda_i(t) = \sum_{i=1}^{N} (\lambda_i + t\delta_i) = \sum_{i=1}^{N} \lambda_i + t \sum_{i=1}^{N} \delta_i = 1.$$

Moreover, when $t = 0$, $\lambda_i(0) = \lambda_i$ for all i, and so this is a convex combination as all $\lambda_i(0) \geq 0$ for all i. On the other hand, from the selection of δ_i, we know that $\sum_{i=1}^{N} \delta_i = 0$ and $[\delta_1; \ldots; \delta_N] \neq 0$, and thus at least one entry in δ must be negative. Therefore, when t is large, some of the coefficients $\lambda_i(t)$ will be negative. There exists, of course, the largest $t = t^*$ for which $\lambda_i(t) \geq 0$ for all $i = 1, \ldots, N$ holds and, for $t = t^*$, at least some of the $\lambda_i(t)$ are zero. Specifically, when setting $I^- := \{i : \delta_i < 0\}$,

$$i^* \in \underset{i}{\arg\min} \left\{ \frac{\lambda_i}{|\delta_i|} : i \in I^- \right\}, \quad \text{and} \quad t^* := \min_i \left\{ \frac{\lambda_i}{|\delta_i|} : i \in I^- \right\},$$

we have $\lambda_i(t^*) \geq 0$ for all i and $\lambda_{i^*}(t^*) = 0$. This then implies that we have represented x as a convex combination of fewer than N points from M, which contradicts the definition of N

(as being the smallest number of points x^i needed in the convex combination representation of x). ∎

Remark 2.4 Carathéodory's Theorem is sharp: for every positive integer n, there is a set of $n + 1$ affinely independent points in $\mathbf{R}^n$ (e.g., the origin and the n standard basis vectors or "orths") such that certain convex combination of these $n + 1$ points (specifically, their average) cannot be represented as a convex combination using strictly less than $n + 1$ points from the set. ◇

Let us see an instructive corollary witnessing the power of Carathéodory's Theorem.

Corollary 2.5 *Let $X \subseteq \mathbf{R}^n$ be a closed and bounded set. Then, $\mathrm{Conv}(X)$ is closed and bounded as well.*

Proof There is nothing to prove when X is empty. Now let X be nonempty, closed and bounded, and define $Y := \mathrm{Conv}(X)$. The boundedness of Y is evident. In order to verify that Y is closed, let $\{x^t\}_{t \geq 1}$ be a converging sequence of points from Y. By Carathéodory's Theorem, every one of the vectors x^t, being a convex combination of vectors from X, has a representation of the form $x^t = \sum_{i=1}^{n+1} \lambda_i^t x_i^t$ of at most $n + 1$ vectors x_i^t from X, $i \leq n + 1$, where λ_i^t are the corresponding convex combination weights. Since X is bounded, passing to a subsequence $t_1 < t_2 < \cdots$, we can assume that the sequences $\{x_i^{t_s}\}_{s \geq 1}$ and $\{\lambda_i^{t_s}\}_{s \geq 1}$ converge as $s \to \infty$ for every $i \leq n + 1$, the limits being, respectively, the vectors x_i and the real numbers λ_i. We clearly have $\lim_{t \to \infty} x^t = \lim_{s \to \infty} \sum_{i=1}^{n+1} \lambda_i^{t_s} x_i^{t_s} = \sum_{i=1}^{n+1} \lambda_i x_i$. In addition, as X is closed, $x_i = \lim_{s \to \infty} x_i^{t_s} \in X$. Moreover, clearly $\lambda_i \geq 0$ and $\sum_{i=1}^{n+1} \lambda_i = 1$. The bottom line is that $\lim_{t \to \infty} x^t = \sum_{i=1}^{n+1} \lambda_i x_i$ is a convex combination of points from X and thus belongs to Y as $Y = \mathrm{Conv}(X)$. ∎

Remark 2.6 Note that the convex hull of a closed *unbounded* set is not always closed. For example, consider the set $X = \{[0; 0]\} \cup \{[u; v] \in \mathbf{R}_+^2 : uv = 1\}$, which is closed and unbounded; we have $\mathrm{Conv}(X) = \{[u; v] \in \mathbf{R}_+^2 : u > 0, v > 0\} \cup \{[0; 0]\}$, which is not closed. ◇

Let us consider a concrete illustration, taken from [Nem24], of Carathéodory's Theorem.

2.1.1 Carathéodory's Theorem, Illustration

Suppose that a supermarket sells 99 different market-blend herbal teas, and every herbal tea is a certain blend of 26 herbs A, ..., Z. In spite of such a variety of marketed blends, John is not satisfied with any one of them; the only herbal tea he likes is their mixture, in the proportion

$$1 : 2 : 3 : \cdots : 98 : 99.$$

Once, it occurred to John that in order to prepare his favorite tea, there is no necessity to buy all 99 market blends; a smaller number of them will do. With some arithmetic, John found a combination of 66 marketed blends which still allows him to prepare his tea. Do you believe John's result can be improved?

Answer: In fact, just 26 properly selected market blends are enough.

Explanation: Let us represent a blend by its unit-weight portion, say, 1 gram. Such a portion can be identified with the 26-dimensional vector $x = [x_1; \ldots; x_{26}]$, where x_i is the weight, in grams, of herb i in the portion. Clearly, we have $x \in \mathbf{R}_+^{26}$ and $\sum_{i=1}^{26} x_i = 1$. When mixing market blends $x^1, x^2, \ldots, x^{99}$ to get a unit-weight portion x of the mixture, we take $\lambda_j \geq 0$ grams of market blend x^j, $j = 1, \ldots, 99$, and mix them together, that is, $x = \sum_{j=1}^{99} \lambda_j x^j$. Since the weight of the mixture represented by x is 1 gram and the λ_j correspond to the weight (in grams) of the market blends x^j used in x, we get $\sum_{j=1}^{99} \lambda_j = 1$. The bottom line is that blend x can be obtained by mixing market blends $x^1, \ldots, x^{99}$ if and only if $x \in \text{Conv}\{x^1, \ldots, x^{99}\}$.

Then, by Carathéodory's Theorem, every blend which can be obtained by mixing market blends can be obtained by mixing $m + 1$ of them, where m is the affine dimension of the affine span of $x^1, \ldots, x^{99}$. In our case, this span belongs to the 25-dimensional affine plane $\left\{ x \in \mathbf{R}^{26} : \sum_{i=1}^{26} x_i = 1 \right\}$ that is, $m \leq 25$. ∎

Carathéodory's Theorem admits a "conic analogy" as follows:

Fact 2.7 *[Carathéodory's Theorem in conic form] Let $a \in \mathbf{R}^m$ be a conic combination (a linear combination with nonnegative coefficients) of N vectors $a^1, \ldots, a^N$. Then, a is a conic combination of at most m vectors from the collection $a^1, \ldots, a^N$.*

2.2 Radon's Theorem

As an informal introduction to what follows, draw four arbitrary points on the plane that are distinct from each other and try to color some of them in red and the remainder in blue in such a way that the convex hull of red points will intersect the convex hull of the blue points. Experimentation will suggest that this is always possible. The general fact is as follows.

Theorem 2.8 *[Radon's Theorem] Let $x^i \in \mathbf{R}^n$, $i \leq N$, where $N \geq n + 2$. Then there exists a partition $I \cup J = \{1, \ldots, N\}$ of the index set $\{1, \ldots, N\}$ into two nonempty disjoint $(I \cap J = \emptyset)$ sets I and J such that*

$$\text{Conv}\left(\{x^i : i \in I\}\right) \cap \text{Conv}\left(\{x^j : j \in J\}\right) \neq \emptyset.$$

Proof Consider the following system of homogeneous equations in N variables $\mu_1, \ldots, \mu_N$:

$$\sum_{i=1}^{N} \mu_i x^i = 0,$$

$$\sum_{i=1}^{N} \mu_i = 0.$$

Note that as $x^i \in \mathbf{R}^n$, this system has $n + 1$ scalar linear equations. Moreover, as the premise of the theorem states that $N > n + 1$, we deduce that this system of equations has a nontrivial solution $\lambda_1, \ldots, \lambda_N$:

$$\sum_{i=1}^{N} \lambda_i x^i = 0, \quad \sum_{i=1}^{N} \lambda_i = 0, \quad \text{and} \quad [\lambda_1; \ldots; \lambda_N] \neq 0.$$

Let $I := \{i : \lambda_i \geq 0\}$ and $J := \{i : \lambda_i < 0\}$. Then, I and J are nonempty and form a partition of $\{1, \ldots, N\}$ (since the sum of all the λ_i is zero and not all λ_i are zero). Moreover, we have

$$a := \sum_{i \in I} \lambda_i = \sum_{j \in J} (-\lambda_j) > 0.$$

Then, by setting

$$\alpha_i := \frac{\lambda_i}{a}, \text{ for } i \in I, \quad \text{and} \quad \beta_j := \frac{-\lambda_j}{a}, \text{ for } j \in J,$$

we get

$$\alpha_i \geq 0, \forall i; \quad \beta_j \geq 0, \forall j; \quad \sum_{i \in I} \alpha_i = 1, \quad \sum_{j \in J} \beta_j = 1.$$

In addition, we also have

$$\sum_{i \in I} \alpha_i x^i - \sum_{j \in J} \beta_j x^j = \frac{1}{a} \left(\sum_{i \in I} \lambda_i x^i - \sum_{j \in J} (-\lambda_j) x^j \right) = \frac{1}{a} \sum_{i=1}^{N} \lambda_i x^i = 0.$$

We conclude that the vector $\sum_{i \in I} \alpha_i x^i = \sum_{j \in J} \beta_j x^j$ is the desired common point of Conv $(\{x^i : i \in I\})$ and Conv $(\{x^j : j \in J\})$. ∎

Remark 2.9 As for Carathéodory's Theorem, the bound in Radon's Theorem is sharp: for every positive integer n, there exist $n+1$ points in $\mathbf{R}^n$ (e.g., the origin and the n standard basic orths) which cannot be split into two disjoint subsets with intersecting convex hulls. ◇

2.3 Helly's Theorem

What follows is a multidimensional extension of a nearly evident fact:

> Given finitely many segments $[a_i, b_i]$ on the line such that every two of the segments intersect, we can always find a point common to all the segments.

The multidimensional extension of this fact is as follows.

Theorem 2.10 [Helly's Theorem, I] *Let $\mathcal{F} := \{S_1, \ldots, S_N\}$ be a finite family of convex sets in $\mathbf{R}^n$. Suppose that, for every collection of at most $n + 1$ sets from this family, the sets from the collection have a point in common. Then, all the sets S_i, $i \leq N$, have a common point.*

Proof We will prove the theorem by induction on the number N of sets in the family. The case $N \leq n + 1$ holds immediately owing to the premise of the theorem. So, suppose that the statement holds for all families with a certain number $N \geq n + 1$ of sets, and let $S_1, \ldots, S_N, S_{N+1}$ be a family of $N + 1$ convex sets which satisfies the premise of Helly's Theorem. We need to prove that the intersection of the sets $S_1, \ldots, S_N, S_{N+1}$ is nonempty.

For each $i \leq N+1$, we construct the following N-set families:

$$\mathcal{F}^i := \{S_1, S_2, \ldots, S_{i-1}, S_{i+1}, \ldots, S_{N+1}\}, \qquad \forall i \leq N+1,$$

where the N-set family $\mathcal{F}^i$ was obtained by deleting from our $(N+1)$-set family the set S_i. Note that each of these N-set families $\mathcal{F}^i$ satisfies the premise of Helly's Theorem, and thus, by the inductive hypothesis, the intersection of the members of $\mathcal{F}^i$ is nonempty:

$$T^i := S_1 \cap S_2 \cap \cdots \cap S_{i-1} \cap S_{i+1} \cap \cdots \cap S_{N+1} \neq \varnothing, \qquad \forall i \leq N+1.$$

For each $i \leq N+1$, choose a point $x^i \in T^i$ (recall that T^i is nonempty!). Then, we have $N+1$ points $x^i \in \mathbf{R}^n$. As $N \geq n+1$, we have $N+1 \geq n+2$ and by Radon's Theorem, we can partition the index set $\{1, \ldots, N+1\}$ into two nonempty disjoint subsets I and J in such a way that a certain convex combination x of the points x^i, $i \in I$, is a convex combination of the points x^j, $j \in J$, as well. Let us verify that x belongs to all the sets $S_1, \ldots, S_{N+1}$, which will complete the inductive step. Indeed, select any index $i^* \leq N+1$ and let us prove that $x \in S_{i^*}$. We have either $i^* \in I$ or $i^* \in J$. Suppose first that $i^* \in I$. Then all the sets T^j, $j \in J$, are contained in S_{i^*} (since S_{i^*} participates in all intersections which give T^i with $i \neq i^*$). Consequently, all the points x^j, $j \in J$, belong to S_{i^*} and therefore x, which is a convex combination of these points, also belongs to S_{i^*} (recall that all the S_i are convex!), as required. Suppose now that $i^* \in J$. In this case, similar reasoning shows that all the points x^i, $i \in I$, belong to S_{i^*} and therefore x, which is a convex combination of these points x^i, $i \in I$, belongs to S_{i^*}. The induction and the proof are complete. ∎

For an alternative proof of Theorem 2.10 which does not utilize Radon's Theorem, see Exercise I.23.

Remark 2.11 Helly's Theorem admits a small and immediate refinement as follows:

Let $\mathcal{F} := \{S_1, \ldots, S_N\}$ be a family of N nonempty convex sets in $\mathbf{R}^n$, and let m be the dimension of $\bigcup_{i \leq N} S_i$. Assume that, for every collection of at most $m+1$ sets from the family, the sets from the collection have a point in common. Then, all sets from the family have a point in common.

The justification of this claim follows from viewing the sets S_i as subsets of $E = \mathrm{Aff}(\cup_i S_i)$ rather than $\mathbf{R}^n$ and applying the standard Helly's Theorem. ◇

Remark 2.12 As for Carathéodory's and Radon's Theorems, Helly's Theorem is sharp: for every positive integer n, there exists a finite family of convex sets in $\mathbf{R}^n$ such that every n of them have a common point but there is no point common to all of them. Indeed, take $n+1$ affinely independent points $x^1, \ldots, x^{n+1}$ in $\mathbf{R}^n$ (say, the origin and the n standard basic orths in $\mathbf{R}^n$) and $n+1$ convex sets $S_1, \ldots, S_{n+1}$ with S_i the convex hull of the points $x^1, \ldots, x^{i-1}, x^{i+1}, \ldots, x^{n+1}$. Every n of these sets have a point in common (e.g., the common point of $S_1, S_3, S_4, \ldots, S_{n+1}$ is x^2), but there is no point common to all $n+1$ sets (why?). ◇

2.3.1 Helly's Theorem, Illustration 1

Suppose that we need to design a factory which, mathematically, is described by the following set of constraints in variables $x \in \mathbf{R}^n$:[1]

$$\left.\begin{array}{l} Ax \geq d \quad [d_1,\ldots,d_{1000}\text{: demands}] \\ Bx \leq f \quad [f_1 \geq 0,\ldots,f_{10} \geq 0\text{: amounts of resources of various types}] \\ Cx \leq c, \quad \text{[other constraints]} \end{array}\right\} \quad (F)$$

where $Ax \geq d$ with vector $d \in \mathbf{R}^{1000}$ represents the demand constraints, $Bx \leq f$ with a vector $f \in \mathbf{R}_+^{10}$ corresponds to the availability of the various resources, and the constraint $Cx \leq c$ represents various additional restrictions. The data A, B, C, c are given in advance but the demand $d \in \mathbf{R}^{1000}$ is unknown, and we are asked to determine the resource availability levels $f \in \mathbf{R}_+^{10}$. In particular, we are asked to buy *in advance* the resources $f_i \geq 0$, $i = 1,\ldots,10$, in such a way that the factory will be able to satisfy all demand scenarios d from a given finite set D, that is, (F) should be feasible for every $d \in D$. The unit cost of resource i is given to us as $a_i > 0$, that is, an amount f_i of resource i costs us $a_i f_i$ dollars.

It is known that for every single demand scenario $d \in D$, a proper investment (depending on the scenario d) of at most \$1 in resources suffices to meet the demand d.

How large should the investment in resources be in the following cases, when D contains

1. just one scenario?
2. 3 scenarios?
3. 10 scenarios?
4. 100,000 scenarios?

Answer: We claim that in these scenarios the following investment amounts will suffice:

1. \$1
2. \$3
3. \$10
4. \$10

Explanation: Cases 1–3: In these cases, we know that every scenario d from D can be met by a vector of resources $f_d \in \mathbf{R}_+^{10}$ incurring a cost of at most \$1. We lose nothing if we assume that f_d costs exactly \$1, since if d is met by a vector of resources f of cost $\leq \$1$, it is met by every vector $f' \geq f$, and since the per-unit costs of the resources are positive, we can select f' to cost exactly \$1. Thus, when we are given scenarios $d^1,\ldots,d^k$ from D, we can meet every one of them with the vector of resources $f := f_{d^1} + \cdots + f_{d^k}$, since $f \geq f_{d^i}$ for every $i = 1,\ldots,k$. Then, because of the structure of our model, f meets demand scenario d^i.

Case 4: We claim that \$10 investment is always enough no matter what the cardinality of D is. To see this, for every $d \in D$, we define the sets

---

[1] The illustration is taken from [Nem24].

2.3 Helly's Theorem

$$S[d, f] := \left\{ x \in \mathbf{R}^n : Ax \geq d, \; Bx \leq f, \; Cx \leq c \right\},$$

$$F_d := \left\{ f \in \mathbf{R}^{10} : f \geq 0, \; a^\top f = 10, \; S[d, f] \neq \varnothing \right\}.$$

From these definitions, F_d is precisely the set of vectors $f \in \mathbf{R}_+^{10}$ such that their cost $a^\top f$ is exactly \$10 and the associated polyhedral set $S[f, d]$ is nonempty, that is, the resource vector f allows the demand d to be met. Note that the set F_d is convex as it is the linear image (in fact just the projection) of the convex set

$$\left\{ f \in \mathbf{R}^{10}, \; x \in \mathbf{R}^n : f \geq 0, \; a^\top f = 10, \; x \in S[d, f] \right\}.$$

The punchline in this illustration is that every 10 sets of the form F_d have a common point. Suppose that we are given 10 scenarios $d^1, \ldots, d^{10}$ from D. Then we can meet demand scenario d^i by investing \$1 in a properly selected vector of resources $f_{d^i} \in \mathbf{R}_+^{10}$. In the same way as we proceeded in cases 1–3, by investing \$10 in the single vector of resources $\bar{f} := f_{d^1} + \cdots + f_{d^{10}}$, we can meet every scenario $d^1, \ldots, d^{10}$, whence $\bar{f} \in F_{d^1} \cap \cdots \cap F_{d^{10}}$. Since all 100,000 sets F_d, $d \in D$, belong to the nine-dimensional affine plane in $\mathbf{R}^{10}$ composed of vectors of cost \$10 (i.e., $F_d \subseteq \{f \in \mathbf{R}^{10} : a^\top f = 10\}$ for all $d \in D$) and since every 10 of these sets F_d have a point in common, all the sets have a common point, say f_*; see Remark 2.11. That is, $f_* \in F_d$ for all $d \in D$ and thus, by the definition of F_d, we deduce that every set $S[d, f_*]$, $d \in D$, is nonempty, that is, the vector of resources f_* (which costs \$10) allows us to satisfy every demand scenario $d \in D$. ∎

2.3.2 Helly's Theorem, Illustration 2

Consider the optimization problem

$$\mathrm{Opt}_* := \min_{x \in \mathbf{R}^{11}} \left\{ c^\top x : g_i(x) \leq 0, \; i = 1, \ldots, 1000 \right\}$$

with 11 variables $x_1, \ldots, x_{11}$ and convex constraints, i.e., every one of the sets

$$X_i := \left\{ x \in \mathbf{R}^{11} : g_i(x) \leq 0 \right\}, \quad i = 1, \ldots, 1000,$$

is convex. Suppose also that the problem is solvable with optimal value $\mathrm{Opt}_* = 0$.
Clearly, when dropping one or more constraints, the optimal value can only decrease or remain the same.

Is it possible to find a constraint such that even if we drop it, we preserve the optimal value? Two constraints which can be dropped simultaneously with no effect on the optimal value? Three of them?

Answer: We can in fact drop as many as $1000 - 11 = 989$ appropriately chosen constraints without changing the optimal value!
Explanation: The case $c = 0$ is trivial – for this case one can drop all 1000 constraints without varying the optimal value! Therefore, from now on we assume $c \neq 0$.

Assume for contradiction that every 11-constraint relaxation of the original problem has negative optimal value. Since there are finitely many such relaxations, there exists $\epsilon < 0$ such that every problem of the form

$$\min_x \left\{ c^\top x : g_{i_1}(x) \leq 0, \ldots, g_{i_{11}}(x) \leq 0 \right\}$$

has a feasible solution with objective value $< -\epsilon$. Besides this, such an 11-constraint relaxation of the original problem also has a feasible solution with objective equal to 0 (namely, the optimal solution of the original problem), and since its feasible set is convex (as it is the intersection of the convex feasible sets of the participating constraints), the 11-constraint relaxation has a feasible solution x with $c^\top x = -\epsilon$. In other words, every 11 of the 1000 convex sets

$$Y_i := \left\{x \in \mathbf{R}^{11} : c^\top x = -\epsilon,\ g_i(x) \leq 0\right\}, \quad i = 1, \ldots, 1000,$$

have a point in common. Now, consider the hyperplane $H := \{x \in \mathbf{R}^{11} : c^\top x = -\epsilon\}$. Note that $Y_i \subseteq H$ for all i. Moreover, as $c \neq 0$, $\dim(H) = 10$ and thus $\dim(\bigcup_{i=1}^{1000} Y_i) \leq \dim(H) = 10$. Since every 11 of these sets Y_i have a nonempty intersection and $\dim(\bigcup_{i=1}^{1000} Y_i) \leq 10$, all of them have a point in common. In other words, the original problem would have a feasible solution with negative objective value, which is not possible as $\mathrm{Opt}_* = 0$. ∎

2.3.3 ★ Helly's Theorem for Infinite Families of Convex Sets

In Helly's Theorem as presented above we dealt with a family of finitely many convex sets. To extend the statement to the case of infinite families, we need to slightly strengthen the assumptions, essentially to avoid complications stemming from the following two situations:

- *lack of closedness*: every two (and in fact finitely many) of the convex sets $A_i = \{x \in \mathbf{R} : 0 < x \leq 1/i\}$, $i = 1, 2, \ldots$, have a point in common, while all the sets have no common point;
- *"intersection at ∞"*: every two (and in fact finitely many) of the closed convex sets $A_i = \{x \in \mathbf{R} : x \geq i\}$, $i = 1, 2, \ldots$, have a point in common, while all the sets have no common point.

The resulting refined statement of Helly's Theorem for a family of (possibly) infinitely many sets is as follows:

Theorem 2.13 *[Helly's Theorem, II] Let $\mathcal{F}$ be an arbitrary family of convex sets in $\mathbf{R}^n$. Assume that*

(i) for every collection of at most $n + 1$ sets from the family, the sets from the collection have a point in common; and

(ii) every set in the family is closed, and the intersection of the sets from a certain finite subfamily of the family is bounded (e.g., one set in the family is bounded).

Then all the sets from the family have a point in common.

Proof By (i), Theorem 2.10 implies that all finite subfamilies of $\mathcal{F}$ have nonempty intersections, and also these intersections are convex (since the intersection of a family of convex sets is convex by Proposition 1.12); in view of (ii) these intersections are also closed. Adding to $\mathcal{F}$ the intersections of sets from finite subfamilies of $\mathcal{F}$, we get a larger family $\mathcal{F}'$ composed of closed convex sets, and sets from a finite subfamily of this larger family again have a nonempty intersection. Moreover, from (ii) it follows that this new family contains a bounded set Q. Since all the sets are closed, the family of sets

$$\{Q \cap Q' : Q' \in \mathcal{F}\}$$

forms a *nested family of compact sets* (i.e., a family of compact sets with nonempty intersection of the sets from every finite subfamily). Then, by a well-known theorem from Real Analysis such a family has a nonempty intersection.[2] ∎

2.4 Proof of Fact

Fact 2.7 *[Carathéodory's Theorem in conic form]* Let $a \in \mathbf{R}^m$ be a conic combination (a linear combination with nonnegative coefficients) of N vectors $a^1, \ldots, a^N$. Then a is a conic combination of at most m vectors from the collection $a^1, \ldots, a^N$.

Proof The proof follows the same lines as the proof of the "plain" Carathéodory's Theorem. Consider the minimal representation, in terms of the number of positive coefficients, of a as a conic combination of $a^1, \ldots, a^N$; w.l.o.g., we can assume that this is the representation $a = \sum_{i=1}^{K} \lambda_i a^i$, $\lambda_i > 0$, $i \leq K$. We need to prove that $K \leq m$. Assume for contradiction that $K > m$. Consider the system of m scalar linear equations $\sum_{i=1}^{K} \delta_i a^i = 0$ in variables δ. The number K of unknowns in this system is larger than the number m of equations. Thus, this system has a nontrivial solution $\bar{\delta}$. Passing, if necessary, from $\bar{\delta}$ to $-\bar{\delta}$, we may further assume that some $\bar{\delta}_i$ are strictly negative. Define $\lambda_i(t) := \lambda_i + t\bar{\delta}_i$ for all i. Note that for all $t \geq 0$ we have $a = \sum_{i=1}^{K} \lambda_i(t) a^i$. Let t^* be the largest $t \geq 0$ for which all $\lambda_i(t)$ are nonnegative (t^* is well defined, since for large t some $\lambda_i(t)$, i.e., those corresponding to $\bar{\delta}_i < 0$, will become negative); then we get $a = \sum_{i=1}^{K} \lambda_i(t^*) a^i$ with all coefficients $\lambda_i(t^*)$ nonnegative and at least one of them zero, contradicting the definition of K. ∎

---

[2] The real analysis theorem states that every nested family of compact sets Q_α, $\alpha \in \mathcal{A}$ has a nonempty intersection. Here is the proof of this theorem: Assume for contradiction that the intersection of the compact sets Q_α, $\alpha \in \mathcal{A}$, is empty. Choose a set Q_{α^*} from the family; for every $x \in Q_{\alpha^*}$ there is a set Q^x in the family which does not contain x (otherwise x would be a common point of all our sets). Since Q^x is closed, there is an open ball V_x centered at x which does not intersect Q^x. The balls V_x, $x \in Q_{\alpha^*}$, form an open covering of the compact set Q_{α^*}. Since Q_{α^*} is compact, there exists a finite subcovering $V_{x_1}, \ldots, V_{x_N}$ of Q_{α^*} by the balls from the covering; see Theorem B.19. Since Q^{x_i} does not intersect V_{x_i}, we conclude that the intersection of the finite subfamily $Q_{\alpha^*}, Q^{x_1}, \ldots, Q^{x_N}$ is empty, which is a contradiction to the nestedness property of this family.

3

Polyhedral Representations and Fourier–Motzkin Elimination

3.1 Polyhedral Representations

Recall that by definition a polyhedral set X in $\mathbf{R}^n$ is the solution set of a finite system of nonstrict linear inequalities in variables $x \in \mathbf{R}^n$:

$$X = \left\{x \in \mathbf{R}^n : Ax \leq b\right\} = \left\{x \in \mathbf{R}^n : a_i^\top x \leq b_i, 1 \leq i \leq m\right\}.$$

We call such a representation of X its *polyhedral description*. A polyhedral set is always convex and closed (Proposition 1.2). We next introduce the notion of the *polyhedral representation* of a set $X \subseteq \mathbf{R}^n$.

Definition 3.1 *A set $X \subseteq \mathbf{R}^n$ is called* polyhedrally representable *if it admits a representation of the form*

$$X = \left\{x \in \mathbf{R}^n : \exists u \in \mathbf{R}^k : Ax + Bu \leq c\right\}, \tag{3.1}$$

where A and B are $m \times n$ and $m \times k$ matrices and $c \in \mathbf{R}^m$. A representation of X of the form (3.1) is called a polyhedral representation *of X, and the variables $u \in \mathbf{R}^k$ in such a representation are called* extra variables.

Geometrically, a polyhedral representation of a set $X \subseteq \mathbf{R}^n$ is its representation as the *projection* $\{x \in \mathbf{R}^n : \exists u \in \mathbf{R}^k : [x; u] \in Y\}$ of a *polyhedral* set $Y = \{(x, u) \in \mathbf{R}^n \times \mathbf{R}^k : Ax + Bu \leq c\}$. Here, Y lives in the space of $n + k$ variables $x \in \mathbf{R}^n$ and $u \in \mathbf{R}^k$, and the polyhedral representation of X states that X is obtained by applying to the points of Y the linear mapping (the projection) $[x; u] \mapsto x : \mathbf{R}^{n+k} \to \mathbf{R}^n$ of the $(n + k)$-dimensional space of (x, u)-variables (the space where Y lives) to the n-dimensional space of x-variables where X lives.

Note that every polyhedrally representable set is the image under a linear mapping (even a projection) of a polyhedral, and thus convex, set. It follows that *a polyhedrally representable set is definitely convex* (Proposition 1.21).

Example 3.1 Every polyhedral set $X = \{x \in \mathbf{R}^n : Ax \leq b\}$ is polyhedrally representable: a polyhedral description of X is simply a polyhedral representation with no extra variables ($k = 0$). Vice versa, a polyhedral representation of a set X with no extra variables ($k = 0$) is clearly a polyhedral description of the set (which therefore is polyhedral). ◇

Example 3.2 Consider the set $X = \{x \in \mathbf{R}^n : \sum_{i=1}^n |x_i| \leq 1\}$. Note that this initial description of X is *not* of the form $\{x \in \mathbf{R}^n : Ax \leq b\}$. Thus, from this description of

3.1 Polyhedral Representations

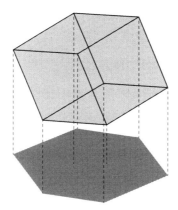

Figure 3.1 Polyhedral representation of a hexagon in the xy-plane as the projection of a rotated three-dimensional cube onto the plane.

X, we cannot immediately say whether it is polyhedral. However, X admits a polyhedral representation, e.g., the following representation:

$$X = \left\{ x \in \mathbf{R}^n : \exists u \in \mathbf{R}^n : \underbrace{-u_i \leq x_i \leq u_i}_{\iff |x_i| \leq u_i}, \ 1 \leq i \leq n, \ \sum_{i=1}^n u_i \leq 1 \right\}. \quad (3.2)$$

Note that the set X in question can be described by a system of linear inequalities *in x-variables only*, namely, as

$$X = \left\{ x \in \mathbf{R}^n : \sum_{i=1}^n \epsilon_i x_i \leq 1, \forall (\epsilon_i \in \{-1, +1\}), \ 1 \leq i \leq n) \right\};$$

thus, X is polyhedral. However, the above polyhedral description of X (which in fact is minimal in terms of the number of inequalities involved) requires 2^n inequalities – an astronomically large number even when n is just few tens. In contrast, the polyhedral representation of the same set given in (3.2) requires only n extra variables u and $2n + 1$ linear inequalities in x and u, and so the "complexity" of this representation is just linear in n. ◇

Example 3.3 Given a finite set of vectors $a^1, \ldots, a^m \in \mathbf{R}^n$, consider their conic hull Cone $\{a^1, \ldots, a^m\} = \{\sum_{i=1}^m \lambda_i a^i : \lambda \geq 0\}$ (see section 1.2.4). From this definition, it is completely unclear whether this set is polyhedral. In contrast to this, its polyhedral representation is immediate:

$$\text{Cone}\{a^1, \ldots, a^m\} = \left\{ x \in \mathbf{R}^n : \exists \lambda \in \mathbf{R}^m_+ : x = \sum_{i=1}^m \lambda_i a_i \right\}$$

$$= \left\{ x \in \mathbf{R}^n : \exists \lambda \in \mathbf{R}^m : \begin{cases} -\lambda \leq 0 \\ x - \sum_{i=1}^m \lambda_i a_i \leq 0 \\ -x + \sum_{i=1}^m \lambda_i a_i \leq 0 \end{cases} \right\}.$$

In other words, the original description of X is simply its polyhedral representation (in a slight disguise), with the λ_i in the role of extra variables. ◇

3.2 Every Polyhedrally Representable Set is Polyhedral (Fourier–Motzkin Elimination)

A surprising and deep fact is that the situation in Example 3.2 above is quite general.

Theorem 3.2 *Every polyhedrally representable set is polyhedral.*

Proof: Fourier–Motzkin elimination Recalling the definition of a polyhedrally representable set, our claim can be rephrased equivalently as follows:

The projection of a polyhedral set Y in a space $\mathbf{R}^{n+k}_{x,u}$ of (x,u)-variables onto the subspace $\mathbf{R}^n_x$ of x-variables is a polyhedral set in $\mathbf{R}^n$.

Note that it suffices to prove this claim in the case of exactly one extra variable since the projection which reduces the dimension by k – "eliminates" k extra variables – is the result of k subsequent projections, each reducing the dimension by 1, so "eliminating" the extra variables one by one.

Thus, consider a polyhedral set with variables $x \in \mathbf{R}^n$ and $u \in \mathbf{R}$, i.e.,

$$Y := \left\{ (x,u) \in \mathbf{R}^{n+1} : a_i^\top x + b_i u \leq c_i, \ 1 \leq i \leq m \right\}.$$

We want to prove that the projection of Y onto the space of x-variables, i.e.,

$$X := \left\{ x \in \mathbf{R}^n : \exists u \in \mathbf{R} : Ax + bu \leq c \right\},$$

is polyhedral. To see this, let us split the indices of the inequalities defining Y into three groups (some of these groups could be empty).

- *Inequalities with $b_i = 0$:* $I_0 := \{i : b_i = 0\}$. These inequalities with index $i \in I_0$ do not involve u at all.
- *Inequalities with $b_i > 0$:* $I_+ := \{i : b_i > 0\}$. An inequality with index $i \in I_+$ can be rewritten equivalently as $u \leq b_i^{-1}[c_i - a_i^\top x]$, and it imposes an *upper bound* on u that depends on x.
- *Inequalities with $b_i < 0$:* $I_- := \{i : b_i < 0\}$. An inequality with index $i \in I_-$ can be rewritten equivalently as $u \geq b_i^{-1}[c_i - a_i^\top x]$, and it imposes a *lower bound* on u that depends on x.

We can now clearly answer the question of when x can be in X, that is, when x can be extended, by some u, to a point (x,u) from Y. This is the case if and only if the following hold: first, x satisfies all inequalities with $i \in I_0$, and, second, the inequalities with $i \in I_+$ giving the upper bounds on u specified by x are compatible with the inequalities with $i \in I_-$ giving the lower bounds on u specified by x, meaning that every lower bound is less than or equal to every upper bound (the latter is necessary and sufficient for finding a value of u which is greater than or equal to all lower bounds and less than or equal to all upper bounds). Thus,

$$X = \left\{ x \in \mathbf{R}^n : \begin{array}{l} a_i^\top x \leq c_i, \quad \forall i \in I_0 \\ b_j^{-1}(c_j - a_j^\top x) \leq b_k^{-1}(c_k - a_k^\top x), \ \forall j \in I_-, \forall k \in I_+ \end{array} \right\}.$$

We see that X is given by finitely many nonstrict linear inequalities in x-variables only, as claimed. ∎

3.2 Every Polyhedrally Representable Set is Polyhedral

The outlined procedure for building polyhedral descriptions (i.e., polyhedral representations not involving extra variables) for projections of polyhedral sets is called *Fourier–Motzkin elimination*.

3.2.1 Some Applications

As an immediate application of Fourier–Motzkin elimination, let us take a linear program (LP) $\min_x\{c^\top x : Ax \leq b\}$ and look at the set T of possible objective values at *feasible solutions* (if any):[1]

$$T := \left\{t \in \mathbf{R} : \exists x \in \mathbf{R}^n : c^\top x = t, \ Ax \leq b\right\}.$$

Rewriting the linear equality $c^\top x = t$ as a pair of opposite inequalities, we see that T is polyhedrally representable, and the above definition of T is just a polyhedral representation of this set, with x in the role of the vector of extra variables. By Fourier–Motzkin elimination, T is polyhedral, that is, T can be expressed as the solution set of a finite system of nonstrict linear inequalities in the variable t only. Thus, we immediately see that T is

1. empty (meaning that the LP in question is infeasible),
2. or is an unbounded-below nonempty set of the form $\{t \in \mathbf{R} : -\infty \leq t \leq b\}$ with $b \in \mathbf{R} \cup \{+\infty\}$ (meaning that the LP is feasible and unbounded),
3. or is a bounded-below nonempty set of the form $\{t \in \mathbf{R} : a \leq t \leq b\}$ with $a \in \mathbf{R}$ and $+\infty \geq b \geq a$. In this case, the LP is feasible and bounded, and its optimal value is a.

Note that given the list of linear inequalities defining T (this list can be built algorithmically by Fourier–Motzkin elimination applied to the original polyhedral representation of T), we can easily detect which of the above cases indeed takes place, i.e., we can identify the feasibility and boundedness status of the LP and find its optimal value. When it is finite (case 3 above), we can use the Fourier–Motzkin elimination backward, starting with $t = a \in T$ and extending this value to a pair (t, x) with $t = a = c^\top x$ and $Ax \leq b$, that is, we can augment the optimal value by an optimal solution. Thus, we can say that Fourier–Motzkin elimination is a finite real arithmetic algorithm, which allows one to check whether an LP is feasible and bounded, and when that is the case, enables one to find the optimal value and an optimal solution.

On the other hand, Fourier–Motzkin elimination is completely impractical since the elimination process can blow up exponentially the number of inequalities. Indeed, from the description of the process it is clear that if a polyhedral set is given by m linear inequalities, when eliminating one variable, we can end up with as many as $m^2/4$ inequalities (this is what happens if there are $m/2$ indices in I_+, $m/2$ indices in I_-, and $I_0 = \varnothing$). Eliminating the next variable, we again can "nearly square" the number of inequalities, and so on. Thus, the number of inequalities in the description of T can become astronomically large even when the dimension of x is something like ten.

The actual importance of Fourier–Motzkin elimination is of a theoretical nature. For example, the linear-programming (LP)-related reasoning we have just carried out shows that

---

[1] See section 4.5.1 for the definitions of feasible/bounded/solvable LPs and their optimal values.

Every feasible and bounded LP problem is solvable, i.e., it has an optimal solution.

(We will revisit this result in more detail in section 7.3.1.) This is a fundamental fact for LP (but in fact is not necessarily true for other types of optimization problems, not even convex problems in general, see Remark 4.6), and the above reasoning (even with the justification of the elimination added to it) is, to the best of our knowledge, the shortest and most transparent way to prove this fundamental fact. Another application of the fact that polyhedrally representable sets are polyhedral is the Homogeneous Farkas' Lemma, to be stated and proved in section 4.1; this lemma will be instrumental in numerous subsequent theoretical developments.

3.3 Calculus of Polyhedral Representations

The fact that polyhedral sets are exactly the same as polyhedrally representable sets does not nullify the notion of a polyhedral representation. The point is that a set can admit "quite compact" polyhedral representation involving extra variables and require an astronomically large – completely meaningless for any practical purpose – number of inequalities in its polyhedral description (think about Example 3.2 and the associated set (3.2) when $n = 100$). Moreover, polyhedral representations admit a kind of fully algorithmic calculus. Specifically, it turns out that all basic convexity-preserving operations (cf. Proposition 1.21) as applied to polyhedral operands preserve polyhedrality. Moreover, polyhedral representations of the results are readily given by polyhedral representations of the operands. Here is the algorithmic polyhedral analogy of Proposition 1.21.

1. *Taking the finite intersection*: Let M_i, $1 \leq i \leq m$, be polyhedral sets in $\mathbf{R}^n$ given by their polyhedral representations

$$M_i = \left\{ x \in \mathbf{R}^n : \exists u^i \in \mathbf{R}^{k_i} : A_i x + B_i u^i \leq c^i \right\}, \quad 1 \leq i \leq m.$$

Then the intersection of the sets M_i is polyhedral with an explicit polyhedral representation, i.e.,

$$\bigcap_{i=1}^m M_i = \left\{ x \in \mathbf{R}^n : \exists u = [u^1; \ldots; u^m] \in \mathbf{R}^{k_1 + \ldots + k_m} : A_i x + B_i u^i \leq c^i, \, 1 \leq i \leq m \right\}.$$

2. *Taking direct product*: Let $M_i \subseteq \mathbf{R}^{n_i}$, $1 \leq i \leq m$, be polyhedral sets given by polyhedral representations:

$$M_i = \left\{ x^i \in \mathbf{R}^{n_i} : \exists u^i \in \mathbf{R}^{k_i} : A_i x^i + B_i u^i \leq c^i \right\}, \quad 1 \leq i \leq m.$$

Then, the direct product

$$M_1 \times \cdots \times M_m := \left\{ x = [x^1; \ldots; x^m] : x^i \in M_i, \, 1 \leq i \leq m \right\}$$

of the sets is a polyhedral set with explicit polyhedral representation, i.e.,

$$M_1 \times \cdots \times M_m = \left\{ x = [x^1; \ldots; x^m] \in \mathbf{R}^{n_1 + \ldots + n_m} : \exists u = [u^1; \ldots; u^m] \in \mathbf{R}^{k_1 + \ldots + k_m} : \\ A_i x^i + B_i u^i \leq c^i, \, 1 \leq i \leq m \right\}.$$

3. *Arithmetic summation and multiplication by reals*: Let $M_i \subseteq \mathbf{R}^n$, $1 \leq i \leq m$, be nonempty polyhedral sets given by polyhedral representations

$$M_i = \left\{x \in \mathbf{R}^n : \exists u^i \in \mathbf{R}^{k_i} : A_i x + B_i u^i \leq c^i\right\}, \ 1 \leq i \leq m,$$

and let $\lambda_1, \ldots, \lambda_m$ be real numbers. Then the set $\lambda_1 M_1 + \cdots + \lambda_m M_m := \{x = \lambda_1 x^1 + \cdots + \lambda_m x^m : x^i \in M_i, \ 1 \leq i \leq m\}$ is polyhedral with explicit polyhedral representation, specifically,

$$\lambda_1 M_1 + \cdots + \lambda_m M_m$$
$$= \left\{ x \in \mathbf{R}^n : \exists (x^i \in \mathbf{R}^n, u^i \in \mathbf{R}^{k_i}, 1 \leq i \leq m) : \atop x \leq \sum_i \lambda_i x^i, \ x \geq \sum_i \lambda_i x^i, \ A_i x^i + B_i u^i \leq c^i, \ 1 \leq i \leq m \right\}.$$

4. *Taking the image under an affine mapping*: Let $M \subseteq \mathbf{R}^n$ be a polyhedral set given by the polyhedral representation

$$M = \left\{x \in \mathbf{R}^n : \exists u \in \mathbf{R}^k : Ax + Bu \leq c\right\},$$

and let $\mathcal{P}(x) = Px + p : \mathbf{R}^n \to \mathbf{R}^m$ be an affine mapping. Then, the image of M under this mapping, i.e., $\mathcal{P}(M) := \{Px + p : x \in M\}$, is a polyhedral set with explicit polyhedral representation given by

$$\mathcal{P}(M) = \left\{ y \in \mathbf{R}^m : \exists (x \in \mathbf{R}^n, u \in \mathbf{R}^k) : \begin{cases} y \leq Px + p \\ y \geq Px + p \\ Ax + Bu \leq c \end{cases} \right\}.$$

5. *Taking the inverse image under affine mapping*: Let $M \subseteq \mathbf{R}^n$ be a polyhedral set given by the polyhedral representation

$$M = \left\{x \in \mathbf{R}^n : \exists u \in \mathbf{R}^k : Ax + Bu \leq c\right\},$$

and let $\mathcal{P}(y) = Py + p : \mathbf{R}^m \to \mathbf{R}^n$ be an affine mapping. Then, the inverse image of M under this mapping, i.e., $\mathcal{P}^{-1}(M) := \{y \in \mathbf{R}^m : Py + p \in M\}$, is a polyhedral set with explicit polyhedral representation given by

$$\mathcal{P}^{-1}(M) = \left\{y \in \mathbf{R}^m : \exists u \in \mathbf{R}^k : A(Py + p) + Bu \leq c\right\}.$$

Note that the rules for taking intersections, direct products, and inverse images, as applied to polyhedral *descriptions* of operands, lead to polyhedral descriptions of the results. In contrast with this, the rules for taking sums with coefficients and images under affine mappings heavily exploit the notion of polyhedral *representation*: even when the operands in these rules are given by polyhedral descriptions, there are no simple ways to point out polyhedral *descriptions* of the results.

An absolutely straightforward justification of the above calculus rules is the subject of Exercise I.27.

Finally, we note that the problem of minimizing a linear form $c^\top x$ over a set M given by its polyhedral representation, i.e.,

$$M = \left\{x \in \mathbf{R}^n : \exists u \in \mathbf{R}^k : Ax + Bu \leq c\right\},$$

can be immediately reduced to an explicit LP problem, namely,

$$\min_{x,u} \left\{ c^\top x : Ax + Bu \leq c \right\}.$$

A reader with some experience in linear programming must have used a lot of the above "calculus of polyhedral representations" when building LPs (perhaps without a clear understanding of what in fact is going on; this is rather similar to Molière's Monsieur Jourdain speaking prose all his life without knowing it).

4

General Theorem of the Alternative and Linear Programming Duality

4.1 Homogeneous Farkas' Lemma

Let $a_1, \ldots, a_N$ be vectors from $\mathbf{R}^n$, and let a be another vector from $\mathbf{R}^n$. Here, we address the question: when does a belong to the cone spanned by the vectors $a_1, \ldots, a_N$, i.e., when can a be represented as a linear combination of the vectors a_i with *nonnegative* coefficients? We immediately observe the following evident *necessary* condition:

$$\text{if } a = \sum_{i=1}^{N} \lambda_i a_i \quad [\text{where } \lambda_i \geq 0, \ i = 1, \ldots, N]$$

then every vector h that has nonnegative inner products with all the a_i should also have a nonnegative inner product with a:

$$\left\{ a = \sum_{i=1}^{N} \lambda_i a_i, \text{ with } \lambda_i \geq 0, \ \forall i, \text{ and } h^\top a_i \geq 0, \ \forall i \right\} \implies h^\top a \geq 0.$$

In fact, this evidently necessary condition is also sufficient. This is given by the Homogeneous Farkas' Lemma.

Lemma 4.1 *[Homogeneous Farkas' Lemma (HFL)] Let $a, a_1, \ldots, a_N$ be vectors from $\mathbf{R}^n$. The vector a is a conic combination of the vectors a_i (a linear combination with nonnegative coefficients), i.e., $a \in \mathrm{Cone}\{a_1, \ldots, a_N\}$, if and only if every vector h satisfying $h^\top a_i \geq 0$, $i = 1, \ldots, N$, satisfies also $h^\top a \geq 0$. In other words, a homogeneous linear inequality*

$$a^\top h \geq 0$$

in the variable h is a consequence of the system

$$a_i^\top h \geq 0, \quad 1 \leq i \leq N,$$

of homogeneous linear inequalities if and only if it can be obtained from the inequalities of the system by "admissible linear aggregation" – taking their weighted sum with nonnegative weights.

Proof The necessity – the "only if" part of the statement – was proved above before we formulated the Homogeneous Farkas' Lemma. Let us prove the "if" part of the lemma. Thus, we assume that $h^\top a \geq 0$ is a *consequence* of the homogeneous system $h^\top a_i \geq 0 \ \forall i$, i.e., every vector h satisfying $h^\top a_i \geq 0 \ \forall i$ satisfies also $h^\top a \geq 0$, and let us prove that a is a conic combination of the vectors a_i.

An "intelligent" proof goes as follows. The set Cone $\{a_1,\ldots,a_N\}$ of all conic combinations of $a_1,\ldots,a_N$ is polyhedrally representable (see Example 3.3) and as such is polyhedral (Theorem 3.2). Hence, we have

$$\text{Cone}\{a_1,\ldots,a_N\} = \left\{x \in \mathbf{R}^n : p_j^\top x \geq b_j, 1 \leq j \leq J\right\}. \tag{4.1}$$

Now, observe that $0 \in \text{Cone}\{a_1,\ldots,a_N\}$, and thus we conclude that $b_j \leq 0$ for all $j \leq J$. Moreover, since $\lambda a_i \in \text{Cone}\{a_1,\ldots,a_N\}$ for every i and every $\lambda \geq 0$, we deduce $\lambda p_j^\top a_i \geq b_j$ for all i, j and all $\lambda \geq 0$, whence $p_j^\top a_i \geq 0$ for all i and j. For every j, the relation $p_j^\top a_i \geq 0$ for all i implies, by the premise of the statement we want to prove, that $p_j^\top a \geq 0$. Then, as $0 \geq b_j$ for all j, we see that $p_j^\top a \geq b_j$ for all j, meaning that a does indeed belong to Cone $\{a_1,\ldots,a_N\}$ due to (4.1). ∎

This very short and elegant proof of the Homogeneous Farkas' Lemma is yet another nice illustration of the power of Fourier–Motzkin elimination.

4.2 Certificates for Feasibility and Infeasibility

Consider a (finite) system of scalar inequalities with n unknowns. To be as general as possible, for the time being we do not assume the inequalities to be linear, and we allow for both non-strict and strict inequalities in the system, and the same for equalities. Since an equality can be represented by a pair of non-strict inequalities, our system can always be written as

$$f_i(x) \, \Omega_i \, 0, \qquad i = 1,\ldots,m, \tag{$\mathcal{S}$}$$

where every Ω_i is either the relation ">" or the relation "$\geq$", and we assume $m \geq 1$, which is the only case of interest here.

The most basic question about $(\mathcal{S})$ is:

($\mathcal{Q}$)   Does $(\mathcal{S})$ have a solution, i.e., is $(\mathcal{S})$ feasible?

Knowing how to answer the question $(\mathcal{Q})$ enables us to answer many other questions. For example, verifying whether a given real number a is a lower bound on the optimal value Opt* of a linear program

$$\min_x \left\{c^\top x : Ax \geq b\right\} \tag{LP}$$

is the same as verifying whether the system

$$-c^\top x + a > 0$$
$$Ax - b \geq 0$$

has no solution.

The general question $(\mathcal{Q})$ above is too difficult, and it makes sense to pass from it to a seemingly simpler one:

($\mathcal{Q}'$)   How do we certify that $(\mathcal{S})$ has, or does not have, a solution?

Imagine that you are very smart and know the correct answer to $(\mathcal{Q})$; how can you convince everyone that your answer is correct? What can be an "evident for everybody" validity certificate for your answer?

4.2 Certificates for Feasibility and Infeasibility

If your claim is that $(\mathcal{S})$ is feasible, a *certificate* can just point out a solution x^* to $(\mathcal{S})$. Given this certificate, one can substitute x^* into the system and check whether x^* is indeed a feasible solution.

Suppose now that your claim is that $(\mathcal{S})$ has no solutions. What can be a "simple certificate" for this claim? How can one certify a *negative* statement? This is a highly nontrivial problem, and not just for mathematics; for example, in criminal law, how should someone accused in a murder prove his innocence? The "real life" answer to the question "how to certify a negative statement" is discouraging: such a statement normally *cannot* be certified.[1] In mathematics, the standard way to justify a negative statement **A**, such as "there is no solution to a particular system of constraints" (e.g., "there is no solution to the equation $x^5 + y^5 = z^5$ with positive integer variables x, y, z") is to obtain a contradiction with the statement opposite to **A**, i.e., $\overline{\mathbf{A}}$ (in our example, $\overline{\mathbf{A}}$ is the statement "the solution exists"). That is, we assume that $\overline{\mathbf{A}}$ is true and derive consequences until a clearly false statement is obtained; when this happens, we know that $\overline{\mathbf{A}}$ is false (since the legitimate consequences of a true statement must be true), and therefore **A** must be true. In general, there is no recipe that leads to a contradiction of something which in fact is false; this is why certifying negative statements is usually difficult.

Fortunately, finite systems of linear inequalities are simple enough to allow for a recipe for certifying their infeasibility: we start with the assumption that a solution exists and then demonstrate a contradiction in a very specific way – *by taking a weighted sum of the inequalities in the system using nonnegative aggregation weights to produce a contradictory inequality*.

Let us start with a simple illustration: we would like to certify the infeasibility of the following system of inequalities in variables u, v, w:

$$\begin{array}{rrrcr} 5u & -6v & -4w & > & 2 \\ & +4v & -2w & \geq & -1 \\ -5u & & +7w & \geq & 1 \end{array}$$

Let us assign to these inequalities "aggregation weights" $2, 3, 2$, multiply the inequalities by the respective weights and sum the resulting inequalities:

$$\begin{array}{r} 2 \times \begin{array}{|rrrcr} 5u & -6v & -4w & > & 2 \end{array} \\ + \\ 3 \times \begin{array}{|rrrcr} & +4v & -2w & \geq & -1 \end{array} \\ + \\ 2 \times \begin{array}{|rrrcr} -5u & & +7w & \geq & 1 \end{array} \\ \hline (*) \quad \begin{array}{|rrrcr} 0 \cdot u & +0 \cdot v & +0 \cdot w & > & 3 \end{array} \end{array}$$

The resulting aggregated inequality $(*)$ is contradictory, it has no solutions at all. At the same time, $(*)$ is a consequence of our system – by the construction of $(*)$, every solution to the original system of three inequalities is also feasible for $(*)$. Taken together, these two observations say that the system has no solutions, and the vector $[2; 3; 2]$ of our aggregation weights can be seen as an *infeasibility certificate* – taking a weighted sum of

---

[1] This is where the law court rule "a person is presumed innocent until proven guilty" comes from – instead of requesting that the accused certifies the negative statement "I did not commit the crime," the court requests the prosecution to certify the positive statement "the accused did commit the crime."

inequalities from the system with corresponding nonnegative weights, we lead the system to a contradiction.

As applied to a general system of inequalities (S), a similar approach to certifying infeasibility would be to assign nonnegative aggregation weights to inequalities, multiply them by these weights, and sum the resulting inequalities, arriving at an aggregated inequality, which, due to its origin, is a *consequence* of system (S), meaning that every solution to the system solves the aggregated inequality as well. It follows that *when the aggregated inequality is contradictory*, i.e., it has no solutions at all, *the original system (S) must be infeasible as well*. When this happens, the collection of weights used to generate the contradictory-consequence inequality can be viewed as an infeasibility certificate for (S).

Let us look at what the outlined approach means when (S) is composed of finitely many *linear* inequalities:

$$a_i^\top x \, \Omega_i \, b_i, \quad i = 1, \ldots, m \qquad [\text{where } \Omega_i \text{ is either ``>'' or ``}\geq\text{''}]. \qquad (S)$$

In this case the "aggregated inequality" is linear as well:

$$\left(\sum_{i=1}^m \lambda_i a_i\right)^\top x \;\; \Omega \;\; \sum_{i=1}^m \lambda_i b_i, \qquad (\text{Comb}(\lambda))$$

where Ω is ">" whenever $\lambda_i > 0$ for at least one i with $\Omega_i = $ " $>$ ", and Ω is "$\geq$" otherwise. Now, when can a *linear* inequality

$$d^\top x \, \Omega \, e$$

be contradictory? Of course, it can happen only when $d = 0$. Furthermore, in this case, whether the inequality is contradictory depends on the relation Ω and the value of e: if Ω is " $>$ " then the inequality is contradictory if and only if $e \geq 0$, and if Ω is " $\geq$ " then the inequality is contradictory if and only if $e > 0$. We have established the following simple result:

Proposition 4.2 *Consider a system of linear inequalities in unknowns $x \in \mathbf{R}^n$:*

$$\begin{cases} a_i^\top x > b_i, & i = 1, \ldots, m_s, \\ a_i^\top x \geq b_i, & i = m_s + 1, \ldots, m. \end{cases} \qquad (S)$$

Let us associate with (S) two systems of linear inequalities and equations with unknowns $\lambda \in \mathbf{R}^m$:

$$\mathcal{T}_\mathrm{I}: \begin{cases} (a) & \lambda \geq 0, \\ (b) & \sum_{i=1}^m \lambda_i a_i = 0, \\ (c_\mathrm{I}) & \sum_{i=1}^m \lambda_i b_i \geq 0, \\ (d_\mathrm{I}) & \sum_{i=1}^{m_s} \lambda_i > 0. \end{cases} \qquad \mathcal{T}_\mathrm{II}: \begin{cases} (a) & \lambda \geq 0, \\ (b) & \sum_{i=1}^m \lambda_i a_i = 0, \\ (c_\mathrm{II}) & \sum_{i=1}^m \lambda_i b_i > 0. \end{cases}$$

If at least one of the systems $\mathcal{T}_\mathrm{I}$, $\mathcal{T}_\mathrm{II}$ is feasible, then the system (S) is infeasible.

4.3 General Theorem of the Alternative

Proposition 4.2 states that in some cases it is easy to certify the infeasibility of a system of linear inequalities: use a "simple certificate" that is a solution to another system of linear

4.3 General Theorem of the Alternative

inequalities. Note, however, that the existence of a certificate of this latter type is so far only a *sufficient*, but not a *necessary*, condition for the infeasibility of $(\mathcal{S})$. A fundamental result in the theory of linear inequalities is that this sufficient condition is in fact also necessary:

Theorem 4.3 *[General Theorem of the Alternative (GTA)] Consider the notation and setting of Proposition 4.2. System $(\mathcal{S})$ has no solutions if and only if at least one of the systems $\mathcal{T}_{\mathrm{I}}$ or $\mathcal{T}_{\mathrm{II}}$ is feasible.*

Proof The GTA is a more or less straightforward corollary of the Homogeneous Farkas' Lemma. Indeed, in view of Proposition 4.2, all we need to prove is that *if $(\mathcal{S})$ has no solution then at least one of the systems $\mathcal{T}_{\mathrm{I}}$, $\mathcal{T}_{\mathrm{II}}$ is feasible*. Thus, assume that $(\mathcal{S})$ has no solutions, and let us look at the consequences. Let us associate with $(\mathcal{S})$ the following system of *homogeneous* linear inequalities in variables x, τ, ϵ:

$$
\begin{array}{lll}
(a) & \tau - \epsilon \geq 0, & \\
(b) & a_i^\top x - b_i \tau - \epsilon \geq 0, & i = 1, \ldots, m_s, \\
(c) & a_i^\top x - b_i \tau \geq 0, & i = m_s + 1, \ldots, m.
\end{array}
\tag{4.2}
$$

First, we claim that *in every solution to* (4.2), *one has $\epsilon \leq 0$*. Indeed, assuming that (4.2) has a solution x, τ, ϵ with $\epsilon > 0$, we conclude from (4.2a) that $\tau > 0$. Then, from (4.2b, c) it will follow that $\tau^{-1} x$ is a solution to $(\mathcal{S})$, while we have assumed $(\mathcal{S})$ is infeasible. Therefore, we must have $\epsilon \leq 0$ in every solution to (4.2).

Now, we have that the homogeneous linear inequality

$$
-\epsilon \geq 0 \tag{4.3}
$$

is a consequence of the system of homogeneous linear inequalities (4.2). Then, by the Homogeneous Farkas' Lemma, there exist nonnegative weights $\nu, \lambda_i, i = 1, \ldots, m$, such that the aggregated inequality from (4.2) using these weights results in precisely the consequence-inequality (4.3), i.e.,

$$
\begin{array}{ll}
(a) & \sum_{i=1}^m \lambda_i a_i = 0, \\
(b) & -\sum_{i=1}^m \lambda_i b_i + \nu = 0, \\
(c) & -\sum_{i=1}^{m_s} \lambda_i - \nu = -1.
\end{array}
\tag{4.4}
$$

Recall that, by their origin, ν and all λ_i are nonnegative. Now, it may happen that $\lambda_1, \ldots, \lambda_{m_s}$ are zero. In this case $\nu = 1$ by (4.4c), and relations (4.4a, b) say that $\lambda_1, \ldots, \lambda_m$ is a solution for $\mathcal{T}_{\mathrm{II}}$. In the remaining case (that is, when not all $\lambda_1, \ldots, \lambda_{m_s}$ are zero, or, which is the same, when $\sum_{i=1}^{m_s} \lambda_i = 1 - \nu > 0$), the same relations (4.4a, b) say that $\lambda_1, \ldots, \lambda_m$ is a solution for $\mathcal{T}_{\mathrm{I}}$. ∎

Remark 4.4 We have derived the GTA from the Homogeneous Farkas' Lemma (HFL). Note that the HFL is just a special case of the GTA. Indeed, identifying when a linear inequality $a^\top x \leq b$ is a consequence of the system $a_i^\top x_i \leq b_i$, $1 \leq i \leq m$ (this is the question answered by the HFL in the case of $b = b_1 = \cdots = b_m = 0$) is exactly the same as identifying when the system of inequalities

$$
a^\top x > b, \quad a_i^\top x \leq b_i, \quad i = 1, \ldots, m \tag{$*$}
$$

in the variables x is infeasible; what in the latter case is said by the GTA is exactly what the HFL states: *when $b = b_1 = \cdots = b_m = 0$, the system $(*)$ is infeasible if and only if*

4.4 Corollaries of the GTA

Let us explicitly state two very useful principles derived from the General Theorem of the Alternative:

A. A system of finitely many linear inequalities

$$a_i^\top x \; \Omega_i \; b_i, \quad i = 1, \ldots, m \qquad [\text{where } \Omega_i \in \{``\geq",``>"\}]$$

has no solutions if and only if one can aggregate the inequalities of the system in a *linear* fashion (i.e., multiplying the inequalities by nonnegative weights, summing the resulting inequalities and passing, if necessary, from an inequality $a^\top x > b$ to the inequality $a^\top x \geq b$) to get a contradictory inequality, namely, either the inequality $0^\top x \geq 1$ or the inequality $0^\top x > 0$.

B. A linear inequality

$$a_0^\top x \; \Omega_0 \; b_0 \qquad [\text{where } \Omega_0 \in \{``\geq",``>"\}]$$

in variables x is a consequence of a *feasible* system of linear inequalities

$$a_i^\top x \; \Omega_i \; b_i, \quad i = 1, \ldots, m \qquad [\text{where } \Omega_i \in \{``\geq",``>"\}]$$

if and only if it can be obtained by *linear* aggregation with nonnegative weights from the inequalities of the system and the trivial identically true inequality $0^\top x > -1$.

In fact, when all Ω_i in **B** are non-strict, **B** can be reformulated equivalently as follows.

Proposition 4.5 *[Inhomogeneous Farkas' Lemma] The linear inequality*

$$a^\top x \leq b$$

is a consequence of the feasible system of linear inequalities

$$a_i^\top x \leq b_i, \quad 1 \leq i \leq m,$$

if and only if there exist nonnegative aggregation weights λ_i, $i = 1, \ldots, m$, such that

$$a = \sum_{i=1}^m \lambda_i a_i \quad \text{and} \quad b \geq \sum_{i=1}^m \lambda_i b_i.$$

We would like to emphasize that the preceding principles are highly nontrivial and very deep. Consider, e.g., the following system of four linear inequalities in two variables u, v:

$$-1 \leq u \leq 1,$$
$$-1 \leq v \leq 1.$$

These inequalities clearly imply that
$$u^2 + v^2 \leq 2, \tag{!}$$
which in turn implies, by the Cauchy–Schwarz inequality, the linear inequality $u + v \leq 2$:
$$u + v = 1 \times u + 1 \times v \leq \sqrt{1^2 + 1^2}\sqrt{u^2 + v^2} \leq (\sqrt{2})^2 = 2. \tag{!!}$$

The concluding inequality $u + v \leq 2$ is linear and is a consequence of the original feasible system, and so we could have simply relied on Principle **B** to derive it. On the other hand, in the preceding demonstration of this linear consequence inequality both the steps (!) and (!!) are "highly nonlinear." It is absolutely unclear a priori why the same consequence-inequality can, as it is stated by Principle **B**, be derived from the system in a "linear" manner as well (of course it can – it suffices just to sum two inequalities, $u \leq 1$ and $v \leq 1$). In contrast, the Inhomogeneous Farkas' Lemma predicts that hundreds of pages of some complicated demonstration that such and such a linear inequality is a consequence of such and such a feasible finite system of linear inequalities can be replaced by a simple demonstration of weights with prescribed sign such that the target inequality is the weighted sum of the inequalities from the system and the identically true linear inequality. One should appreciate the elegance and depth of such a result!

Note that the General Theorem of the Alternative and its corollaries **A** and **B** heavily exploit the fact that we are speaking about *linear* inequalities. For example, consider the following system of two quadratic and two linear inequalities in two variables:

$$(a) \quad u^2 \geq 1,$$
$$(b) \quad v^2 \geq 1,$$
$$(c) \quad u \geq 0,$$
$$(d) \quad v \geq 0,$$

along with the quadratic inequality

$$(e) \quad uv \geq 1.$$

The inequality (e) is clearly a consequence of (a)–(d). However, if we extend the system of inequalities (a)–(b) by all "trivial" (i.e., identically true) linear and quadratic inequalities in two variables, such as $0 > -1$, $u^2 + v^2 \geq 0$, $u^2 + 2uv + v^2 \geq 0$, $u^2 - 2uv + v^2 \geq 0$, etc., and ask whether (e) can be derived in a *linear* fashion from the inequalities of the extended system, the answer will be negative. Thus, Principle **B** fails to be true even for quadratic inequalities (which is a great pity, because otherwise there would be no difficult mathematical problems at all!).

4.5 Application: Linear Programming Duality

We are about to use General Theorem of the Alternative to obtain the basic results of linear programming (LP) duality theory. To do so, we first introduce some basic terminology about mathematical programming problems.

4.5.1 Preliminaries: Mathematical and Linear Programming Problems

A (constrained) mathematical programming problem has the following form:

$$(P) \quad \min_x \left\{ f(x) : \begin{array}{l} x \in X, \\ g(x) \equiv [g_1(x); \ldots ; g_m(x)] \leq 0, \\ h(x) \equiv [h_1(x); \ldots ; h_k(x)] = 0 \end{array} \right\}, \quad (4.5)$$

where

- [domain] X is called the *domain* of the problem,
- [objective] f is called the *objective* (function) of the problem,
- [constraints] g_i, $i = 1, \ldots, m$, are called the (functional) *inequality constraints*, and h_j, $j = 1, \ldots, k$, are called the *equality constraints*.[2]

We always assume that $X \neq \emptyset$ and that the objective and the constraints are well defined on X. Moreover, we typically skip indicating X when $X = \mathbf{R}^n$.

We use the following standard terminology related to (4.5):

- [feasible solution] A point $x \in \mathbf{R}^n$ is called a *feasible solution* to (4.5), if $x \in X$, $g_i(x) \leq 0$, $i = 1, \ldots, m$, and $h_j(x) = 0$, $j = 1, \ldots, k$, i.e., if x satisfies all restrictions imposed by the formulation of the problem.
  - [feasible set] The set of all feasible solutions is called the *feasible set* of the problem.
  - [feasible problem] A problem with a nonempty feasible set (i.e., a problem which admits feasible solutions) is called *feasible* (or consistent).
- [optimal value] The *optimal value* of the problem refers to the quantity

$$\text{Opt} := \begin{cases} \inf_x \{f(x) : x \in X, \, g(x) \leq 0, \, h(x) = 0\}, & \text{if the problem is feasible,} \\ +\infty, & \text{if the problem is infeasible.} \end{cases}$$

  - [below boundedness] The problem is called *below bounded*, if its optimal value is $> -\infty$, i.e., if the objective is bounded from below on the feasible set.
- [optimal solution] A point $x \in \mathbf{R}^n$ is called an *optimal solution* to (4.5) if x is feasible and $f(x) \leq f(x')$ for any other feasible solution x', i.e., if

$$x \in \underset{x'}{\text{Argmin}} \left\{ f(x') : x' \in X, \, g(x') \leq 0, \, h(x') = 0 \right\}.$$

  - [solvable problem] A problem is called *solvable* if it admits optimal solutions.
  - [optimal set] The set of all optimal solutions to a problem is called its *optimal set*.

Remark 4.6 It is crucial to note that to say that a mathematical programming problem is solvable is a much stronger statement than merely to say that it is feasible and below bounded, by definition, a solvable optimization problem with a minimization-type objective is *always* feasible and below bounded. However, the reverse is not true in general, even when the problem is convex. For example, consider the optimization problem given by

$$\min_x \{\exp(x) : x \in \mathbf{R}\}.$$

---

[2] Rigorously speaking, the constraints are not the *functions* g_i, h_j, but the *relations* $g_i(x) \leq 0$, $h_j(x) = 0$. We will use the word "constraints" in both these senses, and it will always be clear what is meant. For example, we will say that "x satisfies the constraints" to refer to the relations and that "the constraints are differentiable" to refer to the underlying functions.

4.5 Application: Linear Programming Duality

Clearly, this problem is feasible, and its optimal value is equal to 0, so it is bounded from below. On the other hand, the problem admits no optimal solution $x \in \mathbf{R}$ achieving the objective value 0.

That said, there is one important class of optimization problems, namely linear programs, where feasibility and below boundedness together imply solvability. In particular, recall from section 3.2.1 that every feasible and below-bounded LP is also solvable. This fundamental fact about LP problems is due to the structure of polyhedral sets, which guarantees that the projection (by Fourier–Motzkin elimination) of a polyhedral set is always closed. ◇

Remark 4.7 In the above description of a mathematical programming problem and related basic notions, such as feasibility, solvability, boundedness, etc., we "standardize" the situation by assuming that the objective is to be minimized, and the inequality constraints are of the form $g_i(x) \leq 0$. Needless to say, we can also speak about problems where the objective should be maximized and/or some of the inequality constraints are of the form $g_i(x) \geq 0$. There is no difficulty in reducing these "more general" forms of optimization problems to our standard form: maximizing $f(x)$ is the same as minimizing $-f(x)$, and the constraint of the form $g_i(x) \geq 0$ is the same as the constraint $-g_i(x) \leq 0$. While this standardization is always possible, from time to time we take the liberty of speaking about maximization problems and/or $\geq$-type constraints. With this in mind, it is worth mentioning that when working with maximization problems, we should update the notions of optimal value, the boundedness of a problem, and optimal solution. For a maximization problem,

- the optimal value is the supremum of the values of the objective at feasible solutions, and is, by definition, $-\infty$ for infeasible problems, and
- boundedness means boundedness from above of the objective on the feasible set (or, which is the same, the fact that the optimal value is $< +\infty$);
- an optimal solution is a feasible solution such that the objective value at this solution is greater than or equal to the objective value at every feasible solution.

Needless to say, when "standardizing" a maximization problem, i.e., replacing the maximization of $f(x)$ with the minimization of $-f(x)$, boundedness and optimal solutions remain intact if the optimal value "is negated," i.e., real number a becomes $-a$, and $\pm\infty$ becomes $\mp\infty$. ◇

Linear programming problems A mathematical programming problem (P) is called a *linear programming* (LP) *problem*, if

- $X = \mathbf{R}^n$ is the entire space,
- $f, g_1, \ldots, g_m$ are *real-valued affine* functions on $\mathbf{R}^m$, that is, functions of the form $a^\top x + b$, and
- there are no equality constraints.

Note that in principle we could allow for linear equality constraints $h_j(x) := a_j^\top x + b_j = 0$. However, a constraint of this type can be equivalently represented as a pair of opposite linear inequalities $a_j^\top x + b_j \leq 0$, $-a_j^\top x - b_j \leq 0$. To save space and words (and, as we have just explained, with no loss in generality), in the sequel we will focus on inequality-constrained linear programming problems.

4.5.2 Dual to an LP Problem: The Origin

Consider the LP problem

$$\text{Opt} = \min_x \left\{ c^\top x : Ax - b \geq 0 \right\} \quad \text{[where } A = \begin{bmatrix} a_1^\top \\ a_2^\top \\ \cdots \\ a_m^\top \end{bmatrix} \in \mathbf{R}^{m \times n}\text{].} \tag{LP}$$

The motivation for constructing the problem *dual* to an LP problem is the desire to generate, in a systematic way, lower bounds on the optimal value Opt of (LP).

Consider the problem

$$\min_x \{ f(x) : g_i(x) \geq b_i, \ i = 1, \ldots, m \}.$$

An evident way to bound from below a given function $f(x)$ in the domain given by a system of inequalities

$$g_i(x) \geq b_i, \quad i = 1, \ldots, m, \tag{4.6}$$

is offered by what is called the *Lagrange duality*. We will discuss the Lagrange duality in full detail for general functions in Part IV. Here, let us do a brief preliminary outline and examine the special case when we are dealing with linear functions only.

Lagrange Duality
- We consider all inequalities which can be obtained from (4.6) by linear aggregation, i.e., inequalities of the form

$$\sum_{i=1}^m y_i g_i(x) \geq \sum_{i=1}^m y_i b_i \tag{4.7}$$

with "aggregation weights" $y_i \geq 0$ for all i. Note that the inequality (4.7), owing to its origin, is valid on the entire set $\mathcal{X}$ of feasible solutions of (4.6).
- Depending on the choice of aggregation weights, it may happen that the left-hand side in (4.7) is $\leq f(x)$ for all $x \in \mathbf{R}^n$. Whenever this is the case, the right-hand side $\sum_{i=1}^m y_i b_i$ of (4.7) is a lower bound on $f(x)$ for any $x \in \mathcal{X}$. It follows that *the optimal value of the problem*

$$\max_y \left\{ \sum_{i=1}^m y_i b_i : \begin{array}{ll} y \geq 0, & (a) \\ \sum_{i=1}^m y_i g_i(x) \leq f(x), \ \forall x \in \mathbf{R}^n & (b) \end{array} \right\} \tag{4.8}$$

is a lower bound on the values of f on the set of feasible solutions to the system (4.6).

Let us now examine what happens to Lagrange duality when f and g_i are homogeneous linear functions, i.e., $f(x) = c^\top x$ and $g_i(x) = a_i^\top x$ for all $i = 1, \ldots, m$. In this case, the requirement (4.8b) merely says that $c = \sum_{i=1}^m y_i a_i$ (or, which is the same, $A^\top y = c$ due to the definition of the matrix A). Thus, problem (4.8) becomes the linear programming problem

$$\max_y \left\{ b^\top y : A^\top y = c, \ y \geq 0 \right\}, \tag{LP*}$$

which is called the LP dual of (LP).

4.5 Application: Linear Programming Duality

By the construction of the dual problem (LP*), we immediately have

[Weak Duality] *The optimal value in* (LP*) *is less than or equal to the optimal value in* (LP).

In fact, "less than or equal to" in the latter statement becomes "equal to" provided that the optimal value Opt in (LP) is a number (i.e., (LP) is feasible and below bounded, in which case Fourier–Motzkin elimination guarantees that Opt is a real number). To see that this indeed is the case, note that a real number α is a lower bound on Opt if and only if $c^\top x \geq \alpha$ holds for all x satisfying $Ax \geq b$, or, which is the same, if and only if the system of linear inequalities

$$-c^\top x > -\alpha, \quad Ax \geq b \tag{S_α}$$

has no solution. Then, by the General Theorem of the Alternative we deduce that at least one of a certain pair of systems of linear inequalities does have a solution. More precisely,

(*) (S_α) *has no solutions if and only if at least one of the following two systems of linear inequalities in* $m+1$ *unknowns has a solution:*

$$\mathcal{T}_\mathrm{I}: \begin{cases} (a) & \lambda = [\lambda_0; \lambda_1; \ldots; \lambda_m] \geq 0, \\ (b) & -\lambda_0 c + \sum_{i=1}^m \lambda_i a_i = 0, \\ (c_\mathrm{I}) & -\lambda_0 \alpha + \sum_{i=1}^m \lambda_i b_i \geq 0, \\ (d_\mathrm{I}) & \lambda_0 > 0; \end{cases}$$

or

$$\mathcal{T}_\mathrm{II}: \begin{cases} (a) & \lambda = [\lambda_0; \lambda_1; \ldots; \lambda_m] \geq 0, \\ (b) & -\lambda_0 c - \sum_{i=1}^m \lambda_i a_i = 0, \\ (c_\mathrm{II}) & -\lambda_0 \alpha - \sum_{i=1}^m \lambda_i b_i > 0. \end{cases}$$

Now assume that (LP) *is feasible*. We first claim that *under this assumption* (S_α) *has no solutions if and only if* $\mathcal{T}_\mathrm{I}$ *has a solution*. The implication "$\mathcal{T}_\mathrm{I}$ has a solution $\Longrightarrow$ (S_α) has no solution" is readily given by the preceding remarks. To verify the inverse implication, assume that (S_α) has no solution and the system $Ax \geq b$ has a solution, and let us prove that then $\mathcal{T}_\mathrm{I}$ has a solution. If $\mathcal{T}_\mathrm{I}$ has no solution then, by (*), $\mathcal{T}_\mathrm{II}$ must have a solution. Moreover, since any solution to $\mathcal{T}_\mathrm{II}$ where $\lambda_0 > 0$ is also a solution to $\mathcal{T}_\mathrm{I}$, we must have $\lambda_0 = 0$ for every solution to $\mathcal{T}_\mathrm{II}$. But, the fact that $\mathcal{T}_\mathrm{II}$ has a solution λ with $\lambda_0 = 0$ is independent of the values of c and α; if this fact were to hold, it would mean, by the same General Theorem of the Alternative, that, e.g., the following instance of (S_α),

$$0^\top x \geq -1, \quad Ax \geq b,$$

has no solution as well. But, then we must have that the system $Ax \geq b$ has no solution – a contradiction to the assumption that (LP) is feasible.

Now, if $\mathcal{T}_\mathrm{I}$ has a solution, this system has a solution with $\lambda_0 = 1$ as well (to see this, pass from a solution λ to the solution λ/λ_0; this construction is well defined, since $\lambda_0 > 0$ for every solution to $\mathcal{T}_\mathrm{I}$). Now, an $(m+1)$-dimensional vector $\lambda = [1; y]$ is a solution to $\mathcal{T}_\mathrm{I}$ if and only if the m-dimensional vector y solves the following system of linear inequalities and equations:

$$[A^\top y \equiv]\ \begin{array}{l} y \geq 0, \\ \sum_{i=1}^m y_i a_i = c, \\ b^\top y \geq \alpha. \end{array} \tag{D}$$

We summarize these observations below.

Proposition 4.8 *If system (D) in unknowns y and α associated with the LP problem (LP) has a solution $(\bar y, \bar\alpha)$ then $\bar\alpha$ is a lower bound on the optimal value of (LP). Vice versa, if (LP) is feasible and $\bar\alpha$ is a lower bound on the optimal value of (LP) then $\bar\alpha$ can be extended by a properly chosen m-dimensional vector $\bar y$ to a solution to (D).*

We see that the entity responsible for lower bounds on the optimal value of (LP) is the system (D): every solution to the latter system induces a bound of this type, and *in the case when (LP) is feasible*, all lower bounds can be obtained from solutions to (D). Now note that if (y,α) is a solution to (D) then the pair $(y, b^\top y)$ also is a solution to the same system, and the lower bound $b^\top y$ on Opt is not worse than the lower bound α. Thus, as far as lower bounds on Opt are concerned, we lose nothing by restricting ourselves to the solutions (y,α) of (D) with $\alpha = b^\top y$. The best lower bound on Opt given by (D) is therefore the optimal value of the problem

$$\max_y \left\{ b^\top y : A^\top y = c,\ y \geq 0 \right\},$$

which is nothing but the dual to the problem (LP), which is given by (LP*). Note that (LP*) is also a linear programming problem.

All we know about the dual problem so far is the following:

Proposition 4.9 *Whenever y is a feasible solution to (LP*), the corresponding value of the dual objective $b^\top y$ is a lower bound on the optimal value Opt in (LP). If (LP) is feasible then for every real number $\alpha \leq$ Opt there exists a feasible solution y of (LP*) with $b^\top y \geq \alpha$.*

4.5.3 Linear Programming Duality Theorem

Proposition 4.9 is in fact equivalent to the following complete statement of the LP Duality Theorem.

Theorem 4.10 *[Duality Theorem in linear programming] Consider a linear programming problem*

$$\min_x \left\{ c^\top x : Ax \geq b \right\}, \tag{LP}$$

along with its dual,

$$\max_y \left\{ b^\top y : A^\top y = c,\ y \geq 0 \right\}. \tag{LP*}$$

Then:

(i) *[Primal–dual symmetry] The dual problem is an LP program, and its dual is equivalent to the primal problem.*

(ii) [Weak duality] *The value of the dual objective at every dual feasible solution is less than or equal to the value of the primal objective at every primal feasible solution, so that the dual optimal value is less than or equal to the primal one.*

(iii) [Strong duality] *The following five properties are equivalent to each other.*
   (a) *The primal is feasible and bounded below.*
   (b) *The dual is feasible and bounded above.*
   (c) *The primal is solvable.*
   (d) *The dual is solvable.*
   (e) *Both primal and dual are feasible.*

Moreover, if any one of these properties (and then, by the equivalence just stated, every one of them) holds, then the optimal values of the primal and the dual problems are equal to each other.

Finally, if at least one of the problems in the primal–dual pair is feasible then the optimal values in both problems are the same, i.e., both are finite and equal to each other, or both are $+\infty$ *(i.e., the primal is infeasible and the dual is not bounded above), or both are* $-\infty$ *(i.e., the primal is unbounded below and the dual is infeasible).*

There is one last remark we should make to complete the story of primal and dual objective values given in Theorem 4.10: in fact it is possible to have both primal and dual problems infeasible simultaneously (see Exercise I.38). This is the only case when the primal and the dual optimal values ($+\infty$ and $-\infty$, respectively) differ from each other.

Proof (i): This part is quite straightforward: writing the dual problem (LP*) in our standard form, we get

$$\min_y \left\{ -b^\top y : \begin{bmatrix} I_m \\ A^\top \\ -A^\top \end{bmatrix} y - \begin{bmatrix} 0 \\ c \\ -c \end{bmatrix} \geq 0 \right\},$$

where I_m is the $m \times m$ identity matrix. Applying the duality transformation to the latter problem, we come to the problem

$$\max_{\xi,\eta,\zeta} \left\{ 0^\top \xi + c^\top \eta + (-c)^\top \zeta : \begin{array}{c} \xi \geq 0 \\ \eta \geq 0 \\ \zeta \geq 0 \\ \xi + A\eta - A\zeta = -b \end{array} \right\},$$

which is clearly equivalent to (LP) (after we set $x = \zeta - \eta$ and eliminate ξ).

(ii): This part follows from the origin of the dual and is thus immediately given by Proposition 4.9.

(iii) We prove the following implications.

(a) $\Longrightarrow$ (d): If the primal is feasible and bounded below, its optimal value Opt (which of course is a lower bound on itself) can, by Proposition 4.9, be (non-strictly) majorized by a quantity $b^\top y^*$, where y^* is a feasible solution to (LP*). Then, of course, $b^\top y^* = \text{Opt}$ by the already proven statement of item (ii). On the other hand, by Proposition 4.9 the optimal value in the dual is less than or equal to Opt. Thus, we conclude that the optimal value in the dual is attained and is equal to the optimal value in the primal.

(d) $\Longrightarrow$ (b): This is evident by the definition of solvability.

(b) $\Longrightarrow$ (c): This implication, in view of the primal–dual symmetry, follows from the already justified implication (a) $\Longrightarrow$ (d).

(c) $\Longrightarrow$ (a): This is evident from the definition of solvability.

We have shown that (a) $\equiv$ (b) $\equiv$ (c) $\equiv$ (d) and that the first (and consequently each) of these four equivalent properties implies that the optimal value in the primal problem is equal to the optimal value in the dual problem. All that remains is to prove the equivalence between (a)–(d) and (e). This is immediate: (a)–(d), of course, imply (e); vice versa, in the case of (e) the primal is not only feasible but also bounded below (this is an immediate consequence of the feasibility of the dual problem, see (ii)), and (a) follows.

It remains to verify that if one problem in the primal–dual pair is feasible then the primal and the dual optimal values are equal to each other. By primal–dual symmetry it suffices to consider the case when the primal problem is feasible. If also the primal is bounded from below then, by what has already been proved, the dual problem is feasible and the primal and dual optimal values coincide with each other. If the primal problem is unbounded from below then the primal optimal value is $-\infty$ and by Weak Duality the dual problem is infeasible, so that the dual optimal value is $-\infty$. ∎

An immediate corollary of the LP Duality Theorem is the following *necessary and sufficient* optimality condition in LP.

Theorem 4.11 *[Necessary and sufficient optimality conditions in Linear Programming] Consider an LP problem* (LP) *along with its dual* (LP*) *as in Theorem 4.10. A pair* (x, y) *of primal and dual feasible solutions is composed of optimal solutions to the respective problems if and only if we have*

$$y_i[Ax - b]_i = 0, \quad i = 1, \ldots, m, \qquad \text{[complementary slackness]}$$

or, equivalently, if and only if

$$c^\top x - b^\top y = 0. \qquad \text{[zero duality gap]}$$

Proof Indeed, the "zero duality gap" optimality condition is an immediate consequence of the fact the optimal values in the primal and the dual are equal to each other whenever one of the problems is feasible; see Theorem 4.10. As a result, denoting by Opt the common optimal value of (LP) and (LP*), the duality gap, when evaluated at a pair (x, y) composed of the primal feasible solution x and the dual feasible solution y, is the sum of two nonnegative quantities $c^\top x - $ Opt and Opt $- b^\top y$, and thus is nonnegative and vanishes if and only if x is a primal optimal solution and y is a dual optimal solution. The equivalence between the "zero duality gap" and the "complementary slackness" optimality conditions is given by the following computation: whenever x is primal feasible and y is dual feasible, we have

$$y^\top(Ax - b) = (A^\top y)^\top x - b^\top y = c^\top x - b^\top y,$$

where the second equality follows from dual feasibility (i.e., $A^\top y = c$). Thus, for a primal–dual feasible pair (x, y), the duality gap vanishes if and only if $y^\top(Ax - b) = 0$, and the latter, since $y \geq 0$ and $Ax - b \geq 0$, happens if and only if $y_i[Ax - b]_i = 0$ for all i, that is, if and only if the complementary slackness takes place. ∎

4.5 Application: Linear Programming Duality

Geometry of primal–dual pair of LP problems. Consider the primal–dual pair of LP problems

$$\min_{x \in \mathbf{R}^n} \{c^\top x : Ax - b \geq 0\}, \qquad \text{(LP)}$$
$$\max_{y \in \mathbf{R}^m} \{b^\top y : A^\top y = c, y \geq 0\}, \qquad \text{(LP*)}$$

as presented in section 4.5.2 and assume that the system of equality constraints in the dual problem is feasible, so that there exists $\beta \in \mathbf{R}^m$ such that $A^\top \beta = c$. It turns out[3] that the pair (LP), (LP*) possesses a nice and transparent geometry. Specifically, the data of the pair give rise to the following geometric entities:

- a pair of linear subspaces $\mathcal{L}, \mathcal{L}_*$ in $\mathbf{R}^m$ which are orthogonal complements to each other, $[\mathcal{L} = \text{Im}(A), \mathcal{L}_* = \text{Ker}(A^\top)]$,
- a pair of shift vectors $d, d_* \in \mathbf{R}^m$ $[d = b, d_* = -\beta]$,

which in turn give rise to

- the pair of convex sets $\mathcal{Q} = [\mathcal{L} - d] \cap \mathbf{R}^m_+$, $\mathcal{Q}_* = [\mathcal{L}_* - d_*] \cap \mathbf{R}^m_+$ [where $\mathbf{R}^m_+ = \{u \in \mathbf{R}^m : u \geq 0\}$].

To solve (LP), (LP*) to optimality is exactly the same as finding vectors $\xi \in \mathcal{Q}$ and $\xi_* \in \mathcal{Q}_*$ that are orthogonal to each other. Such a pair ξ, ξ_* gives rise to primal–dual optimal pair(s) (x^*, y^*) (one can take as x^* any x such that $Ax - b = \xi$ and set $y^* = \xi_*$), and every primal–dual optimal pair (x^*, y^*) can be obtained in this manner from a pair of vectors $\xi \in \mathcal{Q}, \xi_* \in \mathcal{Q}_*$ that are orthogonal to each other. The required pairs ξ, ξ_* exist if and only if both the sets $\mathcal{Q}$ and $\mathcal{Q}_*$ are nonempty.

For an illustration, see Figure 4.1.

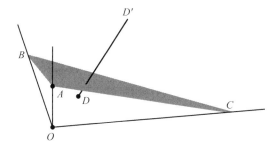

Figure 4.1 Geometry of primal–dual LP pair, $m = 3$: $\mathcal{Q}$, $\triangle ABC$; $\mathcal{Q}_*$, the ray DD'; ξ, the point A; ξ_*, the point D. Note the orthogonality of the ray and the plane of the triangle and also the orthogonality of the vectors $\overrightarrow{OA}$ and $\overrightarrow{OD}$.

---

[3] For derivations, see Exercise IV.7 addressing conic duality, of which LP duality is a special case.

5

Exercises for Part I

5.1 Elementaries

Exercise I.1 Mark in the following list the sets which are convex:

1. $\{x \in \mathbf{R}^2 : x_1 + i^2 x_2 \leq 1, i = 1, \ldots, 10\}$
2. $\{x \in \mathbf{R}^2 : x_1^2 + 2i x_1 x_2 + i^2 x_2^2 \leq 1, i = 1, \ldots, 10\}$
3. $\{x \in \mathbf{R}^2 : x_1^2 + i x_1 x_2 + i^2 x_2^2 \leq 1, i = 1, \ldots, 10\}$
4. $\{x \in \mathbf{R}^2 : x_1^2 + 5 x_1 x_2 + 4 x_2^2 \leq 1\}$
5. $\{x \in \mathbf{R}^{10} : x_1^2 + 2x_2^2 + 3x_3^2 + \cdots + 10 x_{10}^2 \leq 1000 x_1 - 999 x_2 + 998 x_3 - \cdots + 992 x_9 - 991 x_{10}\}$
6. $\{x \in \mathbf{R}^2 : \exp\{x_1\} \leq x_2\}$
7. $\{x \in \mathbf{R}^2 : \exp\{x_1\} \geq x_2\}$
8. $\{x \in \mathbf{R}^n : \sum_{i=1}^n x_i^2 = 1\}$
9. $\{x \in \mathbf{R}^n : \sum_{i=1}^n x_i^2 \leq 1\}$
10. $\{x \in \mathbf{R}^n : \sum_{i=1}^n x_i^2 \geq 1\}$
11. $\{x \in \mathbf{R}^n : \max_{i=1,\ldots,n} x_i \leq 1\}$
12. $\{x \in \mathbf{R}^n : \max_{i=1,\ldots,n} x_i \geq 1\}$
13. $\{x \in \mathbf{R}^n : \max_{i=1,\ldots,n} x_i = 1\}$
14. $\{x \in \mathbf{R}^n : \min_{i=1,\ldots,n} x_i \leq 1\}$
15. $\{x \in \mathbf{R}^n : \min_{i=1,\ldots,n} x_i \geq 1\}$
16. $\{x \in \mathbf{R}^n : \min_{i=1,\ldots,n} x_i = 1\}$

Exercise I.2 Mark with a T those of the following claims which are always true.

1. The linear image $Y = \{Ax : x \in X\}$ of a linear subspace X is a linear subspace.
2. The linear image $Y = \{Ax : x \in X\}$ of an affine subspace X is an affine subspace.
3. The linear image $Y = \{Ax : x \in X\}$ of a convex set X is convex.
4. The affine image $Y = \{Ax + b : x \in X\}$ of a linear subspace X is a linear subspace.

5.1 Elementaries

5. The affine image $Y = \{Ax + b : x \in X\}$ of an affine subspace X is an affine subspace.
6. The affine image $Y = \{Ax + b : x \in X\}$ of a convex set X is convex.
7. The intersection of two linear subspaces in $\mathbf{R}^n$ is always nonempty.
8. The intersection of two linear subspaces in $\mathbf{R}^n$ is a linear subspace.
9. The intersection of two affine subspaces in $\mathbf{R}^n$ is an affine subspace.
10. The intersection of two affine subspaces in $\mathbf{R}^n$, when nonempty, is an affine subspace.
11. The intersection of two convex sets in $\mathbf{R}^n$ is a convex set.
12. The intersection of two convex sets in $\mathbf{R}^n$, when nonempty, is a convex set.

Exercise I.3 [TrYs] Prove that the relative interior of a simplex with vertices $y^0, \ldots, y^m$ is exactly the set

$$\left\{ \sum_{i=0}^{m} \lambda_i y_i : \lambda_i > 0, \sum_{i=0}^{m} \lambda_i = 1 \right\}.$$

Exercise I.4 Which of the following claims is true?

1. The set $X = \{x : Ax \leq b\}$ is a cone if and only if $X = \{x : Ax \leq 0\}$.
2. The set $X = \{x : Ax \leq b\}$ is a cone if and only if $b = 0$.

Exercise I.5 Suppose $\mathbf{K}$ is a closed cone. Prove that the set $X = \{x : Ax - b \in \mathbf{K}\}$ is a cone if and only if $X = \{x : Ax \in \mathbf{K}\}$.

Exercise I.6 [TrYs] Prove that if M is a nonempty convex set in $\mathbf{R}^n$ and $\epsilon > 0$, then, for every norm $\|\cdot\|$ on $\mathbf{R}^n$, the ϵ-neighborhood of M, i.e., the set

$$M_\epsilon = \left\{ y \in \mathbf{R}^n : \inf_{x \in M} \|y - x\| \leq \epsilon \right\},$$

is convex.

Exercise I.7 Which of the following claims are always true? Explain why/why not.

1. The convex hull of a bounded set in $\mathbf{R}^n$ is bounded.
2. The convex hull of a closed set in $\mathbf{R}^n$ is closed.
3. The convex hull of a closed convex set in $\mathbf{R}^n$ is closed.
4. The convex hull of a closed and bounded set in $\mathbf{R}^n$ is closed and bounded.
5. The convex hull of an open set in $\mathbf{R}^n$ is open.

Exercise I.8 [TrYs] [This exercise together with its follow-up, i.e., Exercise II.9 and Exercise I.9, are the most boring exercises ever designed by the authors. Our excuse is that *There is no royal road to geometry* (Euclid of Alexandria, c. 300 BC)].

Let A, B be nonempty subsets of $\mathbf{R}^n$. Consider the following claims. If the claim is always true (i.e., for all data satisfying the premise of the claim), give a proof; otherwise, give a counterexample.

1. If $A \subseteq B$ then $\mathrm{Conv}(A) \subseteq \mathrm{Conv}(B)$.
2. If $\mathrm{Conv}(A) \subseteq \mathrm{Conv}(B)$ then $A \subseteq B$.
3. $\mathrm{Conv}(A \cap B) = \mathrm{Conv}(A) \cap \mathrm{Conv}(B)$.
4. $\mathrm{Conv}(A \cap B) \subseteq \mathrm{Conv}(A) \cap \mathrm{Conv}(B)$.
5. $\mathrm{Conv}(A \cup B) \subseteq \mathrm{Conv}(A) \cup \mathrm{Conv}(B)$.

6. $\text{Conv}(A \cup B) \supseteq \text{Conv}(A) \cup \text{Conv}(B)$.
7. If A is closed, so is $\text{Conv}(A)$.
8. If A is closed and bounded, so is $\text{Conv}(A)$.
9. If $\text{Conv}(A)$ is closed and bounded, so is A.

Exercise I.9 [TrYs] Let A, B, C be nonempty subsets of $\mathbf{R}^n$ and D be a nonempty subset of $\mathbf{R}^m$. Consider the following claims. If the claim is always true (i.e., for all data satisfying the premise of the claim), give a proof; otherwise, give a counterexample.

1. $\text{Conv}(A \cup B) = \text{Conv}(\text{Conv}(A) \cup B)$.
2. $\text{Conv}(A \cup B) = \text{Conv}(\text{Conv}(A) \cup \text{Conv}(B))$.
3. $\text{Conv}(A \cup B \cup C) = \text{Conv}(\text{Conv}(A \cup B) \cup C)$.
4. $\text{Conv}(A \times D) = \text{Conv}(A) \times \text{Conv}(D)$.
5. When A is convex, the set $\text{Conv}(A \cup B)$ (which is always the set of convex combinations of several points from A and several points from B), can be obtained by taking convex combinations of points with *at most one of them* taken from A, and the rest taken from B. Similarly, if A and B are both convex, to get $\text{Conv}(A \cup B)$, it suffices to add to $A \cup B$ all convex combinations of pairs of points, one from A and one from B.
6. Suppose A is a set in $\mathbf{R}^n$. Consider the affine mapping $x \mapsto Px + p : \mathbf{R}^n \to \mathbf{R}^m$, and the image of A under this mapping, i.e., the set $PA + p := \{Px + p : x \in A\}$. Then, $\text{Conv}(PA + p) = P\,\text{Conv}(A) + p$.
7. Consider an affine mapping $y \mapsto P(y) : \mathbf{R}^m \to \mathbf{R}^n$ where $P(y) := Py + p$. Recall that given a set $X \in \mathbf{R}^n$, its inverse image under the mapping $P(\cdot)$ is given by $P^{-1}(X) := \{y \in \mathbf{R}^m : P(y) \in X\}$. Then $\text{Conv}(P^{-1}(A)) = P^{-1}(\text{Conv}(A))$.
8. Consider an affine mapping $y \mapsto P(y) : \mathbf{R}^m \to \mathbf{R}^n$ where $P(y) := Py + p$. Then $\text{Conv}(P^{-1}(A)) \subseteq P^{-1}(\text{Conv}(A))$.

Exercise I.10 [TrYs] Let $X_1, X_2 \in \mathbf{R}^n$ be two nonempty sets, and define $Y := X_1 \cup X_2$ and $Z := \text{Conv}(Y)$. Consider the following claims. If the claim is always true (i.e., for all data satisfying the premise of the claim), give a proof; otherwise, give a counterexample.

1. Whenever X_1 and X_2 are both convex, so is Y.
2. Whenever X_1 and X_2 are both convex, so is Z.
3. Whenever X_1 and X_2 are both bounded, so is Y.
4. Whenever X_1 and X_2 are both bounded, so is Z.
5. Whenever X_1 and X_2 are both closed, so is Y.
6. Whenever X_1 and X_2 are both closed, so is Z.
7. Whenever X_1 and X_2 are both compact, so is Y.
8. Whenever X_1 and X_2 are both compact, so is Z.
9. Whenever X_1 and X_2 are both polyhedral, so is Y.
10. Whenever X_1 and X_2 are both polyhedral, so is Z.
11. Whenever X_1 and X_2 are both polyhedral and bounded, so is Y.
12. Whenever X_1 and X_2 are both polyhedral and bounded, so is Z.

Exercise I.11 Consider two families of convex sets given by $\{F_i\}_{i \in I}$ and $\{G_j\}_{j \in J}$. Prove that the following relation holds:

$$\text{Conv}\left(\bigcup_{i \in I, j \in J}(F_i \cap G_j)\right) \subseteq \text{Conv}\left(\bigcup_{j \in J}[G_j \cap \text{Conv}(\cup_{i \in I} F_i)]\right).$$

Exercise I.12 Let C_1, C_2 be two nonempty conic sets in $\mathbf{R}^n$, i.e., for each $i = 1, 2$, for any $x \in C_i$ and $t \geq 0$, we have $t \cdot x \in C_i$ as well. Note that C_1, C_2 are not necessarily convex. Prove that

1. $C_1 + C_2 \neq \text{Conv}(C_1 \cup C_2)$ may hold if either C_1 or C_2 (or both) is nonconvex.
2. $C_1 + C_2 = \text{Conv}(C_1 \cup C_2)$ always holds if C_1, C_2 are both convex.
3. $C_1 \cap C_2 = \bigcup_{\alpha \in [0,1]} (\alpha C_1 \cap (1-\alpha)C_2)$ always holds if C_1, C_2 are both convex.

Exercise I.13 [TrYs] Let $X \subseteq \mathbf{R}^n$ be a convex set with $\text{int } X \neq \emptyset$, and consider the following set:

$$K := \text{cl}\{[x; t] : t > 0, x/t \in X\}.$$

Prove that the set K is a closed cone with a nonempty interior.

5.2 Around Ellipsoids

Exercise I.14 Verify each of the following statements:

1. Any ellipsoid $E \in \mathbf{R}^n$ is the image of the unit Euclidean ball $B_n = \{x \in \mathbf{R}^n : \|x\|_2 \leq 1\}$ under a one-to-one affine mapping. That is, $E \subset \mathbf{R}^n$ can be represented as $E = \{x : (x-c)^\top C(x-c) \leq 1\}$ with $C \succ 0$ and $c \in \mathbf{R}^n$ if and only if it can be represented as $E = \{c + Du : u \in B_n\}$ with nonsingular D, and in the latter representation D can be selected to be symmetric positive definite.
2. Given $C \succ 0$, $D \succ 0$ and $c, d \in \mathbf{R}^n$, the ellipsoid $E_C := \{x : (x-c)^\top C(x-c) \leq 1\}$ is contained in the ellipsoid $E_D := \{x : (x-c)^\top D(x-c) \leq 1\}$ if and only if $C \succeq D$. If the ellipsoid E_C is contained in the ellipsoid $E'_D = \{x : (x-d)^\top D(x-d) \leq 1\}$ then $C \succeq D$.
3. For a set $U \subset \mathbf{R}^n$, let $\text{Vol}(U)$ be the ratio of the n-dimensional volume of U and the n-dimensional volume of the unit ball B_n. Then, for an n-dimensional ellipsoid E represented as $\{x = c + Du : \|u\|_2 \leq 1\}$ with nonsingular D, we have

$$\text{Vol}(E) = |\text{Det}(D)|,$$

and when E is represented as $\{x : (x-c)^\top C(x-c) \leq 1\}$ with $C \succ 0$, we have

$$\text{Vol}(E) = \text{Det}^{-1/2}(C).$$

Exercise I.15 Given $C \succ 0$, an ellipsoid $\{x : (x-a)^\top C(x-a) \le 1\}$ is the solution set of the quadratic inequality $x^\top C x - 2(Ca)^\top x + (a^\top C a - 1) \le 0$. Prove that the solution set E of any quadratic inequality $f(x) := x^\top C x - c^\top x + \sigma \le 0$ with positive *semi*definite matrix C is convex.

5.3 Truss Topology Design

Exercise I.16 [EdYs] [First acquaintance with truss topology design]

Preamble What follows is the first exercise in a "truss topology design" (TTD) series (other exercises in it are I.18, III.9, IV.11, and IV.28). The underlying "real life" mechanical story is simple enough to be told and rich enough to illustrate numerous constructions and results presented in the main body of our textbook – ranging from Carathéodory's Theorem to semidefinite duality, demonstrating on a real-life example how the theory works.

Trusses A truss is a mechanical construction, such as a railroad bridge, an electric mast, or the Eiffel Tower, composed of thin elastic *bars* linked with each other at *nodes* – points from physical space (three-dimensional space for spatial trusses and two-dimensional space for planar trusses).

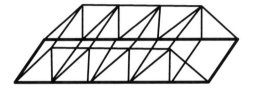

Figure 5.1 Pratt Truss bridge.

When a truss is subject to an external load – a collection of forces acting at the nodes – it starts to deform, so that the nodes move a little bit, leading to elongations or shortenings of the bars, which, in turn, result in reaction forces. At equilibrium, the reaction forces compensate the external forces, and the truss stores a certain potential energy, called the *compliance*. The theory of mechanics models this story as follows.

- The nodes form a finite set $p_1, \ldots, p_K$ of distinct points in physical space $\mathbf{R}^d$ ($d = 2$ for planar and $d = 3$ for spatial constructions). Virtual displacements of the nodes under the load are somewhat restricted by the support conditions; we will focus on the case when some nodes are fixed – they cannot move at all (think about them as being in the wall), and the remaining nodes are free – their virtual displacements form the entire space $\mathbf{R}^d$. A virtual displacement v of the nodal set can be identified with a vector of dimension $M = dm$, where m is the number of free nodes and v is a block vector with m d-dimensional blocks, indexed by the free nodes, representing the physical displacements of these nodes.
- There are N bars; the ith bar links the nodes with indices α_i and β_i (where at least one of these nodes is free) and with volume t_i (three-dimensional or two-dimensional, depending on whether the truss is spatial or planar).
- An external load is a collection of physical forces – vectors from $\mathbf{R}^d$ – acting at the free nodes (the forces acting at the fixed nodes are of no relevance – they are suppressed by the supports). Thus, an external load f can be identified with a block vector of the

same structure as a virtual displacement – the blocks are indexed by the free nodes and represent the external forces acting at these nodes. Thus, the displacements v of the nodal set and the external loads f are vectors from the space $\mathcal{V}$ of *virtual displacements* – they are M-dimensional block vectors with m d-dimensional blocks.

- The bars and the nodes together specify the symmetric positive semidefinite $M \times M$ *stiffness matrix* A of the truss. The role of this matrix is as follows. A displacement $v \in \mathcal{V}$ of the nodal set results in reaction forces at free nodes (as mentioned above, those at the fixed nodes are of no interest – they are compensated by the supports); assembling these forces into an M-dimensional block vector, we get the *reaction*, and this reaction is $-Av$. In other words, the potential energy stored in a truss under a displacement $v \in \mathcal{V}$ of nodes is $\frac{1}{2}v^\top Av$, and the reaction, as it should be, is minus the gradient of the potential energy as a function of v.[1] At equilibrium under an external load f, the total of the reaction and the load should be zero, that is, the equilibrium displacement satisfies

$$Av = f. \tag{5.1}$$

Note that (5.1) may be infeasible, meaning that the truss is crushed by the load in question. Assuming an equilibrium displacement v exists, the truss at equilibrium stores potential energy $\frac{1}{2}v^\top Av$; this energy is the compliance of the truss w.r.t. the load. The compliance is a convenient measure of the rigidity of the truss with respect to the load; the smaller the compliance the better the truss withstands the load.

Let us build the stiffness matrix of a truss. As we have mentioned, the reaction forces originate from elongations or shortenings of the bars under displacement of the nodes. Consider the ith bar linking nodes with initial positions $a_i = p_{\alpha_i}$ and $b_i = p_{\beta_i}$ (i.e., prior to the external load being applied), and let us set

$$d_i = \|b_i - a_i\|_2, \quad e_i = [b_i - a_i]/d_i.$$

Under a displacement $v \in \mathcal{V}$ of the nodal set,

- the positions of the nodes linked by the bar become $a_i + \underbrace{v^{\alpha_i}}_{da}, b_i + \underbrace{v^{\beta_i}}_{db}$, where v^γ is the γth block in v, that is, the displacement of the γth node;
- as a result, the elongation of the bar becomes, in the approximation that is first order in v, $e_i^\top [db - da]$, and the reaction forces caused by this elongation are, by Hooke's Law,[2]

$$\begin{array}{ll} d_i^{-1} S_i e_i e_i^\top [db - da] & \text{at node } \alpha_i \\ -d_i^{-1} S_i e_i e_i^\top [db - da] & \text{at node } \beta_i \\ 0 & \text{at all remaining nodes} \end{array}$$

---

[1] This is called the *linearly elastic* model; it is the linearized-in-displacements approximation to the actual behavior of a loaded truss. This model works better when the nodal displacements are small compared with the inter-nodal distances, and it is accurate enough to be used in typical real-life applications.

[2] Hooke's Law says that the magnitude of the reaction force caused by elongation or shortening of a bar is proportional to $Sd^{-1}\delta$, where S is the bar's cross-sectional size (the area for a spatial truss and the thickness for a planar truss), d is the bar's (pre-deformation) length, and δ is the elongation. With units of length properly adjusted to the bars' material, the proportionality coefficient becomes 1, and this is what we assume from now on.

where $S_i = t_i/d_i$ is the cross-sectional size of the ith bar. It follows that *when both nodes linked by the ith bar are free, the contribution of the ith bar to the reaction is*

$$-t_i \mathsf{b}_i \mathsf{b}_i^\top v,$$

where $\mathsf{b}_i \in \mathcal{V}$ *is the vector with just two nonzero blocks:*

*the block with index α_i – this block is $e_i/d_i = [b_i - a_i]/\|b_i - a_i\|_2^2$ – and
the block with index β_i – this block is $-e_i/d_i = -[b_i - a_i]/\|b_i - a_i\|_2^2$.*

It can immediately be seen that when just one of the nodes linked by the ith bar is free, the contribution of the ith bar to the reaction is given by similar relations but with one block, rather than two blocks, in b_i, the block corresponding to the free node among the nodes linked by the bar.

The bottom line is that *the stiffness matrix of a truss composed of N bars with volumes t_i, $1 \leq i \leq N$, is*

$$A = A(t) := \sum_i t_i \mathsf{b}_i \mathsf{b}_i^\top,$$

where the $\mathsf{b}_i \in \mathcal{V} = \mathbf{R}^M$ are readily given by the geometry of the nodal set and the indices of the nodes linked by bar i.

Truss topology design problem In the simplest truss topology design (TTD) problem, one is given

- a finite *set of tentative nodes* in two or three dimensions along with support conditions indicating which of the nodes are fixed and which are free, and thus specifying the linear space $\mathcal{V} = \mathbf{R}^M$ of virtual displacements of the nodal set,
- a *set of N tentative bars* – unordered pairs of nodes (distinct from each other), which can be linked by bars, and the total volume $W > 0$ of the truss,
- an external load $f \in \mathcal{V}$.

These data specify, as explained above, vectors $\mathsf{b}_i \in \mathbf{R}^M$, $i = 1, \ldots, N$, and the stiffness matrix

$$A(t) = \sum_{i=1}^N t_i \mathsf{b}_i \mathsf{b}_i^\top = B\,\mathrm{Diag}\{t_1, \ldots, t_N\} B^\top \in \mathbf{S}^M \qquad [B = [\mathsf{b}_1, \ldots, \mathsf{b}_N]]$$

of a truss, which for this setup can be identified with the vector $t \in \mathbf{R}_+^N$ of its bar volumes. What we want to find is the truss of given volume capable of best withstanding the given load, that is, the truss that minimizes the corresponding compliance.

When applying the TTD model, one starts with dense grid of tentative nodes and a broad list of tentative bars (e.g., by allowing linkage by a bar of every pair of distinct nodes, with at least one free node in the pair). At the optimal truss, yielded by the optimal solution to the TTD problem, many tentative bars (usually the vast majority) acquire zero volumes, and so a significant part of the tentative nodes become unused. Thus, the TTD problem is in fact not about sizing – it allows one to recover the optimal structure of the construction; this is where the words "Topology Design" in the problem's name come from.

5.3 Truss Topology Design

To illustrate this point, here is a toy example (it will be our guinea pig in the entire series of TTD exercises):

Console design. We want to design a planar truss as follows (see Figure 5.2):

- The set of tentative nodes is the 9×9 grid $\{[p;q] \in \mathbf{R}^2 : p,q \in \{0,1,\ldots,8\}\}$ with the nine left-most nodes fixed and the remaining 72 nodes free, resulting in $M = 144$-dimensional space $\mathcal{V}$ of virtual displacements.
- The external load $f \in \mathcal{V} = \mathbf{R}^{144}$ is a single-force load, with the only nonzero force $[0;-1]$ applied at the fifth node of the right-most column of nodes.
- We allow for all pairwise connections of pairs of nodes that are distinct from each other, with at least one of these nodes free, resulting in $N = 3204$ tentative bars (see upper right panel in Figure 5.2).
- The total volume of the truss is $W = 1000$.

Important: *From now on, speaking about the TTD problem, we always make the following assumption:*

$$\mathfrak{R}: \qquad \sum_{i=1}^{N} \mathsf{b}_i \mathsf{b}_i^\top \succ 0.$$

Under this assumption, the stiffness matrix $A(t) = \sum_i t_i \mathsf{b}_i \mathsf{b}_i^\top$ *associated with a truss* $t > 0$ *is positive definite, so that such a truss can withstand any load* f. You can verify numerically that this is the case in the console design as stated above.

After this lengthy preamble (to justify its length, note that is an investment to a series of exercises, rather than just one), let us pass to the exercise *per se*. Consider a TTD problem.

1. Prove that a truss $t \geq 0$ (recall that we identify a truss with the corresponding vector of bar volumes) is able to carry a load f if and only if the quadratic function

$$F(v) = f^\top v - \frac{1}{2} v^\top A(t) v$$

is bounded from above and that, whenever this is the case,

- the maximum of F over $\mathcal{V}$ is achieved,
- the maximizers of F are exactly the equilibrium displacements v, i.e., those with

$$A(t)v = f,$$

and for such a displacement, one has

$$[\max F =] \ F(v) = \frac{1}{2} v^\top A(t) v = \frac{1}{2} v^\top f,$$

- the maximum value of F is exactly the compliance of the truss w.r.t. the load f.

2. Prove that a real number τ is an upper bound on the compliance of truss $t \geq 0$ w.r.t. load f if and only if the symmetric matrix

$$\mathcal{A} := \left[\begin{array}{c|c} B \, \mathrm{Diag}\{t\} B^\top & f \\ \hline f^\top & 2\tau \end{array} \right], \qquad B = [\mathsf{b}_1, \ldots, \mathsf{b}_N],$$

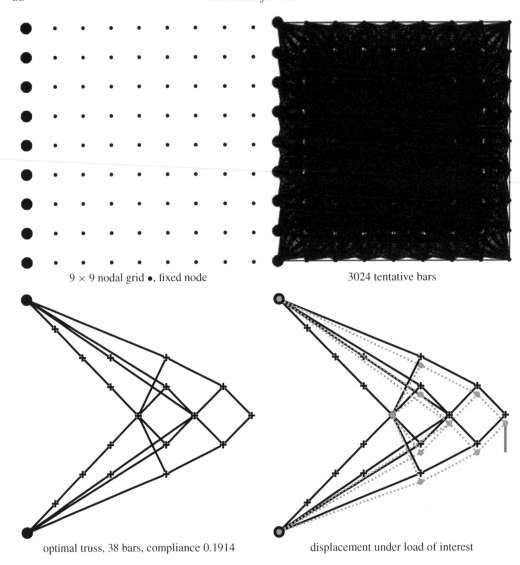

Figure 5.2 Console. Lower figures: Positions of the bars and nodes before and after deformation. The vertical segment starting at the right-most node indicates the external force.

is positive semidefinite. As a result, pose the TTD problem as the optimization problem

$$\text{Opt} = \min_{\tau, t} \left\{ \tau : \left[\begin{array}{c|c} B \operatorname{Diag}\{t\} B^\top & f \\ \hline f^\top & 2\tau \end{array} \right] \succeq 0, t \geq 0, \sum_i t_i = W \right\}. \quad (5.2)$$

Prove that this problem is solvable.

3. [Computational study]

   3.1 Solve the console problem numerically and reproduce the numerical results presented in Figure 5.2.

3.2 Resolve the problem with the set of all possible tentative bars reduced to the subset of "short" bars connecting neighboring nodes only:

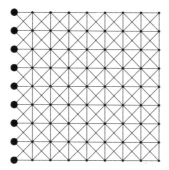

Figure 5.3 Complete set of 262 "short" tentative bars

and compare the resulting design and compliance with those in the item 3.1 above.

5.4 Around Carathéodory's Theorem

Exercise I.17 [EdYs] Prove the following statements:
Let $X \subset \mathbf{R}^n$ be nonempty. Then,

1. if a point x can be represented as a convex combination of a collection of vectors from X, then the collection can be selected to be affinely independent;
2. if a point x can be represented as a conic combination of a collection of vectors from X then the collection can be selected to be linearly independent.

Comment: Note that claims 1 and 2 above are refinements, albeit minor ones, of Carathéodory's Theorem (plain and conic, respectively). Indeed, when $M := \mathrm{Aff}(X)$ and m is the dimension of M, every affinely independent collection of points from X contains at most $m + 1$ points (Proposition A.44), so that the first claim is equivalent to stating that if $x \in \mathrm{Conv}(X)$, then x is a convex combination of at most $m + 1$ points from X. However, the vectors participating in such a convex combination are not necessarily affinely independent, so that the first claim provides a bit more information than the plain Carathéodory's Theorem. Similarly, if $L := \mathrm{Lin}(X)$ and $m := \dim L$, then every linearly independent collection of vectors from X contains at most $m \leq n$ points, that is, the second claim implies Carathéodory's Theorem in conic form, and provides a bit more information than the latter theorem.

Exercise I.18 [EdYs][3] Consider the TTD problem, and let N be the number of tentative bars, M be the dimension of the corresponding space of virtual displacements $\mathcal{V}$, and f be an external load. Prove that if a truss $t \geq 0$ can withstand the load f with compliance $\leq \tau$ for some given real number τ, then there exists a truss $\bar{t}$ of the same total volume as t with compliance w.r.t. f at most τ and at most $M + 1$ bars of positive volume.

---

[3] The previous exercise in the TTD series is Exercise I.16.

Exercise I.19 [EdYs] [Shapley–Folkman Theorem]

1. Prove that if a system of linear equations $Ax = b$ with n variables and m equations has a nonnegative solution, it has a nonnegative solution with at most m positive entries.
2. Let $V_1, \ldots, V_n$ be n nonempty sets in $\mathbf{R}^m$, and define

$$\overline{V} := \mathrm{Conv}(V_1 + V_2 + \cdots + V_n).$$

   1. Prove that
      1. Taking direct product commutes with taking convex hull:
         $$\mathrm{Conv}(V_1 \times \cdots \times V_n) = \mathrm{Conv}(V_1) \times \cdots \times \mathrm{Conv}(V_n).$$
      2. Taking affine image commutes with taking convex hull: if $V \subset \mathbf{R}^n$ is nonempty and $x \mapsto \mathcal{A}(x) = Ax + b : \mathbf{R}^n \to \mathbf{R}^m$ is an affine mapping, then define $\mathcal{A}(V) := \{\mathcal{A}(x) : x \in V\}$ and show that
         $$\mathrm{Conv}(\mathcal{A}(V)) = \mathcal{A}(\mathrm{Conv}(V)).$$
      3. Conclude from the previous two items that taking weighted sum of sets commutes with taking convex hull:
         $$\mathrm{Conv}\left(\lambda_1 V_1 + \cdots + \lambda_n V_n := \{v = \sum_i \lambda_i v_i : v_i \in V_i, i \leq n\}\right)$$
         $$= \lambda_1 \mathrm{Conv}(V_1) + \cdots + \lambda_n \mathrm{Conv}(V_n) \qquad [\lambda_i \in \mathbf{R}, \forall i \leq n].$$
      In particular,
      $$\overline{V} = \mathrm{Conv}(V_1) + \cdots + \mathrm{Conv}(V_n).$$

   2. Prove the *Shapley–Folkman Theorem*:

      Let $x \in \overline{V}$. Then there exists a representation of x such that
      $$x = x_1 + \cdots + x_n, \quad x_i \in \mathrm{Conv}(V_i) \; \forall i \leq n,$$
      in which at least $n - m$ of the x_i belong to the respective sets V_i.

   Comment: The Shapley–Folkman Theorem says, informally, that when $n \gg m$, summing up n nonempty sets in $\mathbf{R}^m$ possesses a certain "convexification property" – every point from $\overline{V}$, i.e., the convex hull of the sum of our sets is the sum of points x_i with all but m of them belonging to V_i rather than to $\mathrm{Conv}(V_i)$, and only $\leq m$ of the points belonging to V_i "fractionally," that is, belonging to $\mathrm{Conv}(V_i)$, but not to V_i. This nice fact has numerous useful applications.

Exercise I.20 [EdYs] Carathéodory's Theorem in its plain and its conic forms represents "existence" statements: if a point $x \in \mathbf{R}^m$ is a convex, respectively conic, combination of points $x^1, \ldots, x^N$ then there exists a representation of x of the same type which involves at most $m + 1$, respectively m, terms. Extract, from the proofs of these theorems, algorithms for finding these "short" representations at the cost of solving at most N solvable systems of m linear equations with at most N variables.

5.4 Around Carathéodory's Theorem

Exercise I.21 [EdYs] Prove *Kirchberger's Theorem*:

Consider two sets of finitely many points $X = \{x^1,\ldots,x^k\}$ and $Y = \{y^1,\ldots,y^m\}$ in $\mathbf{R}^n$ such that $k + m \geq n + 2$ and all the points $x^1,\ldots,x^k, y^1,\ldots,y^m$ are distinct. Assume that, for any subset $S \subseteq X \cup Y$ composed of $n + 2$ points, the convex hulls of the sets $X \cap S$ and $Y \cap S$ do not intersect, i.e., $\mathrm{Conv}(X \cap S) \cap \mathrm{Conv}(Y \cap S) = \emptyset$. Then, the convex hulls of X and Y also do not intersect, i.e., $\mathrm{Conv}(X) \cap \mathrm{Conv}(Y) = \emptyset$.

Hint: Assume that, on the contrary, the convex hulls of X and Y intersect, so that

$$\sum_{i=1}^{k} \lambda_i x^i = \sum_{j=1}^{m} \mu_j y^j$$

for certain nonnegative λ_i, $\sum_{i=1}^{k} \lambda_i = 1$, and certain nonnegative μ_j, $\sum_{j=1}^{m} \mu_j = 1$, and look at the expression of this type with the minimum possible total number of nonzero coefficients λ_i, μ_j.

Exercise I.22 [EdYs] [Follow-up to the Shapley–Folkman Theorem]

1. Let $X_1,\ldots,X_K$ be nonempty convex sets in $\mathbf{R}^n$ and let $X := \bigcup_{k \leq K} X_k$. Prove that

$$\mathrm{Conv}(X) = \left\{ x = \sum_{k=1}^{K} \lambda_k x^k : \lambda_k \geq 0,\ x^k \in X_k,\ \forall k \leq K,\ \sum_{k=1}^{K} \lambda_k = 1 \right\}.$$

2. Let X_k, $k \leq K$, be nonempty bounded polyhedral sets in $\mathbf{R}^n$ given by polyhedral representations:

$$X_k = \left\{ x \in \mathbf{R}^n : \exists u^k \in \mathbf{R}^{n_k} \colon P_k x + Q_k u^k \leq r_k \right\}.$$

Define $X := \bigcup_{k \leq K} X_k$. Prove that the set $\mathrm{Conv}(X)$ is a polyhedral set given by the polyhedral representation

$$\mathrm{Conv}(X) = \left\{ x \in \mathbf{R}^n : \begin{array}{ll} \exists x^k \in \mathbf{R}^n,\ u^k \in \mathbf{R}^{n_k},\ \lambda_k \in \mathbf{R},\ \forall k \leq K : & \\ P_k x^k + Q_k u^k - \lambda_k r_k \leq 0,\ k \leq K & (a) \\ \lambda_k \geq 0,\ \sum_{k=1}^{K} \lambda_k = 1 & (b) \\ x = \sum_{k=1}^{K} x^k & (c) \end{array} \right\}. \quad (*)$$

Does the claim remain true when the assumption of boundedness of the sets X_ks is lifted?

After the two preliminary items above, let us pass to the essence of the matter. Consider the situation as follows. We are given n nonempty and bounded polyhedral sets $X_j \subset \mathbf{R}^r$, $j = 1,\ldots,n$. We will think of X_j as the "resource set" of the jth production unit: entries in $x \in X_j$ are amounts of various resources, and X_j describes the set of vectors of resources available, in principle, for the jth unit. Each production unit j can use any one of its $K_j < \infty$ different production plans. For each $j = 1,\ldots,n$, the vector $y_j \in \mathbf{R}^p$ representing the production of the jth unit depends on the vector x_j of resources consumed by the unit and also on the production plan utilized in the unit. In particular, the production vector $y_j \in \mathbf{R}^p$ stemming from the resources x_j under the k-th plan can be chosen from the set

$$Y_j^k[x_j] := \left\{ y_j \in \mathbf{R}^p : z_j := [x_j; -y_j] \in V_j^k \right\},$$

where V_j^k, $k \leq K_j$, are given bounded polyhedral "technological sets" of the units with projections onto the x_j-plane equal to X_j, so that for every $k \leq K_j$ it holds that

$$x_j \in X_j \iff \exists y_j : [x_j; -y_j] \in V_j^k. \tag{5.3}$$

We assume that all the sets V_j^k are given by polyhedral representations, and we define

$$V_j := \bigcup_{k \leq K_j} V_j^k.$$

Let $R \in \mathbf{R}^r$ be the vector of total resources available to all n units and let $P \in \mathbf{R}^p$ be the vector of total demands for the products. For $j \leq n$, we want to select $x_j \in X_j$, $k_j \leq K_j$, and $y_j \in Y_j^{k_j}[x_j]$ in such a way that

$$\sum_j x_j \leq R \quad \text{and} \quad \sum_j y_j \geq P.$$

That is, we would like to find $z_j = [x_j; v_j] \in V_j$, $j \leq n$, in such a way that $\sum_j z_j \leq [R; -P]$. Note that the presence of a combinatorial part in our decision – the selection of production plans in finite sets – makes the problem difficult.

3. Apply the Shapley–Folkman Theorem (Exercise I.19) to overcome, to some extent, the above difficulty and come up with a good and approximately feasible solution.

5.5 Around Helly's Theorem

Exercise I.23 [TrYs] [Alternative proof of Helly's Theorem] The goal of this exercise is to build an alternative proof of Helly's Theorem without using Radon's Theorem.

1. Consider a system $a_i^\top x \leq b_i$, $i \leq N$, of N linear inequalities in variables $x \in \mathbf{R}^n$. Helly's Theorem applied to the sets $A_i := \{x \in \mathbf{R}^n : a_i^\top x \leq b_i\}$ gives us that

   (!) *If a system $a_i^\top x \leq b_i$, $i \leq N$, of linear inequalities in variables $x \in \mathbf{R}^n$ is infeasible, so is a properly selected subsystem composed of at most $n + 1$ inequalities from the system.*

   Find an alternative proof of (!) without relying on Helly's or Radon's theorems.
2. Extract, from item 1, Helly's Theorem for polyhedral sets: *If $A_1, \ldots, A_N$, $N \geq n + 1$, are polyhedral sets in $\mathbf{R}^n$ and every $n + 1$ of these sets have a point in common then all the sets have a point in common.*
3. Extract, from item 2, Helly's Theorem (Theorem 2.10).

Exercise I.24 [TrYs] $A_0, A_1, \ldots, A_m$, $m = 2025$, are nonempty convex subsets of $\mathbf{R}^{2000}$, and A_0 is a triangle (the convex hull of three affinely independent vectors). Which of the claims below are always true (that is, for any $A_0, \ldots, A_m$ satisfying the above assumptions)?

1. If every three among the sets $A_0, \ldots, A_m$ have a point in common, all $m + 1$ sets have a point in common.

2. If every four among the sets $A_0, \ldots, A_m$ have a point in common, all $m + 1$ sets have a point in common.
3. If every 2001 among the sets $A_0, \ldots, A_m$ have a point in common, all $m + 1$ sets have a point in common.

Exercise I.25 [TrYs] Let $P_i := \{x \in \mathbf{R}^n : A_i x \leq b_i\}$ for $i \in \{1, \ldots, m\}$ and $C := \{x \in \mathbf{R}^n : Dx \geq d\}$ be nonempty polyhedral sets. Suppose that for any $n + 1$ sets $P_{i_1}, \ldots, P_{i_{n+1}}$ there is a translate of C, i.e., the set $C + u$ for some $u \in \mathbf{R}^n$, which is contained in all $P_{i_1}, \ldots, P_{i_{n+1}}$. Prove that there is a translate of C which is contained in all the sets $P_1, \ldots, P_m$.

Exercise I.26 [TrYs] A cake contains 300 grams of raisins (you may think of every one of them as a three-dimensional ball of positive radius). John and Jill are about to divide the cake according to the following rules:

- first, Jill chooses a point a in the cake;
- second, John makes a *cut* through a, that is, chooses a two-dimensional plane Π passing through a, and takes the part of the cake on one side of the plane (both Π and which side are up to John, with the only restriction that the plane should pass through a); all the rest goes to Jill.

1. Prove that it may be the case that Jill cannot guarantee herself 76 grams of the raisins.
2. Prove that Jill always can choose a in a way which guarantees her at least 74 grams of the raisins.
3. Consider an n-dimensional version of the problem, where the raisins are n-dimensional balls, the cake is a domain in $\mathbf{R}^n$, and a "cut" taken by John is defined as the part of the cake contained in the half-space
$$\left\{x \in \mathbf{R}^n : e^\top (x - a) \geq 0\right\},$$
where $e \neq 0$ is the vector (the inner normal to the cutting hyperplane) chosen by John. Prove that, for every $\epsilon > 0$, Jill can guarantee to herself at least $\frac{300}{n+1} - \epsilon$ grams of raisins, but in general cannot guarantee to herself $\frac{300}{n+1} + \epsilon$ grams.

Remarks:

1. With some minor effort, you can prove that Jill can find a point which guarantees her $\frac{300}{n+1}$ grams of raisins instead of $\frac{300}{n+1} - \epsilon$ grams.
2. If, instead of dividing raisins, John and Jill were to divide in the same fashion a *uniform and convex* cake (that is, a closed and bounded convex body X with a nonempty interior in $\mathbf{R}^n$, the reward being the n-dimensional volume of the part a person gets), the results would change dramatically: choosing as the above point the center of mass of the cake
$$\bar{x} := \frac{\int_X x\, dx}{\int_X dx},$$
Jill would guarantee herself a fraction equal to at least $\left(\frac{n}{n+1}\right)^n \approx \frac{1}{e}$ of the cake. This is a not particularly easy corollary of the following extremely important and deep result:

Brunn–Minkowski Symmetrization Theorem: *Let X be as above, and let $[a, b]$ be the projection of X onto an axis ℓ, say, the last coordinate axis. Consider the "symmetrization" Y of X, i.e., Y is the set with the same projection $[a, b]$ on ℓ and, for every hyperplane orthogonal to the axis ℓ and crossing $[a, b]$, the intersection of Y with this hyperplane is an $(n-1)$-dimensional ball centered at the axis with precisely the same $(n-1)$-dimensional volume as that of the intersection of X with the same hyperplane:*

$$\left\{z \in \mathbf{R}^{n-1} : [z; c] \in Y\right\} = \left\{z \in \mathbf{R}^{n-1} : \|z\|_2 \leq \rho(c)\right\}, \quad \forall c \in [a, b], \text{ and}$$

$$\text{Vol}_{n-1}\left(\left\{z \in \mathbf{R}^{n-1} : [z; c] \in Y\right\}\right) = \text{Vol}_{n-1}\left(\left\{z \in \mathbf{R}^{n-1} : [z; c] \in X\right\}\right),$$

$\forall c \in [a, b]$.

Then, Y is a closed convex set.

5.6 Around Polyhedral Representations

Exercise I.27 [TrYs] Justify the calculus rules for polyhedral representations presented in section 3.3.

Exercise I.28 Given two sets $U, V \subseteq \mathbf{R}^m$, define

$$U + V = \left\{x \in \mathbf{R}^m : \exists u \in U, \exists v \in V : x = u + v\right\}.$$

Let $D := \{x \in \mathbf{R}^n : Ax + b + Q_s \subseteq P, \forall s \in S\}$ where the set $\emptyset \neq P \subset \mathbf{R}^m$ admits a polyhedral representation, the set $\emptyset \neq S \subset \mathbf{R}^k$ is given but arbitrary, and the sets $\emptyset \neq Q_s \subset \mathbf{R}^m$ are indexed by $s \in S$.

1. Suppose that S is a finite set and for each $s \in S$ we have $Q_s = \{q_s\}$, i.e., it is a single point. Then will the set D be polyhedrally representable?
2. State sufficient conditions on the structure of sets Q_s and S that will guarantee that the resulting set D is polyhedral. Here, the goal is to have conditions as general as possible. Among your sufficient conditions, can you identify at least some of those that are necessary?

Exercise I.29 [EdYs] For $x \in \mathbf{R}^n$ and integer k, $1 \leq k \leq n$, let $s_k(x)$ be the sum of the k largest entries in x. For example, $s_1(x) = \max_i\{x_i\}$, $s_n(x) = \sum_{i=1}^n x_i$, $s_3([3; 1; 2; 2]) = 3 + 2 + 2 = 7$. Now let $1 \leq k \leq n$ be two integers. For any integer $k = 1, \ldots, n$, define

$$X_{k,n} := \left\{[x; t] \in \mathbf{R}^n \times \mathbf{R} : s_k(x) \leq t\right\}.$$

Observe that $X_{k,n}$ is a polyhedral set. Indeed, $s_k(x) \leq t$ holds if and only if for every k indices $i_1 < i_2 < \cdots < i_k$ from $\{1, 2, \ldots, n\}$ we have $x_{i_1} + x_{i_2} + \cdots + x_{i_k} \leq t$, which is nothing but a linear inequality in the variables x, t. Since there are $\binom{n}{k}$ possible ways of selecting k indices from $\{1, 2, \ldots, n\}$, the number of linear inequalities describing $X_{k,n}$ is $\binom{n}{k}$, and these linear inequalities give the polyhedral description of $X_{k,n}$. The point of this exercise is to demonstrate that $X_{k,n}$ admits a "short" polyhedral representation, specifically,

$$X_{k,n} = \left\{ [x;t] \in \mathbf{R}^n \times \mathbf{R} : \exists z \in \mathbf{R}^n, \exists s \in \mathbf{R}: x_i \leq z_i + s, \forall i, z \geq 0, \sum_{i=1}^{n} z_i + ks \leq t \right\}.$$

Exercise I.30 [TrYs] [Computational study: Fourier–Motzkin elimination as an LP algorithm] It was mentioned in section 3.2.1 that Fourier–Motzkin elimination provides us with an algorithm that terminates in finitely many steps for solving LP problems. This algorithm, however, is of no computational value owing to the potential rapid growth of the number of inequalities one may need to handle when eliminating more and more variables. The goal of this exercise is to get an impression of this phenomenon.

Our "guinea pig" will be a transportation problem with n unit capacity suppliers and n unit demand customers:

$$\min_{x,t} \left\{ t : t \geq \sum_{i=1}^{n} \sum_{i=1}^{n} c_{ij} x_{ij}, \sum_i x_{ij} \geq 1, \forall j, \sum_j x_{ij} \leq 1, \forall i, x_{ij} \geq 0, \forall i,j \right\}.$$

This problem has $n^2 + 1$ variables and $(n+1)^2$ linear inequality constraints; let us solve it by applying Fourier–Motzkin elimination to project the feasible set of the problem onto the axis of the t-variable, that is, to build a finite system S of univariate linear inequalities specifying this projection.

How many inequalities do you think there will be in S when $n = 1, 2, 3, 4$? Check your intuition by implementing and running the F–M elimination, assuming, for the sake of definiteness, that $c_{ij} = 1$ for all i, j, and noting that the F–M elimination presented here does not have a step that eliminates redundant inequalities.

5.7 Around the General Theorem of the Alternative

Exercise I.31

1. Prove Gordan's Theorem of the Alternative: A system of strict homogeneous linear inequalities $Ax < 0$ in variables x has a solution if and only if the system $A^\top \lambda = 0$, $\lambda \geq 0$ in variables λ has only the trivial solution $\lambda = 0$.
2. Prove Motzkin's Theorem of the Alternative: A system $Ax < 0$, $Bx \leq 0$ of strict and nonstrict homogeneous linear inequalities has a solution if and only if the system $A^\top \lambda + B^\top \mu = 0$, $\lambda \geq 0$, $\mu \geq 0$ in variables λ, μ has no solution with $\lambda \neq 0$.

Exercise I.32 For the systems of constraints to follow, write them down equivalently in the standard form $Ax < b, Cx \leq d$ and point out their feasibility status (feasible or infeasible) along with the corresponding certificates (a certificate of feasibility is a feasible solution to the system; a certificate of infeasibility is a collection of weights of constraints which leads to a contradictory consequence inequality, as explained in GTA).

1. $x \leq 0$ ($x \in \mathbf{R}^n$)
2. $x \leq 0$, and $\sum_{i=1}^{n} x_i > 0$ ($x \in \mathbf{R}^n$)
3. $-1 \leq x_i \leq 1$, $1 \leq i \leq n$, $\sum_{i=1}^{n} x_i \geq n$ ($x \in \mathbf{R}^n$)
4. $-1 \leq x_i \leq 1$, $1 \leq i \leq n$, $\sum_{i=1}^{n} x_i > n$ ($x \in \mathbf{R}^n$)
5. $-1 \leq x_i \leq 1$, $1 \leq i \leq n$, $\sum_{i=1}^{n} i x_i \geq \frac{n(n+1)}{2}$ ($x \in \mathbf{R}^n$)
6. $-1 \leq x_i \leq 1$, $1 \leq i \leq n$, $\sum_{i=1}^{n} i x_i > \frac{n(n+1)}{2}$ ($x \in \mathbf{R}^n$)
7. $x \in \mathbf{R}^2$, $|x_1| + x_2 \leq 1$, $x_2 \geq 0$, $x_1 + x_2 = 1$

8. $x \in \mathbf{R}^2$, $|x_1| + x_2 \leq 1$, $x_2 \geq 0$, $x_1 + x_2 > 1$
9. $x \in \mathbf{R}^4$, $x \geq 0$, sum of two largest entries in x does not exceed 2, and $x_1 + x_2 + x_3 \geq 3$
10. $x \in \mathbf{R}^4$, $x \geq 0$, sum of two largest entries in x does not exceed 2, and $x_1 + x_2 + x_3 > 3$

Exercise I.33 Let $(\mathcal{S})$ be the following system of linear inequalities in variables $x \in \mathbf{R}^3$:

$$x_1 \leq 1, \; x_1 + x_2 \leq 1, \; x_1 + x_2 + x_3 \leq 1 \qquad (\mathcal{S})$$

In the following list, point out which inequalities are or are not consequences of this system, and certify your claims. To certify that a given inequality is a consequence of the given system, you need to provide nonnegative aggregation weights $\lambda \in \mathbf{R}_+^3$ for the inequalities in $(\mathcal{S})$ such that the resulting consequence inequality implies the given inequality. To certify that a given inequality is not a consequence of the given system $(\mathcal{S})$, you need to find a point $x \in \mathbf{R}^3$ that satisfies the given system but violates the given inequality.

1. $3x_1 + 2x_2 + x_3 \leq 4$
2. $3x_1 + 2x_2 + x_3 \leq 2$
3. $3x_1 + 2x_2 \leq 3$
4. $3x_1 + 2x_2 \leq 2$
5. $3x_1 + 3x_2 + x_3 \leq 3$
6. $3x_1 + 3x_2 + x_3 \leq 2$

Make a generalization: prove that a linear inequality $px_1 + qx_2 + rx_3 \leq s$ is a consequence of $(\mathcal{S})$ if and only if $s \geq p \geq q \geq r \geq 0$.

Exercise I.34 Is the inequality $x_1 + x_2 \leq 1$ a consequence of the system $x_1 \leq 1$, $x_1 \geq 2$? If so, can it be obtained by taking a legitimate weighted sum of inequalities from the system and an equality that is always true, $0^\top x \leq 1$, as is suggested by the Inhomogeneous Farkas' Lemma?

Exercise I.35 Certify the correct statements in the following list:

1. The polyhedral set $X = \left\{x \in \mathbf{R}^3 : x \geq [1/3; 1/3; 1/3], \; \sum_{i=1}^3 x_i \leq 1\right\}$ is nonempty.
2. The polyhedral set $X = \left\{x \in \mathbf{R}^3 : x \geq [1/3; 1/3; 1/3], \; \sum_{i=1}^3 x_i \leq 0.99\right\}$ is empty.
3. The linear inequality $x_1 + x_2 + x_3 \geq 2$ is violated somewhere on the polyhedral set $X = \left\{x \in \mathbf{R}^3 : x \geq [1/3; 1/3; 1/3], \; \sum_{i=1}^3 x_i \leq 1\right\}$.
4. The linear inequality $x_1 + x_2 + x_3 \geq 2$ is violated somewhere on the polyhedral set $X = \left\{x \in \mathbf{R}^3 : x \geq [1/3; 1/3; 1/3], \; \sum_{i=1}^3 x_i \leq 0.99\right\}$.
5. The linear inequality $x_1 + x_2 \leq 3/4$ is satisfied everywhere on the polyhedral set $X = \left\{x \in \mathbf{R}^3 : x \geq [1/3; 1/3; 1/3], \; \sum_{i=1}^3 x_i \leq 1.05\right\}$.
6. The polyhedral set $Y = \left\{x \in \mathbf{R}^3 : x_1 \geq 1/3, x_2 \geq 1/3, x_3 \geq 1/3\right\}$ is not contained in the polyhedral set $X = \left\{x \in \mathbf{R}^3 : x \geq [1/3; 1/3; 1/3], \; \sum_{i=1}^3 x_i \leq 1\right\}$.
7. The polyhedral set $Y = \left\{x \in \mathbf{R}^3 : x \geq [1/3; 1/3; 1/3], \; \sum_{i=1}^3 x_i \leq 1\right\}$ is contained in the polyhedral set $X = \left\{x \in \mathbf{R}^3 : x_1 + x_2 \leq 2/3, \; x_2 + x_3 \leq 2/3, \; x_1 + x_3 \leq 2/3\right\}$.

5.8 Around Linear Programming Duality

Exercise I.36 [EdYs] Let the polyhedral set $P = \{x \in \mathbf{R}^n : Ax \leq b\}$, where $A = [a_1^\top; \ldots; a_m^\top]$, be nonempty. Prove that P is bounded if and only if every vector from $\mathbf{R}^n$ can be represented as a linear combination of the vectors a_i with nonnegative coefficients, where at most n coefficients are positive. As a result, given A, all nonempty sets of the form $\{x \in \mathbf{R}^n : Ax \leq b\}$ simultaneously are or are not bounded.

Exercise I.37 Consider the linear program

$$\text{Opt} = \max_{x \in \mathbf{R}^2} \{x_1 : x_1 \geq 0, x_2 \geq 0, ax_1 + bx_2 \leq c\} \quad (P)$$

where a, b, c are parameters. Answer the following questions concerning the problem (P):

1. Let $c = 1$. Is the problem feasible?
2. Let $a = b = 1$, $c = -1$. Is the problem feasible?
3. Let $a = b = 1$, $c = -1$. Is the problem bounded?[4]
4. Let $a = b = c = 1$. Is the problem bounded?
5. Let $a = 1$, $b = -1$, $c = 1$. Is the problem bounded?
6. Let $a = b = c = 1$. Is it true that Opt ≥ 0.5?
7. Let $a = b = 1$, $c = -1$. Is it true that Opt ≤ 1?
8. Let $a = b = c = 1$. Is it true that Opt ≤ 1?
9. Let $a = b = c = 1$. Is it true that $x_* = [1; 1]$ is an optimal solution of (P)?
10. Let $a = b = c = 1$. Is it true that $x_* = [1/2; 1/2]$ is an optimal solution of (P)?
11. Let $a = b = c = 1$. Is it true that $x_* = [1; 0]$ is an optimal solution of (P)?

Exercise I.38 Consider the LP program

$$\max_{x_1, x_2} \left\{ -x_2 : \begin{array}{l} x_1 \leq 0 \\ -x_1 \leq -1 \\ x_2 \leq 1 \end{array} \right\}.$$

Write down the dual problem and check whether the optimal values are equal to each other.

Exercise I.39 Write down the problems dual to the following linear programs:

1. $\max\limits_{x \in \mathbf{R}^3} \left\{ x_1 + 2x_2 + 3x_3 : \begin{array}{l} x_1 - x_2 + x_3 = 0, \\ x_1 + x_2 - x_3 \geq 100, \\ x_1 \leq 0, \\ x_2 \geq 0, \\ x_3 \geq 0 \end{array} \right\}$
2. $\max\limits_{x \in \mathbf{R}^n} \{c^\top x : Ax = b, x \geq 0\}$
3. $\max\limits_{x \in \mathbf{R}^n} \{c^\top x : Ax = b, \underline{u} \leq x \leq \overline{u}\}$
4. $\max\limits_{x,y} \{c^\top x : Ax + By \leq b, x \leq 0, y \geq 0\}$

---

[4] Recall that a maximization problem is called *bounded*, if the objective is bounded from above on the feasible set, which is the same as its optimal value being $< \infty$.

Exercise I.40 [TrYs] Consider the primal–dual pair of linear programs

$$\text{Opt}(P) = \min_x \left\{ c^\top x : Ax \geq b \right\}, \quad (P)$$

$$\text{Opt}(D) = \max_y \left\{ b^\top y : y \geq 0, A^\top y = c \right\}. \quad (D)$$

Suppose that both are feasible. Prove that the feasible set of at least one of these problems is unbounded.

Exercise I.41 [TrYs] Consider the following linear program:

$$\text{Opt} = \min_{\{x_{ij}\}_{1 \leq i < j \leq 4}} \left\{ 2 \sum_{1 \leq i < j \leq 4} x_{ij} : x_{ij} \geq 0, \forall 1 \leq i < j \leq 4, \sum_{j > i} x_{ij} + \sum_{j < i} x_{ji} \geq i, 1 \leq i \leq 4 \right\}.$$

1. Show that the optimum objective value is at most 20.
2. Show that the optimum objective value is at least 10.

Exercise I.42 [EdYs] We say that an $n \times n$ matrix P is *stochastic* if all of its entries are nonnegative and the sum of the entries of each row is equal to 1. Show that if P is a stochastic matrix then there is a nonzero vector $a \in \mathbf{R}^n$ such that $P^\top a = a$ and $a \geq 0$.

Exercise I.43 [TrYs] Let $A \in \mathbf{R}^{n \times n}$ be a symmetric matrix, i.e., $A^\top = A$. Consider the linear program

$$\min_x \left\{ c^\top x : Ax \geq c, x \geq 0 \right\}.$$

Prove that if $\bar{x}$ satisfies $A\bar{x} = c$ and $\bar{x} \geq 0$ then $\bar{x}$ is optimal.

Exercise I.44 [TrYs] Let $c \in \mathbf{R}^n$, and let $A \in \mathbf{R}^{n \times n}$ be a *skew-symmetric* matrix, i.e., $A^\top = -A$. Consider the following linear program:

$$\text{Opt}(P) = \min_{x \in \mathbf{R}^n} \left\{ c^\top x : Ax \geq -c, x \geq 0 \right\}.$$

Suppose that the problem is solvable. Provide a closed analytical expression for Opt(P).

Exercise I.45 [TrYs] [Separation Theorem, polyhedral version] Let P and Q be two nonempty polyhedral sets in $\mathbf{R}^n$ such that $P \cap Q = \emptyset$. Suppose that the polyhedral descriptions of these sets are given as

$$P := \left\{ x \in \mathbf{R}^n : Ax \leq b \right\} \text{ and } Q := \left\{ x \in \mathbf{R}^n : Dx \geq f \right\}.$$

Using LP duality show that there exists a vector $c \in \mathbf{R}^n$ such that

$$c^\top x < c^\top y \quad \text{for all } x \in P \text{ and } y \in Q.$$

Exercise I.46 [TrYs] Suppose we are given the linear program

$$\min_x \left\{ c^\top x : Ax = b, x \geq 0 \right\} \quad (P)$$

and its associated *Lagrangian* function

$$L(x, \lambda) := c^\top x + \lambda^\top (b - Ax).$$

The LP dual to (P) is (replacing $Ax = b$ with $Ax \geq b$, $-Ax \geq -b$)

$$\text{Opt}(D) = \max_{\lambda_\pm, \mu} \left\{ b^\top [\lambda_+ - \lambda_-] : A^\top [\lambda_+ - \lambda_-] + \mu = c, \lambda_\pm \geq 0, \mu \geq 0 \right\},$$

or, after eliminating μ and setting $\lambda = \lambda_+ - \lambda_-$,

$$\text{Opt}(D) = \max_\lambda \left\{ b^\top \lambda : A^\top \lambda \leq b \right\}. \quad (D)$$

Now, let us consider the following game: Player 1 chooses some $x \geq 0$, and player 2 chooses some λ simultaneously; then, player 1 pays to player 2 the amount $L(x, \lambda)$. In this game, player 1 would like to minimize $L(x, \lambda)$ and player 2 would like to maximize $L(x, \lambda)$.

A pair (x^*, λ^*) with $x^* \geq 0$, is called an *equilibrium* point (or *saddle point* or *Nash equilibrium*) if

$$L(x^*, \lambda) \leq L(x^*, \lambda^*) \leq L(x, \lambda^*), \quad \forall x \geq 0 \text{ and } \forall \lambda. \quad (*)$$

(That is, in an equilibrium no player is able to improve his performance by unilaterally modifying his choice.)

Show that x^* and λ^* are optimal solutions to the problem (P) and to its dual, respectively, if and only if (x^*, λ^*) is an equilibrium point.

Exercise I.47 [TrYs] Given a polyhedral set $X = \{x \in \mathbf{R}^n : a_i^\top x \leq b_i, \forall i = 1, \ldots, m\}$, consider the associated optimization problem

$$\max_{x, t} \{t : B_1(x, t) \subseteq X\},$$

where $B_1(x, t) := \{y \in \mathbf{R}^n : \|y - x\|_\infty \leq t\}$. Is it possible to pose this optimization problem as a linear program with a polynomial in m variables and n constraints? If it is possible, give such a representation explicitly. If not, argue why.

Exercise I.48 [TrYs] Consider the optimization problem

$$\min_{x \in \mathbf{R}^n} \left\{ c^\top x : \tilde{a}_i^\top x \leq b_i \text{ for some } \tilde{a}_i \in P_i, i = 1, \ldots, m, x \geq 0 \right\}, \quad (*)$$

where $P_i = \{\bar{a}_i + \epsilon_i : \|\epsilon_i\|_\infty \leq \rho\}$ for $i = 1, \ldots, m$ and $\|u\|_\infty := \max_{j=1,\ldots,n}\{|u_j|\}$. In this problem, we basically mean that the constraint coefficient $\tilde{a}_{ij}$ (the jth component of the ith constraint vector $\tilde{a}_i$) belongs to the interval uncertainty set $[\bar{a}_{ij} - \rho, \bar{a}_{ij} + \rho]$, where $\bar{a}_{ij}$ is its nominal value. That is, in $(*)$, we are seeking a solution x such that each constraint is satisfied for *some* coefficient vector from the corresponding uncertainty set.

Note that in its current form $(*)$, this problem is not a linear program (LP). Prove that it can be written as an *explicit* linear program and give the corresponding LP formulation.

Exercise I.49 [EdYs] Let $S = \{a_1, a_2, \ldots, a_n\}$ be a finite set composed of n distinct elements, and let f be a real-valued function defined on the set of all subsets of S. We say that f is *submodular* if, for every $X, Y \subseteq S$, the following inequality holds:

$$f(X) + f(Y) \geq f(X \cup Y) + f(X \cap Y).$$

1. Give an example of a submodular function f.
2. Let $f : 2^S \to \mathbf{Z}$ be an integer-valued submodular function such that $f(\emptyset) = 0$. Consider the polyhedron

$$P_f := \left\{ x \in \mathbf{R}^{|S|} : \sum_{t \in T} x_t \leq f(T), \forall T \subseteq S \right\}.$$

Consider

$$\bar{x}_{a_k} := f(\{a_1, \ldots, a_k\}) - f(\{a_1, \ldots, a_{k-1}\}), \quad k = 1, \ldots, n.$$

Show that $\bar{x}$ is feasible to P_f.

3. Consider the following optimization problem associated with P_f:

$$\max_x \left\{ c^\top x : x \in P_f \right\}.$$

Write down the dual of this LP.

4. Assume without loss of generality that $c_{a_1} \geq c_{a_2} \geq \cdots \geq c_{a_n}$. Identify a dual feasible solution and using the LP Duality Theorem show that the solution $\bar{x}$ specified in item 2 is optimal to the primal maximization problem associated with P_f.

Remark. Note that when the submodular function f is integer-valued, we immediately see from the characterization of the optimal primal solution $\bar{x}$ that for all integer vectors $c \in \mathbf{Z}^n$ such that there exists an optimum solution to the primal problem, there exists an optimum solution (e.g. $\bar{x}$) where all variables take integer values. A system of linear inequalities $Ax \leq b$ with $b \in \mathbf{Z}^m$ and $A \in \mathbf{Q}^{m \times n}$ satisfying such a property (i.e., whenever $c \in \mathbf{Z}^n$ is such that there is an optimal solution to $\max_x \{c^\top x : Ax \leq b\}$ then there is an integer optimum solution) is called *totally dual integral* (TDI). Thus, we conclude that the polyhedron P_f associated with an integer-valued submodular function f is TDI. The TDI property is a well-known sufficient condition that guarantees that every extreme point (see section 6.4) of the associated polyhedron is integral. In particular, the TDI property generalizes *total unimodularity* (TU), i.e., the other well-known sufficient condition for the integrality of a polyhedron, which plays a key role in network-flow based optimization.

Part II

Separation Theorem, Extreme Points, Recessive Directions, and Geometry of Polyhedral Sets

6

Separation Theorem and Geometry of Convex Sets

We next investigate the *Separation Theorem*, which is as indispensable when one is studying general convex sets as is the General Theorem of the Alternative when one is investigating properties of polyhedral sets.

6.1 Separation: Definition

Recall that a *hyperplane* M in $\mathbf{R}^n$ is, by definition, an affine subspace of dimension $n-1$. Then, by Proposition A.47, hyperplanes are precisely the level sets of nontrivial linear forms. That is,

$$M \subset \mathbf{R}^n \text{ is a hyperplane}$$
$$\iff \exists a \in \mathbf{R}^n,\ a \neq 0,\ \exists b \in \mathbf{R} \text{ such that } M = \left\{ x \in \mathbf{R}^n : a^\top x = b \right\}.$$

We can associate with the hyperplane M or, equivalently, with the associated pair a, b (defined by the hyperplane up to multiplication of a, b by a nonzero real number) the following sets:

- the "upper" and "lower" open half-spaces

$$M^{++} := \left\{ x \in \mathbf{R}^n : a^\top x > b \right\} \quad \text{and} \quad M^{--} := \left\{ x \in \mathbf{R}^n : a^\top x < b \right\}.$$

These sets clearly are convex, and since a linear form is continuous, and the sets are given by strict inequalities on the value of a continuous function, they are indeed open.
These open half-spaces are uniquely defined by the hyperplane, up to swapping the "upper" and the "lower" (this is what happens when one passes from a particular pair a, b specifying M to a negative multiple of this pair).
- the "upper" and "lower" closed half-spaces

$$M^+ := \left\{ x \in \mathbf{R}^n : a^\top x \geq b \right\} \quad \text{and} \quad M^- := \left\{ x \in \mathbf{R}^n : a^\top x \leq b \right\}.$$

These are also convex sets. Moreover, these two sets are polyhedral and thus closed. It is easily seen that the closed upper or lower half-space is the closure of the corresponding open half-space, and M itself is the common boundary of all four half-spaces.

Also, note that our half-spaces and M itself partition $\mathbf{R}^n$, i.e.,

$$\mathbf{R}^n = M^{--} \cup M \cup M^{++}$$

(partitioning by disjoint sets), and

$$\mathbf{R}^n = M^- \cup M^+$$

(where M is the intersection of the sets M^- and M^+).

We are now ready to define the basic notion of the *separation* of two convex sets T and S by a hyperplane.

Definition 6.1 *[Separation] Let S, T be two nonempty convex sets in $\mathbf{R}^n$.*

(i) A hyperplane

$$M = \left\{ x \in \mathbf{R}^n : a^\top x = b \right\} \quad [\text{where } a \neq 0]$$

is said to separate *S and T, if it satisfies both of the following properties:*
$S \subseteq \{ x \in \mathbf{R}^n : a^\top x \leq b \}$, $T \subseteq \{ x \in \mathbf{R}^n : a^\top x \geq b \}$ *(i.e., S and T belong to the opposite closed half-spaces into which M splits $\mathbf{R}^n$), and,*
at least one of the sets S, T is not contained in M itself, i.e.,

$$S \cup T \not\subseteq M.$$

(ii) The separation is called strong, *if there exist $b', b'' \in \mathbf{R}$ satisfying $b' < b < b''$ such that*

$$S \subseteq \left\{ x \in \mathbf{R}^n : a^\top x \leq b' \right\}, \quad T \subseteq \left\{ x \in \mathbf{R}^n : a^\top x \geq b'' \right\}.$$

(iii) A linear form $a \neq 0$ is said to separate (strongly separate) *S and T, if for properly chosen b the hyperplane $\{ x \in \mathbf{R}^n : a^\top x = b \}$ separates (strongly separates) S and T.*[1]

(iv) We say that S and T can be (strongly) separated *if there exists a hyperplane which (strongly) separates S and T.*

Let us examine the separation concept on a few simple examples.

Example 6.1 See Figure 6.1 for an illustration of the following examples.

1. The hyperplane $\{ x \in \mathbf{R}^2 : x_2 - x_1 = 1 \}$ strongly separates the polyhedral sets $S = \{ x \in \mathbf{R}^2 : x_2 = 0, x_1 \geq -1 \}$ and $T = \{ x \in \mathbf{R}^2 : 0 \leq x_1 \leq 1, 3 \leq x_2 \leq 5 \}$.
2. The hyperplane $\{ x \in \mathbf{R} : x = 1 \}$ separates (but not strongly separates) the convex sets $S = \{ x \in \mathbf{R} : x \leq 1 \}$ and $T = \{ x \in \mathbf{R} : x \geq 1 \}$.
3. The hyperplane $\{ x \in \mathbf{R}^2 : x_1 = 0 \}$ separates (but not strongly separates) the convex sets $S = \{ x \in \mathbf{R}^2 : x_1 < 0, x_2 \geq -1/x_1 \}$ and $T = \{ x \in \mathbf{R}^2 : x_1 > 0, x_2 > 1/x_1 \}$.
4. The hyperplane $\{ x \in \mathbf{R}^2 : x_2 - x_1 = 1 \}$ does *not* separate the convex sets $S = \{ x \in \mathbf{R}^2 : x_2 \geq 1 \}$ and $T = \{ x \in \mathbf{R}^2 : x_2 = 0 \}$.
5. The hyperplane $\{ x \in \mathbf{R}^2 : x_2 = 0 \}$ does *not* separate the polyhedral sets $S = \{ x \in \mathbf{R}^2 : x_2 = 0, x_1 \leq -1 \}$ and $T = \{ x \in \mathbf{R}^2 : x_2 = 0, x_1 \geq 1 \}$. ◇

---

[1] Here and in what follows we identify a linear form with the vector of its coefficient. Thus, "a linear form $a \in \mathbf{R}^n$" stands for the linear function $x \mapsto a^\top x : \mathbf{R}^n \to \mathbf{R}$.

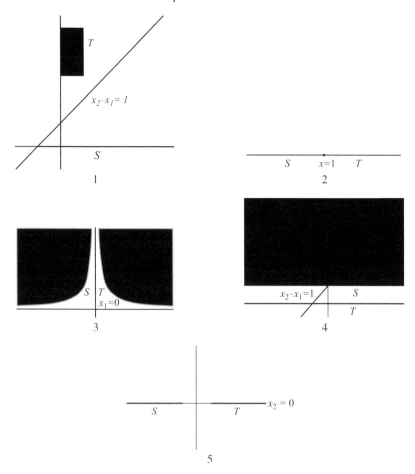

Figure 6.1 Separation figures corresponding to Example 6.1.

The following equivalent description of separation is often used as well.

Fact 6.2 *Let S, T be nonempty convex sets in $\mathbf{R}^n$. A linear form $a^\top x$ separates S and T if and only if*

$$(a) \quad \sup_{x \in S} a^\top x \leq \inf_{y \in T} a^\top y, \quad \text{and}$$

$$(b) \quad \inf_{x \in S} a^\top x < \sup_{y \in T} a^\top y.$$

This separation is strong if and only if (a) holds as a strict inequality:

$$\sup_{x \in S} a^\top x < \inf_{y \in T} a^\top y.$$

6.2 Separation Theorem

One of the most fundamental results in convex analysis is the following Separation Theorem.

Theorem 6.3 *[Separation Theorem] Let S and T be nonempty convex sets in $\mathbf{R}^n$.*

(i) S and T can be separated if and only if their relative interiors do not intersect, i.e., $\operatorname{rint} S \cap \operatorname{rint} T = \emptyset$.

(ii) S and T can be strongly separated if and only if the sets are at a positive distance from each other, i.e.,

$$\text{dist}(S, T) := \inf \{\|x - y\|_2 : x \in S, \ y \in T\} > 0.$$

In particular, if S and T are nonempty non-intersecting closed convex sets and one of these sets is compact then S and T can be strongly separated.

We will use the following simple and important lemma in the proof of the Separation Theorem.

Lemma 6.4 *A point $x \in \text{rint } Q$ of a convex set Q can be the minimizer (or maximizer) of a linear function $f(x) = a^\top x$ if and only if the function is constant on Q.*

Proof The "if" part is evident. To prove the "only if" part, let $\bar{x} \in \text{rint } Q$ be, say, a minimizer of f over Q; then for any $y \in Q$ we need to prove that $f(\bar{x}) = f(y)$. There is nothing to prove if $y = \bar{x}$, so let us assume that $y \neq \bar{x}$. Since Q is convex and $\bar{x}, y \in Q$, the segment $[\bar{x}, y]$ belongs to Q. Moreover, as $\bar{x} \in \text{rint } Q$ we can extend this segment a little further away from $\bar{x}$ and still remain in Q. That is, there exists $z \in Q$ such that $\bar{x} = (1 - \lambda)y + \lambda z$ for a certain $\lambda \in [0, 1)$. As $y \neq \bar{x}$, we have in fact $\lambda \in (0, 1)$. Since f is linear, we deduce

$$f(\bar{x}) = (1 - \lambda)f(y) + \lambda f(z).$$

Because $\bar{x}$ is a minimizer of f over Q and $y, z \in Q$, we have $\min\{f(y), f(z)\} \geq f(\bar{x}) = (1 - \lambda)f(y) + \lambda f(z)$. Then, from $\lambda \in (0, 1)$ we conclude that this relation can be satisfied only when $f(\bar{x}) = f(y) = f(z)$. ∎

Proof of Theorem 6.3 We will prove the Separation Theorem in several steps. We will first focus on the usual separation, i.e., case (i) of the theorem.

Case (i): Necessity. Assume that S, T can be separated. Then, for certain $a \neq 0$ we have

$$\sup_{x \in S} a^\top x \leq \inf_{y \in T} a^\top y, \quad \text{and} \quad \inf_{x \in S} a^\top x < \sup_{y \in T} a^\top y. \tag{6.1}$$

Assume for contradiction that rint S and rint T have a common point $\bar{x}$. Then, from the first inequality in (6.1) and $\bar{x} \in S \cap T$, we deduce

$$a^\top \bar{x} \leq \sup_{x \in S} a^\top x \leq \inf_{y \in T} a^\top y \leq a^\top \bar{x}.$$

Thus, $\bar{x}$ maximizes the linear function $f(x) = a^\top x$ on S and simultaneously minimizes this function on T. Then, as $\bar{x} \in \text{rint } S$ and also $\bar{x} \in \text{rint } T$ using Lemma 6.4, we conclude that $f(x) = f(\bar{x})$ on S and on T, so that $f(\cdot)$ is constant on $S \cup T$. This then yields the desired contradiction to the second inequality in (6.1).

Case (i): Sufficiency. The proof of the sufficiency part of the Separation Theorem is much more instructive. There are several ways to prove it. Below, we present a proof based on Theorem 3.2.

Case (i): Sufficiency, Step 1: Separation of a nonempty polytope and a point outside the polytope. We start with a seemingly very particular case of the Separation Theorem –

6.2 Separation Theorem

one where $S = \mathrm{Conv}\{x^1, \ldots, x^N\}$ and T is a singleton $T = \{x\}$ which does not belong to S. We will prove that in this case there exists a linear form which strongly separates $T = \{x\}$ and S.

The set $S = \mathrm{Conv}\{x^1, \ldots, x^N\}$ is given by the polyhedral representation

$$S = \left\{ z \in \mathbf{R}^n : \exists \lambda \text{ such that } \lambda \geq 0, \sum_{i=1}^{N} \lambda_i = 1, z = \sum_{i=1}^{N} \lambda_i a_i \right\},$$

and thus S is polyhedral (Theorem 3.2). Therefore, for a properly selected $k, a_1, \ldots, a_k$, and $b_1, \ldots, b_k$ we have:

$$S = \left\{ z \in \mathbf{R}^n : a_i^\top z \leq b_i, \, i \leq k \right\}.$$

Since $x \notin S$, there exists $i \leq k$ such that $a_i^\top x > b_i$, and thus the corresponding a_i clearly strongly separates our S and $T = \{x\}$.

Case (i): Sufficiency, Step 2: Separation of a nonempty convex set and a point outside the set. Now consider the case when S is an arbitrary nonempty convex set and $T = \{x\}$ is a singleton outside S (here the difference from Step 1 is that now S is not assumed to be a polytope).

Without loss of generality we may assume that S contains 0 (if that is not the case, by taking any $p \in S$ we may translate S and T to the sets $S \mapsto -p + S$, $T \mapsto -p + T$; clearly, a linear form which separates the shifted sets also separates the original sets as well). Let L be the linear span of S.

If $x \notin L$, the separation is easy: we can write $x = e + f$, where $e \in L$ and f is from the subspace orthogonal to L, and thus

$$f^\top x = f^\top f > 0 = \max_{y \in S} f^\top y,$$

so that f strongly separates S and $T = \{x\}$.

Now, we consider the case when $x \in L$. Since $x \in L$, and $x \notin S$ as well as $\emptyset \neq S \subseteq L$, we deduce that L contains at least two points and so $L \neq \{0\}$. Without loss of generality, we can assume that $L = \mathbf{R}^n$.

Let $\mathcal{B} := \{h \in \mathbf{R}^n : \|h\|_2 = 1\}$ be the unit sphere in $\mathbf{R}^n$. This is a closed and bounded set in $\mathbf{R}^n$ (the boundedness is evident, and the closedness follows from the fact that $\|\cdot\|_2$ is continuous). Thus, $\mathcal{B}$ is a compact set. Let us prove that there exists $f \in \mathcal{B}$ that separates x and S in the sense that

$$f^\top x \geq \sup_{y \in S} f^\top y. \tag{6.2}$$

Assume for contradiction that no such f exists. Then, for every $h \in \mathcal{B}$ there exists $y_h \in S$ such that

$$h^\top y_h > h^\top x.$$

Since the inequality is strict, it immediately follows that there exists an open neighborhood U_h of the vector h such that

$$(h')^\top y_h > (h')^\top x, \quad \forall h' \in U_h. \tag{6.3}$$

Note that the family of open sets $\{U_h\}_{h \in \mathcal{B}}$ covers $\mathcal{B}$. As $\mathcal{B}$ is compact, we can find a finite subfamily $U_{h_1}, \ldots, U_{h_N}$ of this family which still covers $\mathcal{B}$. Let us take the corresponding points $y^1 := y_{h_1}$, $y^2 := y_{h_2}, \ldots, y^N := y_{h_N}$ and define the polytope $\widehat{S} :=$ Conv$\{y^1, \ldots, y^N\}$. Due to the definition of y^i, all these points are in S and thus $S \supseteq \widehat{S}$ (recall that S is convex). Since $x \notin S$, we deduce $x \notin \widehat{S}$. Then, by Step 1, x can be strongly separated from $\widehat{S}$, i.e., there exists $a \neq 0$ such that

$$a^\top x > \sup_{y \in \widehat{S}} a^\top y = \max\left\{a^\top y^i : 1 \leq i \leq N\right\}. \tag{6.4}$$

By normalization, we may also assume that $\|a\|_2 = 1$, so that $a \in \mathcal{B}$. Recall that $U_{h_1}, \ldots, U_{h_N}$ form a covering of $\mathcal{B}$, and as $a \in \mathcal{B}$ we have that a belongs to a certain U_{h_i}. By the construction of U_{h_i} (see (6.3)), we have

$$a^\top y^i \equiv a^\top y_{h_i} > a^\top x,$$

which contradicts (6.4) as $y^i \in \widehat{S}$.

Thus, we conclude that there exists $f \in \mathcal{B}$ satisfying (6.2). We claim that f separates S and $\{x\}$. Given that we have already established (6.2), all we need to verify for establishing that f does indeed separate S and $\{x\}$ is to show that the linear form $f(y) = f^\top y$ is non-constant on $S \cup T$. This is evident as we are in the situation when $0 \in S$ and $L = \text{Lin}(S) = \mathbf{R}^n$ and $f \neq 0$, so that $f(y)$ is non-constant already on S (indeed, otherwise we would have $f^\top y = 0$ for $y \in S$ due to $0 \in S$, whence $f^\top y = 0$ for $y \in \text{Lin}(S) = \mathbf{R}^n$, contradicting $f \neq 0$).

Case (i): Sufficiency, Step 3: Separation of two nonempty and non-intersecting convex sets. Now we are ready to prove that two nonempty and non-intersecting convex sets S and T can be separated. To this end consider the arithmetic difference of the sets S and T, i.e.,

$$\Delta := S - T = \{x - y : x \in S, y \in T\}.$$

As S and T are nonempty and convex, Δ is nonempty and convex (by Proposition 1.21(iii)). Also, as $S \cap T = \emptyset$, we have $0 \notin \Delta$. Then, by Step 2, we can separate Δ and $\{0\}$, i.e., there exists $f \neq 0$ such that

$$f^\top 0 = 0 \geq \sup_{z \in \Delta} f^\top z \quad \text{and} \quad f^\top 0 > \inf_{z \in \Delta} f^\top z.$$

In other words,

$$0 \geq \sup_{x \in S, y \in T} \left\{f^\top x - f^\top y\right\} \quad \text{and} \quad 0 > \inf_{x \in S, y \in T} \left\{f^\top x - f^\top y\right\},$$

which clearly means that f separates S and T.

Case (i): Sufficiency, Step 4: Separation of nonempty convex sets with non-intersecting relative interiors. Now we are ready to complete the proof of the "if" part of part (i) of the Separation Theorem. Let S and T be two nonempty convex sets such that rint $S \cap$ rint $T = \emptyset$; then we will prove that S and T can be separated. Recall from Theorem 1.29 that the sets $S' :=$ rint S and $T' :=$ rint T are nonempty and convex. Moreover, we are given that S' and T' do not intersect, thus they can be separated by Step 3. That is, there exists f such that

$$\inf_{x \in T'} f^\top x \geq \sup_{y \in S'} f^\top x \quad \text{and} \quad \sup_{x \in T'} f^\top x > \inf_{y \in S'} f^\top x. \tag{6.5}$$

It can immediately be seen that in fact f separates S and T. Indeed, the quantities on the left- and the right-hand sides of the first inequality in (6.5) clearly remain unchanged when we replace S' with cl S' and T' with cl T'. Moreover, by Theorem 1.29, cl $S' =$ cl $S \supseteq S$ and cl $T' =$ cl $T \supseteq T$, and we get $\inf_{x \in T} f^\top x = \inf_{x \in T'} f^\top x$, and similarly $\sup_{y \in S} f^\top y = \sup_{y \in S'} f^\top y$. Thus, we obtain from (6.5)

$$\inf_{x \in T} f^\top x \geq \sup_{y \in S} f^\top y.$$

It remains to note that $T' \subseteq T$, $S' \subseteq S$, so that the second inequality in (6.5) implies that

$$\sup_{x \in T} f^\top x > \inf_{y \in S} f^\top x.$$

Case (ii): Necessity. The reader is invited to prove this.

Case (ii): Sufficiency. Define $\rho := \text{dist}(S, T) = \inf\{\|x - y\|_2 : x \in S, y \in T\}$. In case (ii), we are given that $\rho > 0$. Consider the set $\widehat{S} := \left\{ x \in \mathbf{R}^n : \inf_{y \in S} \|x - y\|_2 \leq \rho/2 \right\}$. As S is convex, the set $\widehat{S}$ is convex (recall Example 1.3). Moreover, $\widehat{S} \cap T = \emptyset$ (why?). Then, by part (i), $\widehat{S}$ and T can be separated. Let f be any linear form that separates $\widehat{S}$ and T. Then, the same form strongly separates S and T (why?). The last statement of (ii), i.e., the "in particular" part, readily follows from the statement just proved, owing to the fact that if two closed nonempty sets in $\mathbf{R}^n$ do not intersect and one of them is compact then the sets are at a positive distance from each other (why?). This concludes the proof of Theorem 6.3. ∎

Remark 6.5 In Theorem 6.3, a careful reader might notice that the considerations in the proof part (i): Sufficiency, Step 1, i.e., the separation of a polytope and a point not in the polytope, are based solely on Theorem 3.2, and this is a purely arithmetic statement: when proving it, we did not use convergence, compactness, square roots, etc., just rational arithmetics. Therefore, the result stated at Step 1 remains valid if we replace our universe $\mathbf{R}^n$ with the space $\mathbf{Q}^n$ of n-dimensional *rational* vectors (those with rational coordinates; of course, multiplication by reals in this space should be restricted to multiplication by rationals). The possibility of separating a rational vector from a "rational" polytope by a *rational* linear form, which is the "rational" version of the result of Step 1, is definitely of interest (e.g., for Integer Programming).

In fact, all the results in Part I of the book are derived from Fourier–Motzkin elimination and Theorem 3.2, i.e., the existence of optimal solution(s) to feasible and bounded LP problems, Farkas' Lemmas, the General Theorem of the Alternative, the plain and conic Carathéodory Theorems and (the finite-family version of) Helly's Theorems, etc., remain valid when $\mathbf{R}^n$ is replaced with $\mathbf{Q}^n$ (provided, of course, that the related data are rational). In particular, any feasible and bounded LP problem *with rational data* admits a rational optimal solution, which definitely is worth knowing.

In contrast with these "purely arithmetic" considerations at Step 1, at Step 2, concerning the separation of a closed convex set and a point outside the set, we used compactness, which heavily exploits the fact that our universe is $\mathbf{R}^n$ and not, say, $\mathbf{Q}^n$ (in the latter space bounded and closed sets are not necessarily compact).

In fact, we could not avoid things like compactness arguments at Step 2, since the very fact we are proving is true in $\mathbf{R}^n$ but not in $\mathbf{Q}^n$. Indeed, consider the "rational plane," i.e., the universe composed of all two-dimensional vectors with rational entries, and let S be the half-plane in this rational plane given by the linear inequality

$$x_1 + \alpha x_2 \leq 0,$$

where α is irrational. Clearly S is a "convex set" in $\mathbf{Q}^2$. But, it can immediately be seen that a point outside this set cannot be separated from S by a rational linear form. ◇

The Separation Theorem admits a number of important consequences. In this chapter, we will discuss these.

6.3 Supporting Hyperplanes

By the Separation Theorem, we can immediately deduce that a nonempty closed convex set M is precisely the intersection of all closed half-spaces containing M. Among these half-spaces, the most interesting are the "extreme" ones, i.e., those with boundary hyperplanes touching M. Such extreme hyperplanes are called supporting hyperplanes. While this notion of extreme does make sense for an arbitrary (not necessary closed) convex set, we will use it for closed convex sets only, and include the requirement of closedness in the definition:

Definition 6.6 *[Supporting hyperplane] Let M be a closed convex set in $\mathbf{R}^n$, and let $x \in$ rbd M. A hyperplane*

$$\Pi := \left\{ y \in \mathbf{R}^n : a^\top y = a^\top x \right\} \quad [\text{where } a \neq 0]$$

is called supporting *to M at x, if it separates M and $\{x\}$, i.e., if*

$$a^\top x \geq \sup_{y \in M} a^\top y \quad \text{and} \quad a^\top x > \inf_{y \in M} a^\top y. \tag{6.6}$$

Independently of whether M is or is not closed, a point $x \in$ rbd M is a limit of points from M, and thus the first inequality in (6.6) cannot be strict. As a result, we arrive at an equivalent definition of a supporting hyperplane for convex sets, as follows.

Given a closed convex set $M \in \mathbf{R}^n$ and a point $x \in$ rbd M, a hyperplane

$$\left\{ y \in \mathbf{R}^n : a^\top y = a^\top x \right\}$$

is supporting to M at x if and only if the linear form $a(y) := a^\top y$ attains its maximum on M at the point x and is non-constant on M.

Example 6.2 The hyperplane $\{x \in \mathbf{R}^n : x_1 = 1\}$ clearly is supporting to the unit Euclidean ball $\{x \in \mathbf{R}^n : \|x\|_2 \leq 1\}$ at the point $x = e_1 = [1; 0; \ldots; 0]$. ◇

The most important property of a supporting hyperplane is its existence:

Proposition 6.7 *[Existence of supporting hyperplanes] Let M be a closed convex set in $\mathbf{R}^n$, and let $x \in$ rbd M. Then,*

(i) there exists at least one hyperplane which is supporting to M at x;
(ii) if a hyperplane Π is supporting to M at x, then $\Pi \cap M$ is a nonempty closed convex set, $x \in \Pi \cap M$, and $\dim(\Pi \cap M) < \dim(M)$.

Proof To see (i) consider any $x \in \operatorname{rbd} M$. Then, $x \notin \operatorname{rint} M$, and therefore the point $\{x\}$ and $\operatorname{rint} M$ can be separated by the Separation Theorem. The associated separating hyperplane is exactly the desired hyperplane supporting to M at x.

To prove (ii), note that if $\Pi = \{y \in \mathbf{R}^n : a^\top y = a^\top x\}$ is supporting to M at $x \in \operatorname{rbd} M$ then the set $M' := M \cap \Pi$ is nonempty (as it contains x) and is closed and convex since both Π and M are. Moreover, the linear form $a^\top y$ is constant on M' and therefore (why?) on $\operatorname{Aff}(M')$. At the same time, this form is non-constant on M by the definition of a supporting plane. Thus, $\operatorname{Aff}(M')$ is a proper (less than the entire $\operatorname{Aff}(M)$) subset of $\operatorname{Aff}(M)$, and therefore the affine dimension of $\operatorname{Aff}(M')$, which is by definition just $\dim(M')$, is less than the affine dimension of $\operatorname{Aff}(M)$, which is precisely $\dim(M)$.[2] ∎

6.4 Extreme Points and Krein–Milman Theorem

Supporting hyperplanes are useful in proving the existence of *extreme points* of convex sets. Geometrically, an extreme point of a convex set is a point in the set which cannot be written as a convex combination of other points from the set. The importance of this notion originates from the following fact, which we will soon prove: any "good enough" (in fact, just nonempty compact) convex set M is just the convex hull of its extreme points, and the set of extreme points of such a set M is the *smallest* set whose convex hull is equal to M. That is, every extreme point of a nonempty compact convex set M is essential.

6.4.1 Extreme Points: Definition

The exact definition of an extreme point is as follows:

Definition 6.8 *[Extreme points] Let M be a nonempty convex set in $\mathbf{R}^n$. A point $x \in M$ is called an* extreme point *of M if there is no nontrivial (of positive length) segment $[u, v] \in M$ for which x is an interior point. That is, x is an* extreme point *of M if the relation*

$$x = \lambda u + (1 - \lambda) v$$

with $\lambda \in (0, 1)$ and $u, v \in M$ holds if and only if

$$u = v = x.$$

The set of all extreme points of M is denoted by $\operatorname{Ext}(M)$.

In the case of polyhedral sets, extreme points are also referred to as *vertices*.

Example 6.3

- The extreme points of a segment $[x, y] \in \mathbf{R}^n$ are exactly its endpoints $\{x, y\}$.
- The extreme points of a triangle are its vertices.

---

[2] For the dimension of a subset in $\mathbf{R}^n$, see Definition 2.1 and/or section A.4.3. We have used the following immediate observation: *If $M \subseteq M'$ are two affine planes, then $\dim M \leq \dim M'$, with equality implying that $M = M'$.* Readers are encouraged to prove this fact for themselves.

- The extreme points of a (closed) circle on the two-dimensional plane are the points of the circumference.
- The convex set $M := \{x \in \mathbf{R}_+^2 : x_1 > 0, x_2 > 0\}$ does not have any extreme points.
- The only extreme point of the convex set $M := \{[0;0]\} \cup \{x \in \mathbf{R}_+^2 : x_1 > 0, x_2 > 0\}$ is the point $[0;0]$.
- The closed convex set $\{x \in \mathbf{R}^2 : x_1 = 0\}$ does not have any extreme points. $\diamond$

An equivalent definition of an extreme point is as follows:

Fact 6.9 *Let M be a nonempty convex set and let $x \in M$. Then x is an extreme point of M if and only if any (and then all) of the following hold:*

(i) *the only vector h such that $x \pm h \in M$ is the zero vector;*
(ii) *in every representation $x = \sum_{i=1}^m \lambda_i x^i$ of x as a convex combination, with positive coefficients, of points $x^i \in M$, $i \leq m$, one has $x^1 = \cdots = x^m = x$;*
(iii) *the set $M \setminus \{x\}$ is convex.*

Fact 6.9(iii) also admits the following immediate corollary.

Fact 6.10 *All extreme points of the convex hull $\mathrm{Conv}(Q)$ of a set Q belong to Q:*

$$\mathrm{Ext}(\mathrm{Conv}(Q)) \subseteq Q.$$

6.4.2 Krein–Milman Theorem

There are convex sets that do not necessarily possess extreme points; as an example you may take the open unit ball in $\mathbf{R}^n$. This example is not particularly interesting as the set in question is not closed, and when we replace it with its closure the resulting set is the closed unit ball, which has plenty of extreme points, i.e., all points of the boundary. There are, however, *closed* convex sets which do not possess extreme points. Consider for example, a line, or an affine subspace of larger dimension, as the convex set. Indeed, a nonempty *closed* convex set will have no extreme points *only when* it contains a line.

We will next prove that any nonempty closed convex set M that does not contain lines possesses extreme points. Furthermore, if M is a nonempty convex compact set, it possesses a quite representative set of extreme points, i.e., their convex hull is the entire M.

Theorem 6.11 *Let M be a nonempty closed convex set in $\mathbf{R}^n$. Then,*

(i) *the set of extreme points of M, i.e., $\mathrm{Ext}(M)$, is nonempty if and only if M does not contain lines;*
(ii) *if M is bounded then $M = \mathrm{Conv}(\mathrm{Ext}(M))$, i.e., every point of M is a convex combination of the points of $\mathrm{Ext}(M)$.*

Remark 6.12 Part (ii) of this theorem is the finite-dimensional version of the famous *Krein–Milman Theorem* (1940). In fact, Hermann Minkowski (1911) established part (ii) for the case $n = 3$, and Ernst Steinitz (1916) showed it for any (finite) n. $\diamond$

We will use a number of lemmas in the proof of Theorem 6.11. The first states that in the case of a closed convex set, we can add any "recessive" direction to any point in the set and still remain in the set.

6.4 Extreme Points and Krein–Milman Theorem

Lemma 6.13 *Let $M \in \mathbf{R}^n$ be a nonempty closed convex set. Then, whenever M contains a ray*

$$\{\bar{x} + th : t \geq 0\}$$

starting at $\bar{x} \in M$ with the direction $h \in \mathbf{R}^n$, M also contains all parallel rays starting at the points of M, i.e., for all $x \in M$

$$\{x + th : t \geq 0\} \subseteq M.$$

As a consequence, if M contains a certain line, then it contains also all parallel lines passing through the points of M.

Proof Suppose $\bar{x} + th \in M$ for all $t \geq 0$. Consider any point $x \in M$. Since M is convex, for any fixed $\tau \geq 0$ we have

$$\epsilon \left(\bar{x} + \frac{\tau}{\epsilon} h \right) + (1 - \epsilon) x \in M, \qquad \forall \epsilon \in (0, 1).$$

By taking the limit as $\epsilon \to +0$ and noting that M is closed, we deduce that $x + \tau h \in M$ for every $\tau \geq 0$. ∎

Note that Lemma 6.13 admits a corollary as follows:

Lemma 6.14 *Let $M \in \mathbf{R}^n$ be a nonempty convex set, not necessarily closed. Suppose cl M contains a ray*

$$\{\bar{x} + th : t \geq 0\}$$

starting at $\bar{x} \in \text{cl } M$ with the direction $h \in \mathbf{R}^n$, and let $\widehat{x} \in \text{rint } M$. Then rint M contains the ray

$$\{\widehat{x} + th : t \geq 0\}.$$

In particular, cl M contains a ray (a straight line) if and only if M contains a ray (respectively, a straight line) with the same direction.

Proof With $\bar{x}$, h, and $\widehat{x}$ as above, for every $t \geq 0$ the point $x^t := \widehat{x} + 2th$ belongs to cl M by Lemma 6.13. Taking into account that $\widehat{x} \in \text{rint } M$ and invoking Lemma 1.30, we conclude that $\widehat{x} + th = \frac{1}{2}[\widehat{x} + x^t] \in \text{rint } M$. Thus, $\widehat{x} + th \in \text{rint } M$ for all $t \geq 0$. ∎

Our last ingredient in the proof of Theorem 6.11 is a lemma stating a nice transitive property of extreme points: that is, the extreme points of subsets of nonempty closed convex sets obtained from the intersection with a supporting hyperplane of the set are also extreme for the original set.

Lemma 6.15 *Let $M \subset \mathbf{R}^n$ be a nonempty closed convex set. Then, for any $\bar{x} \in \text{rbd } M$ and any hyperplane Π that is supporting to M at $\bar{x}$, we have that the set $\Pi \cap M$ is nonempty, closed, and convex, and $\text{Ext}(\Pi \cap M) \subseteq \text{Ext}(M)$.*

Proof The first statement, i.e., that $\Pi \cap M$ is nonempty, closed, and convex, follows from Proposition 6.7(ii). Moreover, by Proposition 6.7(ii) we have $\bar{x} \in \Pi \cap M$.

Next, let $a \in \mathbf{R}^n$ be the linear form associated with Π, i.e.,

$$\Pi = \left\{ y \in \mathbf{R}^n : a^\top y = a^\top \bar{x} \right\},$$

so that
$$\inf_{x \in M} a^\top x < \sup_{x \in M} a^\top x = a^\top \bar{x} \tag{6.7}$$

(see Proposition 6.7). Consider any extreme point y of $\Pi \cap M$. Assume for contradiction that $y \notin \mathrm{Ext}(M)$. Then, there exists two distinct points $u, v \in M$ and $\lambda \in (0, 1)$ such that
$$y = \lambda u + (1 - \lambda) v.$$

As $y \in \Pi \cap M$ we have $a^\top y = a^\top \bar{x}$ and also, as $u, v \in M$, from (6.7) we deduce that
$$a^\top y = a^\top \bar{x} \geq \max\left\{a^\top u, a^\top v\right\}.$$

On the other hand, from the relation $y = \lambda u + (1 - \lambda) v$ we have
$$a^\top y = \lambda a^\top u + (1 - \lambda) a^\top v.$$

Combining these last two observations and taking into account that $\lambda \in (0, 1)$, we conclude that
$$a^\top y = a^\top u = a^\top v.$$

Then, by the definition of Π, these equalities imply that $u, v \in \Pi$. As $u, v \in M$ as well, this contradicts the fact that $y \in \mathrm{Ext}(\Pi \cap M)$, as we have written $y = \lambda u + (1 - \lambda) v$ using distinct points $u, v \in \Pi \cap M$ and some $\lambda \in (0, 1)$. ∎

We are now ready to prove Theorem 6.11.

Proof of Theorem 6.11 Let us start with (i). The "only if" part of (i) follows from Lemma 6.13. Indeed, for the "only if" part we need to prove that if M possesses extreme points then M does not contain lines. That is, we need to prove that if M contains lines then it has no extreme points. But this is indeed immediate: if M contains a line then, by Lemma 6.13, there is a line in M passing through every given point of M, so that no point can be extreme.

Now let us prove the "if" part of (i). Thus, from now on we assume that M does not contain lines, and our goal is to prove that then M possesses extreme points. Equipped with Lemma 6.15 and Proposition 6.7, we will prove this by induction on $\dim(M)$.

There is nothing to do if $\dim(M) = 0$, i.e., if M is a single point – then, of course, $M = \mathrm{Ext}(M)$. Now, for the induction hypothesis, for some integer $k \geq 0$, we assume that all nonempty closed convex sets T that do not contain lines and have $\dim(T) = k$ satisfy $\mathrm{Ext}(T) \neq \varnothing$. To complete the induction, we will show that this statement is valid for such sets of dimension $k+1$ as well. Let M be a nonempty closed convex set that does not contain lines and has $\dim(M) = k + 1$. Since M does not contain lines and $\dim(M) > 0$, we have $M \neq \mathrm{Aff}(M)$. We claim that M possesses a relative boundary point $\bar{x}$. To see this, note that there exists $z \in \mathrm{Aff}(M) \setminus M$, and thus for any fixed $x \in M$ the point
$$x_\lambda := x + \lambda(z - x)$$

does not belong to M for some $\lambda > 0$ (and then, by the convexity of M, for all larger values of λ), while $x_0 = x$ belongs to M. The set of those $\lambda \geq 0$ for which $x_\lambda \in M$ is therefore nonempty and bounded from above; this set is clearly closed (since M is closed). Thus, there exists the largest $\lambda = \lambda^*$ for which $x_\lambda \in M$. We claim that $x_{\lambda^*} \in \text{rbd } M$. Indeed, by construction, $x_{\lambda^*} \in M$. If x_{λ^*} were to be in rint M then all the points x_λ with λ values greater than λ^* yet close to λ^* would also belong to M, which would contradict the definition of λ^*.

Thus, there exists $\bar{x} \in \text{rbd } M$. Then, by Proposition 6.7(i), there exists a hyperplane $\Pi = \{x \in \mathbf{R}^n : a^\top x = a^\top \bar{x}\}$ which is supporting to M at $\bar{x}$:

$$\inf_{x \in M} a^\top x < \max_{x \in M} a^\top x = a^\top \bar{x}.$$

Moreover, by Proposition 6.7(ii), the set $T := \Pi \cap M$ is nonempty, closed, and convex and it satisfies $\dim(T) < \dim(M)$, i.e., $\dim(T) \leq k$. As M does not contain lines, $T \subset M$ clearly does not contain lines either. Then, by the inductive hypothesis, T possesses extreme points, i.e., $\text{Ext}(T) \neq \emptyset$. Moreover, by Lemma 6.15, $\text{Ext}(M) \supseteq \text{Ext}(\Pi \cap M) = \text{Ext}(T) \neq \emptyset$. This completes the inductive step, and hence (i) is proved.

Now let us prove (ii). Let M be nonempty, closed, convex, and bounded. We need to prove that

$$M = \text{Conv}(\text{Ext}(M)).$$

As M is convex, we immediately observe that $M \supseteq \text{Conv}(\text{Ext}(M))$. Thus, all we need is to prove that every $x \in M$ is a convex combination of points from $\text{Ext}(M)$. Here, we again use induction on $\dim(M)$. The case of $\dim(M) = 0$, i.e., when M is a single point, is trivial. Assume that the statement holds for all k-dimensional closed, convex, and bounded sets. Let M be a closed convex bounded set with $\dim(M) = k+1$. Consider any $x \in M$. To represent x as a convex combination of points from $\text{Ext}(M)$, let us pass through x an arbitrary line $\ell = \{x + \lambda h : \lambda \in \mathbf{R}\}$ (where $h \neq 0$) in the affine span $\text{Aff}(M)$ of M. Moving along this line from x in each of the two possible directions, we eventually leave M (since M is bounded). Then, there exist nonnegative λ^+ and λ^- such that the points

$$\bar{x}_+ := x + \lambda^+ h, \qquad \bar{x}_- := x - \lambda^- h$$

both belong to rbd M. We claim that $\bar{x}_\pm$ admit convex-combination representation using points from $\text{Ext}(M)$ (this will complete the proof, since x is clearly a convex combination of the two points $\bar{x}_\pm$). Indeed, by Proposition 6.7(i) there exists a hyperplane Π supporting to M at $\bar{x}_+$, and by Proposition 6.7(ii) the set $\Pi \cap M$ is nonempty, closed, and convex with $\dim(\Pi \cap M) < \dim(M) = k+1$. Moreover, as M is bounded, $\Pi \cap M$ is bounded as well. Then, by the inductive hypothesis, $\bar{x}_+ \in \text{Conv}(\text{Ext}(\Pi \cap M))$. Moreover, since by Lemma 6.15 we have $\text{Ext}(\Pi \cap M) \subseteq \text{Ext}(M)$, we conclude that $\bar{x}_+ \in \text{Conv}(\text{Ext}(M))$. Analogous reasoning is valid for $\bar{x}_-$ as well. ∎

6.5 Recessive Directions and Recessive Cone

Lemma 6.13 states that if M is a nonempty closed convex set then the set of all directions h such that $x + th \in M$ for *some* x and all $t \geq 0$ is exactly the same as the set of all directions h such that $x + th \in M$ for *all* $x \in M$ and all $t \geq 0$. Directions of this type play an important role in the theory of convex sets, and consequently they have a name – they are called *recessive directions* of M.

Definition 6.16 *[Recessive directions and recessive cone]* Given a nonempty closed convex set $M \subseteq \mathbf{R}^n$, a direction $h \in \mathbf{R}^n$ is called a *recessive direction of* M if we have $x + th \in M$ for any $x \in M$ and any $t \geq 0$. The set of all recessive directions is called the *recessive cone of* M [notation: $\text{Rec}(M)$].

Remark 6.17 Given a closed convex set M, we immediately deduce that $\text{Rec}(M)$ indeed is a closed cone (prove it!) and that

$$M + \text{Rec}(M) = M. \tag{6.8}$$

$\diamond$

Let us see some examples of recessive cones of sets.

Example 6.4

- The recessive cone of $\mathbf{R}^n_+$ is itself. In fact, the recessive cone of any closed cone is itself.
- Consider the set $M := \{x \in \mathbf{R}^n : \sum_{i=1}^n x_i = 1\}$; then $\text{Rec}(M) = \{h \in \mathbf{R}^n : \sum_{i=1}^n h_i = 0\}$.
- Consider the set $M := \{x \in \mathbf{R}^n_+ : \sum_{i=1}^n x_i = 1\}$; then $\text{Rec}(M) = \{0\}$.
- Consider the set $M := \{x \in \mathbf{R}^n_+ : \sum_{i=1}^n x_i \geq 1\}$; then $\text{Rec}(M) = \mathbf{R}^n_+$.
- Consider the set $M := \{x \in \mathbf{R}^2_+ : x_1 x_2 \geq 1\}$; then $\text{Rec}(M) = \mathbf{R}^2_+$.
- Consider the set $M := \{x \in \mathbf{R}^n : x_n - a_n \geq \|(x_1, \ldots, x_{n-1}) - (a_1, \ldots, a_{n-1})\|_2\}$, where $a = (a_1, \ldots, a_n)$ is a given point. Then, $\text{Rec}(M) = \mathbf{L}^n$.

$\diamond$

Fact 6.18 *Let M be a nonempty closed convex set in $\mathbf{R}^n$. Then,*

(i) $\text{Rec}(M) \neq \{0\}$ *if and only if M is unbounded.*
(ii) *If M is unbounded, then all nonzero recessive directions of M are positive multiples of recessive directions of unit Euclidean length, and the latter are* asymptotic directions *of M, i.e., a unit vector $h \in \mathbf{R}^n$ is a recessive direction of M if and only if there exists a sequence $\{x^i \in M\}_{i \geq 1}$ such that $\|x^i\|_2 \to \infty$ as $i \to \infty$ and $h = \lim_{i \to \infty} x^i / \|x^i\|_2$.*
(iii) *M does not contain lines if and only if the cone $\text{Rec}(M)$ does not contain lines.*

Here is how we can "visualize" (or compute) the recessive cone of a nonempty closed convex set:

Fact 6.19 *Let $M \subseteq \mathbf{R}^n$ be a nonempty closed convex set. Recall that its closed conic transform is given by*

$$\overline{\text{ConeT}}(M) = \text{cl}\left\{[x; t] \in \mathbf{R}^n \times \mathbf{R} : t > 0, \ x/t \in M\right\}$$

(see section 1.5). Then

$$\text{Rec}(M) = \left\{h \in \mathbf{R}^n : [h; 0] \in \overline{\text{ConeT}}(M)\right\}.$$

Finally, the recessive cones of nonempty polyhedral sets in fact admit a much simpler characterization.

Fact 6.20 *For any nonempty polyhedral set $M = \{x \in \mathbf{R}^n : Ax \leq b\}$, its recessive cone is given by*

$$\text{Rec}(M) = \left\{h \in \mathbf{R}^n : Ah \leq 0\right\},$$

i.e., $\text{Rec}(M)$ is given by a homogeneous version of the linear constraints specifying M.

6.5 Recessive Directions and Recessive Cone

We have seen in Theorem 6.11 that if M is a nonempty convex compact set, it possesses a quite representative set of extreme points, i.e., their convex hull is the entire M. We close this section by extending this result as follows.

Theorem 6.21 *Let $M \subset \mathbf{R}^n$ be a nonempty closed convex set.*

(i) If M does not contain lines then the set $\mathrm{Ext}(M)$ of extreme points of M is nonempty, and

$$M = \mathrm{Conv}(\mathrm{Ext}(M)) + \mathrm{Rec}(M). \tag{6.9}$$

(ii) In every representation, if there are any, of M as $M = V + K$ with a nonempty bounded set V and a closed cone K, the cone K is $\mathrm{Rec}(M)$ and V contains $\mathrm{Ext}(M)$.

Proof (i): By Theorem 6.11(i) we already know that any nonempty closed convex set that does not contain lines must possess extreme points. We will prove the rest of part (i) by induction on $\dim(M)$. There is nothing to prove when $\dim(M) = 0$, that is, M is a singleton. So, suppose that the claim holds true for all sets of dimension k. Let M be any nonempty closed convex set that does not contain lines and has $\dim(M) = k + 1$. To complete the induction step, we will show that M satisfies the relation (6.9). Consider $x \in M$ and let e be a nonzero direction parallel to $\mathrm{Aff}(M)$ (such a direction exists, since $\dim(M) = k + 1 \geq 1$). Recalling that M does not contain lines and replacing, if necessary, e with $-e$, we can assume that $-e$ is not a recessive direction of M. As in the proof of Theorem 6.11, x admits a representation $x = x^- + t_- e$ with $t_- \geq 0$ and $x^- \in \mathrm{rbd}(M)$. Define M_- to be the intersection of M with the plane Π_- supporting to M at x^-. Then M_- is a nonempty closed convex subset of M and $\dim(M_-) \leq k$. Also, M_- does not contain lines, as $M_- \subset M$ and M does not contain lines. Thus, by the inductive hypothesis, x^- is the sum of a point from the nonempty set $\mathrm{Conv}(\mathrm{Ext}(M_-))$ and a recessive direction h_- of M_-. As in the proof of Theorem 6.11, $\mathrm{Ext}(M_-) \subseteq \mathrm{Ext}(M)$, and of course $h_- \in \mathrm{Rec}(M)$ since $\mathrm{Rec}(M_-) \subseteq \mathrm{Rec}(M)$ (why?). Thus, $x = v_- + h_- + t_- e$ with $v_- \in \mathrm{Conv}(\mathrm{Ext}(M))$ and $h_- \in \mathrm{Rec}(M)$.

Now, there are two possibilities: $e \in \mathrm{Rec}(M)$ and $e \notin \mathrm{Rec}(M)$. In the first case, $x = v_- + h$ with $h = h_- + t_- e \in \mathrm{Rec}(M)$ (recall that $h_- \in \mathrm{Rec}(M)$ and in this case we also have $e \in \mathrm{Rec}(M)$), that is, $x \in \mathrm{Conv}(\mathrm{Ext}(M)) + \mathrm{Rec}(M)$. In the second case, we can apply the above construction to the vector $-e$ in the role of e, ending up with a representation of x of the form $x = v_+ + h_+ - t_+ e$ where $v_+ \in \mathrm{Conv}(\mathrm{Ext}(M))$, $h_+ \in \mathrm{Rec}(M)$, and $t_+ \geq 0$. Taking an appropriate convex combination of the resulting pair of representations of x, we can cancel the terms with e and arrive at $x = \lambda v_- + (1-\lambda)v_+ + \lambda h_- + (1-\lambda)h_+$, resulting in $x \in \mathrm{Conv}(\mathrm{Ext}(M)) + \mathrm{Rec}(M)$. This reasoning holds true for every $x \in M$, hence we deduce that $M \subseteq \mathrm{Conv}(\mathrm{Ext}(M)) + \mathrm{Rec}(M)$. The opposite inclusion is given by (6.8) since $\mathrm{Conv}(\mathrm{Ext}(M)) \subseteq M$. This then completes the proof of the inductive hypothesis, and thus part (i) is proved.

(ii): Now assume that M, in addition to being nonempty, closed, and convex, is represented as $M = V + K$, where K is a closed cone and V is a nonempty bounded set, and let us prove that $K = \mathrm{Rec}(M)$ and $V \supseteq \mathrm{Ext}(M)$. Indeed, every vector from K is clearly a recessive direction of $V + K$, so that $K \subseteq \mathrm{Rec}(M)$. To prove the opposite inclusion, $K \supseteq \mathrm{Rec}(M)$, consider any $h \in \mathrm{Rec}(M)$, and let us prove that $h \in K$. Fix any point $v \in M$. The vectors $v + ih$, $i = 1, 2, \ldots$, belong to M and therefore $v + ih = v^i + h^i$ for some $v^i \in V$

and $h^i \in K$ since $M = V + K$. It follows that $h = i^{-1}[v^i - v] + i^{-1}h^i$ for $i = 1, 2, \ldots$ Thus, $h = \lim_{i \to \infty} i^{-1} h^i$ (recall that V is bounded). As $h^i \in K$ and K is a cone, $i^{-1} h^i \in K$ and so h is the limit of a sequence of points in K. Since K is closed, we deduce that $h \in K$, as claimed. Thus, $K = \text{Rec}(M)$. It remains to prove that $\text{Ext}(M) \subseteq V$. This is immediate: consider any $w \in \text{Ext}(M)$; then as $M = V + K = V + \text{Rec}(M)$ and $w \in M$, we have $w = v + e$ with some $v \in V \subseteq M$ and $e \in \text{Rec}(M)$, implying that $w - e = v \in M$. Besides this, $w + e \in M$, as $w \in M$ and $e \in \text{Rec}(M)$. Thus, $w \pm e \in M$. Since w is an extreme point of M, we conclude that $e = 0$, that is, $w = v \in V$. ∎

Finally, let us consider what happens to the recessive directions after the projection operation.

Proposition 6.22 *Let $M^+ \in \mathbf{R}_x^n \times \mathbf{R}_u^k$ be a nonempty closed convex set such that its projection*

$$M = \left\{ x \in \mathbf{R}^n : \exists u \colon [x;u] \in M^+ \right\}$$

is closed. Then,

$$[h_x; h_u] \in \text{Rec}(M^+) \implies h_x \in \text{Rec}(M).$$

Proof Consider any recessive direction $[h_x; h_u] \in \text{Rec}(M^+)$. Then, for any $[\bar{x}; \bar{u}] \in M^+$, the ray $\{[\bar{x}; \bar{u}] + t[h_x; h_u] : t \geq 0\}$ is contained in M^+. The projection of this ray onto the x-plane is the ray $\{\bar{x} + t h_x : t \geq 0\}$, which is contained in M. Thus, $h_x \in \text{Rec}(M)$. ∎

While Proposition 6.22 states that $[h_x; h_u] \in \text{Rec}(M^+) \implies h_x \in \text{Rec}(M)$, in general $\text{Rec}(M)$ can be much larger than the projection of $\text{Rec}(M^+)$ onto the x-plane. Our next example illustrates this.

Example 6.5 Consider the sets $M^+ = \{[x;u] \in \mathbf{R}^2 : u \geq x^2\}$ and $M = \{x \in \mathbf{R} : \exists u \in \mathbf{R} : [x;u] \in M^+\}$. Then M is the entire x-axis and $\text{Rec}(M) = M$ is the entire x-axis. On the other hand, $\text{Rec}(M^+) = \{[0; h_u] : h_u \geq 0\}$ and the projection of $\text{Rec}(M^+)$ onto the x-axis is just the origin. ◇

In fact, the pathology highlighted in Example 6.5 can be eliminated when we have that the set of extreme points of the convex representation M^+ of a convex set M is bounded and the projection of $\text{Rec}(M^+)$ is closed.

Proposition 6.23 *Let $M^+ \subset \mathbf{R}_x^n \times \mathbf{R}_u^k$ be a nonempty closed convex set such that $M^+ = V + \text{Rec}(M^+)$ for some bounded and closed set V. Let M be the projection of M^+ onto the x-plane, i.e.,*

$$M = \left\{ x \in \mathbf{R}_x^n : \exists u \in \mathbf{R}_u^k \colon [x;u] \in M^+ \right\}.$$

Assume that the cone

$$K = \left\{ h_x \in \mathbf{R}_x^n : \exists h_u \in \mathbf{R}_u^k \colon [h_x; h_u] \in \text{Rec}(M^+) \right\}$$

is closed. Then, M is closed and $K = \text{Rec}(M)$.

Proof Let M^+ satisfy the premise of the proposition. Define

$$W := \left\{ x \in \mathbf{R}_x^n : \exists u \in \mathbf{R}_u^k \colon [x;u] \in V \right\},$$

that is, W is the projection of V onto the x-space. As V is a closed and bounded (therefore compact) set, its projection W is compact as well (recall that the image of a compact set under a continuous mapping is compact). Note that M is nonempty and satisfies $M = W + K$ (why?). Then M is the sum of a compact set W and a closed set K, and thus M is closed itself (why?). Besides this, M is convex (recall that the projection of a convex set is convex). Thus, the nonempty closed convex set M satisfies $M = W + K$ with nonempty bounded W and closed cone K, implying by Theorem 6.21 that $K = \mathrm{Rec}(M)$. ∎

Recall that we have investigated the relation between the recessive directions of a closed convex set $M \in \mathbf{R}_x^n$ and its closed convex representation $M^+ \in \mathbf{R}_x^n \times \mathbf{R}_u^k$ in Proposition 6.22. We have also observed that, while $[h_x; h_u] \in \mathrm{Rec}(M^+)$ implies $h_x \in \mathrm{Rec}(M)$, the recessive direction of M stemming from those of M^+ can form a small part of $\mathrm{Rec}(M)$, as seen in Example 6.5.

A surprising (and not completely trivial) fact is that *for polyhedral sets M^+, the projection of $\mathrm{Rec}(M^+)$ onto the x-plane is $\mathrm{Rec}(M)$.*

Proposition 6.24 *Let $M \in \mathbf{R}_x^n$ be a nonempty set admitting a polyhedral representation $M^+ \in \mathbf{R}_x^n \times \mathbf{R}_u^k$, i.e.,*

$$M^+ := \left\{ [x; u] \in \mathbf{R}_x^n \times \mathbf{R}_u^k : Ax + Bu \leq c \right\} \quad \text{and}$$

$$M := \left\{ x \in \mathbf{R}_x^n : \exists u \in \mathbf{R}_u^k : [x; u] \in M^+ \right\}.$$

Then

$$\begin{aligned} \mathrm{Rec}(M) &= \left\{ h_x : \exists h_u : [h_x; h_u] \in \mathrm{Rec}(M^+) \right\} \\ &= \{ h_x : \exists h_u : Ah_x + Bh_u \leq 0 \}. \end{aligned} \quad (6.10)$$

That is, a polyhedral representation of M naturally induces a polyhedral representation of $\mathrm{Rec}(M)$.

Proposition 6.24 is an immediate consequence of Proposition 6.23. To derive Proposition 6.24 from Proposition 6.23, it suffices to note that a nonempty polyhedral set is the sum of the convex hull of a finite set and a polyhedral cone, see section 7.3.

6.6 Dual Cone

We start with the definition of the dual of a cone.

Definition 6.25 *[Dual cone] Let $M \subseteq \mathbf{R}^n$ be a cone. The set of all vectors which have nonnegative inner products with all vectors from M, i.e., the set*

$$M_* := \left\{ a \in \mathbf{R}^n : a^\top x \geq 0, \ \forall x \in M \right\}, \quad (6.11)$$

is called the cone dual to M.

From its definition, it is clear that the dual cone M_* of *any* cone M is a closed cone.

Example 6.6 The cone dual to the nonnegative orthant $\mathbf{R}_+^n$ is composed of all n-dimensional vectors y making nonnegative inner products with all entrywise nonnegative

n-dimensional vectors x. As can immediately be seen, the vectors y with this property are exactly entrywise nonnegative vectors: $[\mathbf{R}^n_+]_* = \mathbf{R}^n_+$. ◇

Note that in the preceding example, $\mathbf{R}^n_+$ is given by finitely many homogeneous linear inequalities:
$$\mathbf{R}^n_+ = \left\{ x \in \mathbf{R}^n : e_i^\top x \geq 0, \, i = 1, \ldots, n \right\},$$
where e_i are the standard basic orths; and we observe that the dual cone is the conic hull of these basic orths. This is indeed a special case of the following general fact:

Proposition 6.26 *For any $F \subseteq \mathbf{R}^n$, the set*
$$M := \left\{ x \in \mathbf{R}^n : f^\top x \geq 0, \, \forall f \in F \right\}$$
is a closed cone, and its dual cone is
$$M_* = \operatorname{cl} \operatorname{Cone}(F),$$
where $\operatorname{Cone}(F)$, *as always, is the conic hull of F, see Definition 1.19. In addition, M remains intact when F is extended to its closed conic hull:*
$$M := \left\{ x \in \mathbf{R}^n : f^\top x \geq 0, \, \forall f \in F \right\} = \left\{ x \in \mathbf{R}^n : f^\top x \geq 0, \, \forall f \in \operatorname{cl} \operatorname{Cone}(F) \right\}.$$

Proof The inclusion $\operatorname{Cone}(F) \subseteq M_*$ is evident, and, since M_* is closed, we have also $\operatorname{cl} \operatorname{Cone}(F) \subseteq M_*$. Let us define $\overline{F} := \operatorname{cl} \operatorname{Cone}(F)$, so now we need to prove the inclusion $\overline{F} \supseteq M_*$. Consider $z \in M_*$, and assume for contradiction that $z \notin \overline{F}$. Note that $\overline{F}$ is convex, nonempty, and closed, so that by the Separation Theorem, part (ii), there exists g such that
$$g^\top z < \inf_{f \in \overline{F}} g^\top f.$$
Because $\overline{F}$ is a closed cone (and so $0 \in \overline{F}$), the right-hand side infimum, being finite, must be 0. Then $g^\top f \geq 0$ for all $f \in \overline{F}$ and $g^\top z < 0$. Since $f^\top g \geq 0$ for all $f \in \overline{F}$ and also $\overline{F} \supseteq F$, we deduce $f^\top g \geq 0$ for all $f \in F$, that is, $g \in M$ by the definition of M. But then the inclusion $g \in M$ together with $z \in M_*$ contradicts the relation $z^\top g < 0$. Finally, we clearly have $f^\top x \geq 0$ for all $x \in F$ if and only if $f^\top x \geq 0$ for all $x \in \operatorname{cl} \operatorname{Cone}(F)$. ∎

Remark 6.27 Note that, in contrast with Proposition 6.26, in the concluding expression of the chain
$$[\mathbf{R}^n_+]_* = \left\{ x \in \mathbf{R}^n : e_i^\top x \geq 0, \, i = 1, \ldots, n \right\}_* = \operatorname{Cone}(\{e_i : i = 1, \ldots, n\})$$
we did *not* need to take the closure. This is so because the conic hull of a *finite* set F is polyhedrally representable and is therefore a polyhedral cone (by Theorem 3.2), and as such it is automatically closed.

This fact (i.e., there is no need to take the closure in Proposition 6.26) holds true for the dual of *any* polyhedral cone: consider the set $\{x \in \mathbf{R}^n : a_i^\top x \geq 0, \, i = 1, \ldots, I\}$ given by finitely many homogeneous nonstrict linear inequalities. This set is clearly a polyhedral cone, and its dual is the conic hull of the a_i, i.e., $\operatorname{Cone}(\{a_i : i = 1, \ldots, I\}) = \left\{ \sum_{i=1}^I \lambda_i a_i : \lambda \geq 0 \right\}$. Moreover, clearly this dual cone is also polyhedrally representable as

6.6 Dual Cone

$$\text{Cone}\,\{a_1,\ldots,a_I\} = \left\{ x \in \mathbf{R}^n \,:\, \exists \lambda \geq 0\colon\, x = \sum_{i=1}^{I} \lambda_i a_i \right\},$$

and thus $\text{Cone}\,\{a_1,\ldots,a_I\}$ is polyhedral as well. ◇

In the case of cones of the form $\{x \in \mathbf{R}^n \,:\, f^\top x \geq 0,\, \forall f \in F\}$ stemming from infinite sets F (in fact, *every* closed cone in $\mathbf{R}^n$ can be represented in this way using a properly selected countable set $F = \{f_i \,:\, i = 1, 2, \ldots\}$) (why?), the closure operation in the computation of the dual cone in general cannot be omitted. This is so even when the set F itself is closed, convex, and bounded.

We illustrate this in the next example.

Example 6.7 Consider the cone given by $K := \{x \in \mathbf{R}^2 \,:\, f^\top x \geq 0,\, \forall f \in F\}$, where $F = \{f = [u; v] \in \mathbf{R}^2_+ \,:\, v \geq u^2,\, u \leq 1,\, v \leq 1\}$. Note that F is a compact convex set contained in $\mathbf{R}^2_+$. Moreover, every vector $[u; v] \in \mathbf{R}^2$ with positive entries is a positive multiple of a vector from F (draw a picture!). Thus, the set of vectors K that have nonnegative inner products with all vectors from F is exactly the same as the set of vectors that have nonnegative inner products with all vectors from $\mathbf{R}^2_+$. Hence, we arrive at $K = \{x \in \mathbf{R}^2 \,:\, f^\top x \geq 0,\, \forall f \in F\} = \mathbf{R}^2_+$, so $K_* = \mathbf{R}^2_+$ as well. Now observe that $K_* = \mathbf{R}^2_+$, as it should be by Proposition 6.26, is the *closure* of $\text{Cone}(F)$; nevertheless $K_* = \text{cl}\,\text{Cone}(F)$ is larger than $\text{Cone}(F)$ as $\text{Cone}(F)$ is not closed! Note that $\text{Cone}(F)$ is precisely the set obtained from $\mathbf{R}^2_+$ by eliminating all points $[u; 0]$ with $u > 0$. ◇

Fact 6.28 *Let M be a closed cone in $\mathbf{R}^n$, and let M_* be the cone dual to M. Then*

(i) *Duality does not distinguish between a cone and its closure: whenever $M = \text{cl}\,M'$ for a cone M', we have $M_* = M'_*$.*
(ii) *Duality is symmetric: the cone dual to M_* is M.*
(iii) *One has*

$$\text{int}\,M_* = \left\{ y \in \mathbf{R}^n \,:\, y^\top x > 0,\, \forall x \in M \setminus \{0\} \right\},$$

and $\text{int}\,M_*$ is nonempty if and only if M is pointed (i.e., $M \cap [-M] = \{0\}$).
Moreover, when M, in addition to being closed, is pointed and nontrivial ($M \neq \{0\}$), one has

$$\text{int}\,M_* = \left\{ y \in \mathbf{R}^n \,:\, M_y := \{x \in M \,:\, x^\top y = 1\} \text{ is nonempty and compact} \right\}. \quad (6.12)$$

(iv) *The cone dual to the direct product $M_1 \times \cdots \times M_m$ of cones M_i is the direct product of their duals: $[M_1 \times \cdots \times M_m]_* = [M_1]_* \times \cdots \times [M_m]_*$.*

Let us see some examples of dual cones.

Example 6.8 Consider the epigraph of $\|\cdot\|_\infty$ on $\mathbf{R}^n$ given by

$$K_\infty := \left\{ [x; t] \in \mathbf{R}^{n+1} \,:\, t \geq \|x\|_\infty \right\}.$$

Note that K_∞ is a polyhedral cone (why?). The cone dual to K_∞ is $K_1 = \{[x;t] \in \mathbf{R}^{n+1} : t \geq \|x\|_1\}$, which is just the epigraph of $\|\cdot\|_1$:

$$[K_\infty]_* = \left\{[g;s] \in \mathbf{R}^{n+1} : st + g^\top x \geq 0, \forall([x;t] : \|x\|_\infty \leq t)\right\}$$

$$= \left\{[g;s] \in \mathbf{R}^{n+1} : s + \min_{x:\|x\|_\infty \leq 1} g^\top x \geq 0\right\}$$

$$= \left\{[g;s] \in \mathbf{R}^{n+1} : s \geq \|g\|_1\right\}. \qquad \diamond$$

Definition 6.29 *[Self-dual cone] A cone* $\mathbf{K} \subset \mathbf{R}^n$ *is called* self-dual *if it is equal to its dual cone, i.e.,* $\mathbf{K}_* = \mathbf{K}$.

We next examine a number of very important self-dual cones.

Example 6.9 Let us compute the dual of the Lorentz cone

$$\mathbf{L}^n := \left\{[x;t] \in \mathbf{R}^{n-1} \times \mathbf{R} : t \geq \|x\|_2\right\}.$$

When $n = 1$, $\mathbf{L}^1$ is the nonnegative ray, and thus $\mathbf{L}^1 = \mathbf{R}_+$ and therefore $[\mathbf{L}^1]_* = \mathbf{R}_+ = \mathbf{L}^1$. When $n \geq 2$, we have

$$[\mathbf{L}^n]_* = \left\{[g;s] \in \mathbf{R}^{n-1} \times \mathbf{R} : g^\top x + ts \geq 0, \forall([x;t] : \|x\|_2 \leq t)\right\}$$

$$= \left\{[g;s] \in \mathbf{R}^{n-1} \times \mathbf{R} : g^\top x + s \geq 0, \forall(x : \|x\|_2 \leq 1)\right\}$$

$$= \left\{[g;s] \in \mathbf{R}^{n-1} \times \mathbf{R} : s + \min_{x:\|x\|_2 \leq 1} g^\top x \geq 0\right\}$$

$$= \left\{[g;s] \in \mathbf{R}^{n-1} \times \mathbf{R} : s \geq \|g\|_2\right\},$$

where the concluding equality is due to the Cauchy–Schwarz inequality; see Theorem B.1. Thus, $[\mathbf{L}^n]_* = \mathbf{L}^n$. $\qquad \diamond$

Example 6.10 The cone dual to the semidefinite cone $\mathbf{S}_+^n$ is itself, see Theorem D.32:

$$[\mathbf{S}_+^n]_* := \left\{y \in \mathbf{S}^n : \langle y,x \rangle := \mathrm{Tr}(xy) \geq 0, \forall x \in \mathbf{S}_+^n\right\} = \mathbf{S}_+^n. \qquad \diamond$$

On the basis of these examples, we have arrived at the following conclusion.

The cones $\mathbf{R}_+$, $\mathbf{L}^n$, *and* $\mathbf{S}_+^n$ *are self-dual.*

By Fact 6.28 the direct product of finitely many self-dual cones is self-dual, implying that *finite direct products of nonnegative orthants, Lorentz cones, and semidefinite cones are self-dual.*

6.7 ★ Dubovitski–Milutin Lemma

In this section, we deal with the following basic yet important question. Let $M^1, \ldots, M^k$ be cones (not necessarily closed) in $\mathbf{R}^n$, and M be their intersection. Of course, M also is a cone. But how can we compute M_*, i.e., the cone dual to M? To this end, we first examine the relationship between M_* and the cone $\widetilde{M}$ that is defined as the sum of the dual cones M_*^i.

Proposition 6.30 *Let $k \geq 2$, and let $M^1, \ldots, M^k$ be cones in $\mathbf{R}^n$. Define $M := \bigcap_{i=1}^{k} M^i$, and let M_* be the dual cone of M. Let M_*^i denote the dual cone of M^i, for $i = 1, \ldots, k$, and define $\widetilde{M} := M_*^1 + \cdots + M_*^k$. Then $M_* \supseteq \widetilde{M}$.*

Moreover, if all the cones $M^1, \ldots, M^k$ are closed then $M_* = \operatorname{cl} \widetilde{M}$. In particular, for closed cones $M^1, \ldots, M^k$, $M_* = \widetilde{M}$ holds if and only if $\widetilde{M}$ is closed.

Proof For any $i = 1, \ldots, k$, any $a_i \in M_*^i$, and any $x \in M$, we have $a_i^\top x \geq 0$, and hence $(a_1 + \cdots + a_k)^\top x \geq 0$. Since the latter relation is valid for all $x \in M$, we conclude that $a_1 + \cdots + a_k \in M_*$. Thus, $\widetilde{M} \subseteq M_*$.

Now assume that the cones $M^1, \ldots, M^k$ are closed, and let us define $\widehat{M} := \operatorname{cl} \widetilde{M}$ so that we need to prove $M_* = \widehat{M}$. Recall that we have already seen that $\widetilde{M} \subseteq M_*$, and as M_* is closed we deduce $\widehat{M} = \operatorname{cl} \widetilde{M} \subseteq M_*$. Thus, all we need to prove is that if $a \in M_*$ then $a \in \widehat{M}$ as well. Assume for contradiction that there exists $a \in M_* \setminus \widehat{M}$. As $\widetilde{M}$ is clearly a cone, its closure $\widehat{M}$ is a closed cone. Then, as $a \notin \widehat{M}$, by the Separation Theorem, part (ii), a can be strongly separated from $\widehat{M}$ and thus also from $\widetilde{M} \subseteq \widehat{M}$. Therefore, for some $x \neq 0$ we have

$$a^\top x < \inf_{b \in \widetilde{M}} b^\top x = \inf_{a_i \in M_*^i, i=1,\ldots,k} (a_1 + \cdots + a_k)^\top x = \sum_{i=1}^{k} \inf_{a_i \in M_*^i} a_i^\top x. \tag{6.13}$$

As $a^\top x$ is a finite number, this inequality implies that $\inf_{a_i \in M_*^i} a_i^\top x > -\infty$ holds for all $i = 1, \ldots, k$. Since M_*^i is a cone, this is possible if and only if $\inf_{a_i \in M_*^i} a_i^\top x = 0$. Then, we deduce that $x \in [M_*^i]_* = M^i$, where the last equality follows from the fact that each cone M^i is closed and the use of Fact 6.28(ii). Thus, $x \in M^i$ for all i, and $\sum_{i=1}^{k} \inf_{a_i \in M_*^i} a_i^\top x = 0$. Therefore, $x \in M = \bigcap_{i=1}^{k} M^i$, and (6.13) reduces to $a^\top x < 0$. But this then contradicts the inclusion $a \in M_*$. ∎

Remark 6.31 Note that in general $\widetilde{M}$ can be non-closed even when all the cones $M^1, \ldots, M^k$ are closed. Indeed, take $k = 2$, and let $M^1 = M_*^1$ be the second-order cone $\left\{ (x, y, z) \in \mathbf{R}^3 : z \geq \sqrt{x^2 + y^2} \right\}$, and M_*^2 be the following ray in $\mathbf{R}^3$:

$$\left\{ (x, y, z) \in \mathbf{R}^3 : x = z, \ y = 0, \ x \leq 0 \right\}.$$

Observe that the points from $\widetilde{M} \equiv M_*^1 + M_*^2$ are exactly the points of the form $(x-t, y, z-t)$ with $t \geq 0$ and $z \geq \sqrt{x^2 + y^2}$. In particular, for any $\alpha > 0$, the points $(0, 1, \sqrt{\alpha^2 + 1} - \alpha) = (\alpha - \alpha, 1, \sqrt{\alpha^2 + 1} - \alpha)$ belong to $\widetilde{M}$. As $\alpha \to \infty$, these points converge to $\xi := (0, 1, 0)$, and thus $\xi \in \operatorname{cl} \widetilde{M}$. On the other hand, we clearly cannot find x, y, z, t with $t \geq 0$ and $z \geq \sqrt{x^2 + y^2}$ such that $(x - t, y, z - t) = (0, 1, 0)$, that is, $\xi \notin \widetilde{M}$. ◇

The Dubovitski–Milutin Lemma presents a simple sufficient condition for $\widetilde{M}$ to be closed and thus to coincide with M_*:

Proposition 6.32 *[Dubovitski–Milutin Lemma in finite dimensions]* Let $k \geq 2$, and let $M^1, \ldots, M^k$ be cones such that

$$M^k \cap \text{int } M^1 \cap \text{int } M^2 \cap \cdots \cap \text{int } M^{k-1} \neq \varnothing.$$

Define $M := \bigcap_{i=1}^{k} M^i$. Let M_*^i be the cones dual to M^i. Then

(i) $\text{cl } M = \bigcap_{i=1}^{k} \text{cl } M^i$; and

(ii) the cone $\widetilde{M} := M_*^1 + \cdots + M_*^k$ is closed, and thus, by Proposition 6.30, $M_* = M_*^1 + \cdots + M_*^k$. In other words, every linear form which is nonnegative on M can be represented as a sum of k linear forms which are nonnegative on the respective cones $M^1, \ldots, M^k$.

Proof (i): This is given by Proposition 1.33(ii).

(ii): First, we claim that under the premise of the proposition, without loss of generality we can assume that $M^1, \ldots, M^k$ are closed cones. This is so because when replacing the cones $M^1, \ldots, M^k$ with their closures, we preserve the premise of the proposition, and also $\widetilde{M} = M_*^1 + \cdots + M_*^k = [\text{cl } M^1]_* + \cdots + [\text{cl } M^k]_*$ holds (recall that, by definition of the dual cone, $M_*^i = [\text{cl } M^i]_*$) as well as $M_* = [\text{cl } M]_* = \left[\bigcap_{i=1}^{k} \text{cl } M^i\right]_*$, where the last equality holds by part (i).

To prove part (ii) of the proposition all we need is to show that given closed cones $M^1, \ldots, M^k$ such that $M^K \cap \text{int } M^1 \cap \cdots \cap \text{int } M^{k-1} \neq \varnothing$ we have that the cone $\widetilde{M} := M_*^1 + \cdots + M_*^k$ is closed. To this end, we will use induction on $k \geq 2$.

Base case: Suppose $k = 2$. Consider a sequence $\{f_t + g_t\}_{t=1}^{\infty}$ with $f_t \in M_*^1$, $g_t \in M_*^2$, and $(f_t + g_t) \to h$ as $t \to \infty$. We need to prove that $h = f + g$ for some appropriate $f \in M_*^1$ and $g \in M_*^2$. To this end, it suffices to verify that for an appropriate subsequence t_j of indices there exists $f := \lim_{j \to \infty} f_{t_j}$. Indeed, if this is the case, then $g = \lim_{j \to \infty} g_{t_j}$ also exists since $f_t + g_t \to h$ as $t \to \infty$ and $f + g = h$, and also in this case we will have $f \in M_*^1$ and $g \in M_*^2$ (recall that M_*^1 and M_*^2 are closed cones). Let us verify the existence of the desired subsequence. Assume for contradiction that $\|f_t\|_2 \to \infty$ as $t \to \infty$. Passing to a subsequence, we may assume that the unit vectors $\phi_t := f_t/\|f_t\|_2$ have a limit ϕ as $t \to \infty$. Since M_*^1 is a closed cone, ϕ is a unit vector from M_*^1. Now, since $f_t + g_t \to h$ as $t \to \infty$, we have $\phi = \lim_{t \to \infty} f_t/\|f_t\|_2 = -\lim_{t \to \infty} g_t/\|f_t\|_2$ (recall that $\|f_t\|_2 \to \infty$ as $t \to \infty$, whence $h/\|f_t\|_2 \to 0$ as $t \to \infty$). Then, the vector $(-\phi) \in M_*^2$ as well (recall that M_*^2 is a closed cone). Now, consider any $\bar{x} \in M^2 \cap \text{int } M^1$ (by the premise of the proposition this set is nonempty). We have $\phi^\top \bar{x} \geq 0$ (since $\bar{x} \in M^1$ and $\phi \in M_*^1$) and $\phi^\top \bar{x} \leq 0$ (since $-\phi \in M_*^2$ and $\bar{x} \in M^2$). We conclude that $\phi^\top \bar{x} = 0$, which contradicts the facts that $0 \neq \phi$ (as $\|\phi\|_2 = 1$), $\phi \in M_*^1$, and $\bar{x} \in \text{int } M^1$ (since when $\phi \neq 0$ and $\phi^\top \bar{x} = 0$, $\phi^\top x$ cannot be nonnegative for all x from a neighhborhood of $\bar{x}$).

Inductive step: Assume that the statement is valid for $k - 1 \geq 2$ cones, and let $M^1, \ldots, M^k$ be k cones satisfying the premise of the proposition. By this premise, the cone $M_1 := M^1 \cap \cdots \cap M^{k-1}$ has a nonempty interior, and M^k intersects this interior. Applying to the pair of cones M_1, M^k the two-cone version of the lemma that has already been proved, we see that the set $[M_1]_* + M_*^k$ is closed; here $[M_1]_*$ is the cone dual to M_1. Moreover, the cones

$M^1, \ldots, M^{k-1}$ satisfy the premise of the $(k-1)$-cone version of the lemma. Then, by the inductive hypothesis, the set $M_*^1 + \cdots + M_*^{k-1}$ is closed. Thus, as $M_1 = M^1 \cap \cdots \cap M^{k-1}$, Proposition 6.30 implies that $[M_1]_* = M_*^1 + \cdots + M_*^{k-1}$, and so $M_*^1 + \cdots + M_*^k = [M_1]_* + M_*^k$. As $[M_1]_* + M_*^k$ is closed, we deduce that $M_*^1 + \cdots + M_*^k$ is closed, as desired. This concludes the induction step. ∎

Alternative proof of Proposition 6.32 Here, we present an alternative proof of Proposition 6.32, part (ii), without relying on induction.

Let us start with the following fact, which is important in its own right.

Fact 6.33 *Let $M \subseteq \mathbf{R}^n$ be a cone and M_* be its dual cone. Then, for any $x \in \operatorname{int} M$, there exists a properly selected $C_x < \infty$ such that*

$$\|f\|_2 \leq C_x f^\top x, \quad \forall f \in M_*.$$

Now, as explained at the beginning of part (ii) of the above proof of Proposition 6.32, we can assume without loss of generality that the cones $M^1, \ldots, M^k$ satisfying the premise of the proposition are closed, and all we need to prove is that the cone $M_*^1 + \cdots + M_*^k$ is closed. The latter is the same as verifying that whenever vectors $f_t^i \in M_*^i$, $i \leq k$, $t = 1, 2, \ldots$ are such that $f_t := \sum_{i=1}^k f_t^i \to h$ as $i \to \infty$, it holds that $h \in M_*^1 + \cdots + M_*^k$. Indeed, in the situation in question, selecting $\bar{x} \in M^k \cap \operatorname{int} M^1 \cap \cdots \cap \operatorname{int} M^{k-1}$ (by the premise this intersection is nonempty!) we have $\bar{x}^\top f_t^i \geq 0$ for all i, t and $\sum_{i=1}^k \bar{x}^\top f_t^i \to \bar{x}^\top h$ as $t \to \infty$, implying that for all $i \leq k$ the sequences $\{\bar{x}^\top f_t^i\}_{t=1,2,\ldots}$ are bounded. Moreover, for any $i < k$ we have $\bar{x} \in \operatorname{int} M^i$ and $f_t^i \in M_*^i$, and so Fact 6.33 guarantees that the sequence $\{f_t^i\}_{t=1,2,\ldots}$ is bounded. Thus, as the sequences $\{f_t^i\}_{t=1,2,\ldots}$ are bounded for any $i < k$ and the sequence $\sum_{i=1}^k f_t^i$ has a limit as $t \to \infty$, we conclude that the sequence $\{f_t^k\}_{t=1,2,\ldots}$ is bounded as well. Hence, all k sequences $\{f_t^i\}_{t=1,2,\ldots}$, where $1 \leq i \leq k$, are bounded, so that passing to a subsequence $t_1 < t_2 < \cdots$ we can assume that $f^i := \lim_{j \to \infty} f_{t_j}^i$ is well defined for every $i \leq k$. Since $f_t^i \in M_*^i$ and the cones M_*^i are closed, we have $f^i \in M_*^i$ for all $i \leq k$. Finally, as $h = \lim_{t \to \infty} \sum_i f_t^i = \lim_{j \to \infty} \sum_i f_{t_j}^i = \sum_i f^i$, we conclude that $h \in M_*^1 + \cdots + M_*^k$, as claimed. ∎

6.8 Extreme Rays and Conic Krein–Milman Theorem

The story about extreme points of closed convex sets has a conic analogy, with nontrivial closed *pointed* cones playing the role of nonempty closed convex sets that do not contain straight lines and *extreme rays* of these cones playing the role of extreme points.

Once again, in order to define the extreme rays of nontrivial closed cones, we need to work with cones that do not contain straight lines. In the case of cones, this notion of not containing straight lines has a special name, i.e., *pointed*.

Definition 6.34 *[Pointed cone] A cone $M \subseteq \mathbf{R}^n$ is called* pointed *if $M \cap [-M] = \{0\}$, i.e., the zero vector is the only vector x that satisfies $x \in M$ and $-x \in M$.*

Remark 6.35 Note that a cone $M \subseteq \mathbf{R}^n$ is pointed if and only if M does not contain a straight line passing through the origin. Invoking Lemma 6.13, we see that a *closed* cone is pointed if and only if it does not contain straight lines. ◇

In our discussion, we will focus on *nontrivial* cones M, i.e., $M \neq \{0\}$. For nontrivial closed cones, let us formally introduce the definition of extreme directions and extreme rays.

Definition 6.36 *[Extreme directions and extreme rays] Let $M \subset \mathbf{R}^n$ be a nontrivial closed cone. A direction $d \in \mathbf{R}^n$ is called an* extreme direction *of M if it possesses the following two properties:*

- $d \in M \setminus \{0\}$, *and*
- *in every representation of d as the sum of two vectors from M, i.e., $d = d^1 + d^2$ with $d^1, d^2 \in M$, both d^1 and d^2 are nonnegative multiples of d.*

It is clear that when d is an extreme direction of M, so are all positive multiples of d, i.e., all nonzero vectors on the ray $\mathbf{R}_+(d)$ generated by d are also extreme directions of M. A ray generated by an extreme direction of M is called an extreme ray *of M.*

The sets of all extreme directions and all extreme rays of M are denoted by $\mathrm{ExtD}(M)$ and $\mathrm{ExtR}(M)$, respectively.

Example 6.11 The simplest example of a nontrivial closed cone is the nonnegative orthant $\mathbf{R}^n_+$. From the definition of an extreme direction, the extreme directions of $\mathbf{R}^n_+$ should be the nonzero n-dimensional entrywise nonnegative vectors d such that, whenever $d = d^1 + d^2$ with $d^1 \geq 0$ and $d^2 \geq 0$, both d^1 and d^2 must be nonnegative multiples of d. Such a vector d has all entries nonnegative and at least one of them positive. In fact, the number of positive entries in d is exactly one, since if there were at least two positive entries, say, d_1 and d_2, we would have $d = [d_1; 0; \ldots; 0] + [0; d_2; d_3; \ldots; d_n]$ and both of the vectors on the right-hand side would be nonzero and not proportional to d. Thus, any extreme direction of $\mathbf{R}^n_+$ must be a positive multiple of a standard basic orth. It is immediately seen that every vector of the latter type is an extreme direction of $\mathbf{R}^n_+$. Hence, the extreme directions of $\mathbf{R}^n_+$ are positive multiples of the standard basic orths, and the extreme rays of $\mathbf{R}^n_+$ are the nonnegative parts of the coordinate axes. ◇

We next introduce the concept of a *base*, which is an important type of cross-section of a nontrivial closed pointed cone. Moreover, we will see that a base is a compact convex set and will establish a direct connection between the extreme rays of the underlying cone and the extreme points of its base.

Definition 6.37 *[Base of a cone] Let $M \subseteq \mathbf{R}^n$ be a nontrivial closed cone, and M_* be its dual cone. A set of the form*

$$B := \left\{ x \in M : f^\top x = 1 \right\} \tag{6.14}$$

is called a base *of M if this set intersects every nontrivial ray in M emanating from the origin. Equivalently, a set B of the form (6.14) is a base of M if for every $x \in M \setminus \{0\}$ there exists $t > 0$ such that $f^\top(tx) = 1$, or equivalently, if $f^\top x > 0$ whenever $x \in M \setminus \{0\}$.*

The last "equivalently" above implies that *if a nontrivial closed cone has a base, then the cone is pointed.*

Fact 6.38 *Let $M \subseteq \mathbf{R}^n$ be a nontrivial closed cone, and M_* be its dual cone. Then:*

(i) *M is pointed*
  (i.1) *if and only if M does not contain straight lines,*
  (i.2) *if and only if M_* has a nonempty interior,*
  (i.3) *if and only if M has a base.*
(ii) *The set (6.14) is a base of M*
  (ii.1) *if and only if $f^\top x > 0$ for all $x \in M \setminus \{0\}$,*
  (ii.2) *if and only if $f \in \operatorname{int} M_*$.*
As a result,
  (ii.3) *$f \in \operatorname{int} M_*$ if and only if $f^\top x > 0$ for all $x \in M \setminus \{0\}$.*
(iii) *Every base of M is nonempty, closed, and bounded. Moreover, whenever M is pointed, for any $f \in M_*$ such that the set (6.14) is nonempty (note that this set is always closed for any f), this set is bounded if and only if $f \in \operatorname{int} M_*$, in which case (6.14) is a base of M.*
(iv) *M has extreme rays if and only if M is pointed. Furthermore, when M is pointed, there is a one-to-one correspondence between the extreme rays of M and the extreme points of a base B of M: specifically, the ray $R := \mathbf{R}_+(d)$, $d \in M \setminus \{0\}$, is extreme if and only if $R \cap B$ is an extreme point of B.*

See Figure 6.2 for an illustration of Fact 6.38(iv).

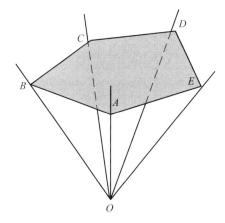

Figure 6.2 Cone and its base (grey pentagon). Extreme rays of the cone are $OA, OB, \ldots, OE$, intersecting the base at its extreme points $A, B, \ldots, E$.

In Example 6.11 we observed that every vector from $\mathbf{R}_+^n$ is the sum of finitely many extreme directions of $\mathbf{R}_+^n$. This observation is indeed a special case of the following result.

Theorem 6.39 *[Krein–Milman Theorem, conic form] Let $M \subset \mathbf{R}^n$ be a nontrivial closed and pointed cone. Then its set of extreme directions, i.e. $\operatorname{ExtD}(M)$, is nonempty and every nonzero vector $d \in M$ is the sum of finitely many extreme directions of M.*

Proof Under the premise of the theorem, Fact 6.28(iii) states that the cone M_*, i.e., the dual cone of M, has a nonempty interior, implying by Fact 6.38 that M has a base B which is a nonempty convex compact set. Then, by Krein–Milman Theorem (Theorem 6.11), the set $V = \text{Ext}(B)$ is nonempty, and $B = \text{Conv}(V)$. By Fact 6.38(iv), the extreme rays of B are exactly the rays generated by points from V, so that the extreme rays do exist. Besides this, every nonzero vector $d \in M$ has a positive multiple, i.e., αd with $\alpha > 0$, that belongs to B (as B is a base of M) and thus αd is a convex combination of points from V, so that the vector d itself is the sum of finitely many positive multiples of points from V. By Fact 6.38(iv), these multiples are extreme directions of M, implying that every nonzero vector from M is the sum of finitely many extreme directions. ∎

The Krein–Milman Theorem in conic form states that a nontrivial closed pointed cone has extreme directions – in fact enough of them to make their conic hull the entire cone. Recall that a trivial cone M has no extreme directions (by definition, these are nonzero vectors from M satisfying certain additional requirements, and the only vector in a trivial cone is the origin). A closed nontrivial *non-pointed* cone M has no extreme directions either. Indeed, if there exists a vector $0 \neq h \in M \cap [-M]$, then for every $d \in M$ one has $\{d + th : t \in \mathbf{R}\} \subseteq M$ by Remark 6.35. Since for all $t \in \mathbf{R}$ and all $d \in M$ it holds $d = d_t^+ + d_t^-$ with $d_t^\pm = \frac{1}{2}[d \pm th] \in M$, we conclude that a nonzero $d \in M$ which is not proportional to h cannot be an extreme direction, since here $d_1^\pm$ are not multiples of d. And a nonzero $d \in M$ proportional to h cannot be an extreme direction either, since now one of the vectors $d_t^\pm$ for large t is not a *nonnegative* multiple of d. We have arrived at the following result:

Corollary 6.40 *A closed cone $M \subseteq \mathbf{R}^n$ possesses extreme rays if and only if M is nontrivial and pointed.*

6.9 ★ Polar of a Convex Set

We next study the *polar* of a convex set, a concept closely related to the duals of cones.

Definition 6.41 *[Polar of a convex set] For any nonempty convex set $M \subseteq \mathbf{R}^n$, we define its* polar *[notation: Polar(M)] to be the set of all vectors $a \in \mathbf{R}^n$ such that $a^\top x \leq 1$ for all $x \in M$, i.e.,*

$$\text{Polar}(M) := \left\{ a \in \mathbf{R}^n : a^\top x \leq 1, \forall x \in M \right\}.$$

Let us see some basic examples.

Example 6.12

1. Polar$(\mathbf{R}^n) = \{0\}$.
2. Polar$(\{0\}) = \mathbf{R}^n$.
3. Given a linear subspace in $L \subseteq \mathbf{R}^n$, we have Polar$(L) = L^\perp$ (why?).
4. Let B be the unit Euclidean ball, i.e., $B := \{x \in \mathbf{R}^n : \|x\|_2 \leq 1\}$. Then Polar$(B) = B$ (by the Cauchy–Schwarz inequality).
5. Let $X \subset \mathbf{R}^n$ be nonempty and D be a nonsingular $n \times n$ matrix. Then Polar$(DX) = D^{-\top}$Polar(X).
6. Let E be an n-dimensional ellipsoid centered at the origin, i.e., $E := \{x : x^\top Cx \leq 1\}$ where $C \succ 0$. Then Polar$(E) = \{x : x^\top C^{-1} x \leq 1\}$, i.e., its polar is another n-dimensional ellipsoid centered at the origin.
7. Finally, note that passing to polars reverses inclusions: when $\emptyset \neq X \subseteq Y \subset \mathbf{R}^n$, we have Polar$(Y) \subseteq$ Polar(X). ◇

6.9 ★ Polar of a Convex Set

For any nonempty convex set M, the following properties of its polar are evident:

1. $0 \in \text{Polar}(M)$;
2. $\text{Polar}(M)$ is convex;
3. $\text{Polar}(M)$ is closed.

It turns out that these properties characterize polars completely:

Proposition 6.42 *Every closed convex set M containing the origin is a polar set. Specifically, such a set is the polar of its polar:*

$$M \text{ is closed, convex, and } 0 \in M \iff M = \text{Polar}(\text{Polar}(M)).$$

Proof From the evident properties of polars, all we need is to prove that if M is closed and convex and $0 \in M$ then $M = \text{Polar}(\text{Polar}(M))$. By definition, for all $x \in M$ and $y \in \text{Polar}(M)$, we have

$$y^\top x \leq 1.$$

Thus, $M \subseteq \text{Polar}(\text{Polar}(M))$.

To prove that this inclusion is in fact equality, we assume for contradiction that there exists $\bar{x} \in \text{Polar}(\text{Polar}(M)) \setminus M$. Since M is a nonempty closed convex set and $\bar{x} \notin M$, the point $\bar{x}$ can be strongly separated from M (Separation Theorem, part (ii)). Thus, there exists $b \in \mathbf{R}^n$ such that

$$b^\top \bar{x} > \sup_{x \in M} b^\top x.$$

As $0 \in M$, we deduce $b^\top \bar{x} > 0$. Passing from b to a proportional vector $a = \lambda b$ with appropriately chosen positive λ, we ensure that

$$a^\top \bar{x} > 1 \geq \sup_{x \in M} a^\top x.$$

From the relation $1 \geq \sup_{x \in M} a^\top x$ we conclude that $a \in \text{Polar}(M)$. But then the relation $a^\top \bar{x} > 1$ contradicts the assumption that $\bar{x} \in \text{Polar}(\text{Polar}(M))$. Hence, we conclude that indeed $M = \text{Polar}(\text{Polar}(M))$. ∎

We close this section with a few important properties of polars.

Fact 6.43 *Let $M \subseteq \mathbf{R}^n$ be a convex set containing the origin. Then:*

(i) $\text{Polar}(M) = \text{Polar}(\text{cl } M)$;
(ii) M *is bounded if and only if* $0 \in \text{int}(\text{Polar}(M))$;
(iii) $\text{int}(\text{Polar}(M)) \neq \varnothing$ *if and only if M does not contain straight lines;*
(iv) *if M is a cone (not necessarily closed) then*

$$\text{Polar}(M) = \left\{ a \in \mathbf{R}^n : a^\top x \leq 0, \forall x \in M \right\} = -M_*. \tag{6.15}$$

Assume further that M is closed. Then, M is a closed cone if and only if $\text{Polar}(M)$ is a closed cone.

For more information on polars, see Exercise II.38.

6.10 Proofs of Facts

Fact 6.2 *Let S, T be nonempty convex sets in $\mathbf{R}^n$. A linear form $a^\top x$ separates S and T if and only if*

(a) $\sup\limits_{x \in S} a^\top x \leq \inf\limits_{y \in T} a^\top y,$ *and*

(b) $\inf\limits_{x \in S} a^\top x < \sup\limits_{y \in T} a^\top y.$

This separation is strong if and only if (a) holds as a strict inequality:

$$\sup_{x \in S} a^\top x < \inf_{y \in T} a^\top y.$$

Proof When a linear form $a^\top x$ separates S and T, (a) holds true. Given (a), (b) could be violated if and only if $\inf_{x \in S} a^\top x = \sup_{y \in T} a^\top y$. But, together with (a), this can happen only if $a^\top x$ is constant on $S \cup T$, which is not the case as $a^\top x$ separates S and T. The above reasoning clearly can be reversed: given (b), we have $a \neq 0$, and given (a), both $\sup_{x \in S} a^\top x$ and $\inf_{y \in T} a^\top y$ are real numbers. Selecting b inbetween these real numbers, the hyperplane $a^\top x = b$ clearly separates S and T. The "strong separation" claim is evident. ∎

Fact 6.9 *Let M be a nonempty convex set and let $x \in M$. Then, x is an extreme point of M if and only if any (and then all) of the following hold:*

(i) *the only vector h such that $x \pm h \in M$ is the zero vector;*
(ii) *in every representation $x = \sum_{i=1}^m \lambda_i x^i$ of x as a convex combination, with positive coefficients, of points $x^i \in M$, $i \leq m$, one has $x^1 = \cdots = x^m = x$;*
(iii) *the set $M \setminus \{x\}$ is convex.*

Proof (i): If x is an extreme point and $x \pm h \in M$ then $h = 0$, since otherwise $x = \frac{1}{2}(x+h) + \frac{1}{2}(x-h)$ implying that x is an interior point of a nontrivial segment $[x-h, x+h]$, which is impossible. For the other direction, assume for contradiction that $x \pm h = 0$ implies $h = 0$ and that x is not an extreme point of M. Then, as $x \notin \text{Ext}(M)$, there exist $u, v \in M$ where both u, v are not equal to x and $\lambda \in (0, 1)$ such that $x = \lambda u + (1 - \lambda)v$. As $u \neq x$ and $v \neq x$ while $x = \lambda u + (1 - \lambda)v$, we conclude that $u \neq v$. Now, consider any $\delta > 0$ such that $\delta < \min\{\lambda, 1-\lambda\}$ and define $h := \delta(u-v)$. Note that $h \neq 0$, $x + h = (\lambda + \delta)u + (1 - \lambda - \delta)v \in M$, and $x - h = (\lambda - \delta)u + (1 - \lambda + \delta)v \in M$, due to $\lambda \pm \delta \in (0, 1)$, $u, v \in M$, and the convexity of M. This then leads to the desired contradiction with our assumption that $x \pm h \in M$ implies that $h = 0$.

As a byproduct of our reasoning, we see that if $x \in M$ can be represented as $x = \lambda u + (1-\lambda)v$ with $u, v \in M$, $\lambda \in (0, 1]$, and $u \neq x$, then x is *not* an extreme point of M.

(ii): In one direction, when x is *not* an extreme point of M, there exists $h \neq 0$ such that $x \pm h \in M$ so that $x = \frac{1}{2}(x+h) + \frac{1}{2}(x-h)$ is a convex combination with positive coefficients for two points $x \pm h$ that are both in M and are distinct from x. To prove the opposite direction, let x be an extreme point of M and suppose $x = \sum_{i=1}^m \lambda_i x^i$ with $\lambda_i > 0$, $\sum_i \lambda_i = 1$, and let us prove that $x^1 = \cdots = x^m = x$. Indeed, assume for contradiction that at least one x^i, say, x^1, differs from x, hence $m > 1$. Since $\lambda_2 > 0$, we have $0 < \lambda_1 < 1$.

Then, the point $v := (1-\lambda_1)^{-1}\sum_{i=2}^{m}\lambda_i x^i$ is well defined. Moreover, as $\sum_{i=2}^{m}\lambda_i = 1 - \lambda_1$, v is a convex combination of $x^2, \ldots, x^m$ and therefore $v \in M$. Then, $x = \lambda_1 x^1 + (1-\lambda_1)v$ with $x, x^1, v \in M$, $\lambda_1 \in (0, 1]$, and $x^1 \neq x$, which, by the concluding comment in part (i) of the proof, implies that $x \notin \text{Ext}(M)$; this is the desired contradiction.

(iii): In one direction, let x be an extreme point of M; let us prove that the set $M' := M \setminus \{x\}$ is convex. Assume for contradiction that this is not the case. Then, there exist $u, v \in M'$ and $\lambda \in [0, 1]$ such that $\bar{x} := \lambda u + (1-\lambda)v \notin M'$, implying that $0 < \lambda < 1$ (since $u, v \in M'$). As M is convex, we have $\bar{x} \in M$ and, since $\bar{x} \notin M'$ and $M \setminus M' = \{x\}$, we conclude that $\bar{x} = x$. Thus, x is a convex combination, with positive coefficients, of two points from M that are distinct from x, contradicting, by the already proved part (ii), the fact that x is an extreme point of M. For the other direction, suppose that $M \setminus \{x\}$ is convex; we will prove that x must be an extreme point of M. Assume for contradiction that $x \notin \text{Ext}(M)$. Then, there exists $h \neq 0$ such that $x \pm h \in M$. As $h \neq 0$, both $x + h$ and $x - h$ are distinct from x, thus $x \pm h \in M \setminus \{x\}$. We see that $x \pm h \in M \setminus \{x\}$, $x = \frac{1}{2}(x+h) + \frac{1}{2}(x-h)$, and $x \notin M \setminus \{x\}$, contradicting the convexity of $M \setminus \{x\}$. ∎

Fact 6.10 *All extreme points of the convex hull $\text{Conv}(Q)$ of a set Q belong to Q:*

$$\text{Ext}(\text{Conv}(Q)) \subseteq Q.$$

Proof Assume for contradiction that $x \in \text{Ext}(\text{Conv}(Q))$ and $x \notin Q$. As $x \in \text{Ext}(\text{Conv}(Q))$, by Fact 6.9 (iii) the set $\text{Conv}(Q) \setminus \{x\}$ is convex and contains Q, contradicting the fact that $\text{Conv}(Q)$ is the smallest convex set containing Q. ∎

Fact 6.18 *Let M be a nonempty closed convex set in $\mathbf{R}^n$. Then*

(i) $\text{Rec}(M) \neq \{0\}$ if and only if M is unbounded.

(ii) If M is unbounded then all nonzero recessive directions of M are positive multiples of recessive directions of unit Euclidean length, and the latter are asymptotic directions of M, i.e., a unit vector $h \in \mathbf{R}^n$ is a recessive direction of M if and only if there exists a sequence $\{x^i \in M\}_{i \geq 1}$ such that $\|x^i\|_2 \to \infty$ as $i \to \infty$ and $h = \lim_{i \to \infty} x^i / \|x^i\|_2$.

(iii) M does not contain lines if and only if the cone $\text{Rec}(M)$ does not contain lines.

Proof (i): If $\text{Rec}(M) \neq \{0\}$ then M contains a ray and therefore M is unbounded. For the reverse direction, suppose M is unbounded and let us prove that $\text{Rec}(M) \neq \{0\}$. As M is unbounded, there exists a sequence of points $x^i \in M$ such that $\|x^i\|_2 > i$, for all $i = 1, 2, \ldots$ Then, for sufficiently large i, the vectors $h^i := (x^i - x^1)/\|x^i - x^1\|_2$ are well-defined unit vectors. Passing to a subsequence, we can assume that $h^i \to h$ as $i \to \infty$ (Theorem B.15), so that h is a unit vector as well. For every $t \geq 0$, the points $x^1 + th^i$, for all i with $\|x^i - x^1\|_2 > t$, are convex combinations of points x^1 and x^i, and both $x^1, x^i \in M$. Then, as M is convex, $x^1 + th^i \in M$ for all large enough i. As $i \to \infty$, the points $x^1 + th^i$ converge to $x^1 + th$, and, since M is closed, we conclude $x^1 + th \in M$. Because this holds for every $t \geq 0$, the vector $h \neq 0$ is a recessive direction of M.

(ii): Suppose M is unbounded, and consider any $h \in \text{Rec}(M)$ such that h is a unit vector. Pick any $x^0 \in M$ and define $x^i := x^0 + ih$, $i = 1, 2, \ldots$ Then we get a sequence of points from M diverging to infinity, i.e., $\|x^i\|_2 \to \infty$ as $i \to \infty$, and also satisfying $h = \lim_{i \to \infty} \|x^i\|_2^{-1} x^i$. Thus, h is an asymptotic direction of M, as claimed. To prove the reverse direction, if $x^i \in M$ are such that $\|x^i\|_2 \to \infty$ and $h := \lim_{i \to \infty} \|x^i\|_2^{-1} x^i$ exists then h is a recessive direction of M by the same reasoning used in the proof of part (i), and of course h is a unit vector.

(iii): Suppose M contains a line with direction $h \neq 0$, i.e., for some $x \in M$ and for all $t \in \mathbf{R}$, we have $x + th \in M$. Then, by the definition of recessive directions, both h and $-h$ are recessive directions of M, so $\pm h \in \text{Rec}(M)$ and thus $\text{Rec}(M)$ contains a line with the direction h. For the reverse direction suppose $h \neq 0$ and $\text{Rec}(M)$ contains a line with direction h. Since $\text{Rec}(M)$ is a closed convex cone, the line with the same direction passing through the origin is contained in $\text{Rec}(M)$ (by Lemma 6.13). Thus $\pm h \in \text{Rec}(M)$. Then, by the definition of $\text{Rec}(M)$, for any $x \in M$ it holds that $x + th \in M$ for all $t \in \mathbf{R}$. Hence, M contains a line with the direction h. ∎

Fact 6.19 *Let $M \subseteq \mathbf{R}^n$ be a nonempty closed convex set. Recall that its closed conic transform is given by*

$$\overline{\text{ConeT}}(M) = \text{cl}\left\{[x;t] \in \mathbf{R}^n \times \mathbf{R} : t > 0,\ x/t \in M\right\},$$

(see section 1.5). Then

$$\text{Rec}(M) = \left\{h \in \mathbf{R}^n : [h;0] \in \overline{\text{ConeT}}(M)\right\}.$$

Proof Let h be such that $[h;0] \in \overline{\text{ConeT}}(M)$, and let us prove that $h \in \text{Rec}(M)$. There is nothing to prove when $h = 0$; thus assume that $h \neq 0$. Since the vectors g such that $[g;0] \in \overline{\text{ConeT}}(M)$ form a closed cone and both $\overline{\text{ConeT}}(M)$ and $\text{Rec}(M)$ are cones as well, we lose nothing when assuming, in addition to $[h;0] \in \overline{\text{ConeT}}(M)$ and $h \neq 0$, that h is a unit vector. Since $[h;0] \in \overline{\text{ConeT}}(M)$, by definition of the latter set there exists a sequence $[u^i;t_i] \to [h;0]$, $i \to \infty$, such that $t_i > 0$ and $x^i := u^i/t_i \in M$ for all i. Then this, together with $u^i \to h$ and $\|h\|_2 = 1$, implies that $\|x^i\|_2 \to \infty$ and $t_i \|x^i\|_2 \to 1$ as $i \to \infty$. As a result, $\lim_{i \to \infty} \|x^i\|_2^{-1} x^i = \lim_{i \to \infty} u^i = h$. By Fact 6.18(ii), we see that $h \in \text{Rec}(M)$.

For the reverse direction, consider any $h \in \text{Rec}(M)$, and let us prove that $[h;0] \in \overline{\text{ConeT}}(M)$. There is nothing to prove when $h = 0$, so we assume $h \neq 0$. Consider any $\bar{x} \in M$ and define $x^i := \bar{x} + ih$, $i = 1, 2, \ldots$ As $h \in \text{Rec}(M)$, we have $x^i \in M$ for all i. Moreover, $\|x^i\|_2 \to \infty$ as $i \to \infty$ due to $h \neq 0$. We clearly have $\lim_{i \to \infty}[x^i/\|x^i\|_2; 1/\|x^i\|_2] = [h/\|h\|_2; 0]$, and the vectors $[y^i;t_i] := [x^i/\|x^i\|_2, 1/\|x^i\|_2]$ for all large enough i satisfy the requirement $t_i > 0$, $y^i/t_i \in M$, so $[y^i;t_i] \in \overline{\text{ConeT}}(M)$ for all large enough i. As $\overline{\text{ConeT}}(M)$ is closed and $[y^i;t_i] \to [h/\|h\|_2; 0]$ as $i \to \infty$, we deduce $[h/\|h\|_2; 0] \in \overline{\text{ConeT}}(M)$. Finally, $\overline{\text{ConeT}}(M)$ is a cone, so $[h;0] \in \overline{\text{ConeT}}(M)$ as well. ∎

Fact 6.20 *For any nonempty polyhedral set $M = \{x \in \mathbf{R}^n : Ax \leq b\}$, its recessive cone is given by*

$$\text{Rec}(M) = \left\{h \in \mathbf{R}^n : Ah \leq 0\right\},$$

i.e., $\text{Rec}(M)$ is given by the homogeneous version of the linear constraints specifying M.

Proof Consider any h such that $Ah \leq 0$. Then, for any $\bar{x} \in M$ and $t \geq 0$, we have $A(\bar{x} + th) = A\bar{x} + tAh \leq A\bar{x} \leq b$, so $\bar{x} + th \in M$ for all $t \geq 0$. Hence, $h \in \text{Rec}(M)$.

6.10 Proofs of Facts 113

For the reverse direction, suppose $h \in \text{Rec}(M)$ and $\bar{x} \in M$. Then, for all $t \geq 0$ we have $A(\bar{x} + th) \leq b$. This is equivalent to $Ah \leq t^{-1}(b - A\bar{x})$ for all $t > 0$, which implies that $Ah \leq 0$. ∎

Fact 6.28 *Let M be a closed cone in $\mathbf{R}^n$, and let M_* be the cone dual to M. Then:*

(i) *Duality does not distinguish between a cone and its closure: whenever $M = \text{cl } M'$ for a cone M', we have $M_* = M'_*$.*

(ii) *Duality is symmetric: the cone dual to M_* is M.*

(iii) *One has*
$$\text{int } M_* = \left\{ y \in \mathbf{R}^n : y^\top x > 0, \forall x \in M \setminus \{0\} \right\},$$
and $\text{int } M_*$ is nonempty if and only if M is pointed (i.e., $M \cap [-M] = \{0\}$).

Moreover, when M, in addition to being closed, is pointed and nontrivial ($M \neq \{0\}$), one has
$$\text{int } M_* = \left\{ y \in \mathbf{R}^n : M_y := \{x \in M : x^\top y = 1\} \text{ is nonempty and compact} \right\}. \quad (6.12)$$

(iv) *The cone dual to the direct product $M_1 \times \cdots \times M_m$ of cones M_i is the direct product of their duals: $[M_1 \times \cdots \times M_m]_* = [M_1]_* \times \cdots \times [M_m]_*$.*

Proof (i): This is evident.

(ii): By definition, any $x \in M$ satisfies $x^\top y \geq 0$ for all $y \in M_*$, hence $M \subseteq [M_*]_*$. To prove $M = [M_*]_*$, assume for contradiction that there exists $\bar{x} \in [M_*]_* \setminus M$. By Separation Theorem, $\{\bar{x}\}$ can be strongly separated from M, i.e., there exists y such that
$$y^\top \bar{x} < \inf_{x \in M} y^\top x.$$
As M is a conic set and the right-hand side infimum is finite, this infimum must be 0. Thus, $y^\top \bar{x} < 0$ while $y^\top x \geq 0$ for all $x \in M$, implying $y \in M_*$. But then this contradicts $\bar{x} \in [M_*]_*$.

(iii): Let us prove that $\text{int } M_* \neq \emptyset$ if and only if M is pointed. If M is not pointed then $\pm h \in M$ for some $h \neq 0$, implying that $y^\top[\pm h] \geq 0$ for all $y \in M_*$, that is, $y^\top h = 0$ for all $y \in M_*$. Thus, when M is not pointed, M_* belongs to a proper (smaller than the entire $\mathbf{R}^n$) linear subspace of $\mathbf{R}^n$ and thus $\text{int } M_* = \emptyset$. This reasoning can be reversed: when $\text{int } M_* = \emptyset$, the affine hull $\text{Aff}(M_*)$ of M_* cannot be the entire $\mathbf{R}^n$ (since $\text{int } M_* = \emptyset$ and $\text{rint } M_* \neq \emptyset$); taking into account that $0 \in M_*$, we have $\text{Aff}(M_*) = \text{Lin}(M_*)$, so that $\text{Lin}(M_*) \subsetneq \mathbf{R}^n$, and therefore there exists a nonzero h orthogonal to $\text{Lin}(M_*)$. We have $y^\top[\pm h] = 0$ for all $y \in M_*$, implying that h and $-h$ belong to the cone dual to M_*, that is, to M (owing to the already verified part (ii)). Thus, for some nonzero h it holds that $\pm h \in M$, that is, M is not pointed.

Now let us prove that $y \in \text{int } M_*$ if and only if $y^\top x > 0$ for every $x \in M \setminus \{0\}$. In one direction: assume that $y \in \text{int } M_*$, so that for some $r > 0$ it holds that $y + \delta \in M_*$ for all δ with $\|\delta_2\| \leq r$. Then, for any $x \in M$, we have $0 \leq \min_{\delta: \|\delta\|_2 \leq r} [y + \delta]^\top x = y^\top x - r\|x\|_2$. Thus,
$$y \in \text{int } M_* \implies \|x\|_2 \leq \frac{1}{r} y^\top x, \forall x \in M, \quad (*)$$

implying that $y^\top x > 0$ for all $x \in M\setminus\{0\}$, as required. In the opposite direction: assume that $y^\top x > 0$ for all $x \in M\setminus\{0\}$, and let us prove that $y \in \operatorname{int} M_*$. There is nothing to prove when $M = \{0\}$ (and therefore $M_* = \mathbf{R}^n$). Assuming $M \neq \{0\}$, let $\bar{M} = \{x \in M : \|x\|_2 = 1\}$. This set is nonempty (since $M \neq \{0\}$), closed (as M is closed), and clearly bounded, and thus is compact. We are in the situation when $y^\top x > 0$ for $x \in \bar{M}$, implying that $\min_{x \in \bar{M}} y^\top x$ (this minimum is achieved since $\bar{M}$ is a nonempty compact set) is strictly positive. Thus, $y^\top x \geq r > 0$ for all $x \in \bar{M}$, whence $[y + \delta]^\top x \geq 0$ for all $x \in \bar{M}$ and all δ with $\|\delta\|_2 \leq r$. Due to the definition of $\bar{M}$, a positive multiple of every nonzero $x \in M$ belongs to $\bar{M}$, and therefore the inequality $[y + \delta]^\top x \geq 0$ for all $x \in \bar{M}$ implies that $[y + \delta]^\top x \geq 0$ for all $x \in M$. The bottom line is that the Euclidean ball of radius r centered at y belongs to M_*, and therefore $y \in \operatorname{int} M_*$, as claimed.

Now let us prove the "moreover" part of item (iii). Thus, let the cone M be closed, pointed, and nontrivial. Consider any $y \in \operatorname{int} M_*$; then the set M_y, first, contains some positive multiple of every nonzero vector from M (by (*)) and thus is nonempty (since $M \neq \{0\}$) and, second, is bounded (by the same (*)). Since M_y is closed (as M is closed), we conclude that M_y is a nonempty compact set. Thus, the left-hand side set in (6.12) is contained in the right-hand side set. To prove the opposite inclusion, let $y \in \mathbf{R}^n$ be such that M_y is a nonempty compact set, and let us prove that $y \in \operatorname{int} M_*$. By the already proved part of (iii), all we need to do is to verify that if $x \neq 0$ and $x \in M$ then $y^\top x > 0$. Assume for contradiction that there exists $\bar{x} \in M \setminus \{0\}$ such that $\alpha := -y^\top \bar{x} \geq 0$. Then, by selecting any $\widehat{x} \in M_y$ (M_y is nonempty!) and setting $e = \alpha \widehat{x} + \bar{x}$, we get $e \in M$ and $y^\top e = 0$. Note that $e \neq 0$; indeed, $e = 0$ would mean that the nonzero vector $\bar{x} \in M$ is such that $-\bar{x} = \alpha \widehat{x} \in M$, contradicting the pointedness of M. The bottom line is that $e \in M \setminus \{0\}$ and $y^\top e = 0$, whence e is a nonzero recessive direction of M_y. This is the desired contradiction as M_y is compact!

(iv): This is evident. ∎

Fact 6.33 *Let $M \subseteq \mathbf{R}^n$ be a cone and M_* be its dual cone. Then, for any $x \in \operatorname{int} M$, there exists a properly selected $C_x < \infty$ such that*

$$\|f\|_2 \leq C_x f^\top x, \qquad \forall f \in M_*.$$

Proof Since $x \in \operatorname{int} M$, there exists $\rho > 0$ such that $x - \delta \in M$ whenever $\|\delta\|_2 \leq \rho$. Then, as $f \in M_*$, we have $f^\top(x - \delta) \geq 0$ for any $\|\delta\|_2 \leq \rho$, i.e., $f^\top x \geq \sup_\delta\{f^\top \delta : \|\delta\|_2 \leq \rho\} = \rho\|f\|_2$. Taking $C_x := 1/\rho$ (note that $C_x < \infty$ as $\rho > 0$) gives us the desired relation. ∎

Fact 6.38 *Let $M \subseteq \mathbf{R}^n$ be a nontrivial closed cone, and M_* be its dual cone. Then:*

(i) *M is pointed*
  (i.1) *if and only if M does not contain straight lines,*
  (i.2) *if and only if M_* has a nonempty interior,*
  (i.3) *if and only if M has a base.*
(ii) *The set (6.14) is a base of M*
  (ii.1) *if and only if $f^\top x > 0$ for all $x \in M \setminus \{0\}$,*
  (ii.2) *if and only if $f \in \operatorname{int} M_*$.*

As a result,

(ii.3) $f \in \operatorname{int} M_*$ if and only if $f^\top x > 0$ for all $x \in M \setminus \{0\}$.

(iii) *Every base of M is nonempty, closed, and bounded. Moreover, whenever M is pointed, for any $f \in M_*$ such that the set (6.14) is nonempty (note that this set is always closed for any f), this set is bounded if and only if $f \in \operatorname{int} M_*$, in which case (6.14) is a base of M.*

(iv) *M has extreme rays if and only if M is pointed. Furthermore, when M is pointed, there is a one-to-one correspondence between the extreme rays of M and the extreme points of a base B of M: specifically, the ray $R := \mathbf{R}_+(d)$, $d \in M \setminus \{0\}$, is extreme if and only if $R \cap B$ is an extreme point of B.*

Proof (i.1): Since M is closed, convex, and contains the origin, M contains a line if and only if it contains a line passing through the origin, and since M is conic, the latter occurs if and only if M is not pointed.

(i.2): This is precisely Fact 6.28(iii).

(i.3): By the definition of a base, (6.14) is a base of M if and only if $f^\top x > 0$ for all $x \in M \setminus \{0\}$, which, by Fact 6.28(iii), holds if and only if $f \in \operatorname{int} M_*$. Thus, a nontrivial closed cone M has a base if and only if $\operatorname{int} M_* \neq \emptyset$, and the latter, by the already proved part (i.2), is the same as the pointedness of M.

(ii.1): This was explained when defining a base.

(ii.2): This is given by Fact 6.28(iii).

(iii): By (ii.2), the set B given by (6.14) is a base of M if and only if $f \in \operatorname{int} M_*$. By Fact 6.28(iii), $f \in \operatorname{int} M_*$ if and only if B is nonempty and compact (Fact 6.28(iii) is applicable, since under the premise of Fact 6.38 M is nontrivial and closed). Thus, M has a base if and only if $\operatorname{int} M_* \neq \emptyset$, the bases are exactly the sets (6.14) given by $f \in \operatorname{int} M_*$, and every base is nonempty and compact. To complete the proof of (iii), we need to verify that a nonempty set B given by some $f \in M_*$ according to (6.14) is bounded if and only if $f \in \operatorname{int} M_*$. In one direction this has just been proved, and all we need is to verify that if B is nonempty and $f \in \operatorname{bd} M_* = M_* \setminus \operatorname{int} M_*$, then B is unbounded. Indeed, by (ii.3), when $f \in \operatorname{bd} M_*$, then there exists nonzero $d \in M$ such that $f^\top d = 0$. It follows that d is a nontrivial recessive direction of the nonempty convex set B, so that B is unbounded, as claimed. (iii) is proved.

(iv): The fact that a closed cone M which is either trivial or non-pointed has no extreme rays was established before formulating Corollary 6.40, and the corresponding reasoning did not refer to Fact 6.38. Now assume that M is nontrivial, closed, and pointed, and let M_* be the cone dual to M. By Fact 6.28(iii), $\operatorname{int} M_* \neq \emptyset$, so that, by the already proved parts (ii) and (iii) of Fact 6.38, M has a base $B = \{x \in M : f^\top x = 1\}$, and this base is a nonempty convex compact set. By the Krein–Milman Theorem (Theorem 6.11), the set V of extreme point of B is nonempty. To complete the proof of (iv), we need to verify that

(a) when $v \in V$, v is an extreme direction of M (this, as a byproduct, implies that M has extreme rays, since $V \neq \emptyset$), and

(b) when R is an extreme ray of M, then the point v where the ray intersects B (such a point exists by the definition of a base) is an extreme point of B.

(a): As $0 \notin B \subset M$, v is a nonzero vector from M. Let $v = d^1 + d^2$ with d^1, d^2 from M; we should prove that d^1, d^2 are nonnegative multiples of v. There is nothing to prove when $d^1 = 0$ or $d^2 = 0$. Assuming $d^1 \neq 0, d^2 \neq 0$, the rays $\mathbf{R}_+(d^1)$ and $\mathbf{R}_+(d^2)$ intersect B at points v^1, respectively, v^2 (the rays do intersect B by the definition of a base). We have $d^i = \lambda_i v^i$ with $\lambda_i > 0$, $i = 1, 2$, and $v = d^1 + d^2$ reads $v = \lambda_1 v^1 + \lambda_2 v^2$, whence $f^\top v = \lambda_1 f^\top v^1 + \lambda_2 f^\top v^2$. As $v, v^1, v^2 \in B$, we have $f^\top v = f^\top v^1 = f^\top v^2 = 1$, implying that $\lambda_1 + \lambda_2 = 1$, so that v is a convex combination, with positive coefficients, of $v^i \in B$, $i = 1, 2$. By Fact 6.9(ii) we have $v^1 = v^2 = v$, that is, d^i are nonnegative multiples of v, as claimed. (a) is proved.

(b): To prove that $v \in \text{Ext}(B)$, assume that h is such that $v \pm h \in B$, and let us prove that $h = 0$. Indeed, setting $d^\pm = \frac{1}{2}[v \pm h]$, we have $d^\pm \in M$ and $v = d^- + d^+$. As $v \in B$, we have $f^\top v = 1$, and also since $v \pm h \in B$, we deduce that $f^\top h = 0$. We see that when $h \neq 0$, the vectors v and h are linearly independent, implying that $d^\pm$ are *not* multiples of v, which is impossible, since v is an extreme direction of M. Thus, $h = 0$, as required. ∎

Fact 6.43 *Let $M \subseteq \mathbf{R}^n$ be a convex set containing the origin. Then:*

(i) Polar $(M) = $ Polar $(\text{cl } M)$;
(ii) M is bounded if and only if $0 \in \text{int}(\text{Polar}(M))$;
(iii) $\text{int}(\text{Polar}(M)) \neq \emptyset$ if and only if M does not contain straight lines;
(iv) if M is a cone (not necessarily closed) then

$$\text{Polar}(M) = \left\{ a \in \mathbf{R}^n : a^\top x \leq 0, \forall x \in M \right\} = -M_*. \tag{6.15}$$

Assume further that M is closed. Then, M is a closed cone if and only if Polar (M) is a closed cone.

Proof (i): This follows immediately from $\sup_{x \in M} a^\top x = \sup_{x \in \text{cl } M} a^\top x$.

(ii): Suppose M is bounded. Then, by the Cauchy–Schwarz inequality all vectors y with small enough norms satisfy $y \in \text{Polar}(M)$ and so $0 \in \text{int}(\text{Polar}(M))$. To see the reverse direction, suppose that $0 \in \text{int}(\text{Polar}(M))$. Note that cl M is a closed convex set and by part (i) we have Polar $(M) = $ Polar $(\text{cl } M)$, so $0 \in \text{int}(\text{Polar}(\text{cl } M))$, i.e., Polar $(\text{cl } M)$ contains a ball centered at the origin with some radius $\rho > 0$. Then, by Proposition 6.42, we have cl $M = $ Polar $(\text{Polar}(\text{cl } M)) = $ Polar $(\text{Polar}(M))$, which implies that $x \in \text{cl } M = $ Polar $(\text{Polar}(M))$ only if $1 \geq \sup_{y \in \mathbf{R}^n} \{ y^\top x : \|y\|_2 \leq \rho \} = \rho \|x\|_2$. Thus, for all $x \in \text{cl } M$ we have $\|x\|_2 \leq 1/\rho$.

(iii): By part (i), the polar remains intact when passing from M to cl M; by Lemma 6.14, a nonempty convex set M contains a straight line if and only if cl M does so. Thus, we lose nothing when assuming in the rest of the proof that M is closed.

Assume, first, that M contains a straight line, and let us prove that $\text{int}(\text{Polar}(M)) = \emptyset$. Indeed, when the closed convex set M contains a line, since $0 \in M$ by Lemma 6.13, M contains a parallel line ℓ passing through $0 \in M$. Thus, Polar $(M) \subseteq $ Polar (ℓ). Since ℓ is a one-dimensional linear subspace of $\mathbf{R}^n$, Polar (ℓ) is the orthogonal complement $\ell^\perp$ of ℓ, so that $\text{int}(\text{Polar}(\ell)) = \text{int}(\ell^\perp) = \emptyset$, hence $\text{int}(\text{Polar}(M)) = \emptyset$ as well.

Now let $\text{int}(\text{Polar}(M)) = \emptyset$, and let us prove that M contains a straight line. Assume that this is not the case, and let us lead this assumption to a contradiction. Since M does not contain lines, the closed cone $K := \text{Rec}(M)$ is pointed, so that its dual cone K_* has a nonempty interior (Fact 6.38(i.2)). Thus, there exists $r > 0$ and $\overline{f} \in K_*$ such that the ball of

6.10 Proofs of Facts

radius $2r$ centered at $\overline{f}$ is contained in K_*. Then $z^\top(f+e) \geq 0$ whenever $z \in K$, $\|e\|_2 \leq r$, and $\|f - \overline{f}\|_2 \leq r$. As a result,

$$f^\top z \geq r\|z\|_2, \quad \forall(z \in K, \ f \in B := \{f : \|f - \overline{f}\|_2 \leq r\}). \tag{$*$}$$

Now let

$$C := \sup_{f \in -B, z \in M} f^\top z;$$

we claim that $C < \infty$. Taking this claim for granted, observe that $C < \infty$ implies, by homogeneity, that $\sup_{f \in -\epsilon B, z \in M} f^\top z \leq \epsilon C$ for all $\epsilon > 0$, hence for properly selected small positive ϵ the ball $-\epsilon B$ is contained in Polar (M), implying int(Polar $(M)) \neq \emptyset$, which is the desired contradiction.

It remains to justify the above claim. To this end assume that $C = +\infty$, and let us lead this assumption to a contradiction. When $C = +\infty$, there exists a sequence $f_i \in -B$ and $z_i \in M$ such that $f_i^\top z_i \to +\infty$ as $i \to \infty$, implying, due to $f_i \in -B$, that $\|z_i\|_2 \to \infty$ as $i \to \infty$. Passing to a subsequence, we can assume that $z_i/\|z_i\|_2 \to h$ as $i \to \infty$. Then, by its origin, h is an asymptotic direction of M and therefore is a unit vector from K (Fact 6.18(ii)). Assuming w.l.o.g. $z_i \neq 0$ for all i, we have

$$f_i^\top z_i = \|z_i\|_2 \bigg(\underbrace{f_i^\top h}_{:=\alpha_i} + \underbrace{f_i^\top (z_i/\|z_i\|_2 - h)}_{:=\beta_i} \bigg). \tag{!}$$

As $i \to \infty$, $f_i \in -B$ remain bounded and $(z_i/\|z_i\|_2 - h) \to 0$, implying that $\beta_i \to 0$ as $i \to \infty$, while $(*)$ as applied with $z = h$ together with $h \in K$, $\|h\|_2 = 1$, and $f_i \in -B$ implies that $\alpha_i \leq -r < 0$. Thus, $\alpha_i + \beta_i \leq -r/2$ for large enough values of i, so that $(!)$ taken together with $\|z_i\|_2 \to \infty$ as $i \to \infty$ says that $f_i^\top z_i \to -\infty$ as $i \to \infty$, contradicting the origin of f_i and z_i. Thus, $C < \infty$, as claimed. This completes the verification of part (iii).

(iv): Clearly, when M is a nonempty conic set, the relation $y^\top x \leq 1$ for all $x \in M$ is exactly the same as $y^\top x \leq 0$ for all $x \in M$. Hence, when M is a cone, its polar is a closed cone given by (6.15). On the other hand, when M contains the origin and is convex and closed, it is the polar of its polar, so that when this polar is a cone, M itself is a closed cone (by the just proved statement in part (iv) as applied to Polar (M) in the role of M). ∎

7

Geometry of Polyhedral Sets

7.1 Extreme Points of Polyhedral Sets

Consider a polyhedral set

$$M = \{x \in \mathbf{R}^n : Ax \leq b\},$$

where A is an $m \times n$ matrix and $b \in \mathbf{R}^m$. In the sequel, we refer to the scalar linear inequalities participating in the polyhedral description of M as the *constraints* specifying M.

We saw a geometric characterization of extreme points for general convex sets in section 6.4.1. In the case of polyhedral sets M, we can also give an *algebraic* characterization of the extreme points as follows.

Theorem 7.1 *[Characterization of extreme points of polyhedral sets] Let $M = \{x \in \mathbf{R}^n : Ax \leq b\}$. A point $x \in M$ is an extreme point of M if and only if there are n linearly independent (i.e., with linearly independent vectors of coefficients) inequalities of the system $Ax \leq b$ that are active (i.e., held as equalities) at x.*

Proof Let $a_i^\top$, $i = 1, \ldots, m$, be the rows of A.

The "only if" part: let x be an extreme point of M, and define the sets $I := \{i : a_i^\top x = b_i\}$ as the set of indices of the constraints which are active (i.e., are satisfied as equalities) at x and $F := \{a_i : i \in I\}$ as the set of vectors of coefficients of active constraints. We will prove that the set F contains n linearly independent vectors, i.e., $\text{Lin}(F) = \mathbf{R}^n$. Assume for contradiction that this is not the case. Then, as $\dim(F^\perp) = n - \dim(\text{Lin}(F))$, we deduce $\dim(F^\perp) > 0$ and so there exists a nonzero vector $d \in F^\perp$. Consider the segment $\Delta_\epsilon := [x - \epsilon d, x + \epsilon d]$, where $\epsilon > 0$ will be the parameter of our construction. Since d is orthogonal to the active vectors a_i (those with $i \in I$), all points $y \in \Delta_\epsilon$ satisfy the relations $a_i^\top y = a_i^\top x = b_i$. Now, if i is a "nonactive" index (one with $a_i^\top x < b_i$) then $a_i^\top y \leq b_i$ for all $y \in \Delta_\epsilon$, provided that ϵ is small enough. Since there are finitely many nonactive indices, we can choose $\epsilon > 0$ in such a way that all $y \in \Delta_\epsilon$ will satisfy all nonactive inequalities $a_i^\top x \leq b_i$, $i \notin I$, as well. So, we conclude that for the above choice of $\epsilon > 0$ we get $\Delta_\epsilon \subseteq M$. But, this is a contradiction to x being an extreme point of M, as we have expressed x as the midpoint of a nontrivial segment Δ_ϵ (recall that $\epsilon > 0$ and $d \neq 0$).

To prove the "if" part, we assume that $x \in M$ is such that among the inequalities $a_i^\top x \leq b_i$ which are active at x there are n linearly independent ones. Without loss of generality, we assume that the indices of these linearly independent inequalities are $1, \ldots, n$. Given this, we will prove that x is an extreme point of M. Assume for contradiction that x is not an extreme point. Then there exists a vector $d \neq 0$ such that $x \pm d \in M$. In other words, for

$i = 1, \ldots, n$ we would have $b_i \geq a_i^\top (x \pm d) \equiv b_i \pm a_i^\top d$ (where the last equivalence follows from $a_i^\top x = b_i$ for all $i \in \{1, \ldots, n\}$), which is possible only if $a_i^\top d = 0$, $i = 1, \ldots, n$. But the only vector which is orthogonal to n linearly independent vectors in $\mathbf{R}^n$ is the zero vector (why?), and so we get $d = 0$, which contradicts the assumption $d \neq 0$. ∎

Theorem 7.1 states that at every extreme point of a polyhedral set $M = \{x \in \mathbf{R}^n : Ax \leq b\}$ we must have n linearly independent constraints from $Ax \leq b$ holding as equalities. Since a system of n linearly independent equality constraints in n unknowns has a unique solution, such a system can specify *at most one extreme point* of M (exactly one, when the (unique!) solution to the system satisfies the remaining constraints in the system $Ax \leq b$). Moreover, when M is defined by m inequality constraints, the number of such systems, and thus the number of extreme points of M, does not exceed the number C_m^n of $n \times n$ submatrices of the matrix $A \in \mathbf{R}^{m \times n}$. Hence, we arrive at the following corollary.

Corollary 7.2 *Every polyhedral set has finitely many extreme points.*

Recall that there are nonempty polyhedral sets which do not have any extreme points; these are precisely those that contain lines.

Note that C_m^n is just an upper (and typically very conservative) bound on the number of extreme points of a polyhedral set in $\mathbf{R}^n$ defined by m inequality constraints. This is so because some $n \times n$ submatrices of A can be singular, and, what is more important, the majority of the nonsingular submatrices typically produce "candidate" points which do not satisfy the remaining inequalities defining M.

Remark 7.3 Historically, Theorem 7.1 has been instrumental in developing an algorithm to solve linear programs, namely the *Simplex method*. Let us consider an LP in standard form,

$$\min_{x \in \mathbf{R}^n} \left\{ c^\top x : Px = p, x \geq 0 \right\},$$

where $P \in \mathbf{R}^{k \times n}$. Note that we can convert any given LP to this form by adding a small number of new variables and constraints if needed. In the context of this LP, Theorem 7.1 states that the extreme points of the feasible set are exactly the *basic feasible solutions* of the system $Px = p$, i.e., nonnegative vectors x such that $Px = p$ and the set of columns of P associated with positive entries of x is linearly independent. As the feasible set of an LP in standard form clearly does not contain lines (note the constraints $x \geq 0$ which restrict the standard form LP domain to being subset of the pointed cone $\mathbf{R}^n_+$), among its optimal solutions (if they exist) at least one is an extreme point of the feasible set (Theorem 7.12(ii)). This then suggests a simple algorithm to solve a solvable LP in standard form: go through the finite set of all extreme points of the feasible set (or equivalently all basic feasible solutions) and choose the one with the best objective value. This algorithm allows one to find an optimal solution in finitely many arithmetic operations, provided that the LP is solvable, and it underlies the basic idea of the Simplex method. As one can immediately recognize, the number of extreme points, although finite, may be quite large. The Simplex method operates in a smarter way and examines *only a subset* of the basic feasible solutions, in an organized way, and can handle other issues such as infeasibility and unboundedness.

Another useful consequence of Theorem 7.1 is that if all the data in an LP are rational then every one of its extreme points is a vector with rational entries. Thus, a solvable standard-form LP with rational data has at least one rational optimal solution. $\diamond$

Theorem 7.1 has further important consequences in terms of the sizes of the extreme points of polyhedral sets.

To this end, let us first recall a simple fact from linear algebra:

Proposition 7.4 *If a system of linear equations $Ax = b$ (with $A \in \mathbf{R}^{m \times n}$) is feasible then it has a solution $x(b)$ of "magnitude of order of the magnitude of b;" that is, $\|x(b)\|_2 \leq C(A)\|b\|_2$ with a parameter $C(A) < \infty$ which depends solely on A and but not on b.*

Proof Let $r := \mathrm{rank}(A)$. There is nothing to prove when $r = 0$ as in this case A is zero, and if the system $Ax = b$ has a solution, the vector zero is one of its solutions as well. When $r > 0$, we can assume without loss of generality that the first r columns of A are linearly independent, and the remaining columns are linear combinations of these r columns. Then, if there is a solution to $Ax = b$ then there must be a solution for which $x_i = 0$ for all $i > r$. We will take such a solution as $x(b)$. As the first r columns of A are linearly independent, A has an $r \times r$ submatrix $\widehat{A}$ composed of the first r columns of A and r properly selected rows of A. Let $\widehat{b}$ be the subvector of b corresponding to the indices of rows selected for $\widehat{A}$. Define the vector $\widehat{x}(b) \in \mathbf{R}^r$ to be the vector obtained from the first r entries in $x(b)$. Since the entries in $x(b)$ with indices greater than r are zero, we have $x(b) = [\widehat{x}(b); 0]$ and so $\widehat{A}\widehat{x}(b) = \widehat{b}$. As $\widehat{A}$ is a nonsingular matrix, we deduce that

$$\|x(b)\|_2 = \|\widehat{x}(b)\|_2 = \|\widehat{A}^{-1}\widehat{b}\|_2 \leq \|\widehat{A}^{-1}\|\|\widehat{b}\|_2 \leq \|\widehat{A}^{-1}\|\|b\|_2,$$

where $\|\widehat{A}^{-1}\|$ is the spectral norm of $\widehat{A}^{-1}$. Then, setting $C(A) := \|\widehat{A}^{-1}\|$ concludes the proof. ∎

Surprisingly, a similar result holds for the solutions of systems of linear inequalities as well.

Proposition 7.5 *Consider a system of linear inequalities $Ax \leq b$ where $A \in \mathbf{R}^{m \times n}$. Whenever b results in a feasible system $Ax \leq b$, then there exists $C(A) < \infty$ depending solely on A, not on b, such that this system has a solution $x(b)$ with $\|x(b)\|_2 \leq C(A)\|b\|_2$.*

Proof This proof is quite similar to that for Proposition 7.4. Let $r := \mathrm{rank}(A)$. The case of $r = 0$ is trivial: in this case $A = 0$, and the system $Ax \leq b$ if feasible, has the solution $x = 0$. When $r > 0$, we can assume without loss of generality that the first r columns in A are linearly independent. Let $\widehat{A} \in \mathbf{R}^{m \times r}$ be the submatrix of A obtained from the first r columns of A. As $\widehat{A}$ has all the linearly independent columns of A, the image spaces of A and $\widehat{A}$ are the same. Thus, the system $Ax \leq b$ is feasible if and only if the system $\widehat{A}u \leq b$ is feasible. Moreover, given any feasible solution $u \in \mathbf{R}^r$ to $\widehat{A}u \leq b$, we can generate a feasible solution $x := [u; 0] \in \mathbf{R}^n$ by adding $n - r$ zeros at the end and still preserve the norm of the solution. Hence, without loss of generality we can assume that the columns of A are linearly independent and $r = n$.

As $A \in \mathbf{R}^{m \times n}$ has n linearly independent columns and each column is a vector in $\mathbf{R}^m$, we deduce that $m \geq n$ and $\{u : Au = 0\} = \{0\}$. Thus, we conclude that the polyhedral set $\{x : Ax \leq b\}$ does not contain lines. Therefore, when it is nonempty, by the Krein–Milman

Theorem this polyhedral set has an extreme point. Let us take this point as $x(b)$. Then, by Theorem 7.1, at least n of the inequality constraints from the system $Ax \leq b$ will be active at $x(b)$, and out of the vectors of coefficients of these active constraints there will be n vectors a_i (corresponding to the rows of the matrix A) that are linearly independent. That is, $A_b x(b) = b$ holds for a certain nonsingular $n \times n$ submatrix A_b of A. So, we conclude that $\|x(b)\|_2 \leq \|A_b^{-1}\| \|b\|_2$. Since the number of $r \times r$ nonsingular submatrices in A is finite, the maximum $C(A)$ of the spectral norms of the inverses of these submatrices is finite as well, and, as we have seen, for every b for which the system $Ax \leq b$ is feasible, it has a solution $x(b)$ with $\|x(b)\|_2 \leq C(A)\|b\|_2$, as claimed. ∎

7.1.1 Important Polyhedral Sets and Their Extreme Points

In this section, we will examine a number of important polyhedral sets and their extreme points. Throughout this section, we suppose that k and n are positive integers with $k \leq n$.

Example 7.1 Suppose k, n are integers satisfying $0 \leq k \leq n$. Consider the polytope

$$X := \left\{ x \in \mathbf{R}^n : 0 \leq x_i \leq 1, \forall i \leq n, \sum_{i=1}^{n} x_i = k \right\}.$$

The set of extreme points of X is precisely the set of vectors with entries 0 and 1 which have exactly k entries equal to 1. That is,

$$\text{Ext}(X) = \left\{ x \in \{0, 1\}^n : \sum_{i=1}^{n} x_i = k \right\}.$$

In particular, the extreme points of the "flat" (a.k.a. *probabilistic*) simplex $\{x \in \mathbf{R}^n_+ : \sum_{i=1}^{n} x_i = 1\}$ are the standard basic orths (see Figure 7.1 for an illustration of this set with $k = 1$).

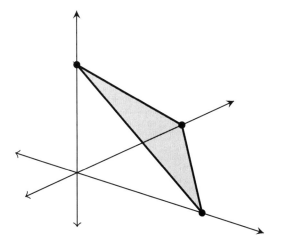

Figure 7.1 Example 7.1 for $n = 3, k = 1$.

Let us justify the claim of this example. For convenience, we define $Y := \{x \in \{0,1\}^n : \sum_{i=1}^n x_i = k\}$; then we need to show that $\text{Ext}(X) = Y$. Clearly, $Y \subseteq X$. Moreover, for any $y \in Y$, for each coordinate $i = 1,\ldots,n$ we have either $y_i = 0$ or $y_i = 1$; thus we have n bound constraints active. Since the vectors of coefficients of these n constraints are linearly independent, we conclude by Theorem 7.1 that $Y \subseteq \text{Ext}(X)$. Now, consider any $w \in \text{Ext}(X)$. Then, by Theorem 7.1, among the constraints specifying X, n constraints with linearly independent vectors of coefficients should be active at w. Thus, we must have at least $n-1$ of the bound constraints $0 \le x_i \le 1$ active at w, i.e., at least $n-1$ of the entries of w must be in $\{0,1\}$. Let i_* be the index of the remaining entry of w. As $w \in X$, it must also satisfy $\sum_{i=1}^n w_i = k$. As k is an integer, we deduce $w_{i_*} = k - \sum_{i \ne i_*} w_i$ must be an integer as well. But then, as we also have $0 \le w_{i_*} \le 1$, we deduce that $w_{i_*} \in \{0,1\}$. Thus $w \in Y$ holds, as desired. $\diamond$

Example 7.2 Suppose k, n are integers satisfying $0 \le k \le n$. Consider the polytope

$$X = \left\{ x \in \mathbf{R}^n : 0 \le x_i \le 1, \forall i \le n, \sum_{i=1}^n x_i \le k \right\}.$$

The set of extreme points of X is precisely the set of vectors with entries 0 and 1 which have at most k entries equal to 1. That is,

$$\text{Ext}(X) = \left\{ x \in \{0,1\}^n : \sum_{i=1}^n x_i \le k \right\}.$$

In particular, the extreme points of the "full-dimensional simplex" $\{x \in \mathbf{R}^n_+ : \sum_{i=1}^n x_i \le 1\}$ are the standard basic orths and the origin (see Figure 7.2 for an illustration of this set with $k = 1$).

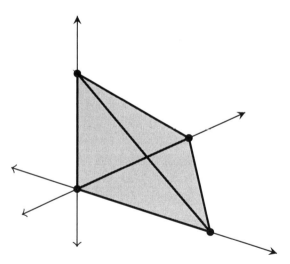

Figure 7.2 Example 7.2 for $n = 3, k = 1$.

The justification of this example follows that for Example 7.1 and is left as an exercise for the reader. $\diamond$

7.1 Extreme Points of Polyhedral Sets

Example 7.3 Suppose k, n are integers satisfying $0 \leq k \leq n$. Consider the polytope

$$X = \left\{ x \in \mathbf{R}^n : |x_i| \leq 1, \forall i \leq n, \sum_{i=1}^n |x_i| \leq k \right\}.$$

The extreme points of X are exactly the vectors with entries $0, 1, -1$ which have exactly k nonzero entries. That is,

$$\mathrm{Ext}(X) = \left\{ x \in \{-1, 0, 1\}^n : \sum_{i=1}^n |x_i| = k \right\}.$$

In particular, the extreme points of the unit $\|\cdot\|_1$-ball $\{x \in \mathbf{R}^n : \|x\|_1 \leq 1\} = \{x \in \mathbf{R}^n : \sum_{i=1}^n |x_i| \leq 1\}$ are exactly the vectors $\{\pm e_i : i = 1, \ldots, n\}$ where e_i is the ith standard basic orth (see Figure 7.3 for an illustration of this set with $k = 1$).

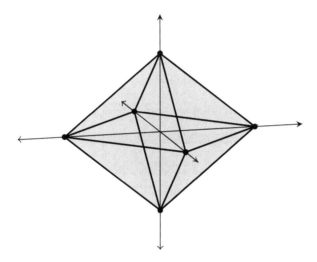

Figure 7.3 Example 7.3 for $n = 3, k = 1$.

Similarly, the extreme points of the unit $\|\cdot\|_\infty$-ball $\{x \in \mathbf{R}^n : \|x\|_\infty \leq 1\} = \{x \in \mathbf{R}^n : |x_i| \leq 1, \forall i = 1, \ldots, n\}$ are the 2^n vectors with all entries ± 1 (see Figure 7.4 for an illustration of this set).

Here is the justification for the claim of this example. For convenience, we define $Y := \{x \in \{-1, 0, 1\}^n : \sum_{i=1}^n |x_i| = k\}$; we need to show that $\mathrm{Ext}(X) = Y$. Clearly, $Y \subseteq X$. Consider any $y \in Y$. Without loss of generality, suppose that the nonzero entries of y are the first k entries. Now, consider any h such that $y \pm h \in X$. Since $y_i \in \{-1, +1\}$ for $i = 1, \ldots, k$ and $y \pm h \in X$, we must have $h_i = 0$ for all $i = 1, \ldots, k$. Also, $y + h \in X$ implies that $k \geq \sum_{i=1}^n |y_i + h_i| = \sum_{i=1}^k |y_i| + \sum_{i=k+1}^n |h_i| = k + \sum_{i=k+1}^n |h_i|$, and thus $|h_i| = 0$ for all $i = k+1, \ldots, n$. This proves that $h = 0$ and thus y must indeed be an extreme point of X. So $Y \subseteq \mathrm{Ext}(X)$. Now, consider any $w \in \mathrm{Ext}(X)$; we will show that $w \in Y$. Note that X has a certain symmetry: for any $x \in X$ and any $d \in \{-1, +1\}^n$, we have $\mathrm{Diag}(d)x \in X$ as well. In particular, any $d \in \{-1, +1\}^n$ maps X onto itself and therefore maps $\mathrm{Ext}(X)$ onto $\mathrm{Ext}(X)$. As a result, we can assume without loss of generality that $w \geq 0$. Then, all we need to prove is that w has k entries equal to 1 and all remaining entries equal

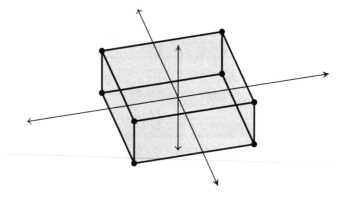

Figure 7.4 Extreme points of the box $\{x \in \mathbf{R}^3 : \|x\|_\infty \leq 1\}$.

to 0. The set $X_+ := \{x \in X : x \geq 0\} = \{x \in \mathbf{R}^n : 0 \leq x_i \leq 1, \forall i \leq n, \sum_{i=1}^k x_i \leq k\}$ is contained in X and $w \in X_+$. As $w \in \mathrm{Ext}(X)$ and $w \in X_+ \subset X$, we must have that w is an extreme point of X_+ as well.

In the preceding reasoning, we have used the following evident fact:

> *Suppose $P \subset Q$ are convex sets. If $\bar{x} \in P$ is an extreme point of Q, then it is an extreme point of P as well.*

(Otherwise $\bar{x}$ would be the midpoint of a nontrivial segment contained in P and therefore contained in Q.)

Then, noting that X_+ is precisely the set from Example 7.2 and $w \in \mathrm{Ext}(X_+)$, we conclude that w has only 0 and 1 entries and at most k of its entries are nonzero, therefore the sum of the entries of w is at most k. It remains to verify that the number of nonzero entries in w is equal to k. Indeed, if w were to have fewer than k nonzero entries, w would have a zero entry, say, $w_1 = 0$, and $\sum_{i=1}^n |w_i| < k$, implying that there would exist $\epsilon \in (0, 1)$ such that the vector $h = [\epsilon; 0; \ldots; 0]$ satisfies $(w \pm h) \in X$, which is impossible since $w \in \mathrm{Ext}(X)$. $\diamondsuit$

As our last example we discuss the so-called *Assignment polytope*, which is closely connected to the very important concept of *doubly stochastic matrices* and the Birkhoff Theorem.

Definition 7.6 *[Doubly stochastic matrix]* A matrix $X = [x_{ij}]_{i,j=1}^n \in \mathbf{R}^{n \times n}$ is called doubly stochastic if $x_{ij} \geq 0$ for all $i, j \in \{1, \ldots, n\}$, $\sum_{i=1}^n x_{ij} = 1$ for all $j \in \{1, \ldots, n\}$ (i.e., all column sums are equal to 1), and $\sum_{j=1}^n x_{ij} = 1$ for all $i \in \{1, \ldots, n\}$ (i.e., all row sums are equal to 1).

The set of all doubly stochastic matrices (treated as elements of $\mathbf{R}^{n^2} = \mathbf{R}^{n \times n}$) form the following bounded polyhedral set:

$$\Pi_n := \left\{ X = [x_{ij}]_{i,j=1}^n : \begin{array}{l} x_{ij} \geq 0, \forall i, j \in \{1, \ldots, n\}, \\ \sum_{i=1}^n x_{ij} = 1, \forall j \in \{1, \ldots, n\}, \\ \sum_{j=1}^n x_{ij} = 1, \forall i \in \{1, \ldots, n\} \end{array} \right\}.$$

The set Π_n is also called the *Assignment (or balanced matching) polytope*. As Π_n is a polytope, by the Krein–Milman Theorem, we have that Π_n is the convex hull of its extreme points. What are these extreme points? The answer is given by the following fundamental result.

Theorem 7.7 [Birkhoff–von-Neumann Theorem] *The extreme points of Π_n are exactly the permutation matrices of order n, which are $n \times n$ Boolean (i.e., with 0/1 entries) matrices with exactly one nonzero element (equal to 1) in every row and every column.*

Proof It is indeed easy to see that every $n \times n$ permutation matrix is an extreme point of Π_n; we give this as Exercise II.42.

We now prove the difficult part, that is, to show that every extreme point of Π_n is a permutation matrix. First, note that the $2n$ linear equations in the definition of Π_n, those saying that all row and column sums are equal to 1, are linearly dependent (observe that the sum of the first group of equalities is exactly the same as the sum of the second group of equalities). Thus, we lose nothing when assuming that there are just $2n - 1$ equality constraints in the description of Π_n. Now, let us prove the claim by induction on n. The base case $n = 1$ is trivial. As the induction hypothesis suppose that the statement holds for Π_{n-1}. Let X be an extreme point of Π_n. By Theorem 7.1, among the constraints defining Π_n (i.e., $2n - 1$ equalities and n^2 inequalities $x_{ij} \geq 0$) there should be n^2 linearly independent constraints which are satisfied at X as equations. Thus, at least $n^2 - (2n - 1) = (n - 1)^2$ entries in X should be zeros. It follows that at least one of the columns of X contains ≤ 1 nonzero entries (since otherwise the number of zero entries in X would be at most $n(n-2) < (n-1)^2$). Thus, there exists at least one column with at most 1 nonzero entry; since the sum of entries in this column is 1, this nonzero entry, say $x_{\bar{i}\bar{j}}$, is equal to 1. As the entries in row $\bar{i}$ are nonnegative and sum up to 1 and as $x_{\bar{i}\bar{j}} = 1$, $x_{\bar{i}\bar{j}} = 1$ is the only nonzero entry in the ith row and also in the jth column. Eliminating from X row $\bar{i}$ and column $\bar{j}$, we get an $(n - 1) \times (n - 1)$ doubly stochastic matrix. By the inductive hypothesis, this matrix is a convex combination of $(n - 1) \times (n - 1)$ permutation matrices. Augmenting every one of these matrices by the column and the row that we have eliminated, we get a representation of X as a convex combination of $n \times n$ permutation matrices P^ℓ, i.e., $X = \sum_\ell \lambda_\ell P^\ell$ with nonnegative λ_ℓ summing up to 1. Since $P^\ell \in \Pi_n$ and X is an extreme point of Π_n, in this representation all terms with nonzero coefficients λ_ℓ must be equal to $\lambda_\ell X$, so that X is one of the permutation matrices P^ℓ and as such is a permutation matrix. ∎

7.2 Extreme Rays of Polyhedral Cones

Recall that for nontrivial closed pointed cones, we defined the concepts of extreme directions and extreme rays in section 6.8. In the case of polyhedral cones, we can also give an algebraic characterization of their extreme directions analogous to Theorem 7.1.

Theorem 7.8 *[Characterization of extreme directions of polyhedral cones] Consider a polyhedral cone $M = \{d \in \mathbf{R}^n : a_i^\top d \leq 0, i = 1, \ldots, m\}$. Suppose that M is nontrivial and pointed. A direction $d \in M \setminus \{0\}$ is an extreme direction of M if and only if there are $n - 1$ linearly independent constraints (i.e., with linearly independent vectors a_i) which are active at d (i.e., such that $a_i^\top d = 0$).*

Proof As M is a nonempty closed (recall that it is polyhedral!) pointed cone, from Fact 6.28(iii) we deduce that its dual cone M_* has a nonempty interior. Let $f \in \operatorname{int} M_*$. Consider the set

$$B := M \cap \left\{ d \in \mathbf{R}^n : f^\top d = 1 \right\}.$$

Then, by Fact 6.38(ii), B is a base of M. Thus B is nonempty and compact (see Fact 6.38(iii)). As M is polyhedral, by the definition of B we have that B is polyhedral as well. Since B is a nonempty bounded polyhedral set, from Theorem 6.11(ii) we have $B = \operatorname{Conv}(\operatorname{Ext}(B))$. Recall from Fact 6.38(iv) that there is a one-to-one correspondence between extreme rays of M and extreme points of a base B of M; specifically, the ray $\mathbf{R}_+(d) = \{td : t \geq 0\}, d \in M \setminus \{0\}$, is extreme if and only if the point x_d where the ray intersects B is an extreme point of B.

Consider an extreme direction d; then x_d is an extreme point of B. Applying Theorem 7.1, we deduce that among the constraints of the system $f^\top x = 1, a_i^\top x \leq 0, i = 1, \ldots, m$ (these are the constraints specifying B), n constraints with linearly independent vectors of coefficients are active at x_d. Among these n constraints, $n - 1$ constraints are of the form $a_i^\top x \leq 0$, and the corresponding a_i are linearly independent. Thus, at an extreme direction of M, $n - 1$ of the constraints $a_i^\top x \leq 0$ with linearly independent vectors of coefficients are active. Vice versa, let a nonzero direction $d \in M$ be such that at least $n - 1$ constraints from $a_i^\top x \leq 0$ are active and the set of vectors a_i corresponding to these active constraints has at least $n - 1$ linearly independent vectors, and let I be the set of indices of these constraints. Note that at $x_d \in B$ we have the constraints $f^\top x = 1$ and $a_i^\top x = 0, i \in I$ active, and f is not a linear combination of $a_i, i \in I$, since otherwise $a_i^\top x_d = 0, i \in I$, would imply that $f^\top x_d = 0$, which is not the case. As the vectors $a_i, i \in I$, are linearly independent, and f is not their linear combination, the n vectors f and $a_i, i \in I$, are linearly independent, and since at $x_d \in B$ the constraints $f^\top x = 1, a_i^\top x \leq 0, i \in I$ are active, by Theorem 7.1 $x_d \in \operatorname{Ext}(B)$. This, in turn, implies that d is an extreme direction of M by Fact 6.38(iii). ■

Analogously to Corollary 7.2, we have the following immediate corollary of Theorem 7.8.

Corollary 7.9 *Every nontrivial pointed polyhedral cone M has finitely many extreme rays. Moreover, the sum of these extreme rays is the entire cone M.*

Proof Let M be a nontrivial pointed polyhedral cone. As any polyhedral cone is defined by finitely many linear inequalities, using the algebraic characterization of the extreme directions given in Theorem 7.8 we deduce that M has finitely many extreme rays. The last claim of the corollary is justified by Theorem 6.39, i.e., the Krein–Milman Theorem in conic form, as this theorem states that M has extreme rays, and their conic hull is the entire cone M. ■

7.3 Geometry of Polyhedral Sets

By definition, a polyhedral set M is the set of all solutions to a finite system of nonstrict linear inequalities:

$$M := \left\{ x \in \mathbf{R}^n : Ax \leq b \right\}, \tag{7.1}$$

where $A \in \mathbf{R}^{m \times n}$ and $b \in \mathbf{R}^m$. This is an "outer" description of a polyhedral set, that is, it explains what we should delete from $\mathbf{R}^n$ to get M (cf. "to create a sculpture, take a big stone and delete everything that is redundant"). We are about to establish an important result on the equivalent "inner" representation of a polyhedral set, that is, one explaining how to build the set starting with simple "building blocks."

Consider the following construction. Let us take two finite sets of vectors V ("vertices;" this set must be nonempty) and R ("rays;" this set can be empty) and build the set

$$M(V, R) := \text{Conv}(V) + \text{Cone}(R)$$

$$= \left\{ \sum_{v \in V} \lambda_v v + \sum_{r \in R} \mu_r r : \begin{array}{l} \lambda_v \geq 0, \; \forall v \in V, \; \sum_{v \in V} \lambda_v = 1, \\ \mu_r \geq 0, \; \forall r \in R \end{array} \right\}.$$

Thus, in the construction of $M(V, R)$ we take all vectors which can be represented as sums of convex combinations of the points from V and conic combinations of the points from R. The set $M(V, R)$ is clearly convex as it is the arithmetic sum of two convex sets $\text{Conv}(V)$ and $\text{Cone}(R)$ (recall our convention that $\text{Cone}(\varnothing) = \{0\}$, see Fact 1.20). We are now ready to present the promised inner description of a polyhedral set.

Theorem 7.10 *[Inner description of a polyhedral set] Sets of the form $M(V, R)$ where V, R are finite sets of vectors and $V \neq \varnothing$, are exactly the nonempty polyhedral sets: $M(V, R)$ is polyhedral, and every nonempty polyhedral set M is $M(V, R)$ for properly chosen finite sets $V \neq \varnothing$ and R. Sets of the type $M(\{0\}, R)$ are exactly the polyhedral cones (sets given by finitely many nonstrict homogeneous linear inequalities).*

Remark 7.11 We will see in section 7.3.2 that the inner characterization of the polyhedral sets given in Theorem 7.10 can be made much more precise. Suppose that we are given a nonempty polyhedral set M. Then, we can select an inner characterization of it in the form $M = \text{Conv}(V) + \text{Cone}(R)$ with finite V and finite R, where the "conic" part $\text{Cone}(R)$ (not the set R itself!) is uniquely defined by M; in fact it will always hold that $\text{Cone}(R) = \text{Rec}(M)$, i.e., R can be taken as the set of generators of the recessive cone of M (see the comment in Lemma 6.13). Moreover, if M does not contain lines then V can be chosen as the set of all extreme points of M, and this choice is "minimal" – it always holds that $\text{Ext}(M) \subseteq V$. ◇

We will prove Theorem 7.10 in section 7.3.2. Before proceeding with its proof, let us understand why this theorem is so important, i.e., why it is so nice to know both the inner and outer descriptions of a polyhedral set.

Consider the following natural questions:

- A. Is it true that the inverse image of a polyhedral set $M \subset \mathbf{R}^n$ under an affine mapping $y \mapsto \mathcal{P}(y) = Py + p : \mathbf{R}^m \to \mathbf{R}^n$, i.e., the set

$$\mathcal{P}^{-1}(M) = \{y \in \mathbf{R}^m : Py + p \in M\},$$

  is polyhedral?
- B. Is it true that the image of a polyhedral set $M \subset \mathbf{R}^n$ under an affine mapping $x \mapsto y = \mathcal{P}(x) = Px + p : \mathbf{R}^n \to \mathbf{R}^m$, i.e., the set

$$\mathcal{P}(M) = \{Px + p : x \in M\},$$

  is polyhedral?

- C. Is it true that the intersection of two polyhedral sets is again a polyhedral set?
- D. Is it true that the arithmetic sum of two polyhedral sets is again a polyhedral set?

The answers to all these questions are positive; one way to see this is to use the calculus of polyhedral representations along with the fact that polyhedrally representable sets are exactly the same as polyhedral sets (see chapter 3). Another very instructive way is to use the results just outlined on the structure of polyhedral sets, which we will do now.

It is very easy to answer affirmatively to A, starting from the original "outer" definition of a polyhedral set: if $M = \{x : Ax \leq b\}$ then, of course,

$$\mathcal{P}^{-1}(M) = \{y : A(Py + p) \leq b\} = \{y : (AP)y \leq b - Ap\}$$

and therefore $\mathcal{P}^{-1}(M)$ is a polyhedral set.

An attempt to answer affirmatively to B via the same "outer" definition fails – there is no easy way to convert the linear inequalities defining a polyhedral set into those defining its image, and it is absolutely unclear why the image is in fact given by finitely many linear inequalities. Note, however, that there is no difficulty in answering affirmatively to B with the inner description of a nonempty polyhedral set: if $M = M(V, R)$, then, evidently,

$$\mathcal{P}(M) = M(\mathcal{P}(V), PR),$$

where $PR := \{Pr : r \in R\}$ is the image of R under the action of the homogeneous part of $\mathcal{P}$.

A positive answer to C becomes evident when we use the outer description of a polyhedral set: taking the intersection of the solution sets to two systems of finitely many nonstrict linear inequalities, we, of course, again get the solution set to a system of this type – we simply put together all inequalities from both of the original systems.

On the other hand, it is very unclear how to give an affirmative answer to D using the outer description of a polyhedral set – what happens to the inequalities when we add the solution sets? In contrast with this, an inner description gives the answer immediately:

$$\begin{aligned} M(V, R) + M(V', R') &= \text{Conv}(V) + \text{Cone}(R) + \text{Conv}(V') + \text{Cone}(R') \\ &= [\text{Conv}(V) + \text{Conv}(V')] + [\text{Cone}(R) + \text{Cone}(R')] \\ &= \text{Conv}(V + V') + \text{Cone}(R \cup R') \\ &= M(V + V', R \cup R'). \end{aligned}$$

Note that in this computation we used two rules which should be justified: $\text{Conv}(V) + \text{Conv}(V') = \text{Conv}(V + V')$ and $\text{Cone}(R) + \text{Cone}(R') = \text{Cone}(R \cup R')$. The second is evident from the definition of the conic hull, and the first follows from very simple reasoning. To see it, note that $\text{Conv}(V) + \text{Conv}(V')$ is a convex set which by its definition contains $V + V'$ and thus also contains $\text{Conv}(V + V')$. The inverse inclusion is proved as follows: if

$$x = \sum_i \lambda_i v_i, \qquad y = \sum_j \lambda'_j v'_j$$

are convex combinations of points from V, respectively, V', then, as it can immediately be seen (please check!),

$$x + y = \sum_{i,j} \lambda_i \lambda'_j (v_i + v'_j)$$

and the right-hand side expression is just a convex combination of points from $V + V'$.

7.3 Geometry of Polyhedral Sets

To conclude, *it is extremely useful to keep in mind both descriptions of polyhedral sets* – what may be difficult to see with one of them is absolutely clear with the other.

As a seemingly "more important" application of the developed theory, let us look at Linear Programming.

7.3.1 Application: Descriptive Theory of Linear Programming

A general linear program is the problem of minimizing a linear objective function over a polyhedral set:

$$\min_x \left\{ c^\top x : x \in M \right\}, \text{ where } M := \{x \in \mathbf{R}^n : Ax \leq b\}. \tag{P}$$

Here, $c \in \mathbf{R}^n$ is the objective, $A \in \mathbf{R}^{m \times n}$ is the constraint matrix, and $b \in \mathbf{R}^m$ is the right-hand side vector. Note that (P) is called a "linear programming problem in the canonical form;" there are other equivalent forms of this problem.

According to the Linear Programming terminology discussed in section 4.5.1, (P) is called

- *feasible*, if it admits a feasible solution, i.e., the system $Ax \leq b$ is feasible, and it is *infeasible* otherwise;
- *bounded*, if its objective is below bounded on the feasible set (e.g., (P) is below bounded whenever its feasible set is empty), and *unbounded* otherwise;
- *solvable*, if it is feasible and the optimal solution exists, i.e., the objective function attains its minimum on the feasible set.

Whenever (P) is feasible, the infimum of the values of the objective function at feasible solutions is called the *optimal value* $\mathrm{Opt}(P)$ of (P). $\mathrm{Opt}(P)$ is finite when the problem (P) is bounded from below and $-\infty$ when (P) is unbounded. In the case of a minimization-type problem, it is convenient to assign the optimal value of $+\infty$ whenever the problem is infeasible.

Note that our terminology is aimed to deal with minimization problems; if the problem is to maximize the objective, the terminology is updated in the natural way: when defining bounded or unbounded programs, we should speak about above-boundedness rather than about the below-boundedness of the objective on the feasible set, etc. As a result, the optimal value of an LP problem

- in the case of a minimization problem is the infimum of the objective over the feasible set, provided the latter is nonempty, and $+\infty$ when the problem is infeasible;
- in the case of a maximization problem is the supremum of the objective over the feasible set, provided the latter is nonempty, and $-\infty$ when the problem is infeasible.

This terminology is consistent with the usual way of converting a minimization problem into an equivalent maximization problem by replacing the original objective c with $-c$: the properties of feasibility, boundedness, solvability remain unchanged, and the optimal value in all cases changes its sign.

When talking about the possible outcomes of solving an LP problem, we talk about three possibilities: (i) an infeasible LP problem, (ii) an unbounded and feasible LP problem, and

(iii) a solvable LP problem. In particular, it seems that in the case of a "bounded and feasible" LP problem, we are jumping straight to the conclusion that the corresponding optimization problem will be "solvable." That a bounded LP program is always solvable is indeed true, although it is absolutely unclear in advance why (note that this statement does not hold for general convex programming problems without further assumptions). We have already established this fact, in fact twice – via Fourier–Motzkin elimination (section 3.2) and via the LP Duality Theorem (Theorem 4.10). In fact yet another proof of this fundamental fact of Linear Programming follows immediately from the inner characterization of polyhedral sets, as shown next.

Theorem 7.12 *Suppose we are given a feasible minimization-type LP problem (P) via an inner representation of its feasible set M:*

$$M = \text{Conv}(V) + \text{Cone}(R),$$

where $V = \{v_i : i = 1, \ldots, I\}$ and $R = \{r_j : j = 1, \ldots, J\}$ are finite and nonempty sets (cf. Theorem 7.10). Then,

(i) *(P) is solvable if and only if it is below-bounded, which is the case if and only if $c^\top r_j \geq 0$ for all $1 \leq j \leq J$.*
In particular, the set C of objectives c for which (P) is below-bounded is a polyhedral cone.

(ii) *When (P) is below-bounded, its optimal value is equal to*

$$\text{Opt}(P) = \min_{v \in V} c^\top v.$$

Thus, Opt(P) is a concave function of the objective vector c, and there is an optimal solution which is the best, in terms of its objective value, among the points in V. In addition, when the feasible set of (P) does not contain lines and (P) is below-bounded, there is at least one optimal solution of (P) which is an extreme point of M.

Proof (i): By Theorem 7.10, we clearly have

$$\text{Opt}(P) = \min_v \left\{ c^\top v : v \in \text{Conv}(V) \right\} + \inf_r \left\{ c^\top r : r \in \text{Cone}(R) \right\}.$$

We see that $\text{Opt}(P)$ is finite if and only if $\inf_r \left\{ c^\top r : r \in \text{Cone}(R) \right\} > -\infty$, and the latter is clearly the case if and only if $c^\top r \geq 0$ for all $r \in R$. Then, in such a case $\inf_r \left\{ c^\top r : r \in \text{Cone}(R) \right\} = 0$, and also $\min_v \left\{ c^\top v : v \in \text{Conv}(V) \right\} = \min_v \left\{ c^\top v : v \in V \right\}$.

(ii): The first claim in (ii) is an immediate byproduct of the proof of (i). The second claim follows from the fact that when M does not contain lines, we can take $V = \text{Ext}(M)$; see Remark 7.11. ∎

7.3.2 Proof of Theorem 7.10

In this section, we will prove Theorem 7.10. To simplify our language let us call VR-sets ("V" from "vertex" and "R" from "rays") sets of the form $M(V, R)$, and P-sets nonempty polyhedral sets, i.e., defined by finitely many linear non-strict inequalities. We need to prove that every P-set is a VR-set, and vice versa.

VR $\Longrightarrow$ P: This is immediate: a VR-set is nonempty and polyhedrally representable (why?) and thus is a nonempty P-set by Theorem 3.2.

P $\Longrightarrow$ VR: Let $M \neq \emptyset$ be a P-set, so that M is the set of all solutions to a feasible system of linear inequalities:
$$M = \{x \in \mathbf{R}^n : Ax \leq b\}, \tag{7.2}$$
where $A \in \mathbf{R}^{m \times n}$.

We will first study the case of P-sets that do not contain lines, and then reduce the general case to this one.

Theorem 7.13 *[Structure of a polyhedral set with no lines] A nonempty polyhedral set $M = \{x \in \mathbf{R}^n : Ax \leq b\}$ which does not contain lines admits a VR-representation given by $M = M(V, R) = \mathrm{Conv}(V) + \mathrm{Cone}(R)$, where V is the set of extreme points of M and $R = \{0\}$ if M is bounded, and R is the set of generators of extreme rays of $\mathrm{Rec}(M)$ if M is unbounded.*

Proof As M is a nonempty closed convex set that does not contain lines, by Theorem 6.11(i) we know that $\mathrm{Ext}(M) \neq \emptyset$, and by Theorem 6.21 we have $M = \mathrm{Conv}(\mathrm{Ext}(M)) + \mathrm{Rec}(M)$. Moreover, by Corollary 7.2, we have that $\mathrm{Ext}(M)$ is a finite set.

If M is bounded then $\mathrm{Rec}(M) = \{0\}$, and thus the result follows. Suppose M is unbounded. Then $\mathrm{Rec}(M)$ is nontrivial. Also, $\mathrm{Rec}(M)$ is pointed as the fact that M does not contain lines implies that $\mathrm{Rec}(M)$ does not contain lines either. Moreover, from Fact 6.20, we deduce that $\mathrm{Rec}(M) = \{h \in \mathbf{R}^n : Ah \leq 0\}$ and thus is a polyhedral cone. Then, by Corollary 7.9 we have that $\mathrm{Rec}(M)$ has finitely many extreme rays and $\mathrm{Rec}(M)$ is the sum of its extreme rays. ∎

Next, we study the case when M contains a line. We start with the following observation.

Lemma 7.14 *A nonempty polyhedral set $M = \{x \in \mathbf{R}^n : Ax \leq b\}$ contains lines if and only if $\mathrm{Ker} A \neq \{0\}$. Moreover, given a vector $h \neq 0$, the set M contains a line with direction h (i.e., $x + th \in M$ for all $x \in M$ and $t \in \mathbf{R}$) if and only if $h \in \mathrm{Ker} A$. That is, the nonzero vectors from $\mathrm{Ker} A$ are exactly the directions of the lines contained in M.*

Proof If $h \neq 0$ is the direction of a line in M then $A(x + th) \leq b$ for some $x \in M$ and all $t \in \mathbf{R}$, which is possible if and only if $Ah = 0$. Vice versa, if $h \neq 0$ is from the kernel of A, i.e., if $Ah = 0$, then the line $x + \mathbf{R}(h)$ with $x \in M$ is clearly contained in M. ∎

Given a nonempty set M as in (7.2), define $L := \mathrm{Ker} A$, let $L^\perp$ be the orthogonal complement to L, and let M' be the intersection of M and $L^\perp$:
$$M' := \{x \in L^\perp : Ax \leq b\}.$$

First, note that as $M \neq \emptyset$ we have $M' \neq \emptyset$, (since when passing from $x \in M$ to the orthogonal projection x' of x onto $L^\perp$, we shift x along $\mathrm{Ker} L$, whence $Ax' = Ax$ and thus $x' \in M$). Besides this, the set M' clearly does not contain lines. This is so because if $h \neq 0$ is the direction of a line satisfying $x + th \in M'$ for all $t \in \mathbf{R}$ and some $x \in M'$, by the definition of M' we must have $x + th \in L^\perp$ for all t and thus $h \in L^\perp$. On the other hand, by Lemma 7.14, we must also have $h \in \mathrm{Ker} A = L$. Then $h \in L \cap L^\perp$ implies $h = 0$, which is a contradiction.

Now, note that $M' \neq \varnothing$ satisfies $M = M' + L$. Indeed, M' contains the orthogonal projections of all points from M onto $L^\perp$ (since to project a point onto $L^\perp$, one moves from this point along a certain line with a direction from L, and all these movements, started in M, keep one in M by Lemma 7.14) and therefore, M' is nonempty, first, and is such that $M' + L \supseteq M$, second. On the other hand, $M' \subseteq M$ and $M + L = M$ (by Lemma 7.14), and so $M' + L \subseteq M$. Thus, $M' + L = M$.

Finally, it is clear that M' is a polyhedral set as the inclusion $x \in L^\perp$ can be represented by $\dim(L)$ linear equations (i.e., by $2\dim(L)$ nonstrict linear inequalities). To this end, all we need is a set of vectors $\xi_1, \ldots, \xi_{\dim(L)}$ forming a basis in L, and then $L^\perp := \{x \in \mathbf{R}^n : \xi_i^\top x = 0, \forall i = 1, \ldots, \dim(L)\}$.

Therefore, with these steps, given an arbitrary nonempty P-set M, we have represented it as the sum of a P-set M' which does not contain lines and a linear subspace L. Then, as M' does not contain lines, by Theorem 7.13 we have $M' = M(V', R')$ where V' is the nonempty set of extreme points of M' and R' is the set of generators of extreme rays of $\mathrm{Rec}(M')$. Let us define R'' to be the finite set of generators for L, i.e., $L = \mathrm{Cone}(R'')$ (one can take as R'' the collection composed of vectors from a basis of L and the negations of these vectors). Then, we arrive at

$$M = M' + L$$
$$= [\mathrm{Conv}(V') + \mathrm{Cone}(R')] + \mathrm{Cone}(R'')$$
$$= \mathrm{Conv}(V') + [\mathrm{Cone}(R') + \mathrm{Cone}(R'')]$$
$$= \mathrm{Conv}(V') + \mathrm{Cone}(R' \cup R'')$$
$$= M(V', R' \cup R'').$$

This then proves that a P-set is indeed a VR-set, as desired. ∎

Finally, let us justify Remark 7.11. Suppose we are given $M = \mathrm{Conv}(V) + \mathrm{Cone}(R)$ with finite sets V, R and $V \neq \varnothing$. Justifying the first claim in this remark requires us to show that $\mathrm{Cone}(R) = \mathrm{Rec}(M)$, but this is readily given by Theorem 6.21(ii).

The last claim of Remark 7.11 states that when a nonempty P-set M does not contain lines, in every representation $M = M(V, R)$ with finite V and R one has $\mathrm{Ext}(M) \subseteq V$, and there is a representation with $V = \mathrm{Ext}(M)$. The latter fact is given by Theorem 7.13. To justify the first fact, suppose x is an extreme point of M. We will first show that $x \in \mathrm{Conv}(V)$. Assume for contradiction that $x \in M \setminus \mathrm{Conv}(V)$. Then, from $x \in M = \mathrm{Conv}(V) + \mathrm{Cone}(R)$, we deduce that $x = \bar{x} + r$ with $\bar{x} \in \mathrm{Conv}(V)$ and $0 \neq r \in \mathrm{Cone}(R)$ (why is $r \neq 0$?). But, this would imply $\bar{x} = x - r \in M$ and $x + r \in M$, as $x \in M, r \in \mathrm{Cone}(R) = \mathrm{Rec}(M)$, and M is convex. Thus, we arrive at $x \pm r \in M$ with $r \neq 0$, contradicting the fact that $x \in \mathrm{Ext}(M)$. Therefore, $x \in \mathrm{Conv}(V)$, and hence $x \in V$ by Fact 6.10.

7.4 ★ Majorization

In this section we will introduce and study the *Majorization Principle*, which describes the convex hull of permutations of a given vector.

For any $x \in \mathbf{R}^n$, we define $X[x]$ to be the set of all convex combinations of $n!$ vectors obtained from x by permuting its coordinates. That is,

$$X[x] := \mathrm{Conv}\left(\{Px : P \text{ is an } n \times n \text{ permutation matrix}\}\right)$$
$$= \{Dx : D \in \Pi_n\},$$

where Π_n is the set of all $n \times n$ doubly stochastic matrices. Here, the equality is due to the Birkhoff–von-Neumann Theorem. Note that $X[x]$ is a *permutationally symmetric* set, that is, given any vector from this set, the vector obtained by permuting its entries is also in the set.

Theorem 7.15 *[Majorization Principle] Given two vectors $x, y \in \mathbf{R}^n$, we have $y \in X[x]$ if and only if y satisfies*

$$\begin{aligned} s_j(y) &\leq s_j(x), \quad j = 1, \ldots, n-1, \\ y_1 + \cdots + y_n &= x_1 + \cdots + x_n, \end{aligned} \tag{7.3}$$

where $s_j(y)$ is the sum of the j largest entries of the vector y.

Proof To ease our notation, let us define the set

$$Y := \left\{ y \in \mathbf{R}^n : \begin{array}{l} s_j(y) \leq s_j(x), \ j = 1, \ldots, n-1 \\ s_n(y) = y_1 + \cdots + y_n = x_1 + \cdots + x_n = s_n(x) \end{array} \right\}.$$

Then, we need to show that $Y = X[x]$.

For any k, define $\mathcal{I}_k$ to be the family of all k-element subsets of the index set $\{1, 2, \ldots, n\}$, and so

$$s_k(y) = \max_{I \in \mathcal{I}_k} \sum_{i \in I} y_i. \tag{7.4}$$

We first prove that $Y \supseteq X[x]$. Consider any $y \in X[x]$. By the definition of $X[x]$, y can be represented as a convex combination,

$$y = \sum_\sigma \lambda_\sigma x^\sigma,$$

where the sum is taken over all permutations σ of indices $1, 2, \ldots, n$, λ_σ are convex combination weights, and x^σ is obtained from x by permutation σ of the entries, i.e.,

$$x^\sigma := [x_{\sigma(1)}; \ldots ; x_{\sigma(n)}].$$

Consequently, for every $I \in \mathcal{I}_k$ we have

$$\sum_{i \in I} y_i = \sum_{i \in I} \sum_\sigma \lambda_\sigma x_{\sigma(i)} = \sum_\sigma \lambda_\sigma \sum_{i \in I} x_{\sigma(i)} \leq \sum_\sigma \lambda_\sigma s_k(x) = s_k(x),$$

where the inequality is due to (7.4). Maximizing both sides of this inequality over $I \in \mathcal{I}_k$ and invoking (7.4) once again, we get $s_k(y) \leq s_k(x)$ for all $k \leq n$. In addition,

$$\sum_{i=1}^n y_i = \sum_{i=1}^n \sum_\sigma \lambda_\sigma x_{\sigma(i)} = \sum_\sigma \lambda_\sigma \sum_{i=1}^n x_{\sigma(i)} = \sum_\sigma \lambda_\sigma \sum_{i=1}^n x_i = \sum_{i=1}^n x_i,$$

that is, $s_n(y) = s_n(x)$. Thus, $y \in Y$, whence $Y \supseteq X[x]$.

We will now prove the difficult part of Majorization Principle, which states that $Y \subseteq X[x]$. Consider any $y \in Y$ and let us prove that $y \in X[x]$. By symmetry, we may assume without loss of generality that the vectors x and y are ordered: $x_1 \geq x_2 \geq \cdots \geq x_n$ and

$y_1 \geq y_2 \geq \cdots \geq y_n$. Assume for contradiction that $y \notin X[x]$. Since $X[x]$ is clearly a convex compact set and $y \notin X[x]$, there exists a linear form $c^\top z$ which strongly separates y and $X[x]$, i.e.,

$$c^\top y > \max_{z \in X[x]} c^\top z.$$

As the set $X[x]$ is permutationally symmetric and the vector y is ordered, without loss of generality we can select the vector c to be ordered as well. This is so because when permuting the entries of c, we preserve $\max_{z \in X[x]} c^\top z$, and arranging the entries of c in nonincreasing order, we do not decrease $c^\top y$: assuming, say, that $c_1 < c_2$, by swapping c_1 and c_2 we do not decrease $c^\top y$: $[c_2 y_1 + c_1 y_2] - [c_1 y_1 + c_2 y_2] = [c_2 - c_1][y_1 - y_2] \geq 0$. Next, by Abel's formula (the discrete analogy of integration by parts) we have

$$c^\top y = \sum_{i=1}^n c_i y_i = \sum_{i=1}^{n-1}(c_i - c_{i+1}) \sum_{j=1}^i y_j + c_n \sum_{j=1}^n y_j$$

$$= \sum_{i=1}^{n-1}(c_i - c_{i+1}) s_i(y) + c_n s_n(y)$$

$$\leq \sum_{i=1}^{n-1}(c_i - c_{i+1}) s_i(x) + c_n s_n(x) = \sum_{i=1}^n c_i x_i = c^\top x.$$

where the inequality follows from the "orderedness" of entries in c and $y \in Y$. Thus, we conclude that $c^\top y \leq c^\top x$, which is the desired contradiction as clearly $x \in X[x]$. ∎

8

Exercises for Part II

8.1 Separation

Exercise II.1 Mark with **Y** those of the cases listed below where the linear form $f^\top x$ separates the sets S and T:

- $S = \{0\} \subset \mathbf{R}$, $T = \{0\} \subset \mathbf{R}$, $f^\top x = x$
- $S = \{0\} \subset \mathbf{R}$, $T = [0, 1] \subset \mathbf{R}$, $f^\top x = x$
- $S = \{0\} \subset \mathbf{R}$, $T = [-1, 1] \subset \mathbf{R}$, $f^\top x = x$
- $S = \{x \in \mathbf{R}^3 : x_1 = x_2 = x_3\}$, $T = \{x \in \mathbf{R}^3 : x_3 \geq \sqrt{x_1^2 + x_2^2}\}$, $f^\top x = x_1 - x_2$
- $S = \{x \in \mathbf{R}^3 : x_1 = x_2 = x_3\}$, $T = \{x \in \mathbf{R}^3 : x_3 \geq \sqrt{x_1^2 + x_2^2}\}$, $f^\top x = x_3 - x_2$
- $S = \{x \in \mathbf{R}^3 : -1 \leq x_1 \leq 1\}$, $T = \{x \in \mathbf{R}^3 : x_1^2 \geq 4\}$, $f^\top x = x_1$
- $S = \{x \in \mathbf{R}^2 : x_2 \geq x_1^2, x_1 \geq 0\}$, $T = \{x \in \mathbf{R}^2 : x_2 = 0\}$, $f^\top x = -x_2$

Exercise II.2 Consider the set

$$M = \left\{ x \in \mathbf{R}^{1000} : \begin{array}{r} x_1 + x_2 + \cdots + x_{1000} \geq 1 \\ x_1 + 2x_2 + 3x_3 + \cdots + 1000 x_{1000} \geq 10 \\ x_1 + 2^2 x_2 + 3^2 x_3 + \cdots + 1000^2 x_{1000} \geq 10^2 \\ \vdots \\ x_1 + 2^{999} x_2 + 3^{999} x_3 + \cdots + 1000^{999} x_{1000} \geq 10^{999} \end{array} \right\}$$

Is it possible to separate this set from the set $\{x_1 = x_2 = \cdots = x_{1000} \leq 0\}$? If it is, what could be a separating plane?

Exercise II.3 Can the sets $S = \{x \in \mathbf{R}^2 : x_1 > 0, x_2 \geq 1/x_1\}$ and $T = \{x \in \mathbf{R}^2 : x_1 < 0, x_2 \geq -1/x_1\}$ be separated? Can they be strongly separated?

Exercise II.4 [EdYs] Let $M \subset \mathbf{R}^n$ be a nonempty closed convex set. The *metric projection* $\mathrm{Proj}_M(x)$ of a point $x \in \mathbf{R}^n$ onto M is the $\|\cdot\|_2$-closest to x point of M, so that

$$\mathrm{Proj}_M(x) \in M \quad \text{and} \quad \|x - \mathrm{Proj}_M(x)\|_2^2 = \min_{y \in M} \|x - y\|_2^2. \tag{$*$}$$

1. Prove that for every $x \in \mathbf{R}^n$ the minimum in the right-hand side of $(*)$ is achieved, and x_+ is a minimizer if and only if

$$x_+ \in M \quad \text{and} \quad \forall y \in M : [x - x_+]^\top [x_+ - y] \geq 0. \tag{8.1}$$

Derive from the latter fact that the minimum in $(*)$ is achieved at a unique point, the bottom line being that $\mathrm{Proj}_M(\cdot)$ is well defined.

2. Prove that when passing from a point $x \in \mathbf{R}^n$ to its metric projection $x_+ := \mathrm{Proj}_M(x)$, the distance to any point of M does not increase, specifically,

$$\forall y \in M : \|x_+ - y\|_2^2 \leq \|x - y\|_2^2 - \mathrm{dist}^2(x, M),$$
$$\mathrm{dist}(x, M) := \min_{u \in M} \|x - u\|_2 = \|x - x_+\|_2. \tag{8.2}$$

3. Let $x \notin M$, so that, denoting $x_+ = \mathrm{Proj}_M(x)$, the vector $e = \frac{x - x_+}{\|x - x_+\|_2}$ is well defined. Prove that the linear form $e^\top z$ strongly separates x and M, specifically,

$$\forall y \in M : \quad e^\top y \leq e^\top x - \mathrm{dist}(x, M).$$

Note: The fact just outlined underlies an alternative proof of the Separation Theorem, where the first step is to prove that a point outside a nonempty closed convex set can be strongly separated from the set. In our proof, the first step was similar, but with M restricted to be polyhedral rather than merely convex and closed.

4. Prove that the mapping $x \mapsto \mathrm{Proj}_M(x) : \mathbf{R}^n \to M$ is a *contraction* in $\|\cdot\|_2$:

$$\forall u, u' \in \mathbf{R}^n : \|\mathrm{Proj}_M(u) - \mathrm{Proj}_M(u')\|_2 \leq \|u - u'\|_2.$$

5. Let M be the probabilistic simplex, i.e., $M = \{x \in \mathbf{R}^n : x \geq 0, \sum_i x_i = 1\}$. Justify the following recipe for computing $\mathrm{Proj}_M(x)$:

Let $\psi(t) = \sum_{i=1}^m [x_i - t]_+$, where $[s]_+ := \max[s, 0]$. The function ψ is piecewise linear, with breakpoints $x_1, x_2, \ldots, x_n$; it is a continuous function of $t \in \mathbf{R}$; $\psi(t) \to +\infty$ as $t \to -\infty$, and $\psi(t) \to 0$ as $t \to +\infty$. Consequently, there exists (and can be easily computed owing to the piecewise linearity of ψ) $\hat{t} \in \mathbf{R}$ such that $\sum_i [x_i - \hat{t}]_+ = 1$. The metric projection of x onto M is just the vector x_+ with coordinates $[x_i - \hat{t}]_+$, $1 \leq i \leq n$.

What is the metric projection of the point $x = [1; 2; 2.5]$ onto the three-dimensional probabilistic simplex?

Exercise II.5 [EdYs] [Follow-up to Exercise II.4] Let $p(z) = z^n + p_{n-1} z^{n-1} + \cdots + p_1 z + p_0$, $n \geq 1$ be a polynomial of complex variable z. By the Fundamental Theorem of Algebra, p has n roots $\lambda_1, \ldots, \lambda_n$. Treating complex numbers as two-dimensional real vectors, prove that all roots of the derivative $p'(z) = n z^{n-1} + (n-1) p_{n-1} z^{n-2} + \cdots + p_1$ belong to the convex hull of $\lambda_1, \ldots, \lambda_n$.

Exercise II.6 [TrYs] Derive the statement in Remark 1.4 from the Separation Theorem.

8.2 Extreme Points

Exercise II.7 Find extreme points of the following sets:

1. $X = \{x \in \mathbf{R}^3 : x_1 + x_2 \leq 1, x_2 + x_3 \leq 1, x_3 + x_1 \leq 1\}$
2. $X = \{x \in \mathbf{R}^4 : x_1 + x_2 \leq 1, x_2 + x_3 \leq 1, x_3 + x_4 \leq 1, x_4 + x_1 \leq 1\}$

Exercise II.8 [EdYs] Let $M \subset \mathbf{R}^n$ be a nonempty closed convex set not containing lines, and $f^\top x$ be a linear function of $x \in \mathbf{R}^n$ achieving its maximum over M. Prove that among maximizers of this function on M there are extreme points of M.

8.2 Extreme Points

Exercise II.9 [Follow-up to Exercise I.8] Let A, B be subsets of $\mathbf{R}^n$. Mark with **T** those claims below which are always true (i.e., for every data satisfying premise of the claim):

1. If $\mathrm{Conv}(A) = \mathrm{Conv}(B)$ then $A = B$.
2. If $\mathrm{Conv}(A) = \mathrm{Conv}(B)$ is nonempty and $A, B, \mathrm{Conv}(A)$ are closed, then $A \cap B \neq \emptyset$.
3. If $\mathrm{Conv}(A) = \mathrm{Conv}(B)$ is nonempty and bounded, $A \cap B \neq \emptyset$.
4. If $\mathrm{Conv}(A) = \mathrm{Conv}(B)$ is nonempty, closed, and bounded, then $A \cap B \neq \emptyset$.

Exercise II.10 As can be immediately seen, the only extreme point of the nonnegative orthant $\mathbf{R}^n_+ = \mathbf{R}_+ \times \mathbf{R}_+ \times \cdots \times \mathbf{R}_+$ is the origin, that is, the vector from $\{0\} \times \{0\} \times \cdots \times \{0\}$; as we know, the extreme points of the n-dimensional unit box $\{x \in \mathbf{R}^n : 0 \leq x_i \leq 1, i \leq n\} = [0, 1] \times [0, 1] \times \cdots \times [0, 1]$ are zero/one vectors, that is, vectors from $\{0, 1\} \times \{0, 1\} \times \cdots \times \{0, 1\}$. Prove the following generalization of these observations:

> Let $X_i \subset \mathbf{R}^{n_i}$, $1 \leq i \leq K$, be closed convex sets. The set of extreme points of the direct product $X = X_1 \times \cdots \times X_K$ of these sets is the direct product of the sets of extreme points of X_i.

Exercise II.11 [EdYs] Looking at the sets of extreme points of closed convex sets such as the unit Euclidean ball, a polytope, the paraboloid $\{[x; t] : t \geq x^\top x\}$, etc., we see that these sets are closed. Do you think this is always the case? Is it true that the set $\mathrm{Ext}(M)$ of extreme points of a closed convex set M is always closed?

Exercise II.12 [TrYs] Derive representation $(*)$ in Exercise I.29 from Example 7.1 in section 7.1.1.

Exercise II.13 [EdYs] By Birkhoff's Theorem, the extreme points of the polytope $\Pi_n := \{[x_{ij}] \in \mathbf{R}^{n \times n} : x_{ij} \geq 0, \sum_i x_{ij} = 1 \, \forall j, \sum_j x_{ij} = 1 \, \forall i\}$ are exactly the Boolean matrices (i.e., with entries 0 and 1) from this set. Prove that the same holds true for the "polytope of doubly sub-stochastic" matrices $\Pi_{m,n} := \{[x_{ij}] \in \mathbf{R}^{m \times n} : x_{ij} \geq 0, \sum_i x_{ij} \leq 1 \, \forall j, \sum_j x_{ij} \leq 1 \, \forall i\}$.

Exercise II.14 [EdYs] [Follow-up to Exercise II.13] Let m, n be two positive integers with $m \leq n$, and $X_{m,n}$ be the set of $m \times n$ matrices $[x_{ij}]$ with $\sum_i |x_{ij}| \leq 1$ for all $j \leq n$ and $\sum_j |x_{ij}| \leq 1$ for all $i \leq m$. Describe the set $\mathrm{Ext}(X_{m,n})$. To get an educated guess, look at the matrices $\begin{bmatrix} 1 & 0 & 0 \\ 0 & 0 & -1 \end{bmatrix}, \begin{bmatrix} 0 & 0 & 0 \\ 0 & 0 & -1 \end{bmatrix}, \begin{bmatrix} 0.5 & -0.5 & 0 \\ -0.5 & 0.5 & 0 \end{bmatrix}$ from $X_{2,3}$.

Exercise II.15 [EdYs] [Follow-up to Exercise II.13] Let x be an $n \times n$ entrywise nonnegative matrix with all row and all column sums ≤ 1. Is it true that, for some doubly stochastic matrix $\overline{x}$, the matrix $\overline{x} - x$ is entrywise nonnegative?

Exercise II.16 [EdYs] [Assignment problem] Consider the following problem:

> There are n jobs and n workers. When worker j is assigned to job i, we get profit c_{ij}. We want to assign every worker to a job in such a way that every worker is assigned to exactly one job and every job is assigned to exactly one worker. Under this restriction, we want to maximize the total profit.

1. Pose this assignment problem as a Boolean linear programming problem (i.e., one where the decision variables are restricted to be zeros and ones).

2. Think how to solve the problem from item 1 via plain linear programming
3. [Computational study] Consider the special case of this assignment problem where all profits c_{ij} are zero or one; you can interpret $c_{ij} = 1/0$ as the fact that worker j knows/does not know how to execute job j. In this situation the assignment problem requires us to find an assignment which maximizes the total number of executed jobs. Assume now that the matrix $C = [c_{ij}]$ is generated at random, with entries taking, independently of each other, the value 1 with probability $\epsilon \in (0, 1)$ and the value 0 with probability $1 - \epsilon$. For $n \in \{4, 8, 16, 32, 64, 128, 256\}$ and $\epsilon \in \{1/2, 1/4, 1/8, 1/16\}$, run 100 simulations per pair n, ϵ to find the empirical mean of the ratio "number of executed jobs in optimal assignment"/n and observe what happens to this ratio as ϵ is fixed but $n \to \infty$.

Exercise II.17 [TrYs] Let $\nu = (\nu_1, \ldots, \nu_K)$ be a vector of positive integers ν_i, and let $\mathbf{S}^\nu = \mathbf{S}^{\nu_1} \times \cdots \times \mathbf{S}^{\nu_K}$ be the space of block-diagonal symmetric matrices, with K diagonal blocks of sizes $\nu_i \times \nu_i$, $i \leq K$; let $\mathbf{S}^\nu_+$ be the cone composed of positive semidefinite matrices from $\mathbf{S}^\nu$, and let E be an m-dimensional affine plane in $\mathbf{S}^\nu$ which intersects $\mathbf{S}^\nu_+$. The intersection $X := E \cap \mathbf{S}^\nu_+$ is a nonempty closed convex set. Assume further that X does not contain lines and thus possesses extreme points. Let W be such a point, W^{ii} be the diagonal blocks of W, and r_i be the ranks of $\nu_i \times \nu_i$ matrices W^{ii}. Prove that

$$\sum_{i=1}^{k} r_i(r_i + 1) \leq \sum_{i=1}^{K} \nu_i(\nu_i + 1) - 2m.$$

What happens in the diagonal case $\nu_1 = \cdots = \nu_K = 1$?

Exercise II.18 [EdYs] Let M be a closed convex set in $\mathbf{R}^n$ and $\bar{x}$ be a point of M.

1. Prove that if there exists a linear form $a^\top x$ such that $\bar{x}$ is the *unique* maximizer of the form on M then $\bar{x}$ is an extreme point of M.
2. Is the inverse of item 1 true, i.e., is it true that every extreme point $\bar{x}$ of M is the unique maximizer, over $x \in M$, of a properly selected linear form?

Exercise II.19 Identify and justify the correct claims in the following list:

1. Let $X \subset \mathbf{R}^n$ be a nonempty closed convex set, P be an $m \times n$ matrix, and $Y = PX := \{Px : x \in X\} \subset \mathbf{R}^m$. Then
   - For every $x \in \text{Ext}(X)$, $Px \in \text{Ext}(Y)$.
   - Every extreme point of Y is of the form Px for some $x \in \text{Ext}(X)$.
   - When X does not contain lines, every extreme point of Y is of the form Px for some $x \in \text{Ext}(X)$.
2. Let X, Y be nonempty closed convex sets in $\mathbf{R}^n$, and let $Z = X + Y$. Then
   - If $w \in \text{Ext}(Z)$ then $w = x + y$ for some $x \in \text{Ext}(X)$ and $y \in \text{Ext}(Y)$.
   - If $x \in \text{Ext}(X)$, $y \in \text{Ext}(Y)$ then $x + y \in \text{Ext}(Z)$.

Exercise II.20 [EdYs] [Faces of a polyhedral set] Let $X = \{x \in \mathbf{R}^n : a_i^\top x \leq b_i, i \leq m\}$ be a nonempty polyhedral set and $f^\top x$ be a linear form of $x \in \mathbf{R}^n$ which is bounded above on X:

$$\text{Opt}(f) = \sup_{x \in X} f^\top x < \infty.$$

8.2 Extreme Points

Prove that

1. Opt(f) is achieved – the set $\text{Argmax}_{x \in X} f^\top x := \{x \in X : f^\top x = \text{Opt}(f)\}$ is nonempty.
2. The set $\text{Argmax}_{x \in X} f^\top x$ is as follows: there exists an index set $I \subseteq \{1, 2, \ldots, m\}$, perhaps empty, such that

$$\text{Argmax}_{x \in X} f^\top x = X_I := \{x : a_i^\top x \leq b_i \,\forall i, \, a_i^\top x = b_i \,\forall i \in I\}.$$

3. Vice versa, if $I \subseteq \{1, \ldots, m\}$ is such that the set $X_I = \{x : a_i^\top x \leq b_i \,\forall i, a_i^\top x = b_i \,\forall i \in I\}$ is nonempty, then $X_I = X_* := \text{Argmax}_{x \in X} f^\top x$ for properly selected f.
   Note: Nonempty sets of the form X_I, $I \subseteq \{1, \ldots, m\}$, are called *faces* of the polyhedral set X. Note that this definition of a face is not geometric – according to it, whether a given set Y is or is not a face in X may depend not on X *per se*, but on its representation as the solution set of a finite system of linear inequalities. Facts 2 and 3, taken together, state that being a face of a polyhedral set is a geometric property – faces are exactly the sets $\text{Argmax}_{x \in X} f^\top x$ of all maximizers of linear forms bounded from above on X.
4. Extreme points of a face of X are extreme points of X.
5. Extreme points of X, if any, are exactly the faces of X which are singletons.
   Note: As a corollary of 1–3, 5, we see that the extreme points of the polyhedral set X are exactly the maximizers of those linear forms which achieve their maximum on X at a unique point.

Exercise II.21 [EdYs] [Follow-up to Exercise II.20]

1. Let $X \subset Y$ be nonempty closed convex sets in $\mathbf{R}^n$. Is it true that $\text{Ext}(Y) \cap X \subseteq \text{Ext}(X)$?
2. Let X be a nonempty closed convex set contained in the polyhedral set $\{x : Ax \leq b\}$. Assuming that the set $\overline{X} = X \cap \{x : Ax = b\}$ is nonempty, is it true that $\text{Ext}(\overline{X}) = \text{Ext}(X) \cap \overline{X}$?
3. By the result of Exercise II.13, the extreme points of the polytope $\Pi_{m,n} = \{[x_{ij}] \in \mathbf{R}^{m \times n} : x_{ij} \geq 0, \sum_i x_{ij} \leq 1 \,\forall j, \sum_j x_{ij} \leq 1 \,\forall i\}$ are exactly the Boolean matrices from this polytope. Now let $\widehat{\Pi}_{m,n}$ be the part of $\Pi_{m,n}$ cut off from $\Pi_{m,n}$ by imposing on prescribed row sums and column sums of the $m \times n$ matrix $x \in \Pi_{m,n}$ the requirement of being equal to 1, rather than ≤ 1. Assuming $\widehat{\Pi}_{m,n}$ to be nonempty, prove that the extreme points of this polytope are exactly the Boolean matrices contained in it.

Exercise II.22 Let $X \subset \mathbf{R}^m$ be a nonempty polyhedral set, $x \mapsto Px + p : \mathbf{R}^n \to \mathbf{R}^m$ be an affine mapping, and Y be the image of X under this mapping. Mark with **T** the statements in the list below which are always true (i.e., for all X, P, p compatible with the above assumptions):

1. Y is a nonempty polyhedral set.
2. If X does not contain lines, neither does Y.
3. If X does contain lines, so does Y.
4. If v is an extreme point of X then $Pv + p$ is an extreme point of Y.
5. If z is an extreme point of Y then $z = Pv + p$ for a certain extreme point v of X.

6. If z is an extreme point of Y and X does not contain lines then $z = Pv + p$ for a certain extreme point v of X.

Exercise II.23 [TrYs] Find extreme points of the following closed convex sets:

1. The set $\mathcal{S}_n = \{X \in \mathbf{S}^n : -I_n \preceq X \preceq I_n\}$.
2. The set $\mathcal{S}_n^+ = \{X \in \mathbf{S}^n : 0 \preceq X \preceq I_n\}$.
3. The set $\mathcal{D}_{k,n} = \{X \in \mathbf{S}^n : I_n \succeq X \succeq 0, \mathrm{Tr}(X) = k\}$, where k is a positive integer $\leq n$.
4. The set $\mathcal{M}_n = \{X \in \mathbf{R}^{n \times n} : \|X\|_{2,2} \leq 1\}$ ($\|\cdot\|_{2,2}$ is the spectral norm).

Exercise II.24 [TrYs] Prove the following fact (which can be considered as a matrix extension of Birkhoff's Theorem):

For positive integers d, n, let $\Pi_{d,n}$ be the set of all $n \times n$ block matrices with $d \times d$ symmetric blocks X^{ij} satisfying

$$X^{ij} \succeq 0, \quad \sum_j \mathrm{Tr}(X^{ij}) = 1 \forall i, \quad \sum_i \mathrm{Tr}(X^{ij}) = 1 \forall j.$$

The extreme points of $\Pi_{d,n}$ are exactly the block matrices $[X^{ij}]_{i,j \leq n}$ as follows: for a certain $n \times n$ permutation matrix P and unit vectors $e_{ij} \in \mathbf{R}^d$, one has

$$X^{ij} = P_{ij} e_{ij} e_{ij}^\top \forall i, j.$$

Exercise II.25 [TrYs] Let k, n be positive integers with $k \leq n$, and let $s_k(\lambda)$ for $\lambda \in \mathbf{R}^n$ be the sum of the k largest entries in λ. From the description of the extreme points of the polytope $X = \{x \in \mathbf{R}^n : 0 \leq x_i \leq 1, i \leq n, \sum_{i=1}^n x_i \leq k\}$, see Example 7.2 in section 7.1.1, it follows that when $\lambda \in \mathbf{R}_+^n$,

$$\max_{x \in X} \sum_{i=1}^n \lambda_i x_i = s_k(\lambda).$$

Prove the following matrix analogy of this fact:

For k, n as above, let $\mathcal{X} = \{(X_1, \ldots, X_n) : X_i \in \mathbf{S}^d, 0 \preceq X_i \preceq I_d, i \leq n, \sum_{i=1}^n X_i \preceq k I_d\}$. Then for $\lambda \in \mathbf{R}_+^n$ one has

$$(X_1, \ldots, X_n) \in \mathcal{X} \implies \sum_{i=1}^n \lambda_i X_i \preceq s_k(\lambda) I_d,$$

with the concluding $\preceq$ becoming an equality for properly selected $(X_1, \ldots, X_n) \in \mathcal{X}$.

8.3 Cones and Extreme Rays

Exercise II.26 [TrYs] Let X be a nonempty closed and bounded set in $\mathbf{R}^n$. Which of the following statements are true?

1. $\mathrm{Conv}(X)$ is a closed convex set.
2. $\mathrm{Cone}(X)$ is a closed cone.
3. When X is convex, $\mathrm{Cone}(X)$ is a closed cone.
4. When $0 \notin X$, $\mathrm{Cone}(X)$ is a closed cone.

5. When $0 \notin X$ and X is convex, Cone(X) is a closed cone.
6. When X is polyhedral, Cone(X) is a closed cone.

Exercise II.27 [EdYs] Let $X \subset \mathbf{R}^n$ be a nonempty polyhedral set given by the polyhedral representation:
$$X = \{x : \exists u : Ax + Bu \leq r\}$$
and let $K = \text{Cone}(X)$ be the conic hull of X.

1. Is it true that K is a closed cone?
2. Prove that $\overline{K} := \text{cl } K$ is a polyhedral cone and find a polyhedral representation of $\overline{K}$.
3. Assume that X is given by a plain – no extra variables – polyhedral representation: $X = \{x : Ax \leq b\}$. Build a plain polyhedral representation of $\overline{K} := \text{cl Cone}(X)$.

Exercise II.28 As we know, the extreme directions of the nonnegative orthant $\mathbf{R}^n_+ = \mathbf{R}_+ \times \mathbf{R}_+ \times \cdots \times \mathbf{R}_+$ are vectors with a single positive entry and remaining entries equal to 0. Prove the following generalization of this observation:

Let $X_i \subset \mathbf{R}^{n_i}$, $1 \leq i \leq K$, be closed, nontrivial, and pointed cones. The extreme directions of the direct product $X = X_1 \times \cdots \times X_K$ of these cones are the block-vectors $d = [d_1; \ldots; d_K]$, with $d_i \in \mathbf{R}^{n_i}$, of the following structure: all but one block in d are zero, and the only nonzero block is an extreme direction of the corresponding cone X_i.

Exercise II.29 Describe all extreme rays of

1. the positive semidefinite cone $\mathbf{S}^n_+$.
2. the Lorentz cone $\mathbf{L}^n$.
3. the Lorentz cone $\mathbf{L}^n$, $n \geq 2$, is a special case of the following construction: given a norm $\|\cdot\|$ on $\mathbf{R}^{n-1}$ ($n \geq 2$), we associate with it the set
$$\mathbf{K}^n_{\|\cdot\|} = \{[x; t] \in \mathbf{R}^n : t \geq \|x\|\},$$
which is a pointed nontrivial cone with a nonempty interior (why?); note that $\mathbf{L}^n = \mathbf{K}^n_{\|\cdot\|_2}$. Describe the extreme directions of $\mathbf{K}^n_{\|\cdot\|}$.

8.4 Recessive Cones

Exercise II.30 [TrYs] Let M be a convex set, and let $\bar{x}$ and h be such that $R_{\bar{x}} := \{\bar{x} + th : t \geq 0\} \subseteq M$.

1. Is it always true that whenever $x \in M$, the set $R_x = \{x + th, t \geq 0\}$ is contained in M?
2. Let h be a recessive direction of $\overline{M} = \text{cl } M$, and let $\bar{x}$ be a point from the relative interior of M. Is it always true that the set $R_{\bar{x}} = \{\bar{x} + th : t \geq 0\}$ is contained in M?

Exercise II.31 [TrYs] Let $M \subset \mathbf{R}^n$ be a cone, not necessary closed; recall that the pointedness of a cone M means that the only vector x such that $x \in M$ and $-x \in M$ is the zero vector. Which of the following statements are always true:

1. M is pointed if and only if the only representation of 0 as the sum of $k \geq 1$ vectors $x_i \in M$ is the representation with $x_i = 0$, $i \leq k$.

2. M is pointed if and only if M does not contain straight lines (one-dimensional affine planes) passing through the origin.
3. Assuming M to be closed, M is pointed if and only if M does not contain straight lines.
4. M is a pointed cone if and only if the closure of M is a pointed cone.
5. The closure of M is a pointed cone if and only if M does not contain straight lines.

Exercise II.32 A literal interpretation of the words "polyhedral cone" is: a polyhedral set $\{x : Ax \leq b\}$ which is a cone. An immediate example is the solution set $\{x : Ax \leq 0\}$ of a *homogeneous* system of linear inequalities. Prove that this example is generic: whenever a polyhedral set $K = \{x : Ax \leq b\}$ is a cone, one has $K = \{x : Ax \leq 0\}$.

Exercise II.33 [EdYs] Prove the following modification of Proposition 6.23:

(!) *Let $X \subset \mathbf{R}^N$ be a nonempty closed convex set such that $X \subseteq V + \text{Rec}(X)$ for some bounded and closed set V, let $x \mapsto \mathcal{A}(x) = Ax + b : \mathbf{R}^N \to \mathbf{R}^n$ be an affine mapping, and let $Y = \mathcal{A}(X) := \{y : \exists x \in X : y = \mathcal{A}(x)\}$ be the image of X under this mapping. Also let*

$$K := \{Ag : g \in \text{Rec}(X)\}.$$

Then the recessive cone of the closure $\overline{Y}$ of Y is the closure $\overline{K}$ of K. In particular, when K is closed (as is definitely the case when $\text{Rec}(X)$ is polyhedral), it holds that $\text{Rec}(\overline{Y}) = K$.

Exercise II.34 [EdYs] [Follow-up to Exercise II.33]
1. Let $K_1 \subset \mathbf{R}^n$, $K_2 \subset \mathbf{R}^n$ be closed cones, and let $K = K_1 + K_2$.
   - Is it always true that K is a cone?
   - Is it always true that K is closed?
   - Let K_2 be polyhedral. Is it always true that K is closed?
   - Let both K_1 and K_2 be polyhedral. Is it always true that K is closed?
2. Let X_i, $i = 1, \ldots, I$, be closed convex sets in $\mathbf{R}^n$ with nonempty intersection. Is it true that $\cap_i \text{Rec}(X_i) = \text{Rec}(\cap_i X_i)$?
3. Let X_1, X_2 be nonempty closed convex sets in $\mathbf{R}^n$, and define $K_1 = \text{Rec}(X_1)$, $K_2 = \text{Rec}(X_2)$, $\overline{X} = \text{cl}(X_1 + X_2)$, $\overline{K} = \text{cl}(K_1 + K_2)$.
   - Is it always true that $\overline{K} \subseteq \text{Rec}(\overline{X})$?
   - Is is always true that $\overline{K} = \text{Rec}(\overline{X})$?
   - Assume that $X_i \subseteq V_i + K_i$ for properly selected closed and bounded sets V_i, $i = 1, 2$. Is it true that $\overline{K} = \text{Rec}(\overline{X})$?

Exercise II.35 [TrYs] Let $f(x) = x^\top C x - c^\top x + \sigma$ be a quadratic form with $C \succeq 0$. By Exercise I.15, the set $E = \{x : f(x) \leq 0\}$ is convex (and of course closed). Assuming $E \neq \emptyset$, describe $\text{Rec}(E)$.

8.5 Around Majorization

Exercise II.36 [EdYs] Let $x \in \mathbf{R}^m$, let $X[x]$ be the convex hull of all permutations of x, and let $X_+[x]$ be the set of all vectors x' dominated by a vector form $X[x]$:

$$X_+[x] = \{y : \exists z \in X[x] : y \leq z\}.$$

1. Prove that $X_+[x]$ is a closed convex set.
2. Prove the following characterization of $X_+[x]$: $X_+[x]$ is exactly the set of solutions of the system of inequalities $s_j(y) \leq s_j(x)$, $j = 1, \ldots, m$, in variables y, where, as always $s_j(z)$ is the sum of the j largest entries in vector z.

8.6 Around Polars

Exercise II.37 Justify the last three claims in Example 6.12.

Exercise II.38 [EdYs] [More on polars]

1. Recall that, for $U \subset \mathbf{R}^n$, Vol(U) stands for the ratio of the n-dimensional volume of U and the volume of the n-dimensional unit Euclidean ball. Check that for the ellipsoid $E = \{x : x^\top C x \leq 1\}$ ($C \succ 0$), centered at the origin, we have Vol(E)Vol(Polar(E)) = 1.
2. Let $C \succ 0$ and let the ellipsoid $E = \{x : (x-c)^\top C(x-c) \leq 1\}$ contain the origin. Compute Polar(E).
3. Let X_k, $k \leq K$, be closed convex sets in $\mathbf{R}^n$ containing the origin. Prove that

$$\text{Polar}(\text{Conv}(\cup_k X_k)) = \cap_k \text{Polar}(X_k), \quad (a)$$
$$\text{Polar}(\cap_k X_k) = \text{cl Conv}(\cup_k \text{Polar}(X_k)). \quad (b)$$

Exercise II.39 [EdYs] Let $X \subset \mathbf{R}^n$ be a cone given by the polyhedral representation

$$X = \{x \in \mathbf{R}^n : \exists u : Ax + Bu \leq r\}.$$

Let X_* be the dual cone of X. Is X_* polyhedral? If yes, build a polyhedral representation of X_*.

Exercise II.40 [EdYs]

1. Let $X \subset \mathbf{R}^n$ be a nonempty polyhedral set given by the polyhedral representation

$$X = \{x \in \mathbf{R}^n : \exists u : Ax + Bu \leq r\}.$$

Is Polar(X) polyhedral? If yes, point out a polyhedral representation of Polar(X). For a non-polyhedral extension, see Exercise IV.36.
2. Compute the polars of
   - the probabilistic simplex $\Delta = \{x \in \mathbf{R}^n : x \geq 0, \sum_i x_i = 1\}$,
   - the convex hull of a nonempty finite set of points $a_1, \ldots, a_N$ from $\mathbf{R}^n$,
   - the set $\{x \in \mathbf{R}^n : x \leq b\}$.

8.7 Miscellaneous Exercises

Exercise II.41 [TrYs] Let $X = \{x \in \mathbf{R}^n : Ax \leq b\}$ be a nonempty polyhedral set.

1. Prove that X is bounded if and only if every one of the vectors $\pm e_i$ (the e_i, $1 \leq i \leq n$, are the standard basic orths) can be represented as conic combination of columns of $A^\top$.
2. Certify the statement:
   - The polyhedral set $X = \{x \in \mathbf{R}^3 : x \geq [1/3; 1/3; 1/3], \sum_{i=1}^3 x_i \leq 1\}$ is bounded.
   - The polyhedral set $X = \{x \in \mathbf{R}^3 : x_1 \geq 1/3, x_2 \geq 1/3, \sum_{i=1}^3 x_i \leq 1\}$ is unbounded.

Exercise II.42 Prove the easy part of Theorem 7.7, specifically, that every $n \times n$ permutation matrix is an extreme point of the polytope Π_n of $n \times n$ doubly stochastic matrices.

Exercise II.43 [EdYs] [Robust LP] Consider an *uncertain* linear programming problem – the family

$$\left\{ \min_{x \in \mathbf{R}^n} \{c^\top x : [A + \sum_{\nu=1}^{N} \zeta_\nu \Delta_\nu] x \leq b + \sum_{\nu=1}^{N} \zeta_\nu \delta_\nu \} : \zeta \in \mathcal{Z} \right\} \qquad (8.3)$$

of LP instances of common size (n variables, m constraints). The associated story is as follows. We want to solve an LP problem with the data not known exactly at the time when the problem is being solved; what we do know at this time is that the "true problem" belongs to the parametric family given, according to (8.3), by the "nominal data" c, A, b, the "basic perturbations Δ_ν, δ_ν" and the *perturbation set* $\mathcal{Z}$ through which run the data perturbations ζ specifying particular instances in the family. In this situation (which is quite typical for real-life applications of LP, where partial data uncertainty is the rule rather than the exception), one way to "immunize" decisions against data uncertainty is to look for *robust solutions* – those remaining feasible for all perturbations of the data from the perturbation set – by solving the *robust counterpart* (RC) of our uncertain problem – the optimization problem

$$\min_x \left\{ c^\top x : [A + \sum_{\nu=1}^{N} \zeta_\nu \Delta_\nu] x \leq b + \sum_{\nu=1}^{N} \zeta_\nu \delta_\nu \; \forall (\zeta \in \mathcal{Z}) \right\}. \qquad \text{(RC)}$$

The robust counterpart is *not* an LP program – it has finitely many decision variables and an infinite (when $\mathcal{Z}$ is "massive") system of linear constraints on these variables. Optimization problems of this type are called *semi-infinite* and are, in general, difficult to solve. However, the RC of an uncertain LP is easy, provided that $\mathcal{Z}$ is a "computation-friendly" set, for example, the nonempty set given by the polyhedral representation

$$\mathcal{Z} = \{\zeta : \exists u : P\zeta + Qu \leq r\}. \qquad (8.4)$$

Now here is the exercise *per se*:
Use LP duality to reformulate (RC) where $\mathcal{Z}$ is given by (8.4) as an explicit LP program.

Exercise II.44 [TrYs] Consider the scalar linear constraint

$$a^\top x \leq b \qquad (1)$$

with uncertain data $a \in \mathbf{R}^n$ (b is certain) varying in the set

$$\mathcal{U} = \left\{ a : |a_i - a_i^*|/\delta_i \leq 1, 1 \leq i \leq n, \sum_{i=1}^{n} |a_i - a_i^*|/\delta_i \leq k \right\}, \qquad (2)$$

where the a_i^* are the given "nominal data," the $\delta_i > 0$ are given quantities, and $k \leq n$ is an integer (in the literature, this is called "budgeted uncertainty"). Rewrite the robust counterpart

$$a^\top x \leq b, \quad \forall a \in \mathcal{U} \qquad \text{(RC)}$$

in a tractable LO form (that is, write down an explicit system (S) of linear inequalities in the variable x and additional variables such that x satisfies (RC) if and only if x can be extended to a feasible solution of (S)).

8.7 Miscellaneous Exercises

Exercise II.45 [TrYs] [Computational study, follow-up to Exercise II.43]

Preliminaries. Consider an oscillator transmitting a harmonic wave with unit wavelength and placed at some point P in $\mathbf{R}^3$. Physics says that the electric field generated by the oscillator, when measured at a remote point A, is

$$e_A(t) \approx r^{-1} \underbrace{\alpha \cos\left(\omega t - 2\pi r + \theta + 2\pi d \cos(\phi)\right)}_{E_A(t)} \tag{$*$}$$

where

- t is time, ω is the frequency,
- r is the distance from A to the origin O, d is the distance from P to the origin, $\phi \in [0, \pi]$ is the angle between the directions $\overrightarrow{OP}$ and $\overrightarrow{OA}$,
- α and θ are the actuation parameters of the oscillator.

The difference between the left- and the right-hand sides in $(*)$ is of order r^{-2} and in all our subsequent considerations can be completely ignored.

It is convenient to assemble α and θ into the *actuation weight* – the complex number $w = \alpha e^{\iota\theta}$ (ι is the imaginary unit); with this convention, we have

$$E_A(t) = \Re\left[wD_P(\phi)e^{\iota\omega t - 2\pi r}\right], \quad D_P(\phi) = e^{2\pi \iota d \cos(\phi)},$$

where $\Re[\cdot]$ stands for the real part of a complex number. The complex-valued function $D_P(\phi) : [0, \pi] \to \mathbf{C}$, called the *diagram* of the oscillator, is responsible for the directional density of the energy emitted by the oscillator: when evaluated at a certain three-dimensional direction $\vec{e}$, this density is proportional to $|D_p(\phi)|^2$, where ϕ is the angle between the direction $\vec{e}$ and the direction $\overrightarrow{OP}$. Physics says that when our transmitting antenna is composed of K harmonic oscillators located at points $P_1, \ldots, P_K$ and actuated with weights $w_1, \ldots, w_K$, the directional density of the energy transmitted by the resulting *antenna array*, as evaluated at a direction $\vec{e}$, is proportional to $|\sum_k w_k D_k(\phi_k(\vec{e}))|^2$, where $\phi_k(\vec{e})$ is the angle between the directions $\vec{e}$ and $\overrightarrow{OP_k}$.

Consider the design problem as follows. We are given linear array of K oscillators placed at the points $P_k = (k-1)\delta \mathbf{e}$, $k \leq K$, where $\mathbf{e}$ is the first standard basic orth (that is, the unit vector pointing along the positive direction of the x-axis), and $\delta > 0$ is a given distance between consecutive oscillators. Our goal is to specify actuation weights w_k, $k \leq K$ that will maximize the percentage of the energy sent along the directions which make at most a given angle γ with $\mathbf{e}$. To this end, we intend to act as follows.

We want to select actuation weights w_k, $k \leq K$, in such a way that the magnitude $|D^w(\phi)|$ of the complex-valued function

$$D^w(\phi) = \sum_{k=1}^{K} w_k e^{2\pi \iota (k-1)\delta \cos(\phi))}$$

of $\phi \in [0, \pi]$ is "concentrated" on the segment $0 \leq \phi \leq \gamma$. Let us normalize the weights by the requirement

$$D^w(0) = 1$$

and minimize under this restriction the "sidelobe level"
$$\max_{\gamma \leq \phi \leq \pi} |D^w(\phi)|$$
over w.

To get a computation-friendly version of this problem, we replace the full range $[0, \pi]$ of values of ϕ with an M-point equidistant grid
$$\Gamma = \left\{ \phi_\ell = \frac{\ell \pi}{M-1} : 0 \leq \ell \leq M-1 \right\},$$
thus converting our design problem into the optimization problem
$$\text{Opt} = \min_{t,w} \left\{ t : \begin{array}{l} \left| \sum_{k=1}^{K} w_k e^{2\pi i (k-1)\delta \cos(\phi_\ell)} \right| \leq t \, \forall (\ell : \phi_\ell > \gamma) \\ \sum_{k=1}^{K} w_k e^{2\pi i (k-1)\delta} = 1 \end{array}, w_k \in \mathbf{C}, k \leq K \right\}, \quad (P)$$
which is a convex problem in $2k+1$ real variables – the variable $t \in \mathbf{R}$ and the real and imaginary parts of $w_1, \ldots, w_K$.

The tasks of the problem are as follows:

1. Process problem (P) numerically and find the optimal design $w^n = \{w_k^n, k \leq K\}$ along with the optimal value Opt^n. Here, and in what follows, the recommended setup is:
   - the number of oscillators $K = 24$, the distance between consecutive oscillators $\delta = 0.125$;
   - $\gamma = \pi/12$;
   - the cardinality M of the equidistant grid Γ is 512.

   Draw the plot of the modulus of the resulting diagram,
   $$D^n(\phi) = \sum_{k=1}^{K} w_k^n e^{2\pi i (k-1)\delta \cos(\phi)},$$
   and compute the corresponding "energy concentration" $\mathcal{C}^n$, the concentration of a diagram $D(\cdot)$ being defined as
   $$\mathcal{C} = \frac{\sum_{\ell : \phi_\ell \leq \gamma} \sin(\phi_\ell) |D(\phi_\ell)|^2}{\sum_{\ell=1}^{M} \sin(\phi_\ell) |D(\phi_\ell)|^2}.$$
   Up to discretization of ϕ, this is the ratio of the energy emitted in the "cone of interest" (i.e., along the directions making angles at most γ with $\mathbf{e}$) to the total emitted energy. The factors $\sin(\phi_\ell)$ reflect the fact that when computing the energy emitted in a spatial cone, we should integrate $|D(\cdot)|^2$ over the part of the unit sphere in three dimensions cut off the sphere by the cone.

2. Now note that "in reality" the optimal weights w_k^n, $k \leq K$ are used to actuate physical devices and as such cannot be implemented with the same 16-digit accuracy with which they are computed; they will be subject to small implementation errors. We can model these errors by assuming that the "real-life" diagram is
   $$D(\phi) = \sum_{k=1}^{K} w_k^n (1 + \rho \xi_k) e^{2\pi i (k-1)\delta \cos(\phi)}$$

where $\rho \geq 0$ is some (perhaps small) perturbation level and $\xi_k \in \mathbf{C}$ are "primitive" perturbations responsible for the implementation errors and running through the unit disk $\{\xi : |\xi| \leq 1\}$. For the sake of definiteness, we can assume that the ξ_k are independent across k random variables uniformly distributed on the unit circumference in $\mathbf{C}$. Now the diagram becomes random and can violate the constraints of (P), unless $\rho = 0$; in the latter case, the diagram is the "nominal" one given by the optimal weights w^n, so that it satisfies the constraints of (P), with t set to Opt^n.

Now, what happens when $\rho > 0$? In this case, the diagram $D(\cdot)$ and its deviation v from the prescribed value 1 at the origin, its sidelobe level $\mathfrak{l} = \max_{\ell:\phi_\ell > \gamma} |D(\phi_\ell)|$, and its energy concentration become random. A crucial "real-life" question is how large are "typical values" of these quantities. To get an impression of what happens, carry out a numerical experiment as follows:

- select a perturbation level $\rho \in \{10^{-\ell} : 1 \leq \ell \leq 6\}$;
- for this selected ρ, simulate and plot 100 realizations of the modulus of the actual diagram, and find empirical averages $\bar{v}$ of v, $\bar{\mathfrak{l}}$ of $\mathfrak{l}$, and $\bar{\mathcal{C}}$ of $\mathcal{C}$.

3. Apply the robust optimization methodology from Exercise II.43 to build solutions to (P) that are "immunized against implementation errors", compute these solutions for perturbation levels $\rho = 10^{-\ell}$, $1 \leq \ell \leq 6$, and subject the resulting designs to numerical study similar to that outlined in the previous item.

Note: (P) is *not* a linear programming program, so one cannot formally apply the results stated in Exercise II.43; what can be applied is the robust optimization "philosophy."

Exercise II.46 [EdYs] Prove the following statement, "symmetric" to the Dubovitski–Milutin Lemma:

The cone M_* dual to the arithmetic sum of k (closed or not) cones $M^i \subset \mathbf{R}^n$, $i \leq k$, is the intersection of the k cones M_*^i dual to M^i.

Exercise II.47 [EdYs] Prove the following polyhedral version of the Dubovitski–Milutin Lemma:

Let $M^1, \ldots, M^k$ be polyhedral cones in $\mathbf{R}^n$, and let $M = \cap_i M^i$. Then, the cone M_* dual to M is the sum of cones M_*^i, $i \leq k$, dual to M^i, so that a linear form $e^\top x$ is nonnegative on M if and only it can be represented as the sum of linear forms $e_i^\top x$ that are nonnegative on the respective cones M_i.

Exercise II.48 [EdYs] [Follow-up to Exercise II.47] Let $A \in \mathbf{R}^{m \times n}$ be a matrix with trivial kernel, let $e \in \mathbf{R}^n$, and suppose the set

$$X = \{x : Ax \geq 0, e^\top x = 1\} \tag{$*$}$$

is nonempty and bounded. Prove that there exists $\lambda \in \mathbf{R}^m$ such that $\lambda > 0$ and $A^\top \lambda = e$.

Prove the "partial inverse" of this statement: *if* $\mathrm{Ker}\, A = \{0\}$ *and* $e = A^\top \lambda$ *for some* $\lambda > 0$, *the set X is bounded.*

Exercise II.49 [EdYs] Let E be a linear subspace in $\mathbf{R}^n$, K be a closed cone in $\mathbf{R}^n$, and $\ell(x) : E \to \mathbf{R}$ be a linear (linear, not affine!) function which is nonnegative on $K \cap E$. Which of the following claims is/are always true?

1. $\ell(\cdot)$ can be extended from E onto the entire space $\mathbf{R}^n$ to yield a linear function which is nonnegative on K.
2. Assuming $(\text{int } K) \cap E \neq \emptyset$, $\ell(\cdot)$ can be extended from E onto the entire space $\mathbf{R}^n$ to yield a linear function which is nonnegative on K.
3. Assuming, in addition to $\ell(x) \geq 0$ for $x \in K \cap E$, that $K = \{x : Px \leq 0\}$ is a polyhedral cone, $\ell(\cdot)$ can be extended from E onto the entire space $\mathbf{R}^n$ to yield a linear function which is nonnegative on K.

Exercise II.50 Let $n > 1$. Is the unit $\|\cdot\|_2$-ball $B_n = \{x \in \mathbf{R}^n : \|x\|_2 \leq 1\}$ a polyhedral set? Justify your answer.

Exercise II.51 [TrYs] The unit box $\{x \in \mathbf{R}^n : -1 \leq x_i \leq 1, i \leq n\}$ is cut off from $\mathbf{R}^n$ by a system of $m = 2n$ linear inequalities and is a nonempty and bounded polyhedral set. However, when we eliminate any inequality from this system, the solution set of the resulting system becomes unbounded. To see that this situation is in a sense extreme, prove the following claim:

Consider the solution set of a system of m linear inequalities in n variables x, i.e., the set

$$X := \{x \in \mathbf{R}^n : Ax \leq b\},$$

where $A = [a_1^\top; a_2^\top; \ldots; a_m^\top]$. Suppose that X is nonempty and bounded. Then, whenever $m > 2n$, one can eliminate from this system a properly selected inequality in such a way that the solution set of the resulting subsystem remains bounded.

A provocative follow-up: Is it possible to cut off from $\mathbf{R}^{1000}$ a bounded set by using *only* a single linear inequality?

Exercise II.52 [TrYs] [Computational study] Let $\omega^N = (\omega_1, \ldots, \omega_N)$ be an N-element i.i.d. sample drawn from the standard Gaussian distribution (zero mean, unit covariance) on $\mathbf{R}^d$. How many extreme points are there in the convex hull of the points from the sample?

1. Consider the planar case $d = 2$ and think how to list extreme points of $\text{Conv}\{\omega_1, \ldots, \omega_N\}$. Fill in the following table:

N	2	4	8	16	32	64	128
U							
M							
L							

where U is the maximal number, M is the mean, and L is the minimal number of extreme points observed when processing 100 samples ω^N of a given cardinality.

2. Think how to upper-bound the expected number of extreme points in $W = \text{Conv}(\omega^N)$.

Exercise II.53 [TrYs] [Computational study] Given positive integers m, n, with $n \geq 2$, consider a randomly generated system $Ax \leq b$ of m linear inequalities with n variables. We assume that A, b are generated by drawing the entries, independently of each other, from $\mathcal{N}(0, 1)$.

8.7 Miscellaneous Exercises

1. Consider the planar case, i.e., $n = 2$. For $m = 2, 4, 8, 16$, generate 100 samples of $m \times 2$ systems and fill in the following table:

m	2	4	8	16
F				
B				

where F is the number of feasible systems and B is the number of feasible systems with bounded solution sets.

An aside: Related theoretical results originating from [Nem24, Exercise 2.23] are as follows. Given positive integers m, n with $n \geq 2$, consider a homogenous system $Ax \leq 0$ of m inequalities with n variables. We call this system *regular* if its matrix A is regular. Here, a matrix Q is regular if all square submatrices of Q are nonsingular. Clearly, the entries of a regular matrix are nonzero, and when a $p \times q$ matrix Q is drawn at random from a probability distribution on $\mathbf{R}^{p \times q}$ which has a density w.r.t. the Lebesgue measure, Q is regular with probability 1.

Given a regular $m \times n$ homogeneous system of inequalities $Ax \leq 0$, let $g_i(x) = \sum_{j=1}^{n} A_{ij} x_j$, $i \leq m$, so that the g_j are nonconstant linear functions. Setting $\Pi_i = \{x : g_i(x) = 0\}$, we get a collection of m hyperplanes in $\mathbf{R}^n$ passing through the origin. For a point $x \in \mathbf{R}^n$, the *signature* of x is, by definition, the m-dimensional vector $\sigma(x)$ of signs of the reals $g_i(x)$, $1 \leq i \leq m$. Denoting by Σ the set of all m-dimensional vectors with entries ± 1, for $\sigma \in \Sigma$ the set $\mathcal{C}_\sigma = \{x : \sigma(x) = \sigma\}$ is either empty, or is a nonempty open convex set; when it is nonempty, let us call it a *cell* associated with A, and the corresponding σ an A-feasible signature. Clearly, for a regular system, $\mathbf{R}^n$ is the union of all hyperplanes Π_i and all cells associated with A. It turns out that

The number $N(m, n)$ of cells associated with a regular homogeneous $m \times n$ system $Ax \leq 0$ is independent of the system and is given by a simple recurrence:

$$N(1, 2) = 2$$
$$m \geq 2, n \geq 2 \implies N(m, n) = N(m-1, n) + N(m-1, n-1)$$
$$[N(m, 1) = 2, m \geq 1].$$

Next, when A is drawn at random from a probability distribution P on $\mathbf{R}^{m \times n}$ which possesses a *symmetric* density p, that is, such that $p([a_1^\top; a_2^\top; \ldots; a_m^\top]) = p([\epsilon_1 a_1^\top; \epsilon_2 a_2^\top; \ldots; \epsilon_m a_m^\top])$ for all $A = [a_1^\top; a_2^\top; \ldots; a_m^\top]$ and all $\epsilon_i = \pm 1$, then *the probability that a vector $\sigma \in \Sigma$ is an A-feasible signature is*

$$\pi(m, n) = N(m, n)/2^m.$$

In particular, the probability that the system $Ax \leq 0$ has a solution set with a nonempty interior (this is simply the A-feasibility of the signature $[-1; \ldots; -1]$) is $\pi(m, n)$.

The inhomogeneous version of these results is as follows. An $m \times n$ system of linear inequalities $Ax \leq b$ is called regular if the matrix $[A, -b]$ is regular. Setting $g_i(x) = \sum_{j=1}^{n} A_{ij} x_j - b_i$, $i \leq n$, the $[A, b]$-signature of x is, as above, the vector of signs of the reals $g_i(x)$. For $\sigma \in \Sigma$, the set $\mathcal{C}_\sigma = \{x : \sigma(x)) = \sigma\}$ is either empty or is a nonempty open convex set; in the latter case, we call $\mathcal{C}_\sigma$ an $[A, b]$-cell and σ an $[A, b]$-feasible signature. Setting $\Pi_i = \{x : g_i(x) = 0\}$, we get m hyperplanes in $\mathbf{R}^n$, and the entire space $\mathbf{R}^n$ is the union of those hyperplanes and all $[A, b]$-cells. It turns out that

The number $N(m, n)$ of cells associated with a regular $m \times n$ system $Ax \leq b$ is independent of the system and is equal to $\frac{1}{2}N(m+1, n+1)$.

In addition, when an $m \times (n+1)$ matrix $[A, b]$ is drawn at random from a probability distribution on $\mathbf{R}^{m \times (n+1)}$ possessing a symmetric density w.r.t. the Lebesgue distribution, *the probability that a given $\sigma \in \Sigma$ is an $[A, b]$-feasible signature is*

$$\overline{\pi}(m, n) = N(m+1, n+1)/2^{m+1}.$$

In particular, the probability that the system $Ax \leq b$ is strictly feasible (i.e., the system $Ax < b$ is feasible) is $\overline{\pi}(m, n)$.

2. Accompanying exercise: Prove that if A is an $m \times n$ regular matrix then the system $Ax \leq 0$ has a nonzero solution if and only if the system $Ax < 0$ is feasible. Derive from this fact that if $[A, b]$ is regular then the system $Ax \leq b$ is feasible if and only if it is strictly feasible, and that when the system $Ax \leq 0$ has a nonzero solution, the system $Ax \leq b$ is strictly feasible for every b.
3. Use the results from the Aside to compute the expected values of F and B; see item 1.

Exercise II.54 [TrYs] [Computational study]

1. For $v = 1, 2, \ldots, 6$, generate 100 systems of linear inequalities $Ax \leq b$ with $n = 2^v$ variables and $m = 2n$ inequalities, the entries in A, b being drawn, independently of each other, from $\mathcal{N}(0, 1)$. Fill in the following table:

n	2	4	8	16	32	64
F						
$\mathbf{E}\{F\}$						
B						

F is the number of feasible systems in the sample;
B is the number of feasible systems with bounded solution sets

To compute the expected value of F, use the results from [Nem24, Exercise 2.23] cited in item 2 of Exercise II.53.

2. Carry out an experiment similar to that in item 1, but with $m = n+1$ rather than $m = 2n$.

n	2	4	8	16	32	64
F						
$\mathbf{E}\{F\}$						
B						
$\mathbf{E}\{B\}$						

F is the number of feasible systems in the sample;
B is the number of feasible systems with bounded solution sets

Part III

Convex Functions

9

First Acquaintance with Convex Functions

9.1 Definition and Examples

Definition 9.1 *[Convex function] A function* $f : Q \to \mathbf{R}$ *defined on a subset Q of $\mathbf{R}^n$ and taking real values is called* convex *if*

- *the domain Q of the function is convex, and*
- *for every $x, y \in Q$ and every $\lambda \in [0, 1]$ one has*

$$f(\lambda x + (1 - \lambda)y) \leq \lambda f(x) + (1 - \lambda)f(y). \tag{9.1}$$

The function f is called strictly convex *whenever the above inequality holds strictly for every $x \neq y$ and $0 < \lambda < 1$.*

A function f such that $-f$ is convex is called *concave*. In particular, the domain Q of a concave function f is convex, and the function f itself satisfies the inequality opposite to (9.1), i.e.,

$$f(\lambda x + (1 - \lambda)y) \geq \lambda f(x) + (1 - \lambda)f(y), \quad \forall x, y \in Q \text{ and } \forall \lambda \in [0, 1].$$

Example 9.1 The simplest example of a convex function is an *affine function*

$$f(x) = a^\top x + b,$$

which is simply the sum of a linear form and a constant. This function is clearly convex on the entire space, and in this case the convexity inequality (9.1) holds as an equality everywhere. In fact, an affine function is both convex and concave. Moreover, it is easily seen that a function which is both convex and concave on the entire space must be affine. ◇

Example 9.2 Here are several elementary examples of nonlinear convex functions of one variable:

- functions convex on the entire axis:
  x^{2p}, where p is a positive integer;
  $\exp(x)$;
  $\exp(-x)$;
- functions convex on the nonnegative ray:
  x^p, where $p \geq 1$;
  $-x^p$, where $0 \leq p \leq 1$;
  $x \ln x$;

- functions convex on the positive ray:
  $1/x^p$, where $p > 0$;
  $-\ln x$.

At the moment it is not clear why these functions are convex. We will soon derive a simple analytic criterion for detecting convexity which will immediately demonstrate that the above functions are indeed convex. ◊

A very convenient equivalent definition of a convex function is in terms of its *epigraph*. Given a real-valued function f defined on a subset Q of $\mathbf{R}^n$, we define its epigraph as the set

$$\mathrm{epi}\{f\} := \left\{[x;t] \in \mathbf{R}^{n+1} : x \in Q, \, t \geq f(x)\right\}.$$

Geometrically, to define the epigraph, we plot the *graph* of the function, i.e., the surface $\{(x,t) \in \mathbf{R}^{n+1} : x \in Q, \, t = f(x)\}$ in $\mathbf{R}^{n+1}$, and add to this surface all points which are "above" it. The epigraph allows us to give an equivalent, more geometric, definition of a convex function as follows.

Proposition 9.2 *[Epigraph-based definition of convex function] A function $f : Q \to \mathbf{R}$ defined on a subset Q of $\mathbf{R}^n$ is convex if and only if its epigraph is a convex set in $\mathbf{R}^{n+1}$.*

Proof Let $f : Q \to \mathbf{R}$ be convex; we will show that $\mathrm{epi}(f)$ is convex. Consider any two points $[x';t']$, $[x'';t'']$ from $\mathrm{epi}\{f\}$ and any $\lambda \in [0,1]$. Then $x', x'' \in Q$ and

$$\lambda[x';t'] + (1-\lambda)[x'';t''] = [\underbrace{\lambda x' + (1-\lambda)x''}_{:=x}; \underbrace{\lambda t' + (1-\lambda)t''}_{:=t}],$$

and since f is convex, we see that Q is convex and $x \in Q$. Moreover, by definition, $t = \lambda f(x') + (1-\lambda)f(x'') \geq f(x)$, where the inequality follows from the convexity of f. Then $[x;t] \in \mathrm{epi}\{f\}$. Thus $\mathrm{epi}\{f\}$ is convex.

For the other direction, suppose that $\mathrm{epi}\{f\}$ is convex. Consider any $x', x'' \in Q$ and $\lambda \in [0,1]$. Then we have $[x'; f(x')] \in \mathrm{epi}\{f\}$ and $[x''; f(x'')] \in \mathrm{epi}\{f\}$ by the definition of $\mathrm{epi}\{f\}$. Since $\mathrm{epi}\{f\}$ is convex, the point

$$\lambda[x'; f(x')] + (1-\lambda)[x''; f(x'')] = [\lambda x' + (1-\lambda)x''; \lambda f(x') + (1-\lambda)f(x'')]$$

is in $\mathrm{epi}\{f\}$ as well. This, by definition of $\mathrm{epi}\{f\}$, implies the relations $\lambda x' + (1-\lambda)x'' \in Q$ and $f(\lambda x' + (1-\lambda)x'') \leq \lambda f(x') + (1-\lambda)f(x'')$, so that f is convex. ∎

9.1.1 More Examples of Convex Functions: Norms

Equipped with Proposition 9.2, we can extend our initial list of convex functions (affine functions and several one-dimensional functions) to more examples, namely *norms*. Let $\|x\|$ be a norm on $\mathbf{R}^n$ (see section 1.1.2). So far, we encountered three examples of norms: the Euclidean (ℓ_2-) norm $\|x\|_2 = \sqrt{x^\top x}$, the ℓ_1-norm $\|x\|_1 = \sum_i |x_i|$, and the ℓ_∞-norm $\|x\|_\infty = \max_i |x_i|$. It has also been claimed (although not proved) that these are three members from an infinite family of norms

$$\|x\|_p := \left(\sum_{i=1}^n |x_i|^p\right)^{1/p}, \text{ where } 1 \leq p \leq \infty$$

(the right-hand side of the latter relation for $p = \infty$ is, by definition, $\max_i |x_i|$).

We say that a function $f : \mathbf{R}^n \to \mathbf{R}$ is *positively homogeneous of degree* 1 if it satisfies

$$f(tx) = tf(x), \quad \forall x \in \mathbf{R}^n, \ t \geq 0.$$

Also, we say that the function $f : \mathbf{R}^n \to \mathbf{R}$ is *subadditive* if it satisfies

$$f(x + y) \leq f(x) + f(y), \quad \forall x, y \in \mathbf{R}^n.$$

Note that every norm is positively homogeneous of degree 1 and subadditive. We are about to prove that all such functions (in particular, all norms) are convex:

Proposition 9.3 *Let $\pi(x)$ be a real-valued function on $\mathbf{R}^n$ which is positively homogeneous of degree 1. Then π is convex if and only if it is subadditive.*

Proof Note that the epigraph of a function π that is positively homogeneous of degree 1 is a conic set since for any $\lambda \geq 0$ we have that $[x; t] \in \text{epi}\{\pi\}$ implies $\lambda[x; t] \in \text{epi}\{\pi\}$. Moreover, by Proposition 9.2, π is convex if and only if $\text{epi}\{\pi\}$ is convex. It is clear that a conic set is convex if and only if it contains the sum of every pair of its elements (why?). This latter property is satisfied for the epigraph of a real-valued function if and only if the function is subadditive (this should be evident). ∎

9.2 Jensen's Inequality

The following basic observation is, we believe, one of the most useful observations ever made.

Proposition 9.4 *[Jensen's inequality] Let $f : Q \to \mathbf{R}$ be convex. Then, for every convex combination of points x^i from Q, i.e.,*

$$\sum_{i=1}^N \lambda_i x^i,$$

for some $\lambda \in \mathbf{R}_+^N$ satisfying $\sum_{i=1}^N \lambda_i = 1$, we have

$$f\left(\sum_{i=1}^N \lambda_i x^i\right) \leq \sum_{i=1}^N \lambda_i f(x^i).$$

Proof Note that the points $[x^i; f(x^i)]$ belong to the epigraph of f. As f is convex, its epigraph is a convex set. Then, for any $\lambda \in \mathbf{R}_+^N$ satisfying $\sum_{i=1}^N \lambda_i = 1$, we have that the corresponding convex combination of the points given by

$$\sum_{i=1}^N \lambda_i [x^i; f(x^i)] = \left[\sum_{i=1}^N \lambda_i x^i; \sum_{i=1}^N \lambda_i f(x^i)\right]$$

also belongs to epi$\{f\}$. By definition of the epigraph, this means exactly that

$$\sum_{i=1}^{N} \lambda_i f(x^i) \geq f\left(\sum_{i=1}^{N} \lambda_i x^i\right).$$

∎

Note that the definition of convexity of a function f is exactly the requirement on f to satisfy the Jensen inequality for the case $N = 2$. We see that to satisfy this inequality for $N = 2$ is the same as to satisfy it for *all* $N \geq 2$.

Remark 9.5 An instructive interpretation of Jensen's inequality is as follows: Given a convex function $f : \text{Dom } f \to \mathbf{R}$, consider a discrete random variable x taking values $x^i \in \text{Dom } f$, $i \leq N$, with probabilities λ_i. Then,

$$f(\mathbf{E}[x]) \leq \mathbf{E}[f(x)],$$

where $\mathbf{E}[\cdot]$ stands for the expectation operator. The resulting inequality, under mild regularity conditions, holds true for general-type random vectors x taking values in Dom f with probability 1. ◇

9.3 Convexity of Sublevel Sets

The *sublevel set* of a function $f : Q \to \mathbf{R}$ given by $\alpha \in \mathbf{R}$ is defined as

$$\text{lev}_\alpha(f) := \{x \in Q : f(x) \leq \alpha\}.$$

We have the following simple yet useful observation on the sublevel sets of convex functions.

Proposition 9.6 *[Convexity of sublevel sets] Let $f : Q \to \mathbf{R}$ be convex. Then, for every $\alpha \in \mathbf{R}$, the sublevel set of f given by $\alpha \in \mathbf{R}$, i.e., $\text{lev}_\alpha(f)$, is convex.*

Proof Suppose $x, y \in \text{lev}_\alpha(f)$ and $\lambda \in [0, 1]$. Then, from the convexity of f, we have $f(\lambda x + (1 - \lambda)y) \leq \lambda f(x) + (1 - \lambda)f(y) \leq \lambda \alpha + (1 - \lambda)\alpha = \alpha$, so that $\lambda x + (1 - \lambda)y \in \text{lev}_\alpha(f)$. ∎

It is important to note that the convexity of sublevel sets does *not* characterize convex functions; there are nonconvex functions which possess this property (e.g., every monotone function on the axis has all of its sublevel sets convex). Thus, the convexity of sublevel sets specifies a wider family of functions, the so-called *quasiconvex* functions. The "proper" characterization of convex functions in terms of convex sets is given by Proposition 9.2 – convex functions are exactly the functions with convex epigraphs.

9.4 Value of a Convex Function Outside Its Domain

In its literal meaning, a function is not defined outside its domain and thus does not have any associated "value" outside its domain. Nevertheless, when speaking of *convex* functions, it is extremely convenient to think that the function also has a value outside its domain, namely, it takes the value $+\infty$. With this convention, we revise our definition of convex functions as follows.

9.4 Value of a Convex Function Outside Its Domain

A convex function f on $\mathbf{R}^n$ is a function taking values in the extended real axis $\mathbf{R} \cup \{+\infty\}$ and such that for all $x, y \in \mathbf{R}^n$ and all $\lambda \in [0, 1]$ one has

$$f(\lambda x + (1-\lambda)y) \leq \lambda f(x) + (1-\lambda)f(y). \tag{9.2}$$

When f takes values in the extended real axis, some terms in the inequality (9.2) may involve infinities. In such a case, the left- and right-hand side values of this inequality as well as its validity status are determined according to the standard conventions on the operations of summation, multiplication, and comparison in the "extended real line" $\mathbf{R} \cup \{+\infty\} \cup \{-\infty\}$. These conventions are as follows:

- Operations with real numbers are understood in their usual sense.
- The sum of $+\infty$ and a real number, like the sum of $+\infty$ and $+\infty$, is $+\infty$. Similarly, the sum of a real number and $-\infty$, like the sum of $-\infty$ and $-\infty$, is $-\infty$. The sum of $+\infty$ and $-\infty$ is undefined.
- The product of a real number and $+\infty$ is $+\infty$, 0, or $-\infty$, depending on whether the real number is positive, zero, or negative, and similarly for the product of a real number and $-\infty$. The product of two "infinities" is again infinity, with the usual rule for assigning the sign to the product.
- Finally, any real number is $< +\infty$ and $> -\infty$, and of course $-\infty < \infty$.

For a function f taking values in the extended real axis, the set of points where the function value is finite is called the *domain* of f and is denoted by $\mathrm{Dom}\, f$. On the basis of our revised definition of convex functions on the extended real axis, the function f that is defined to be identically equal to $+\infty$ is a legitimate convex function. A convex function with nonempty domain (that is, a convex function which is not identically $+\infty$) is called *proper*.

When f takes all its values in $\mathbf{R} \cup \{+\infty\}$, the inequality (9.2) is automatically valid when $\lambda = 0$ or $\lambda = 1$; when $0 < \lambda < 1$, it is automatically valid when $x = y$, or when at least one of the points x, y is not in $\mathrm{Dom}\, f$. Thus, we arrive at the following equivalent definition of convex functions:

A function $f : \mathbf{R}^n \to \mathbf{R} \cup \{+\infty\}$ is convex if and only if the inequality (9.2) holds for every $x \neq y$, $x, y \in \mathrm{Dom}\, f$, and for every $\lambda \in (0, 1)$.

Note that our initial definition of a convex function included the requirement for the domain of the function to be convex; our new, equivalent, definition of a convex function does not include such a requirement – after f is extended outside of its domain by $+\infty$, the inequality (9.2) automatically takes care of the convexity of $\mathrm{Dom}\, f$.

The simplest function with a given domain Q is identically zero on Q and identically $+\infty$ outside of Q. This function, called the *characteristic* (a.k.a. *indicator*) *function of* Q [1] is convex if and only if Q is a convex set.

It is convenient to think of a convex function as of something which is defined everywhere, since it saves a lot of words. For example, with this convention we can write $f + g$ (f and g are convex functions on $\mathbf{R}^n$), and everybody will understand what is meant. Without this convention, we would need to add to this expression the following explanation: "$f + g$ is a function whose domain is the intersection of the domains of f and g, and in this intersection it is defined as $(f + g)(x) = f(x) + g(x)$."

---

[1] This terminology is standard in convex analysis; in other areas of mathematics, the characteristic, a.k.a. indicator, function of a set $Q \subset \mathbf{R}^n$ is defined as the function equal to 1 on the set and to 0 outside the set.

10

How to Detect Convexity

In an optimization problem

$$\min_x \{f(x) : g_j(x) \leq 0, \, j = 1, \ldots, m\}$$

the convexity of the objective function f and the constraint functions g_i is crucial. Indeed, convex problems – those with convex f and g_j – possess nice theoretical properties with important practical implications. For example, the local *necessary* optimality conditions for these problems are *sufficient for global optimality*. Moreover, much more importantly, convex problems can be efficiently solved (both in the theoretical and, to some extent, in the practical meaning of the word), which is not, unfortunately, the case for general nonconvex problems. This is why it is so important to know how to detect the convexity of a given function.

The scheme of our investigation is typical for mathematics. Let us start with a well-known example from analysis. How does one detect the continuity of a function? Of course, there is a definition of continuity in terms of ϵ and δ, but it would be an actual disaster if each time we needed to prove the continuity of a function, we were supposed to write down the proof that "for every positive ϵ there exists positive δ such that ...". In fact, we use another approach: we list once and for all a number of standard operations which preserve continuity (such as addition, multiplication, taking superpositions, etc.) and point out a number of standard examples of continuous functions (such as the power function or the exponent, etc.). Note that both steps, proving that the operations in the list preserve continuity and proving that the standard functions are continuous, take a certain amount of effort and indeed are done in $\epsilon - \delta$ terms. But, after investing in this effort once, typically proving the continuity of a given function becomes a much simpler task: it suffices to demonstrate that the function can be obtained, in finitely many steps, from our "raw materials" (standard functions which are known to be continuous) by applying our "machinery" (the combination rules which preserve continuity). Normally, this demonstration is given by a single word "evident" or is even understood by default.

This is exactly the case with convexity. We will next point out the list of operations which preserve convexity and a number of standard convex functions.

10.1 Operations Preserving Convexity of Functions

We start with the following basic operations preserving the convexity of functions:

- *Stability under taking nonnegative weighted sums:* If f, g are convex functions on $\mathbf{R}^n$ then their linear combination $\lambda f + \mu g$ with *nonnegative* coefficients $\lambda, \mu \in \mathbf{R}_+$ is also convex. [This can be verified straightforwardly using the convex function definition.]
- *Stability under affine substitutions of the argument:* Given a convex function f on $\mathbf{R}^n$ and an affine mapping $x \mapsto Ax + b$ from $\mathbf{R}^m$ into $\mathbf{R}^n$, the superposition $f(Ax + b)$ is convex. [This can be proved directly by verifying the convex function definition or by noting that the epigraph of the superposition is the inverse image of the epigraph of f under an affine mapping.]
- *Stability under taking pointwise supremum:* Given any nonempty (and possibly infinite!) family of convex functions $\{f_\alpha(\cdot)\}_{\alpha \in \mathcal{A}}$ on $\mathbf{R}^n$, their supremum $\sup_{\alpha \in \mathcal{A}} f_\alpha(\cdot)$ is convex.

[Note that the epigraph of the supremum is clearly the intersection of the epigraphs of the functions from the family. Then, recall that the intersection of any family of convex sets is convex.]

- *Convex monotone superposition:* Let $f(x) := [f_1(x), \ldots, f_K(x)]$ be a vector-valued map with convex component functions $f_i : \mathbf{R}^n \to \mathbf{R} \cup \{+\infty\}$, and let F be a convex function on $\mathbf{R}^K$. Suppose F is *monotone nondecreasing*, i.e., for any $z, z' \in \mathbf{R}^K$ satisfying $z \leq z'$ we always have $F(z) \leq F(z')$. Then, the superposition function given by

$$\phi(x) := F(f(x)) = F(f_1(x), \ldots, f_K(x)) : \mathbf{R}^n \to \mathbf{R} \cup \{+\infty\}$$

is convex.

Here, note that the expression $F(f_1(x), \ldots, f_K(x))$ makes no evident sense at a point x where some f_i take the value of $+\infty$. At a such point, *by definition*, we assign the value of $+\infty$ to the superposition function.

Let us now justify the convex monotone composition rule. Consider any $x, x' \in \mathrm{Dom}\,\phi$. Then $z := f(x)$ and $z' := f(x')$ are vectors from $\mathbf{R}^K$ which belong to Dom F. Owing to the convexity of the components of f, for any $\lambda \in (0, 1)$ we have the vector inequality

$$f(\lambda x + (1 - \lambda)x') \leq \lambda z + (1 - \lambda)z'.$$

In particular, the left-hand side in this inequality is a vector from $\mathbf{R}^K$, i.e., it has no "infinite entries," and we may further use the monotonicity of F to arrive at

$$\phi(\lambda x + (1 - \lambda)x') = F(f(\lambda x + (1 - \lambda)x')) \leq F(\lambda z + (1 - \lambda)z').$$

Moreover, using the convexity of F we deduce

$$F(\lambda z + (1 - \lambda)z') \leq \lambda F(z) + (1 - \lambda)F(z').$$

Then, combining these two inequalities and noting that $F(z) = F(f(x)) = \phi(x)$ and $F(z') = F(f(x')) = \phi(x')$, we arrive at the desired convexity relation

$$\phi(\lambda x + (1 - \lambda)x') \leq \lambda \phi(x) + (1 - \lambda)\phi(x').$$

Imagine how many extra words would be necessary here if there were no convention on the value of a convex function outside its domain!

In the convex monotone superposition rule, the monotone nondecreasing property of F is crucial. (Look what happens when $n = K = 1$, $f_1(x) = x^2$, $F(z) = -z$). This rule, however, admits the following two useful variants where the monotonicity requirement is somewhat relaxed (the justifications of these variants are left to the reader):

- *Convex affine superposition:* Let $F(z)$ be a convex function on $\mathbf{R}^K$, and let the functions $f_i(x)$, $i \leq K$, be convex functions on $\mathbf{R}^n$. Suppose that for some $k \leq K$ the functions $f_1, \ldots, f_k$ are affine, and the function $F(z)$ is nondecreasing in the entries z_s of z with indices $s > k$. Then, the function $F(f_1(x), \ldots, f_K(x))$ is convex.
- Let $F(z)$ be a convex function on $\mathbf{R}^K$, and let the functions $f_i(x)$, $i \leq K$, be convex functions on $\mathbf{R}^n$. Define $f(x) := [f_1(x); \ldots; f_K(x)]$. Let Y be a convex set in $\mathbf{R}^K$ such that $f(x) \in Y$ whenever all entries in $f(x)$ are finite. Suppose that for some $k \leq K$ the functions $f_1, \ldots, f_k$ are affine. Assume, next, that $F(z)$ is nondecreasing in only the entries z_s, $s > k$, of z on Y, i.e., $F(z') \geq F(z)$ whenever z', z are such that $z', z \in Y$ and $z'_s = z_s$ for all $s \leq k$ and $z'_s \geq z_s$ for all $s > k$. Then $F(f(x))$ is convex on $\mathbf{R}^n$.
Example: Let $f_i(x)$ be convex, $i \leq K$, and $F(z) := \|z\|_1$. Note that in general the function $F(f(x))$ is *not* necessarily convex (look what happens when $n = K = 1$ and $f_1(x) = x^2 - 1$). However, $F(f(x))$ is convex provided that the $f_i(x)$ *are not just convex, but are nonnegative as well* (set $Y = \mathbf{R}^K_+$). More generally, the functions $\|f(x)\|_p$, $1 \leq p \leq \infty$, are convex provided that each component $f_i : \mathbf{R}^n \to \mathbf{R} \cup \{+\infty\}$, $i \leq K$ of the vector function $f(x) = [f_1(x); \ldots; f_K(x)]$ is convex *and nonnegative*.

We close this section with two more convexity-preserving operations:

- *Stability under partial minimization:* If $f(x, y) : \mathbf{R}^n_x \times \mathbf{R}^m_y \to \mathbf{R} \cup \{+\infty\}$ is convex (as a function of $z = [x; y]$; this is called *joint convexity*) and the function

$$g(x) := \inf_y f(x, y)$$

is greater than $-\infty$ everywhere, then g is convex.

The justification of this is as follows. First, the only values that convex functions can take are real numbers and $+\infty$, and in the case of g this is assumed. Now, consider any $x, x' \in \text{Dom } g$ and any $\lambda \in [0, 1]$. Define $x'' := \lambda x + (1 - \lambda)x'$. We need to show that $g(x'') \leq \lambda g(x) + (1 - \lambda)g(x')$. There is clearly nothing to prove when $\lambda = 0$ or $\lambda = 1$, same as when $0 < \lambda < 1$ and x, or x', or both do not belong to Dom g. Thus, we assume that $x, x' \in \text{Dom } g$. For any positive ϵ, we can find y_ϵ and y'_ϵ such that $[x; y_\epsilon] \in \text{Dom } f$, $[x'; y'_\epsilon] \in \text{Dom } f$, and $g(x) + \epsilon \geq f(x, y_\epsilon)$, $g(x') + \epsilon \geq f(x', y'_\epsilon)$. Taking a weighted sum of these two inequalities, we get

$$\begin{aligned} \lambda g(x) + (1 - \lambda)g(x') + \epsilon &\geq \lambda f(x, y_\epsilon) + (1 - \lambda)f(x', y'_\epsilon) \\ &\geq f(\lambda x + (1 - \lambda)x', \lambda y_\epsilon + (1 - \lambda)y'_\epsilon) \\ &= f(x'', \lambda y_\epsilon + (1 - \lambda)y'_\epsilon), \end{aligned}$$

where the last inequality follows from the convexity of f. By the definition of $g(x'')$ we have $f(x'', \lambda y_\epsilon + (1 - \lambda)y'_\epsilon) \geq g(x'')$, and thus we get $\lambda g(x) + (1 - \lambda)g(x') + \epsilon \geq g(x'')$. In particular, $x'' \in \text{Dom } g$ (recall that $x, x' \in \text{Dom}(g)$ and thus $g(x), g(x') \in \mathbf{R}$). Moreover, since the resulting inequality is valid for all $\epsilon > 0$, we come to $g(x'') \leq \lambda g(x) + (1 - \lambda)g(x')$, as required.

- *Perspective transform of a convex function:* Given a convex function f on $\mathbf{R}^n$, we define the function $g(x,y) := yf(x/y)$ with domain $\{[x;y] \in \mathbf{R}^{n+1} : y > 0, x/y \in \mathrm{Dom}\, f\}$ to be its *perspective function*. The perspective function of a convex function is convex.

  Let us first examine a direct justification of this. Consider any $[x';y']$ and $[x'';y'']$ from Dom g and any $\lambda \in [0,1]$. Define $x := \lambda x' + (1-\lambda)x''$, $y := \lambda y' + (1-\lambda)y''$. Then, $y > 0$. We also define $\lambda' := \lambda y'/y$ and $\lambda'' := (1-\lambda)y''/y$, so that $\lambda', \lambda'' \geq 0$ and $\lambda' + \lambda'' = 1$. As f is convex, we deduce that $x/y = \lambda x'/y + (1-\lambda)x''/y = \lambda' x'/y' + \lambda'' x''/y'' = \lambda' x'/y' + (1-\lambda')x''/y'' \in \mathrm{Dom}\, f$ and $f(x/y) \leq \lambda' f(x'/y') + (1-\lambda')f(x''/y'')$. Thus, as $y > 0$, we arrive at $yf(x/y) \leq y\lambda' f(x'/y') + y(1-\lambda')f(x''/y'') = \lambda[y' f(x'/y')] + (1-\lambda)[y'' f(x''/y'')]$, that is, $g(x,y) \leq \lambda g(x',y') + (1-\lambda)g(x'',y'')$.

  Here is an alternative, smarter, justification. There is nothing to prove when Dom $f = \emptyset$. So, suppose that Dom $f \neq \emptyset$. Consider the epigraph $\mathrm{epi}(f) = \{[x;s] : s \geq f(x)\}$ and the perspective transform of this nonempty convex set, which is given by (see section 1.5)

  $$\begin{aligned}
  \mathrm{Persp}(\mathrm{epi}\{f\}) &:= \left\{[[x;s];t] \in \mathbf{R}^{n+2} : t > 0,\ [x/t;s/t] \in \mathrm{epi}\{f\}\right\} \\
  &= \left\{[x;s;t] \in \mathbf{R}^{n+2} : t > 0,\ s/t \geq f(x/t)\right\} \\
  &= \left\{[x;s;t] \in \mathbf{R}^{n+2} : t > 0,\ s \geq tf(x/t)\right\} \\
  &= \left\{[x;s;t] \in \mathbf{R}^{n+2} : t > 0,\ x/t \in \mathrm{Dom}\, f,\ s \geq tf(x/t)\right\} \\
  &= \left\{[x;s;t] \in \mathbf{R}^{n+2} : [x;t] \in \mathrm{Dom}\, g,\ s \geq g(x,t)\right\} \\
  &= \left\{[x;s;t] \in \mathbf{R}^{n+2} : [x;t;s] \in \mathrm{epi}\{g\}\right\},
  \end{aligned}$$

  where the second from last equality follows from the fact that by the definition of $g(x,t)$, whenever $t > 0$ the inclusion $x/t \in \mathrm{Dom}\, f$ holds if and only if $[x;t] \in \mathrm{Dom}\, g$. Thus, we observe that $\mathrm{Persp}(\mathrm{epi}\{f\})$ is just the image of $\mathrm{epi}\{g\}$ under the one-to-one linear transformation $[x;t;s] \mapsto [x;s;t]$. As $\mathrm{Persp}(\mathrm{epi}\{f\})$ is a convex set (recall from section 1.5 that the perspective transform of a nonempty convex set is convex), we conclude that g is convex.

Now that we have presented the basic operations preserving the convexity of a function, let us look at the standard convex functions to which these operations can be applied. We have already seen several cases in Example 9.2; but we still do not know why these functions are convex. The usual way to check the convexity of a "simple" function (i.e., a function given by a simple formula) is based on *differential criteria of convexity*, which we will examine in the next section.

10.2 Criteria of Convexity

The definition of the convexity of a function immediately reveals that *convexity is a one-dimensional property*: a function f on $\mathbf{R}^n$ taking values in $\mathbf{R} \cup \{+\infty\}$ is convex if and only if its restriction on every line, i.e., every function $g : \mathbf{R} \to \mathbf{R} \cup \{+\infty\}$ of the type $g(t) := f(x + th)$ with $x, h \in \mathbf{R}^n$ is convex.

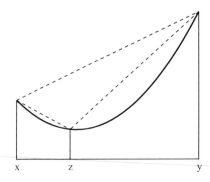

Figure 10.1 Univariate convex function $f : [x, y] \to \mathbf{R}$. The average rate of change of f on the entire segment $[x, y]$ is in between the average rates of change "at the beginning," i.e., when passing from x to z, and "at the end," i.e., when passing from z to y.

We are about to show that an univariate function $f : \mathbf{R} \to \mathbf{R} \cup \{+\infty\}$ is convex if and only if for every three real numbers $x < z < y$ such that $x, y \in \text{Dom } f$ we have $z \in \text{Dom } f$ and the average rate $\frac{f(y)-f(x)}{y-x}$ at which f varies when moving from x to y is in between the average rate $\frac{f(z)-f(x)}{z-x}$ at which f changes "at the beginning," i.e., when moving from x to z, and the average rate $\frac{f(y)-f(z)}{y-z}$ at which f changes "at the end," i.e., when moving from z to y, see Figure 10.1:

$$\frac{f(z)-f(x)}{z-x} \leq \frac{f(y)-f(x)}{y-x} \leq \frac{f(y)-f(z)}{y-z},$$

so that

$$\begin{aligned}\frac{f(z)-f(x)}{z-x} &\leq \frac{f(y)-f(x)}{y-x}, \\ \frac{f(y)-f(x)}{y-x} &\leq \frac{f(y)-f(z)}{y-z}, \\ \frac{f(z)-f(x)}{z-x} &\leq \frac{f(y)-f(z)}{y-z}.\end{aligned} \qquad (10.1)$$

As is immediately seen, every one of the three inequalities in (10.1) implies the other two.

Here is a justification for the above characterization of the convexity of a univariate function f. Note that this convexity is simply the requirement that, for any real numbers $x, y \in \text{Dom } f$ with $x < y$ and every $\lambda \in (0, 1)$, for $z_\lambda := (1 - \lambda)x + \lambda y$ it holds that $f(z_\lambda) \leq (1 - \lambda)f(x) + \lambda f(y)$, or, which is the same,

$$f(z_\lambda) - f(x) \leq \lambda(f(y) - f(x)). \qquad (10.2)$$

When $\lambda \in (0, 1)$, the pair $(1, \lambda)$ is a positive multiple of the pair $(y - x, z_\lambda - x)$, thus (10.2) is equivalent to $(y-z)(f(z_\lambda) - f(x)) \leq (z_\lambda - x)(f(y) - f(x))$. Note that this inequality is the same as $\frac{f(z_\lambda)-f(x)}{z_\lambda - x} \leq \frac{f(y)-f(x)}{y-x}$. As λ runs through the interval $(0, 1)$ the point z_λ runs through the entire set $\{z : x < z < y\}$, and so we conclude that f is convex if and only if for every triple $x < z < y$ with $x, y \in \text{Dom } f$ the first inequality in (10.1) holds true. As every one of the inequalities in (10.1) implies the other two, this justifies our "average rate of change" characterization of univariate convexity.

In the case of multivariate convex functions, we have the following immediate consequence of the preceding observations.

Lemma 10.1 *A function $f : \mathbf{R}^n \to \mathbf{R} \cup \{+\infty\}$ is convex if and only if for every three distinct points w, w', w'' in $\mathbf{R}^n$ with $w' \in [w, w'']$ and $w, w'' \in \text{Dom } f$ it holds that $w' \in \text{Dom } f$ and*

$$\frac{f(w') - f(w)}{\|w' - w\|_2} \leq \frac{f(w'') - f(w')}{\|w'' - w'\|_2}. \tag{10.3}$$

Proof As we have already seen, f is convex if and only if for every line $\ell = \{w_\ell(t) := u + th : t \in \mathbf{R}\}$ the univariate function $\phi_\ell(t) := f(u + th)$ is convex; the latter automatically holds when $h = 0$, thus, it suffices to require the convexity of ϕ_ℓ only when $h \neq 0$, which we assume from now on. Next, we have seen that ϕ_ℓ is convex if and only if for every real number $x < z < y$ with $x, y \in \text{Dom } \phi_\ell$ one has $z \in \text{Dom } \phi_\ell$ and

$$\frac{\phi_\ell(z) - \phi_\ell(x)}{z - x} \leq \frac{\phi_\ell(y) - \phi_\ell(z)}{y - z}. \tag{$*$}$$

Setting $w = w_\ell(x)$, $w' = w_\ell(z)$, $w'' = w_\ell(y)$, we have

$$\frac{\|w' - w\|_2}{\|w'' - w'\|_2} = \frac{z - x}{y - z},$$

so that ($*$) is just (10.3). It remains to note that the triples w, w', w'' of distinct vectors with $w' \in [w, w'']$ and $w, w'' \in \text{Dom } f$ are exactly the triples $w_\ell(x), w_\ell(z), w_\ell(y)$ stemming from different lines ℓ and triples of reals $x < z < y$ with $x, y \in \text{Dom } \phi_\ell$. ∎

To sum up, to detect the convexity of a function, in principle it suffices to know how to detect the convexity of functions of a single variable. Moreover, this latter question can be resolved by the standard calculus tools.

10.2.1 Differential Criteria of Convexity

We have from calculus the following simple and complete characterization of the convexity of smooth univariate functions.

Proposition 10.2 *[Necessary and sufficient condition for the convexity of smooth univariate functions] Let (a, b) be an interval on the axis (where the cases of $a = -\infty$ and/or $b = +\infty$ are also possible). Then,*

(i) *A function f that is differentiable everywhere on (a, b) is convex on (a, b) if and only if its derivative f' is monotonically nondecreasing on (a, b).*
(ii) *A function f that is twice differentiable everywhere on (a, b) is convex on (a, b) if and only if its second derivative f'' is nonnegative everywhere on (a, b).*

Proof (i): We start by proving the necessity of the stated condition. Suppose that f is differentiable and convex on (a, b). We will prove that then f' is monotonically nondecreasing. Let $x < y$ be two points from the interval (a, b), and let us prove that $f'(x) \leq f'(y)$. Consider any $z \in (x, y)$. Invoking the convexity of f and applying (10.1), we have

$$\frac{f(z) - f(x)}{z - x} \leq \frac{f(y) - f(z)}{y - z}.$$

Passing to the limit as $z \to x + 0$, we get

$$f'(x) \leq \frac{f(y) - f(x)}{y - x},$$

and passing to the limit in the same inequality as $z \to y - 0$, we arrive at

$$\frac{f(y) - f(x)}{y - x} \leq f'(y),$$

and so $f'(x) \leq f'(y)$, as claimed.

Let us now prove the sufficiency of the condition (i). Thus, we assume that f' exists and is nondecreasing on (a, b), and we will verify that f is convex on (a, b). By the "average rate of change" description of the convexity of a univariate function, all we need to do is to verify that if $x < z < y$ and $x, y \in (a, b)$ then

$$\frac{f(z) - f(x)}{z - x} \leq \frac{f(y) - f(z)}{y - z}.$$

This is indeed evident: by the Lagrange Mean Value Theorem, the left-hand side ratio is $f'(u)$ for some $u \in (x, z)$, and the right-hand side ratio is $f'(v)$ for some $v \in (z, y)$. Since $v > u$ and f' is nondecreasing on (a, b), we conclude that the left-hand side ratio is indeed less than or equal to the right-hand side ratio.

(ii): This part is an immediate consequence of (i) as we know from calculus that a differentiable function – in our present case this is the function f' – is monotonically nondecreasing on an interval if and only if its derivative is nonnegative on this interval. ∎

Proposition 10.2 immediately allows us to verify the convexity of the functions listed in Example 9.2. To this end, the only difficulty which we may encounter is that some of these functions (e.g., x^p with $p \geq 1$, and $-x^p$ with $0 \leq p \leq 1$) are claimed to be convex on the half-interval $[0, +\infty)$, while Proposition 10.2 talks about the convexity of functions on open intervals. This difficulty can be addressed with the following simple result, which allows us to extend the convexity of continuous functions beyond open sets.

Proposition 10.3 *Let M be a convex set and let f be a function with $\mathrm{Dom}\, f = M$. Suppose that f is convex on $\mathrm{rint}\, M$ and is continuous on M, i.e.,*

$$f(x_i) \to f(x), \quad \text{as } i \to \infty,$$

whenever $x_i, x \in M$ and $x_i \to x$ as $i \to \infty$. Then f is convex on M.

Proof Consider any $x, y \in M$ and any $\lambda \in [0, 1]$. Define $z := \lambda x + (1 - \lambda) y$. We need to prove that

$$f(z) \leq \lambda f(x) + (1 - \lambda) f(y).$$

As $x, y \in M$ and M is convex, by Theorem 1.29 (iii) there exist sequences $\{x_i\}_{i \geq 1} \in \mathrm{rint}\, M$ and $\{y_i\}_{i \geq 1} \in \mathrm{rint}\, M$ converging to x and to y, respectively. Then the sequence

$z_i := \lambda x_i + (1 - \lambda) y_i$ is in rint M and it converges to z as $i \to \infty$. Since f is convex on rint M, for all $i \geq 1$ we have

$$f(z_i) \leq \lambda f(x_i) + (1 - \lambda) f(y_i).$$

By taking the limits of both sides of this inequality, noting that f is continuous on M, and that as $i \to \infty$ the sequences x_i, y_i, z_i converge to $x, y, z \in M$, respectively, we obtain the desired inequality. ∎

Example 10.1 We are now able to justify the claim, made as early as in section 1.1.2, that the functions $\|\cdot\|_p$, $1 \leq p \leq \infty$, are norms.

So far, we have verified this only for $p = 1, 2, \infty$, in Remark 1.7. Now consider the case $p \in (1, \infty)$. In order to prove that $\|\cdot\|_p$ is indeed a norm, using Fact 1.8 all we need to do is to show that the set $V := \{x \in \mathbf{R}^n : \|x\|_p \leq 1\}$ is closed, bounded, symmetric with respect to the origin, contains a neighborhood of the origin, and is convex. All these except the convexity are evident. So, we will prove that V is indeed convex. Define the function $f(x) := \|x\|_p^p = \sum_{i=1}^n |x_i|^p$. Then $V = \{x \in \mathbf{R}^n : f(x) \leq 1\}$. By Proposition 10.2, the univariate function $|x|^p$ is convex (recall that $p \in (1, \infty)$). Moreover, by the "calculus of convexity" (section 10.1), f is convex (as it is the sum of convex functions). Thus, V is the sublevel set of a convex function, and by Proposition 9.6, it is convex. ◇

In the preceding illustration of the convexity of norms we started to examine multivariate functions. Let us continue our discussion of multivariate functions by examining a differential criterion for their convexity. Propositions 10.2(ii) and 10.3 give us the following convenient *necessary and sufficient condition* for the convexity of *smooth multivariate* functions.

Corollary 10.4 *[Convexity criterion for smooth multivariate functions] Let $f : \mathbf{R}^n \to \mathbf{R} \cup \{+\infty\}$ be a function whose domain $Q := \mathrm{Dom}\, f$ is a convex set with a nonempty interior. Suppose that f is*

- *continuous on Q, and*
- *twice differentiable on $\mathrm{int}\, Q$.*

Then f is convex on Q if and only if for all $x \in \mathrm{int}\, Q$ its Hessian matrix $f''(x)$ is positive semidefinite, i.e.,

$$h^\top f''(x) h \geq 0, \quad \forall h \in \mathbf{R}^n.$$

That is, f is convex on Q if and only if the second-order directional derivative of f taken at any point $x \in \mathrm{int}\, Q$ along any direction $h \in \mathbf{R}^n$ is nonnegative, i.e., for all $x \in \mathrm{int}\, Q$, and $h \in \mathbf{R}^n$ we have

$$h^\top f''(x) h = \frac{d^2}{dt^2}\Big|_{t=0} f(x + th) \geq 0.$$

Proof The "only if" part is evident: if f is convex on Q and $x \in \mathrm{int}\, Q$ then for any fixed direction $h \in \mathbf{R}^n$ the function $g : \mathbf{R} \to \mathbf{R} \cup \{+\infty\}$, defined as

$$g(t) := f(x + th),$$

is convex on the axis (recall that affine substitutions of argument preserve convexity). Since f is twice differentiable in a neighborhood of x, the function g is twice differentiable in a neighborhood of $t = 0$. Thus, by Proposition 10.2, we have $0 \leq g''(0) = h^\top f''(x) h$.

In order to prove the "if" part we need to show that every function $f : Q \to \mathbf{R} \cup \{+\infty\}$ that is continuous on Q and that satisfies $h^\top f''(x) h \geq 0$ for all $x \in \mathrm{int}\, Q$ and all $h \in \mathbf{R}^n$ is convex on Q.

Let us first prove that f is convex on $\mathrm{int}\, Q$. By Theorem 1.29, $\mathrm{int}\, Q$ is a convex set. Since the convexity of a function on a convex set is a one-dimensional property, all we need to prove is that for any $x, y \in \mathrm{int}\, Q$ the univariate function $g : [0, 1] \to \mathbf{R} \cup \{+\infty\}$ given by

$$g(t) := f(x + t(y - x))$$

is convex on the segment $[0, 1]$. As f is twice differentiable on $\mathrm{int}\, Q$, g is continuous and twice differentiable on the segment $[0, 1]$ and its second derivative is given by

$$g''(t) = (y - x)^\top f''(x + t(y - x))(y - x) \geq 0,$$

where the inequality follows from the premise on f. Then, by Propositions 10.2(ii) and 10.3, g is convex on $[0, 1]$. Thus, f is convex on $\mathrm{int}\, Q$. As f is convex on $\mathrm{int}\, Q$ and is continuous on Q, by Proposition 10.3 we conclude that f is convex on Q. ∎

Corollary 10.5 *[Sufficient condition for strict convexity of smooth functions] Consider the setting of Corollary 10.4. Suppose, in addition, that $Q := \mathrm{Dom}\, f$ is open and the Hessian of f is positive definite on Q, i.e.,*

$$\frac{d^2}{dt^2} f(x + th) > 0, \quad \forall x \in Q \text{ and } \forall h \neq 0.$$

Then f is strictly convex.

Proof Consider any $x, y \in Q$ with $x \neq y$ and any $\lambda \in (0, 1)$. We need to show that $f(\lambda x + (1 - \lambda) y) < \lambda f(x) + (1 - \lambda) f(y)$. Consider the function $\phi : [0, 1] \to \mathbf{R}$ given by $\phi(t) := f(tx + (1 - t)y)$. Then, as f is twice differentiable on Q, ϕ is twice differentiable on $[0, 1]$. Moreover, from the premise on f, we have $\phi''(t) > 0$ for all $t \in [0, 1]$. Note that our target inequality is simply the relation $\phi(\lambda) < \lambda \phi(1) + (1 - \lambda) \phi(0)$. Since $0 < \lambda < 1$, we can rewrite this target inequality as

$$\frac{\phi(\lambda) - \phi(0)}{\lambda} < \frac{\phi(1) - \phi(\lambda)}{1 - \lambda}.$$

Finally, by the Lagrange Mean Value Theorem and the strict monotonicity of ϕ' we conclude that the desired target inequality holds. ∎

We conclude this section by highlighting that the convexity of many "complicated" functions can be proved easily by the application of a combination of the "calculus of convexity" rules to simple functions which pass the "infinitesimal" convexity tests.

Example 10.2 Consider the following *exponential "posynomial"* function $f : \mathbf{R}^n \to \mathbf{R}$, given by

$$f(x) = \sum_{i=1}^{N} c_i \exp(a_i^\top x),$$

where the coefficients c_i are positive (this is why the function is called *posynomial*). This function is in fact convex on $\mathbf{R}^n$. How can we prove this?

An immediate proof is as follows:

1. The function $\exp(t)$ is convex on $\mathbf{R}$ as its second-order derivative is positive, as required by the infinitesimal convexity test for smooth univariate functions.
2. Thus, by the stability of convexity under affine substitutions of the argument, we deduce that all functions $\exp(a_i^\top x)$ are convex on $\mathbf{R}^n$.
3. Finally, by the stability of convexity under taking linear combinations with nonnegative coefficients, we conclude that f is convex on $\mathbf{R}^n$.

And if we wanted to prove that the maximum of three exponential posynomials is convex? Then, we would just need to add to our three steps above a fourth, which refers to the stability of convexity under taking a pointwise supremum. ◇

10.3 Important Multivariate Convex Functions

Let us start with some simple yet important multivariate convex functions with the convexity justified by solely the "calculus of convexity" presented in section 10.1.

Example 10.3 Let k, n be two positive integers such that $k \leq n$. Consider the function $s_k : \mathbf{R}^n \to \mathbf{R}$ given by

$$s_k(x) := \sum_{i=1}^{k} x_{[i]},$$

where $x_{[i]}$ denotes the ith largest entry in the vector x, where $i = 1, \ldots, n$. That is, given any vector $x \in \mathbf{R}^n$, we have $x_{[1]} \geq x_{[2]} \geq \cdots \geq x_{[n]}$. By definition, $s_k(x)$ is simply the sum of the k largest elements in x. We claim that $s_k(x)$ is a convex function of x. Given any index set I, the function $\ell_I(x) := \sum_{i \in I} x_i$ is a linear function of x and thus it is convex. Now, $s_k(x)$ is clearly the maximum of the linear functions $\ell_I(x)$ over all index sets I with exactly k elements from $\{1, \ldots, n\}$, and as such is convex.

Note also that $s_k(x)$ is a *permutation symmetric* function of x, that is, the value of the function $s_k(x)$ remains the same when entries in its argument x are permuted. Taking together the convexity and the permutation symmetry of $s_k(x)$ will be very useful in our development regarding functions of eigenvalues of symmetric matrices in chapter 14. ◇

Remark 10.6 An immediate implication of Example 10.3 is that, given a vector $x \in \mathbf{R}^n$, the function $\max_i \{x_i\}$, i.e., the value of its largest element, is convex in x. Thus, the function corresponding to the value of the minimum element in x, i.e., $\min_i \{x_i\} = -\max_i \{-x_i\}$, is concave in x. That said, for $1 < k < n$, the "intermediate element" function, i.e., the function given by $x_{[k]}$, which stands for the kth largest element in x, is neither a convex nor a concave function of x. ◇

While the "calculus of convexity" presented in section 10.1 is sufficient and quite practical in proving the convexity of many functions, there are still several multivariate functions for which convexity seemingly cannot be extracted from this calculus and needs to be verified via Corollary 10.4. Here are some important examples.

Example 10.4 The function $f : \mathbf{R}^n \to \mathbf{R}$ given by

$$f(x) := \ln\left(\sum_{i=1}^{n} \exp(x_i)\right)$$

is convex.

Let us first verify the convexity of this function via direct computation using Corollary 10.4. To this end, we define $p_i := \frac{\exp(x_i)}{\sum_j \exp(x_j)}$. Then, the second-order directional derivative of f along the direction $h \in \mathbf{R}^n$ is given by

$$\omega := \left.\frac{d^2}{dt^2}\right|_{t=0} f(x+th) = \sum_i p_i h_i^2 - \left(\sum_i p_i h_i\right)^2.$$

Observing that $p_i > 0$ and $\sum_i p_i = 1$, we see that ω is the variance (the expectation of the square minus the squared expectation) of a discrete random variable taking values h_i with probabilities p_i, and it is well known that the variance of any random variable is always nonnegative. Here is a direct verification of this fact:

$$\left(\sum_i p_i h_i\right)^2 = \left(\sum_i \sqrt{p_i}(\sqrt{p_i} h_i)\right)^2 \le \left(\sum_i p_i\right)\left(\sum_i p_i h_i^2\right) = \sum_i p_i h_i^2,$$

where the inequality follows from the Cauchy–Schwarz inequality and the last equality holds since $\sum_i p_i = 1$.

In fact, in this example we can prove the convexity of f via the calculus of convexity as well. Recall that $\ln(z) = \min_{s \in \mathbf{R}}\{z \exp(s) - s - 1\}$. Thus, $\ln\left(\sum_i \exp(x_i)\right) = \min_{s \in \mathbf{R}}\{(\sum_i \exp(x_i + s)) - s + 1\}$. Then, by the stability of convexity under partial minimization we conclude that f is convex.

An even simpler proof is as follows:

$$\mathrm{epi}\left\{\ln\left(\sum_i \exp(x_i)\right)\right\} = \left\{[x;t] : t \ge \ln\left(\sum_i \exp(x_i)\right)\right\}$$

$$= \left\{[x;t] : \exp(t) \ge \sum_i \exp(x_i)\right\}$$

$$= \left\{[x;t] : \sum_i \exp(x_i - t) \le 1\right\},$$

and the concluding set is convex (as a sublevel set of a convex function). ◇

Example 10.5 The function $f : \mathbf{R}^n_+ \to \mathbf{R}$ given by

$$f(x) = -\prod_{i=1}^n x_i^{\alpha_i},$$

where $\alpha_i > 0$ for all $1 \leq i \leq n$ and satisfy $\sum_i \alpha_i \leq 1$, is convex.

To prove the convexity of f via Corollary 10.4 all we need to do is to verify that, for any $x \in \mathbf{R}^n$ satisfying $x > 0$ and for any $h \in \mathbf{R}^n$, we have $\frac{d^2}{dt^2}\big|_{t=0} f(x + th) \geq 0$. Let $\eta_i := h_i/x_i$; then direct computation shows that

$$\frac{d^2}{dt^2}\bigg|_{t=0} f(x+th) = -\left[\left(\sum_i \alpha_i \eta_i^2\right) - \left(\sum_i \alpha_i \eta_i\right)^2\right] f(x)$$

and, as we have seen in Example 10.4, the premise on α implies that

$$\left(\sum_i \alpha_i \eta_i\right)^2 = \left(\sum_i \sqrt{\alpha_i}(\sqrt{\alpha_i}\eta_i)\right)^2 \leq \left(\sum_i \alpha_i\right)\left(\sum_i \alpha_i \eta_i^2\right) \leq \sum_i \alpha_i \eta_i^2,$$

where the first inequality follows from the Cauchy–Schwarz inequality and the second inequality holds since $\sum_i \alpha_i \leq 1$. Noting that $f(x) \leq 0$ whenever $x > 0$ completes the proof. ◇

Example 10.5 admits the following immediate extension:

Example 10.6 The function $f : \text{int } \mathbf{R}^n_+ \to \mathbf{R}$ given by

$$f(x) = \prod_{i=1}^n x_i^{-\alpha_i},$$

where $\alpha_i > 0$ for all $1 \leq i \leq n$ is convex.

Here the "calculus of convexity" already works: the function

$$\ln(f(x)) = -\sum_{i=1}^n \alpha_i \ln(x_i)$$

is convex on int $\mathbf{R}^n_+$ as the function $g(y) = -\ln y$ is convex on the positive ray. It remains to note that taking the exponent preserves convexity by the convex monotone superposition rule. ◇

10.4 Gradient Inequality

We next present an extremely important property of convex functions:

Proposition 10.7 *[Gradient inequality]* Let $f : \mathbf{R}^n \to \mathbf{R} \cup \{+\infty\}$, $x \in \text{Dom}(f)$ and let Q be a convex set containing x. Suppose that

(i) f is convex on Q, and
(ii) f is differentiable at x (see Definition C.1 and item 3 at the end of section C.1.2).

Let $\nabla f(x)$ be the gradient of the function f at x. Then the following inequality holds:
$$f(y) \geq f(x) + (y-x)^\top \nabla f(x), \quad \forall y \in Q. \tag{10.4}$$

Geometrically, this relation states that the graph of the function f restricted the set Q, i.e.,
$$\left\{(y,t) \in \mathbf{R}^{n+1} : y \in \mathrm{Dom}\, f \cap Q,\ t = f(y)\right\}$$

lies above the graph
$$\{(y,t) \in \mathbf{R}^{n+1} : t = f(x) + (y-x)^\top \nabla f(x)\}$$

of the linear form tangent to f at x.

Proof Let $y \in Q$. There is nothing to prove if $y \notin \mathrm{Dom}\, f$ (since then the left-hand side of the gradient inequality is $+\infty$). Similarly, there is nothing to prove when $y = x$. Thus, we can assume that $y \neq x$ and $y \in \mathrm{Dom}\, f$. Let us set
$$y_\tau := x + \tau(y-x), \quad \text{where } 0 \leq \tau \leq 1,$$

so that $y_0 = x$, $y_1 = y$, and y_τ is an interior point of the segment $[x,y]$ for $0 < \tau < 1$. Applying Lemma 10.1 to the triple (w, w', w'') taken as (x, y_τ, y), we get
$$\frac{f(x + \tau(y-x)) - f(x)}{\tau \|y-x\|_2} \leq \frac{f(y) - f(x)}{\|y-x\|_2};$$

as $\tau \to +0$, the left-hand side of this inequality, by the definition of the gradient, tends to $\frac{(y-x)^\top \nabla f(x)}{\|y-x\|_2}$, and so we get
$$\frac{(y-x)^\top \nabla f(x)}{\|y-x\|_2} \leq \frac{f(y) - f(x)}{\|y-x\|_2},$$

and, as $\|y-x\|_2 > 0$, this is equivalent to
$$(y-x)^\top \nabla f(x) \leq f(y) - f(x).$$

Note that this inequality is exactly the same as (10.4). ∎

Corollary 10.8 *Suppose Q is a convex set with a nonempty interior and f is continuous on Q and differentiable on $\mathrm{int}\, Q$. Then, f is convex on Q if and only if the gradient inequality (10.4) is valid for every pair $x \in \mathrm{int}\, Q$ and $y \in Q$.*

Proof Indeed, the "only if" part, i.e., that the convexity of f on Q implies the gradient inequality for all $x \in \mathrm{int}\, Q$ and all $y \in Q$, is given by Proposition 10.7. Let us prove the "if" part, i.e., establish the reverse implication. Suppose that f satisfies the gradient inequality for all $x \in \mathrm{int}\, Q$ and all $y \in Q$, and let us verify that f is convex on Q. As f is continuous on Q and Q is convex, by Proposition 10.3 it suffices to prove that f is convex on $\mathrm{int}\, Q$. Recall also that by Theorem 1.29 $\mathrm{int}\, Q$ is convex. Moreover, owing to the gradient inequality, on $\mathrm{int}\, Q$ the function f is the supremum of the family of affine functions, i.e., for all $y \in \mathrm{int}\, Q$ we have
$$f(y) = \sup_{x \in \mathrm{int}\, Q} f_x(y), \quad \text{where } f_x(y) := f(x) + (y-x)^\top \nabla f(x).$$

10.5 Boundedness and Lipschitz Continuity of a Convex Function

In this section we will discuss some local properties of convex functions such as boundedness and Lipschitz continuity. Recall that given a function f and a set $Q \subseteq \text{Dom } f$, we say that f is *Lipschitz continuous* on Q if there exists a constant $L > 0$ (referred to as the *Lipschitz constant* of f on Q) such that

$$|f(x) - f(y)| \leq L\|x - y\|_2, \quad \forall x, y \in Q.$$

Theorem 10.9 *[Local boundedness and Lipschitz continuity of convex functions] Let f be a convex function and let K be a closed and bounded set contained in* $\text{rint}(\text{Dom } f)$. *Then, f is Lipschitz continuous on K, i.e.,*

$$|f(x) - f(y)| \leq L\|x - y\|_2, \quad \forall x, y \in K, \qquad (10.5)$$

for properly selected $L < \infty$. In particular, f is bounded on K.

We shall prove this theorem later in this section, after some preliminary effort.

Remark 10.10 In Theorem 10.9, all three assumptions on K, (1) closedness, (2) boundedness, and (3) $K \subseteq \text{rint}(\text{Dom } f)$, are essential. The following three examples illustrate their importance:

- Suppose $f(x) = 1/x$; then $\text{Dom } f = (0, +\infty)$. Consider $K = (0, 1]$. Assumptions (2) and (3) are satisfied, but not (1). Note that f is neither bounded nor Lipschitz continuous on K.
- Suppose $f(x) = x^2$; then $\text{Dom } f = \mathbf{R}$. Consider $K = \mathbf{R}$. Assumptions (1) and (3) are satisfied, but not (2). Note that f is neither bounded nor Lipschitz continuous on K.
- Suppose $f(x) = -\sqrt{x}$; then $\text{Dom } f = [0, +\infty)$. Consider $K = [0, 1]$. Assumptions (1) and (2) are satisfied, but not (3). Note that f is not Lipschitz continuous on K (indeed, we have $\lim_{t \to +0} \frac{f(0) - f(t)}{t} = \lim_{t \to +0} t^{-1/2} = +\infty$, while for a Lipschitz continuous f the ratios $t^{-1}(f(0) - f(t))$ are bounded). On the other hand, f is bounded on K. With a properly chosen convex function f of two variables and having a nonpolyhedral compact domain (e.g., with $\text{Dom } f$ the unit circle), we can demonstrate also that the lack of (3), even in the presence of (1) and (2), may cause the unboundedness of f on K as well – take the function f which is zero everywhere on the unit circle $K = \text{Dom } f$, except for a sequence $\{x_i\}$ of points on the boundary, where $f(x_i) = i$. ◇

Theorem 10.9 says that a convex function f is bounded on every compact (i.e., closed and bounded) subset of $\text{rint}(\text{Dom } f)$. In fact, in the case of convex functions f we can make a much stronger statement on their *below boundedness*: any convex function f is below bounded on *any* bounded subset of $\mathbf{R}^n$.

Proposition 10.11 *A convex function f is bounded from below on every bounded subset of $\mathbf{R}^n$.*

Proof It is clear that f is bounded from below at any point outside the domain of f. Thus, without loss of generality we may assume that Dom f is full-dimensional (with nonempty interior) and that $0 \in \text{int}(\text{Dom } f)$. By Theorem 10.9, there exists a neighborhood U of the origin – which can be thought of as a ball of some radius $r > 0$ centered at the origin – where f is bounded from above by some C. Now, consider an arbitrary $R > 0$ and an arbitrary point x satisfying $\|x\|_2 \leq R$. Then, the point

$$y := -\frac{r}{R}x$$

is in U, and we have

$$0 = \frac{r}{r+R}x + \frac{R}{r+R}y.$$

As f is convex and $\|y\|_2 \leq r$ implies $f(y) \leq C$, we conclude that

$$f(0) \leq \frac{r}{r+R}f(x) + \frac{R}{r+R}f(y) \leq \frac{r}{r+R}f(x) + \frac{R}{r+R}C.$$

By reorganizing the terms in this inequality, we get the lower bound

$$f(x) \geq \frac{r+R}{r}f(0) - \frac{R}{r}C.$$

Thus, f is bounded below for any x in the ball that is centered at 0 and has radius R. As $R > 0$ is arbitrary, for any bounded set, by selecting $R > 0$ large enough we can find such a ball with radius R covering it. ∎

Our proof of Theorem 10.9 relies on the following "local" version of the theorem.

Proposition 10.12 *Let f be a convex function. Then, for any $\bar{x} \in \text{rint}(\text{Dom } f)$, we have*

(i) *f is bounded at $\bar{x}$, i.e., there exists $r > 0$ such that f is bounded in the r-neighborhood $U_r(\bar{x})$ of $\bar{x}$ in the affine span of Dom f:*

$$\exists r > 0 \text{ and } C \text{ such that } |f(x)| \leq C, \quad \forall x \in U_r(\bar{x}),$$
$$\text{where } U_r(\bar{x}) := \{x \in \text{Aff}(\text{Dom } f) : \|x - \bar{x}\|_2 \leq r\}.$$

(ii) *f is Lipschitz continuous at $\bar{x}$, i.e., there exist constants $\rho > 0$ and L such that*

$$|f(x) - f(x')| \leq L\|x - x'\|_2, \quad \forall x, x' \in U_\rho(\bar{x}).$$

Proof (i): We start with proving the *above boundedness* of f in a neighborhood of $\bar{x}$. This is immediate: by the premise of the proposition we have $\bar{x} \in \text{rint}(\text{Dom } f)$, so there exists $\bar{r} > 0$ such that the neighborhood $U_{\bar{r}}(\bar{x})$ is contained in Dom f. Now, we can find a small simplex Δ of dimension $m := \dim(\text{Aff}(\text{Dom } f))$ with vertices $x^0, \ldots, x^m$ in $U_{\bar{r}}(\bar{x})$ such that $\bar{x}$ will be a convex combination of the vectors x^i with *positive* coefficients, even with the coefficients $1/(m+1)$, i.e.,

$$\bar{x} = \sum_{i=0}^{m} \frac{1}{m+1}x^i.$$

Here is the justification of this claim that such a simplex Δ exists: First, when Dom f is a singleton, the claim is evident. So, we assume that $\dim(\text{Dom } f) = m \geq 1$. Without loss of

10.5 Boundedness and Lipschitz Continuity of a Convex Function

generality, we may assume that $\bar{x} = 0$, so that $0 \in \text{Dom } f$ and therefore $\text{Aff}(\text{Dom } f) = \text{Lin}(\text{Dom } f)$. Then, by Linear Algebra, we can find m vectors $y^1, \ldots, y^m$ in $\text{Dom } f$ which form a basis of $\text{Lin}(\text{Dom } f) = \text{Aff}(\text{Dom } f)$. Setting $y^0 := -\sum_{i=1}^{m} y^i$ and taking into account that $0 = \bar{x} \in \text{rint}(\text{Dom } f)$, we can find $\epsilon > 0$ such that the vectors $x^i := \epsilon y^i, i = 0, \ldots, m$, belong to $U_{\bar{r}}(\bar{x})$. By construction, $\bar{x} = 0 = \frac{1}{m+1} \sum_{i=0}^{m} x^i$.

Note that $\bar{x} \in \text{rint}(\Delta)$ (see Exercise I.3). Since Δ spans the same affine subspace as $\text{Dom } f$, we can find a sufficiently small $r > 0$ such that $r \leq \bar{r}$ and $U_r(\bar{x}) \subseteq \Delta$. Now, by definition,

$$\Delta = \left\{ \sum_{i=0}^{m} \lambda_i x^i \; : \; \lambda_i \geq 0, \forall i, \; \sum_{i=0}^{m} \lambda_i = 1 \right\},$$

so that, by Jensen's inequality, for all $x \in \Delta$ we have

$$f\left(\sum_{i=0}^{m} \lambda_i x^i \right) \leq \sum_{i=0}^{m} \lambda_i f(x^i) \leq \max_{i=0,\ldots,m} f(x^i) =: C_u.$$

Thus, as $\Delta \supseteq U_r(\bar{x})$, we conclude that $f(x) \leq C_u$ for all $x \in U_r(\bar{x})$ as well.

Now let us prove that if f is above bounded, by some C_u in $U_r := U_r(\bar{x})$, then in fact it is also below bounded in this neighborhood (and, consequently, f is bounded in U_r). Consider any $x \in U_r$, so that $x \in \text{Aff}(\text{Dom } f)$ and $\|x - \bar{x}\|_2 \leq r$. Define $x' := \bar{x} - [x - \bar{x}] = 2\bar{x} - x$. Then, $x' \in \text{Aff}(\text{Dom } f)$ and $\|x' - \bar{x}\|_2 = \|x - \bar{x}\|_2 \leq r$, and so $x' \in U_r$. As $\bar{x} = \frac{1}{2}[x + x']$ and f is convex, we have

$$2 f(\bar{x}) \leq f(x) + f(x').$$

Note that this inequality holds for all $x \in U_r$, hence

$$f(x) \geq 2 f(\bar{x}) - f(x') \geq 2 f(\bar{x}) - C_u =: C_\ell, \quad \forall x \in U_r(\bar{x}).$$

Thus, f is indeed bounded below by some C_ℓ in U_r.

Setting $C := \max\{|C_u|, |C_\ell|\}$ then concludes the proof of part (i).

(ii): Part (ii) is indeed an immediate consequence of part (i) and Lemma 10.1. From part (i) we already know that there exist positive constants r, C such that $|f(x)| \leq C$ for all $x \in U_r(\bar{x})$. We will prove that f is Lipschitz continuous in the neighborhood $U_{r/2}(\bar{x})$. Consider any $x, x' \in U_{r/2}(\bar{x})$ such that $x \neq x'$. Let us extend the segment $[x, x']$ through the point x' until it reaches a certain point x'' on the (relative) boundary of $U_r(\bar{x})$. Then

$$x' \in (x, x'') \quad \text{and} \quad \|x'' - \bar{x}\|_2 = r.$$

Then, by (10.3) we have

$$f(x') - f(x) \leq \|x' - x\|_2 \frac{f(x'') - f(x)}{\|x'' - x\|_2}.$$

As $|f(y)| \leq C$ for any $y \in U_r(\bar{x})$ and $x, x'' \in U_r(\bar{x})$, we observe that $|f(x'') - f(x)| \leq 2C$. Moreover, by the triangle inequality we have $\|x'' - x\|_2 \geq \|x'' - \bar{x}\|_2 - \|\bar{x} - x\|_2 \geq r - (r/2) = r/2$, where the last inequality holds since $\|x'' - \bar{x}\|_2 = r$ and $\|\bar{x} - x\|_2 \leq r/2$ (as

$x \in U_{r/2}(\bar{x})$). Hence, the term $\frac{f(x'')-f(x)}{\|x''-x\|_2}$ is bounded above by the quantity $(2C)/(r/2) = 4C/r$. Thus, we have

$$f(x') - f(x) \leq \frac{4C}{r}\|x' - x\|_2, \quad \forall x, x' \in U_{r/2}.$$

By swapping the roles of x and x', we arrive at

$$f(x) - f(x') \leq \frac{4C}{r}\|x' - x\|_2,$$

and hence

$$|f(x) - f(x')| \leq \frac{4C}{r}\|x - x'\|_2, \quad \forall x, x' \in U_{r/2},$$

as required in part (ii). ∎

We are now ready to prove Theorem 10.9.

Proof of Theorem 10.9 We can extract Theorem 10.9 from Proposition 10.12 by standard analysis reasoning. All we need is to prove that if K is a bounded and closed (i.e., a compact) subset of rint (Dom f) then f is Lipschitz continuous on K (the boundedness of f on K is an evident consequence of its Lipschitz continuity on K and the boundedness of K). Assume for contradiction that f is not Lipschitz continuous on K. Then, for every integer i, there exists a pair of distinct points $x^i, y^i \in K$ such that

$$f(x^i) - f(y^i) \geq i\|x^i - y^i\|_2. \tag{10.6}$$

Since K is compact, passing to a subsequence we can ensure that $x^i \to x \in K$ and $y^i \to y \in K$.

First, we claim that it is not possible to have $x = y$. To see this, recall that by Proposition 10.12 f is Lipschitz continuous in a neighborhood B of x. If $x = y$ were to hold, then since $x^i \to x$, $y^i \to y$, this neighborhood B would contain all x^i and y^i with large enough indices i; but then, from the Lipschitz continuity of f in B, the ratios $(f(x^i) - f(y^i))/\|x^i - y^i\|_2$ would form a bounded sequence, which we know is not the case. Thus, the case $x = y$ is impossible.

Next, we claim that the case $x \neq y$ is not possible, either. Proposition 10.12 implies that f is continuous on Dom f at both the points x and y (note that Lipschitz continuity at a point clearly implies the usual continuity at the point), so that $f(x^i) \to f(x)$ and $f(y^i) \to f(y)$ as $i \to \infty$. Thus, the left-hand side of (10.6) remains bounded as $i \to \infty$. On the other hand, as $i \to \infty$ the right-hand side of (10.6) tends to ∞ since, when i tends to ∞, $\|x^i - y^i\|_2$ tends to the positive limit $\|x - y\|_2$ (recall that $x \neq y$ in this case).

This is the desired contradiction since precisely one of the cases $x = y$ and $x \neq y$ must hold. ∎

11

Minima and Maxima of Convex Functions

11.1 Minima of Convex Functions

11.1.1 Unimodality

As has already been mentioned, optimization problems involving convex functions possess nice theoretical properties. One of the most important of these properties is given by the following result which states that any local minimizer of a convex function is also its *global* minimizer.

Theorem 11.1 *[Unimodality] Let $f : \mathbf{R}^n \to \mathbf{R} \cup \{+\infty\}$ be a convex function and let $Q \subseteq \mathbf{R}^n$ be a nonempty convex set. Suppose $x^* \in Q \cap \mathrm{Dom}\, f$ is a local minimizer of f on Q, i.e.,*

$$\exists r > 0 \text{ such that } f(y) \geq f(x^*) \; \forall (y \in Q : \|y - x^*\|_2 < r). \tag{11.1}$$

Then x^* is a global minimizer of f on Q, i.e.,

$$f(x^*) \leq f(y) \; \forall y \in Q. \tag{11.2}$$

Moreover, the set $\mathrm{Argmin}_Q f$ of all local ($\equiv$ global) minimizers of f on Q is convex.

If f is strictly convex, then the set of its global minimizers $\mathrm{Argmin}_Q f$ is either empty or a singleton.

Proof If f is the convex function that is identical to $+\infty$, $\mathrm{Dom}\, f = \emptyset$ and there is nothing to prove. So, we assume that $\mathrm{Dom}\, f \neq \emptyset$.

We will first show that any local minimizer of f is also a global minimizer of f. Let x^* be a local minimizer of f on Q. Consider any $y \in Q$ such that $y \neq x^*$. We need to prove that $f(y) \geq f(x^*)$. If $f(y) = +\infty$, this relation is automatically satisfied. So, we assume that $y \in \mathrm{Dom}\, f$. Note that by the definition of a local minimizer, we certainly also have $x^* \in \mathrm{Dom}\, f$. Now, for any $\tau \in (0, 1)$, by Lemma 10.1 we have

$$\frac{f(x^* + \tau(y - x^*)) - f(x^*)}{\tau \|y - x^*\|_2} \leq \frac{f(y) - f(x^*)}{\|y - x^*\|_2}.$$

Since x^* is a local minimizer of f, the left-hand side of this inequality is nonnegative for all small enough values of $\tau > 0$. Thus, we conclude that the right-hand side is nonnegative as well, i.e., $f(y) \geq f(x^*)$.

Note that $\mathrm{Argmin}_Q f$ is nothing other than the sublevel set $\mathrm{lev}_\alpha(f)$ of f, associated with α, taken as the minimal value $\min_Q f$ of f on Q. Recall that by Proposition 9.6 any sublevel set of a convex function is convex, so this sublevel set $\mathrm{Argmin}_Q f$ is convex.

Finally, let us prove that the set $\text{Argmin}_Q f$ associated with a strictly convex f is, if nonempty, a singleton. Assume for contradiction that there are two distinct minimizers x', x'' in $\text{Argmin}_Q f$. Then, from the strict convexity of f, we would have

$$f\left(\frac{1}{2}x' + \frac{1}{2}x''\right) < \frac{1}{2}(f(x') + f(x'')) = \min_Q f,$$

where the equality follows since $x', x'' \in \text{Argmin}_Q f$. But this strict inequality is impossible since $\frac{1}{2}x' + \frac{1}{2}x'' \in Q$ as Q is convex and, by definition of $\min_Q f$, we cannot have a point in Q with objective value strictly smaller than $\min_Q f$. ∎

11.1.2 Necessary and Sufficient Optimality Conditions

Another pleasant fact regarding differentiable convex functions is that the calculus-based necessary optimality condition (a.k.a. the Fermat rule) is *sufficient* for global optimality.

Theorem 11.2 *[Necessary and sufficient optimality condition for differentiable convex functions] Let $f : \mathbf{R}^n \to \mathbf{R} \cup \{+\infty\}$ be a convex function and let $Q \subseteq \mathbf{R}^n$ be a nonempty convex set. Consider any $x^* \in \text{int } Q$. Suppose that f is differentiable at x^*. Then, x^* is a minimizer of f on Q if and only if*

$$\nabla f(x^*) = 0.$$

Proof The *necessity* of the condition $\nabla f(x^*) = 0$ for local optimality is due to calculus, and it has nothing to do with convexity. The essence of the matter is, of course, the *sufficiency* of the condition $\nabla f(x^*) = 0$ for the *global optimality* of x^* in the case of a *convex* function f. In fact, this sufficiency is readily given by the gradient inequality (10.4). In particular, when $\nabla f(x^*) = 0$ holds, (10.4) becomes

$$f(y) \geq f(x^*) + (y - x^*)^\top \nabla f(x^*) = f(x^*), \quad \forall y \in Q.$$

∎

A natural question is what happens if x^* in Theorem 11.2 is not necessarily an interior point of $Q \subseteq \mathbf{R}^n$. Let us now assume that x^* is an *arbitrary* point of a convex set Q and that f is convex on Q and differentiable at x^* (what the differentiability of $f : \text{Dom } f \to R$ at a point $x \in \text{Dom } f$ means, is explained in item 3 at the end of section C.1.2). Under these assumptions, our goal is to characterize the condition for which x^* will be a minimizer of f on Q.

In order to give such a characterization, given a convex set Q and a point $x^* \in Q$, we will look at the *radial cone* of Q at x^* [notation: $T_Q(x^*)$] defined as the set

$$T_Q(x^*) := \{h \in \mathbf{R}^n : x^* + th \in Q, \quad \forall \text{ small enough } t > 0\}.$$

Geometrically, this is the set of all directions "looking" from x^* towards Q, so that a small enough positive step from x^* along the direction, i.e., adding to x^* a small enough positive multiple of the direction, keeps the point in Q. That is, $T_Q(x^*)$ is the set of all "feasible" directions at x^* for which, starting from x^*, we can go a positive distance along the direction and remain in Q. Equivalently, $T_Q(x^*) = \text{Cone}(Q - \{x^*\})$ (check equivalence!). From the convexity of Q it immediately follows that the radial cone is indeed a cone (not necessary closed). For example, when $x^* \in \text{int } Q$, we have $T_Q(x^*) = \mathbf{R}^n$. Let us examine a more interesting example, e.g., the polyhedral set

$$Q = \left\{ x \in \mathbf{R}^n : a_i^\top x \le b_i, i = 1, \ldots, m \right\}, \tag{11.3}$$

and its radial cone. For any $x^* \in Q$, we define $\mathcal{I}(x^*) := \{i : a_i^\top x^* = b_i\}$ as the set of indices of constraints that are *active* at x^* (i.e., those that are satisfied at x^* as equalities rather than as strict inequalities). Then

$$T_Q(x^*) = \left\{ h \in \mathbf{R}^n : a_i^\top h \le 0, \quad \forall i \in \mathcal{I}(x^*) \right\}. \tag{11.4}$$

The radial cone $T_Q(x^*)$ of a convex set Q taken at $x^* \in Q$ gives rise to another cone, namely the *normal cone* [notation: $N_Q(x^*)$]. By definition, the normal cone is the negative of the dual of the radial cone $T_Q(x^*)$, i.e.,

$$N_Q(x^*) := \left\{ h \in \mathbf{R}^n : h^\top (x' - x^*) \le 0, \forall x' \in Q \right\}. \tag{11.5}$$

Now we are ready to present the necessary and sufficient condition for x^* to be a minimizer of a convex function f on Q.

Proposition 11.3 *Let Q be a convex set and $x^* \in Q$, and suppose f is a convex function on Q which is differentiable at x^*. The necessary and sufficient condition for x^* to be a minimizer of f on Q is that the derivative of f taken at x^* along every direction from $T_Q(x^*)$ should be nonnegative, i.e.,*

$$h^\top \nabla f(x^*) \ge 0, \quad \forall h \in T_Q(x^*).$$

By the duality relation between $T_Q(x^*)$ and $N_Q(x^*)$, this is precisely the same condition as the inclusion

$$\nabla f(x^*) \in -N_Q(x^*).$$

Thus, with Q, x^*, and f as above, x^* minimizes f over Q if and only if $-\nabla f(x^*)$ belongs to the normal cone $N_Q(x^*)$ of Q at x^*.

Proof The necessity of this condition is an evident fact which has nothing to do with the convexity of f. Suppose that x^* is a local minimizer of f on Q. Assume for contradiction that there exists $h \in T_Q(x^*)$ such that $h^\top \nabla f(x^*) < 0$. Then, by $h \in T_Q(x^*)$, we get $x^* + th \in Q$ for all small enough positive t. Moreover, as $h^\top \nabla f(x^*) < 0$, we would also get

$$f(x^* + th) < f(x^*),$$

for all small enough positive t. These two observations together thus imply that in every neighborhood of x^* there are points x from Q with values $f(x)$ strictly smaller than $f(x^*)$. This clearly contradicts the assumption that x^* is a local minimizer of f on Q.

Once again, the sufficiency of this condition is given by the gradient inequality, exactly as in the case when $x^* \in \text{int } Q$, discussed in the proof of Theorem 11.2. ∎

Proposition 11.3 states that under its premise the necessary and sufficient condition for x^* to minimize f on Q is the inclusion $\nabla f(x^*) \in -N_Q(x^*)$. What does this condition actually mean? The answer depends on what the normal cone is: whenever we have an explicit description of it, we have an explicit form of the optimality condition. For example,

- Consider the case $T_Q(x^*) = \mathbf{R}^n$, i.e., $x^* \in \text{int } Q$. Then the normal cone $N_Q(x^*)$ is the cone of all the vectors h that have nonpositive inner products with every vector in $\mathbf{R}^n$, i.e., $N_Q(x^*) = \{0\}$. Consequently, in this case the necessary and sufficient optimality condition of Proposition 11.3 becomes the Fermat rule $\nabla f(x^*) = 0$, which we already know.

- When Q is an affine plane given by linear equalities $Ax = b$, $A \in \mathbf{R}^{m \times n}$, the radial cone at every point $x \in Q$ is the linear subspace $\{d : Ad = 0\}$, the normal cone is the orthogonal complement $\{u = A^\top v : v \in \mathbf{R}^m\}$ to this linear subspace, and the optimality condition reads:

    Given $Q = \{x : Ax = b\}$, given a point $x^* \in Q$, and given a function f that is convex on Q and differentiable at x^*, we have that x^* is a minimizer of f on Q if and only if
    $$\exists v \in \mathbf{R}^m : \nabla f(x^*) + A^\top v = 0.$$

- When Q is the polyhedral set (11.3), the radial cone is the polyhedral cone (11.4), i.e., it is the set of all directions which have nonpositive inner products with all a_i for $i \in \mathcal{I}(x^*)$ (recall that these a_i come from the constraints $a_i^\top x \leq b_i$ specifying Q, which are satisfied as equalities at x^*). The corresponding normal cone is thus the set of all vectors which have nonpositive inner products with all these directions in $T_Q(x^*)$, i.e., the set of vectors a such that the inequality $h^\top a \leq 0$ is a consequence of the inequalities $h^\top a_i \leq 0$, $i \in \mathcal{I}(x^*) = \{i : a_i^\top x^* = b_i\}$. From the Homogeneous Farkas' Lemma we conclude that the normal cone is simply the conic hull of the vectors a_i, $i \in \mathcal{I}(x^*)$. Thus, in this case our necessary and sufficient optimality condition becomes:

    Given $Q = \{x \in \mathbf{R}^n : a_i^\top x \leq b_i, i = 1, \ldots, m\}$, given a point $x^* \in Q$, and given a function f that is convex on Q and differentiable at x^*, we have that x^* is a minimizer of f on Q if and only if there exist nonnegative reals λ_i^* ("Lagrange multipliers") associated with "active" indices i (those from $\mathcal{I}(x^*)$) such that
    $$\nabla f(x^*) + \sum_{i \in \mathcal{I}(x^*)} \lambda_i^* a_i = 0,$$
    or, equivalently, there exist nonnegative λ_i^*, $1 \leq i \leq m$, such that
    $$\nabla f(x^*) + \sum_{i=1}^{m} \lambda_i^* a_i = 0 \quad \text{[Karush–Kuhn–Tucker equation]}$$
    $$\text{and} \quad \lambda_i^* [a_i^\top x^* - b_i] = 0, \quad 1 \leq i \leq m \quad \text{[complementary slackness]}$$

This is our first acquaintance with the famous *Karush–Kuhn–Tucker* (KKT) optimality conditions. We will eventually extend them to the case where Q is given by convex, rather than just linear, constraints (see Theorem 19.4). Indeed, KKT conditions are also *necessary* for *local* optimality in the case of nonconvex mathematical programming problems.

Remark 11.4 Let us give an informal explanation of the preceding results on first-order optimality conditions in terms of physics (see Figure 11.1 for a graphical illustration). Consider the optimization problem given by

11.1 Minima of Convex Functions

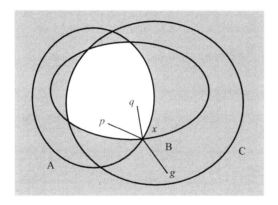

Figure 11.1 Physical illustration of KKT optimality onditions for the optimization problem $\min_{x \in \mathbf{R}^2}\{f(x) : a_i(x) \leq 0, i = 1, 2, 3\}$. The white area represents the feasible domain Q, while the ellipses **A**, **B**, **C** represent the sets $a_1(x) \leq 0, a_2(x) \leq 0, a_3(x) \leq 0$. The point x is a candidate for a feasible solution located at the intersection $\{u \in \mathbf{R}^2 : a_1(u) = a_2(u) = 0\}$ of the boundaries of **A** and **B**. The external force $g = -\nabla f(x)$ acts on a particle located at x and p and q are reaction forces created by obstacles **A** and **B**. The condition for x to be an equilibrium reduces to $g + p + q = 0$, as in the picture. The equilibrium condition $g + p + q = 0$ translates to the KKT equation $\nabla f(x) + \lambda_1 a_1(x) + \lambda_2 \nabla a_2(x) = 0$ holding for some nonnegative λ_1, λ_2.

$$\min_{x \in \mathbf{R}^n} \{f(x) : a_i(x) \leq 0, \quad i = 1, \ldots, m\}.$$

This problem can be interpreted as locating the equilibrium position of a particle that is moving through $\mathbf{R}^n$ while being affected by an external force (such as gravity) with potential f, meaning that when the position of the particle is $x \in \mathbf{R}^n$, the force acting on the particle is $-\nabla f(x)$. The domain in which the particle can actually travel is $Q := \{x \in \mathbf{R}^n : a_i(x) \leq 0, i \leq m\}$; think about the areas $a_i(x) > 0$ as rigid obstacles into which the particle cannot penetrate. When the particle touches the ith obstacle (i.e., it is in position x with $a_i(x) = 0$), the obstacle produces a reaction force directed along the inward normal $-\nabla a_i(x)$ to the boundary of the obstacle, so that the reaction force is $-\lambda_i \nabla a_i(x)$; here $\lambda_i \geq 0$ depends on the pressure on the obstacle exerted by the particle. At an equilibrium x^* (which, by laws of Physics, should minimize, at least locally, the potential f over Q), the sum of the forces acting at the particle should be zero, that is, for properly selected $\lambda_i \geq 0$ one should have $-\nabla f(x^*) - \sum_{i:a_i(x^*)=0} \lambda_i \nabla a_i(x^*) = 0$, which is exactly what is stated by our Karush–Kuhn–Tucker (KKT) optimality condition as applied to the problem where the functions $a_i(x) = a_i^\top x - b_i$ are affine. ◇

11.1.3 ★ Symmetry Principle

We close this section by discussing a simple characterization of minimizers of convex functions admitting a certain symmetry. When applicable, this characterization is extremely useful. This result is indeed an almost immediate consequence of Proposition 9.6.

Proposition 11.5 *Let f be a convex function with domain $Q := \mathrm{Dom}\, f \subseteq \mathbf{R}^n$. Suppose that f and Q admit a group of symmetries. That is, there exists a finite collection $\mathcal{G} = \{G_0, \ldots, G_m\}$ of $n \times n$ nonsingular matrices, distinct from each other, such that*

(i) $\mathcal{G}$ *is a finite group, i.e.,* $G \in \mathcal{G}$ *implies* $G^{-1} \in \mathcal{G}$ *as well, and, for any* $G', G'' \in \mathcal{G}$, *we also have* $G'G'' \in \mathcal{G}$;

(ii) *all matrices* $G \in \mathcal{G}$ *are symmetries of* Q *and* f, *i.e., for any* $G \in \mathcal{G}$, *we have* $Gx \in Q$ *and* $f(x) = f(Gx)$ *for all* $x \in Q$.

Suppose also that the set of minimizers of f *on* Q, *i.e.,* $Q_* := \mathrm{Argmin}_{x \in Q} f(x)$, *is nonempty. Then* f *admits a* $\mathcal{G}$-*symmetric minimizer, that is, there exists some* $x_* \in Q_*$ *such that* $Gx_* = x_*$ *for all* $G \in \mathcal{G}$.

Proof Consider any fixed $x \in Q_*$. Since the matrices $G \in \mathcal{G}$ are symmetries of f and Q, they are symmetries of Q_*, i.e., $Gx \in Q_*$ for all $G \in \mathcal{G}$ (why?) as well. Now, because f is convex, Proposition 9.6 implies that the set $Q_* = \mathrm{Argmin}_{u \in Q} f(u) = \{u \in Q : f(u) \le \min_Q f\}$ is convex and thus contains, along with the point x, the point $x_* := \frac{1}{m} \sum_{G \in \mathcal{G}} Gx$. It remains to show that $Hx_* = x_*$ for all $H \in \mathcal{G}$. Indeed, $Hx_* = \frac{1}{m} \sum_{G \in \mathcal{G}} HGx$, and the terms in the latter sum form a permutation of the terms in the sum specifying x_* owing to the fact that $\mathcal{G}$ is a group. Thus, x_* is the desired $\mathcal{G}$-symmetric minimizer of f. ∎

Illustration [Arithmetic-geometric mean inequality] Let us use the Symmetry Principle to justify the following classical inequality:

$$\text{for any } a_1, \ldots, a_m \in \mathbf{R}_+, \text{ we have } \sqrt[m]{a_1 \cdots a_m} \le \frac{a_1 + a_2 + \cdots + a_m}{m}.$$

To prove this inequality, we define $a := a_1 + \cdots + a_m$, $Q := \{x \in \mathbf{R}_+^m : \sum_{i=1}^m x_i = a\}$, and

$$f(x) := \begin{cases} \sqrt[m]{x_1 \cdots x_m}, & \text{if } x \in Q, \\ +\infty, & \text{otherwise.} \end{cases}$$

Note that Q is nonempty and compact, and $f(x)$ is continuous and concave (see Example 10.5). Thus, by Theorem B.31, the set of maximizers of f on Q, i.e., Q_*, is nonempty. Clearly, the $m!$ permutation matrices of size $m \times m$ form a group of symmetries of f and Q, so that Q_* contains a permutationally symmetric point x_* (apply Proposition 11.5 to minimize the convex function $-f$ over Q). Since the sum of all entries in a point from Q is equal to a, Q contains exactly one permutationally symmetric point $\frac{1}{m}[a; a; \ldots; a]$. Then, as $\frac{1}{m}[a; a; \ldots; a]$ is a maximizer of f over Q, we conclude that for every $[a_1; \ldots; a_m] \in Q$ we have

$$\sqrt[m]{a_1 \cdots a_m} = f([a_1; \ldots; a_m]) \le f\left(\frac{1}{m}[a; \ldots; a]\right) = \frac{1}{m}a = \frac{a_1 + a_2 + \cdots + a_m}{m}.$$

∎

11.2 Maxima of Convex Functions

So far we have seen that the fact that a point $x^* \in \mathrm{Dom}\, f$ is a global minimizer of a convex function f depends only on the local behavior of f at x^*. This is not the case with maximizers of a convex function. First of all, in all nontrivial cases, such a maximizer, if one exists at all, must belong to the relative boundary of the domain of the function.

11.2 Maxima of Convex Functions

Theorem 11.6 *Let f be a convex function. If a point $x^* \in \operatorname{rint}(\operatorname{Dom} f)$ is a maximizer of f over $\operatorname{Dom} f$ then f is constant on $\operatorname{Dom} f$.*

Proof Define $Q := \operatorname{Dom} f$, and consider any $y \in Q$. We need to prove that $f(y) = f(x^*)$. There is nothing to prove if $y = x^*$, so we assume that $y \neq x^*$. Since by assumption $x^* \in \operatorname{rint} Q$, there exists a point $y' \in Q$ such that x^* is an interior point of the segment $[y', y]$, i.e., there exists $\lambda \in (0, 1)$ such that

$$x^* = \lambda y' + (1 - \lambda) y.$$

Then, as f is convex, we get

$$f(x^*) \leq \lambda f(y') + (1 - \lambda) f(y).$$

As x^* is a maximizer of f on Q, we have $f(x^*) \geq f(y)$ and $f(x^*) \geq f(y')$. Combining this with the preceding inequality leads to $\lambda f(y') + (1 - \lambda) f(y) = f(x^*)$. Since $\lambda \in (0, 1)$ and $\max\{f(y), f(y')\} \leq f(x^*)$, this equation can hold as an equality only if $f(y') = f(y) = f(x^*)$. ∎

In particular, Theorem 11.6 states that given a convex function f the only way for a point $x^* \in \operatorname{rint}(\operatorname{Dom} f)$ to be a global maximizer of f is if the function f is constant over its domain.

Next, we provide further information on the maxima of convex functions.

Theorem 11.7 *Let f be a convex function on $\mathbf{R}^n$ and $E \subseteq \mathbf{R}^n$ be a nonempty set. Then*

$$\sup_{x \in \operatorname{Conv} E} f(x) = \sup_{x \in E} f(x). \tag{11.6}$$

In particular, if $S \subset \mathbf{R}^n$ is a nonempty convex and compact set, then the supremum of f on S is equal to the supremum of f on the set of extreme points of S, i.e.,

$$\sup_{x \in S} f(x) = \sup_{x \in \operatorname{Ext}(S)} f(x). \tag{11.7}$$

Proof We will first prove (11.6). As $\operatorname{Conv} E \supseteq E$, we have $\sup_{x \in \operatorname{Conv} E} f(x) \geq \sup_{x \in E} f(x)$. To prove the reverse direction, consider any $\bar{x} \in \operatorname{Conv} E$. Then, there exist $x^i \in E$ and convex combination weights $\lambda_i \geq 0$ satisfying $\sum_i \lambda_i = 1$ and

$$\bar{x} = \sum_i \lambda_i x^i.$$

Applying Jensen's inequality (Proposition 9.4), we get

$$f(\bar{x}) \leq \sum_i \lambda_i f(x^i) \leq \sum_i \lambda_i \sup_{x \in E} f(x) = \sup_{x \in E} f(x).$$

Since the preceding inequality holds for any $\bar{x} \in \operatorname{Conv} E$, we conclude that $\sup_{x \in \operatorname{Conv} E} f(x) \leq \sup_{x \in E} f(x)$ holds as well, as desired.

To prove (11.7), note that when S is a nonempty convex compact set, by the Krein–Milman Theorem (Theorem 6.11) we have $S = \operatorname{Conv}(\operatorname{Ext}(S))$. Then (11.7) follows immediately from (11.6). ∎

Our last theorem on the maxima of convex functions is as follows.

Theorem 11.8 *[Maxima of convex functions]* Let f be a proper convex function, and let $Q := \text{Dom } f$. Then:

(i) If Q is closed and does not contain lines and the set of global maximizers of f, i.e.,

$$\text{Argmax}_Q f := \{x \in Q : f(x) \geq f(y), \forall y \in Q\},$$

is nonempty then $\text{Argmax}_Q f \cap \text{Ext}(Q) \neq \emptyset$, i.e., at least one of the maximizers of f is an extreme point of Q.

(ii) If the set Q is polyhedral and f is above bounded on Q then the maximum of f on Q is achieved, i.e., $\text{Argmax}_Q f \neq \emptyset$.

Proof As f is a proper convex function, we have $Q \neq \emptyset$. Let us start with the following immediate observation:

(!) *Let a convex function f be bounded from above on a ray $\ell = \bar{x} + \mathbf{R}_+(e)$. Then, the function does not increase along the ray, i.e., for any $x \in \ell$ and $s \geq 0$, we have $f(x + se) \leq f(x)$.*

Indeed, assuming for contradiction that there exists $x \in \ell$ and $x' = x + se$, $s \geq 0$, with $f(x') > f(x)$, we conclude that $s > 0$ and that the average rate of change $\kappa := \frac{f(x')-f(x)}{s}$ when moving from x to x' is positive: $\kappa > 0$. Since f is convex, the average rate $\frac{f(x+te)-f(x)}{t}$ when moving from x to $x + te$ with $t \geq s$ is at least $\kappa > 0$ (section 10.2), implying that $f(x + te) \to +\infty$ as $t \to \infty$, which is the desired contradiction, since f is bounded from above on ℓ.

As an immediate corollary of (!), we conclude that

(!!) *If Q is a convex set represented as $V + R$, where V is nonempty and R is a cone, and f is a convex function on Q that is bounded from above, then f attains its maximum on Q if and only if it attains its maximum on V, the maxima being equal to each other.*

Indeed, assume that f attains its maximum on Q at some point $\bar{x}$. As $Q = V + R$, we have $\bar{x} = \bar{v} + e$ for some $\bar{v} \in V$ and $e \in R$. By (!) we have $f(\bar{v}) \geq f(\bar{x})$, which combines with $V \subseteq Q$ to imply that $\bar{v}$ is a maximizer of f on both V and Q. Vice versa, assume that f attains its maximum on V at a certain point $\bar{v}$. Every $x \in Q$ can be represented as $v + e$ with $v \in V$ and $e \in R$, so that by (!) we have $f(x) \leq f(v)$ and thus $f(x) \leq f(\bar{v})$, and we again conclude that $\bar{v}$ maximizes f both on V and on Q.

Theorem 11.8 is an immediate corollary of (!) and (!!). Indeed, in case (i), as Q is a nonempty closed convex set that does not contain lines, by Theorem 6.21 we deduce that $\text{Ext}(Q) \neq \emptyset$, and Q admits a representation as

$$Q = \underbrace{\text{Conv}(W)}_{V} + R, \tag{11.8}$$

where $W = \text{Ext}(Q)$, $R = \text{Rec}(Q)$. In the situation of (i), f attains its maximum on Q and therefore, by (!!), f has a maximizer $\bar{v}$ belonging to V. Owing to the origin of V, $\bar{v} = \sum_{i \in I} \lambda_i v^i$ with nonempty finite set I, $\lambda_i \geq 0$, $\sum_i \lambda_i = 1$, and $v^i \in \text{Ext}(Q)$. By the convexity of f, $f(\bar{v}) \leq \sum_{i \in I} \lambda_i f(v^i) \leq \max_{i \in I} f(v^i) =: f(v^{i*})$, implying that the extreme point v^{i*} of Q maximizes f on Q.

In case (ii), by Theorem 7.10 the polyhedral set Q admits representation (11.8) with nonempty finite W, implying that f, which is convex and real-valued on $V = \text{Conv}(W) \subseteq Q = \text{Dom } f$, attains its maximum on V, e.g., at its maximizer on the nonempty and finite set W. By (!!), this maximizer maximizes f on Q as well, so that f achieves its maximum on Q, as claimed in (ii). ∎

12

Subgradients

12.1 Proper Lower Semicontinuous Convex Functions and Their Representation

Recall that an equivalent definition of a convex function f on $\mathbf{R}^n$ is that f is a function taking values in $\mathbf{R} \cup \{+\infty\}$ such that its epigraph

$$\mathrm{epi}\{f\} = \left\{[x;t] \in \mathbf{R}^{n+1} : t \geq f(x)\right\}$$

is a convex set. Thus, there is no essential difference between convex functions and convex sets: a convex function generates a convex set, i.e., its epigraph, which of course remembers everything about the function. And the only specific property of the epigraph as a convex set is that it always possesses a very specific recessive direction, namely $h = [0; 1]$. That is, the ray $\{z + th : t \geq 0\}$ directed by h belongs to the epigraph set whenever the starting point z of the ray is in the set. In addition, the intersection of the epigraph with any line directed by h is closed and differs from the entire line. Whenever a convex set possesses a nonzero recessive direction h such that the intersection of the set with any line directed by h is closed and differs from the entire line, the set in appropriate coordinates becomes the epigraph of a convex function. Thus, a convex function is, basically, simply a way to look, in the literal meaning of the latter verb, at a convex set possessing a specific property.

Now, we know that the convex sets that are "actually nice" are the closed ones: "actually nice" means that they possess a lot of important properties (e.g., they admit a good outer description) which are not shared by arbitrary convex sets. Therefore, among convex functions there are also "actually nice" ones, namely those with closed epigraphs. The closedness of the epigraph of a function can be "translated" to functional language and there it becomes a special kind of continuity, namely *lower semicontinuity*. Before formally defining lower semicontinuity, we present a brief preamble on the convergence of sequences on the extended real line. In the sequel, we will operate with limits of sequences $\{a_i\}_{i \geq 1}$ with terms a_i from the extended real line $\overline{\mathbf{R}} := \mathbf{R} \cup \{+\infty\} \cup \{-\infty\}$. These limits are defined in the natural way: the relation

$$\lim_{i \to \infty} a_i = a \in \overline{\mathbf{R}}$$

means that for every $a' \in \overline{\mathbf{R}}$

- $a_i < a'$ for all but finitely many values of i, when $a' > a$, and
- $a_i > a'$ for all but finitely many values of i, when $a' < a$.

In particular, a sequence of reals $\{a_i\}$ converges, as $i \to \infty$, to $a \in \overline{\mathbf{R}}$ if and only if $a = \lim_{i \to \infty} a_i$, where the limit is understood in the standard calculus way.

12.1 Proper Lower Semicontinuous Convex Functions and Their Representation

An equivalent way to treat the convergence of sequences $\{a_i\}_i \subseteq \overline{\mathbf{R}}$ is as follows: let us consider a strictly monotone continuous function θ on $\mathbf{R}$ which maps the axis onto the interval $(-1, 1)$ (that is, $\lim_{s\to -\infty} \theta(s) = -1$, $\lim_{s\to\infty} \theta(s) = 1$), e.g., $\theta(s) = \frac{2}{\pi}\arctan(s)$, and extend it from $\mathbf{R}$ to $\overline{\mathbf{R}}$ by setting

$$\overline{\theta}(a) = \begin{cases} -1, & \text{if } a = -\infty, \\ \theta(a), & \text{if } a \in \mathbf{R}, \\ 1, & \text{if } a = +\infty. \end{cases}$$

With this "encoding," $\overline{\mathbf{R}}$ becomes the segment $[-1, 1]$, and the relation $a = \lim_{i\to\infty} a_i$ as defined above is the same as $\overline{\theta}(a) = \lim_{i\to\infty} \overline{\theta}(a_i)$, that is, this relation stands for the usual convergence, as $i \to \infty$, of reals $\overline{\theta}(a_i)$ to the real $\overline{\theta}(a)$. Note also that for $a, b \in \overline{\mathbf{R}}$ the relation $a \leq b$ ($a < b$) is exactly the same as the usual arithmetic inequality $\overline{\theta}(a) \leq \overline{\theta}(b)$ (respectively, $\overline{\theta}(a) < \overline{\theta}(b)$).

With the convergence and limits of sequences $\{a_i\}_i \subseteq \overline{\mathbf{R}}$ already defined, we can speak about the upper (lower) limits of these sequences. For example, we can define $\liminf_{i\to\infty} a_i$ as $a \in \overline{\mathbf{R}}$ as being uniquely specified by the relation $\overline{\theta}(a) = \liminf_{i\to\infty} \overline{\theta}(a_i)$. As for the lower limits of sequences of reals, $\liminf_{i\to\infty} a_i$ is the smallest (in terms of the relation $\leq$ on $\overline{\mathbf{R}}$!) of the limits of converging (in $\overline{\mathbf{R}}$!) subsequences of the sequence $\{a_i\}_i$.

It is time to come back to lower semicontinuity.

Definition 12.1 *[Lower semicontinuity] Let $f : \mathbf{R}^n \to \mathbf{R} \cup \{+\infty\}$ be a function (not necessarily convex). The function f is called* lower semicontinuous *(lsc for short) at a point $\bar{x}$, if for every sequence of points $\{x^i\}_i$ converging to $\bar{x}$ one has*

$$f(\bar{x}) \leq \liminf_{i\to\infty} f(x^i).$$

A function f is called lower semicontinuous if it is lower semicontinuous at every point.

A trivial example of an lsc function is a continuous function. Note, however, that an lsc function need not be continuous; what is needed for lower semicontinuity is that the function can make only "jump downs." For example, the function

$$f(x) = \begin{cases} 0, & \text{if } x \neq 0, \\ a, & \text{if } x = 0, \end{cases}$$

is lsc if $a \leq 0$ (the function can "jump down at $x = 0$ or no jump at all"), and is *not* lsc if $a > 0$ (a "jump up"). For more illustrations, see Figure 12.1.

Here is the connection between the lower semicontinuity of a function and the geometry of its epigraph.

Proposition 12.2 *A function $f : \mathbf{R}^n \to \mathbf{R} \cup \{+\infty\}$ is lower semicontinuous if and only if its epigraph is closed.*

Proof First, suppose epi$\{f\}$ is closed, and let us prove that f is lsc. Consider a sequence $\{x^i\}_i$ such that $x^i \to x$ as $i \to \infty$, and let us prove that $f(x) \leq a := \liminf_{i\to\infty} f(x^i)$. There is nothing to prove when $a = +\infty$. Assuming $a < +\infty$, by the definition of $\liminf$ there exists a sequence $i_1 < i_2 < \cdots$ such that $f(x^{i_j}) \to a$ as $j \to \infty$. Let us assume

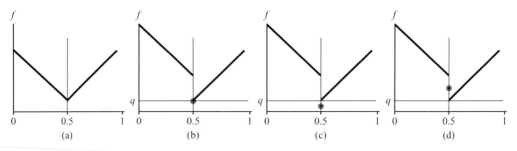

Figure 12.1 Continuity and lower semicontinuity. (a) A continuous function $f(x)$, Dom $f = [0, 1]$. (b)–(d) Functions on $[0, 1]$ that are discontinuous at $x = 0.5$. Given their common restriction on $\{0 \leq x \leq 1, x \neq 0.5\}$ and setting $q = \lim_{\epsilon \to +0} \inf_{\substack{|x-0.5|\leq\epsilon \\ x \neq 0.5}} f(x)$, to be lsc, we should have $f(0.5) \leq q$, as in (b), (c). Function (d) is not lsc.

that $a > -\infty$ (we will verify later on that this is in fact the case). Then, as the points $[x^{i_j}; f(x^{i_j})] \in \text{epi}\{f\}$ converge to $[x; a]$ and $\text{epi}\{f\}$ is closed, we see that $[x; a] \in \text{epi}\{f\}$, that is, $f(x) \leq a$, as claimed. It remains to verify that $a > -\infty$. Indeed, assuming $a = -\infty$, we conclude that for every $t \in \mathbf{R}$ the points $[x^{i_j}; t]$ belong to $\text{epi}\{f\}$ for all large enough values of j, which, as above, implies that $[x; t] \in \text{epi}\{f\}$, that is, $t \geq f(x)$. The latter inequality cannot hold true for all real t, since f does not take the value $-\infty$; thus, $a = -\infty$ is impossible.

Now, for the opposite direction, let f be lsc, and let us prove that $\text{epi}\{f\}$ is closed. So, we need to prove that if $[x^i; t_i] \to [x; t]$ as $i \to \infty$ and $[x^i; t_i] \in \text{epi}\{f\}$, that is, $t_i \geq f(x^i)$ for all i, then $[x; t] \in \text{epi}\{f\}$, that is, $t \geq f(x)$. Indeed, since f is lsc and $f(x^i) \leq t_i$, we have $f(x) \leq \liminf_{i \to \infty} f(x^i) \leq \liminf_{i \to \infty} t_i = \lim_{i \to \infty} t_i = t$. ∎

An immediate consequence of Proposition 12.2 is as follows:

Corollary 12.3 *Given an arbitrary family of lsc functions $f_\alpha : \mathbf{R}^n \to \mathbf{R} \cup \{+\infty\}$, their supremum given by*

$$f(x) := \sup_{\alpha \in \mathcal{A}} f_\alpha(x)$$

is lower semicontinuous.

Proof The epigraph of the function f is the intersection of the epigraphs of all the functions f_α, and the intersection of closed sets is always closed. ∎

Now let us look at *convex, proper, and lower semicontinuous* functions, that is, functions $\mathbf{R}^n \to \mathbf{R} \cup \{+\infty\}$ with closed, convex, and nonempty epigraphs. To save words, let us call these functions *regular*.

What we are about to do is to translate to the functional language several constructions and results related to convex sets. In ordinary life, a translation (e.g., of poetry) typically results in something less rich than the original. In contrast with this, in mathematics this is a powerful source of new ideas and constructions.

"Outer description" of a proper lower semicontinuous convex function. We know that any closed convex set is the intersection of closed half-spaces. What does this fact imply

12.1 Proper Lower Semicontinuous Convex Functions and Their Representation

when the set is the epigraph of a regular function f? First, note that the epigraph is not a completely arbitrary convex set in $\mathbf{R}^{n+1}$: it has the recessive direction $e := [0_n; 1]$, i.e., the standard basic orth along the t-axis in the space of variables $x \in \mathbf{R}^n$, $t \in \mathbf{R}$ where the epigraph lives. This direction, of course, should be recessive for every closed half-space

$$\Pi = \left\{ [x;t] \in \mathbf{R}^{n+1} : \alpha t \geq d^\top x - a \right\} \quad [\text{where } |\alpha| + \|d\|_2 > 0], \qquad (*)$$

containing epi$\{f\}$. Note that in (*) we are adopting a specific form of the nonstrict linear inequality describing the closed half-space Π among many possible forms in the space where the epigraph lives; this form is the most convenient for us at present. Thus, e is a recessive direction of $\Pi \supseteq \text{epi}\{f\}$, and the recessiveness of e for Π means exactly that $\alpha \geq 0$. Thus, when speaking about closed half-spaces containing epi$\{f\}$, we in fact are considering some of the half-spaces (*) with $\alpha \geq 0$.

Now, there are two essentially different possibilities for α to be nonnegative: (A) $\alpha > 0$, and (B) $\alpha = 0$. In case (B) the boundary hyperplane of Π is "vertical," i.e., it is parallel to e, and in fact it "bounds" only x. And, in such cases, Π is the set of all vectors $[x;t]$ with x belonging to a certain half-space in the x-subspace and t an arbitrary real number. These "vertical" half-spaces will be of no interest to us.

The half-spaces which indeed are of interest to us are the "nonvertical" ones: those given by the case (A), i.e., with $\alpha > 0$. For a non-vertical half-space Π, we can always divide the inequality defining Π by α and make $\alpha = 1$. Thus, a "nonvertical" candidate eligible for the role of a closed half-space containing epi$\{f\}$ can always be written as

$$\Pi = \left\{ [x;t] \in \mathbf{R}^{n+1} : t \geq d^\top x - a \right\}. \qquad (**)$$

That is, a "nonvertical" closed half-space containing epi$\{f\}$ *can be represented as the epigraph of an affine function of x.*

Now, when is such a candidate indeed a half-space containing epi$\{f\}$? It is clear that the answer is yes *if and only if the affine function $d^\top x - a$ is less than or equal to $f(x)$ for all $x \in \mathbf{R}^n$*. When this is the case, we call the function $d^\top x - a$ an *affine minorant* of f. In fact, we have a very nice characterization of proper lsc convex functions through their affine minorants!

Proposition 12.4 *A proper lower semicontinuous convex function f is the pointwise supremum of all its affine minorants. Moreover, at every point $\bar{x} \in \text{rint}(\text{Dom } f)$, the function f is not only the supremum but actually the maximum of its affine minorants. That is, at every point $\bar{x} \in \text{rint}(\text{Dom } f)$, there exists an affine function $f_{\bar{x}}(x)$ such that $f_{\bar{x}}(x) \leq f(x)$ for all $x \in \mathbf{R}^n$ and $f_{\bar{x}}(\bar{x}) = f(\bar{x})$.*

Note that Proposition 12.4 essentially gives us an *outer description* of a proper lsc convex function. This outer description is in fact instrumental in developing algorithms for convex optimization. Before proceeding with the proof of Proposition 12.4, we will first prove an intermediate result on proper convex (but not necessarily lower semicontinuous) functions. This result indeed forms the most important step in the proof of Proposition 12.4.

Proposition 12.5 *Let f be a proper convex function. By definition of the affine minorant, we immediately have that the supremum of all affine minorants of f is less than or equal* to f everywhere. Let $\bar{x} \in \operatorname{rint}(\operatorname{Dom} f)$. Then there exists an affine minorant $d^\top x - a$ of f which coincides with f at $\bar{x}$, i.e.,

$$f(x) \geq d^\top x - a, \ \forall x \in \mathbf{R}^n, \quad \text{and} \quad d^\top \bar{x} - a = f(\bar{x}). \tag{12.1}$$

Moreover, this supremum is $+\infty$ outside $\operatorname{cl}(\operatorname{Dom} f)$. Thus, the supremum of all affine minorants of f is equal to f everywhere except, perhaps, in some subset of $\operatorname{rbd}(\operatorname{Dom} f)$.

Proof I. There is nothing to prove when $\operatorname{Dom} f$ is a singleton. Assume from now on that this is not the case. We will first prove that at every $\bar{x} \in \operatorname{rint}(\operatorname{Dom} f)$ there exists an affine function $f_{\bar{x}}(x)$ such that $f_{\bar{x}}(x) \leq f(x)$ for all $x \in \mathbf{R}^n$ and $f_{\bar{x}}(\bar{x}) = f(\bar{x})$.

I(1). First, we can easily reduce the situation to one when $\operatorname{Dom} f$ is full-dimensional. Indeed, by shifting f we can make $\operatorname{Aff}(\operatorname{Dom} f)$ a linear subspace L in $\mathbf{R}^n$; by restricting f to this linear subspace, we clearly get a proper function on L. If we believe that our statement is true for the case when $\operatorname{Dom} f$ is full-dimensional, we can conclude that there exists an affine function *on L*, i.e.,

$$d^\top x - a \quad [\text{where } x \in L \text{ and } d \in L]$$

such that

$$f(x) \geq d^\top x - a, \ \forall x \in L, \quad \text{and} \quad f(\bar{x}) = d^\top \bar{x} - a.$$

This affine function $d^\top x - a$ on L clearly can be extended, by the same formula, from L to the entire space $\mathbf{R}^n$ and is a minorant of f on the entire space $\mathbf{R}^n$ (note that outside $L \supseteq \operatorname{Dom} f$, the function f is simply $+\infty$!). This affine minorant on $\mathbf{R}^n$ is exactly what we need.

I(2). Now let us prove that our statement is valid when $\operatorname{Dom} f$ is full-dimensional. In such a case, $\bar{x} \in \operatorname{int}(\operatorname{Dom} f)$. Consider the point $\bar{y} := [\bar{x}; f(\bar{x})]$. Note that $\bar{y} \in \operatorname{epi}\{f\}$, and in fact we claim further that $\bar{y} \in \operatorname{rbd}(\operatorname{epi}\{f\})$. Assume for contradiction that $\bar{y} \in \operatorname{rint}(\operatorname{epi}\{f\})$. Recall that $e := [0_n; 1]$ is the special recessive direction of $\operatorname{epi}\{f\}$, thus $\bar{y}' := \bar{y} + e$ satisfies $\bar{y}' \in \operatorname{epi}\{f\}$ and so the segment $[\bar{y}', \bar{y}]$ is contained in $\operatorname{epi}\{f\}$. Since $\bar{y} \in \operatorname{rint}(\operatorname{epi}\{f\})$, we can extend this segment a little more through its endpoint $\bar{y}$, without leaving $\operatorname{epi}\{f\}$. But this is clearly impossible, since in such a case the t-coordinate of the new endpoint would be $< f(\bar{x})$ while the x-component of it still would be $\bar{x}$. Thus, $\bar{y} \in \operatorname{rbd}(\operatorname{epi}\{f\})$.

Next, we claim that $\bar{y}'$ is an interior point of $\operatorname{epi}\{f\}$. This is immediate: we know from Theorem 10.9 that f is continuous at $\bar{x}$ (recall that $\bar{x} \in \operatorname{int}(\operatorname{Dom} f)$), so that there exists a neighborhood U of $\bar{x}$ in $\operatorname{Aff}(\operatorname{Dom} f) = \mathbf{R}^n$ such that $f(x) \leq f(\bar{x}) + 0.5$ whenever $x \in U$, or, in other words, the set

$$V := \{[x; t] : x \in U, \ t > f(\bar{x}) + 0.5\}$$

is contained in $\operatorname{epi}\{f\}$; but this set clearly contains a neighborhood of $\bar{y}'$ in $\mathbf{R}^{n+1}$. We see that $\operatorname{epi}\{f\}$ is full-dimensional, so that $\operatorname{rint}(\operatorname{epi}\{f\}) = \operatorname{int}(\operatorname{epi} f)$ and $\operatorname{rbd}(\operatorname{epi}\{f\}) = \operatorname{bd}(\operatorname{epi} f)$.

Now let us look at a hyperplane Π supporting $\operatorname{cl}(\operatorname{epi}\{f\})$ at the point $\bar{y} \in \operatorname{rbd}(\operatorname{epi}\{f\})$. Thus, w.l.o.g., we can represent this hyperplane via a nontrivial (i.e., with $|\alpha| + \|d\|_2 > 0$) linear inequality

$$\alpha t \geq d^\top x - a. \tag{12.2}$$

12.1 Proper Lower Semicontinuous Convex Functions and Their Representation

satisfied everywhere on cl(epi$\{f\}$), specifically, since this is the hyperplane where this inequality holds true as an equality. Now, inequality (12.2) is satisfied everywhere on epi$\{f\}$, and therefore at the point $\bar{y}' := [\bar{x}; f(\bar{x})+1] \in \text{epi}\{f\}$ as well, and is satisfied as an equality at $\bar{y} = [\bar{x}; f(\bar{x})]$ (since $\bar{y} \in \Pi$). These two observations clearly imply that $\alpha \geq 0$. We claim that $\alpha > 0$. Indeed, inequality (12.2) says that the linear form $h^\top [x; t] := \alpha t - d^\top x$ attains its minimum over $y \in \text{cl}(\text{epi}\{f\})$, equal to $-a$, at the point $\bar{y}$. Were $\alpha = 0$, we would have $h^\top \bar{y} = h^\top \bar{y}'$, implying that the set of minimizers of the linear form $h^\top y$ on the set cl(epi$\{f\}$) contains an interior point (namely, $\bar{y}'$) of the set. This is possible only when $h = 0$, that is, $\alpha = 0, d = 0$, which is not the case.

Now, as $\alpha > 0$, by dividing both sides of (12.2) by α, we get a new inequality of the form

$$t \geq d^\top x - a, \tag{12.3}$$

(here we keep the same notation for the right-hand side coefficients as we will not be returning to the old coefficients), which is valid on epi$\{f\}$ and is an equality at $\bar{y} = [\bar{x}; f(\bar{x})]$. Its validity on epi$\{f\}$ implies that, for all $[x; t]$ with $x \in \text{Dom } f$ and $t = f(x)$, we have

$$f(x) \geq d^\top x - a, \quad \forall x \in \text{Dom } f. \tag{12.4}$$

Thus, we conclude that the function $d^\top x - a$ is an affine minorant of f on Dom f and therefore on $\mathbf{R}^n$ ($f = +\infty$ outside Dom f!). Finally, note that the inequality (12.4) becomes an equality at $\bar{x}$, since (12.3) holds as an equality at $\bar{y}$. The affine minorant we have just built justifies the validity of the first claim of the proposition.

II. Let $\mathcal{F}$ be the set of all affine functions which are minorants of f, and define the function

$$\bar{f}(x) := \sup_{\phi \in \mathcal{F}} \phi(x).$$

We have proved that $\bar{f}(x)$ is equal to f on rint (Dom f) (and at any $x \in \text{rint}(\text{Dom } f)$, sup on the right-hand side can be replaced with max). To complete the proof of the proposition, we need to prove that $\bar{f}$ is equal to f outside cl(Dom f) as well. Note that this is the same as proving that $\bar{f}(x) = +\infty$ for all $x \in \mathbf{R}^n \setminus \text{cl}(\text{Dom } f)$. To see this, consider any $\bar{x} \in \mathbf{R}^n \setminus \text{cl}(\text{Dom } f)$. As cl(Dom f) is a closed convex set, $\bar{x}$ can be strongly separated from Dom f; see part (ii) of the Separation Theorem (Theorem 6.3). Thus, there exist $\zeta > 0$ and $z \in \mathbf{R}^n$ such that

$$z^\top \bar{x} \geq z^\top x + \zeta, \quad \forall x \in \text{Dom } f. \tag{12.5}$$

In addition, we already know that there exists at least one affine minorant of f, i.e., there exist a and d such that

$$f(x) \geq d^\top x - a, \quad \forall x \in \text{Dom } f. \tag{12.6}$$

Multiplying both sides of (12.5) by a positive weight λ and then adding it to (12.6), we get

$$f(x) \geq \underbrace{(d + \lambda z)^\top x + [\lambda \zeta - a - \lambda z^\top \bar{x}]}_{=:\phi_\lambda(x)}, \quad \forall x \in \text{Dom } f.$$

This inequality clearly says that $\phi_\lambda(\cdot)$ is an affine minorant of f on $\mathbf{R}^n$ for every $\lambda > 0$. The value of this minorant at $x = \bar{x}$ is equal to $d^\top \bar{x} - a + \lambda \zeta$ and therefore it goes to $+\infty$ as $\lambda \to +\infty$. We see that the supremum of affine minorants of f at $\bar{x}$ is indeed $+\infty$, as claimed. This concludes the proof of Proposition 12.5. ∎

Let us now prove Proposition 12.4.

Proof of Proposition 12.4   Under the premise of the proposition, f is a proper lsc convex function. Let $\mathcal{F}$ be the set of all affine functions which are minorants of f, and let

$$\bar{f}(x) := \sup_{\phi \in \mathcal{F}} \phi(x)$$

be the supremum of all affine minorants of f. Then, by Proposition 12.5, we know that $\bar{f}$ is equal to f everywhere on rint (Dom f) and everywhere outside cl(Dom f). Thus, all we need to prove is that when f is also lsc, $\bar{f}$ is equal to f everywhere on rbd(Dom f) as well.

Consider any $\bar{x} \in$ rbd(Dom f). Recall that by construction $\bar{f}$ is everywhere $\leq f$. So, there is nothing to prove if $\bar{f}(\bar{x}) = +\infty$. Thus, we assume that $\bar{f}(\bar{x}) = c < \infty$ holds, and we will prove that in this case $f(\bar{x}) = c$ holds as well. Since $\bar{f} \leq f$ everywhere, proving $f(\bar{x}) = c$ is the same as proving $f(\bar{x}) \leq c$. In fact $f(\bar{x}) \leq c$ holds owing to the lower semicontinuity of f: pick any $x' \in$ rint (Dom f) and consider a sequence of points $x^i \in [x', \bar{x})$ converging to $\bar{x}$. For all i, by Lemma 1.30, we have $x^i \in$ rint (Dom f) and thus by Proposition 12.5 we conclude that

$$f(x^i) = \bar{f}(x^i).$$

Also, since $x^i \in [x', \bar{x})$, there exists $\lambda_i \in (0, 1]$ such that $x^i = (1 - \lambda_i)\bar{x} + \lambda_i x'$. Note that as $i \to \infty$, we have $x^i \to \bar{x}$ and so $\lambda_i \to +0$. Since $\bar{f}$ is clearly convex (as it is the supremum of a family of affine and thus convex functions), we have

$$\bar{f}(x^i) \leq (1 - \lambda_i)\bar{f}(\bar{x}) + \lambda_i \bar{f}(x').$$

Since $x', x^i \in$ rint Dom f and $\bar{f} = f$ on rint Dom f, we have $f(x') = \bar{f}(x')$, $f(x^i) = \bar{f}(x^i)$, and we get

$$f(x^i) \leq (1 - \lambda_i)\bar{f}(\bar{x}) + \lambda_i f(x').$$

Moreover, as $i \to \infty$, we have $\lambda_i \to +0$ and so the right-hand side of our inequality converges to $\bar{f}(\bar{x}) = c$. In addition, as $i \to \infty$, we have $x^i \to \bar{x}$ and, since f is lower semicontinuous, we get $f(\bar{x}) \leq c$. ■

We see why the "translation of mathematical facts from one mathematical language to another" – in our case, from the language of convex sets to the language of convex functions – may be fruitful: because we invest a lot of effort into the process rather than running it mechanically.

Closure of a convex function. Proposition 12.4 presents a nice result on the outer description of a *proper lower semicontinuous convex* function: it is the supremum of a family of affine functions. Note that the reverse is also true: the supremum of every family of affine functions is a proper lsc convex function, provided that this supremum is finite at least at one point. This is so because we know from section 10.1 that the supremum of every family of convex functions is convex and from Corollary 12.3 that the supremum of lsc functions, e.g., affine functions (these are in fact even continuous), is lower semicontinuous.

Now, what to do with a convex function which is not lower semicontinuous? There is a similar question about convex sets: what to do with a convex set which is not closed?

12.1 Proper Lower Semicontinuous Convex Functions and Their Representation

We can resolve this question very simply by passing from the set to its closure and thus getting a "much easier to handle" object which is very "close" to the original object: the "main part" of the original set – its relative interior – remains unchanged, and the "correction" adds to the set something relatively small, i.e., (part of) its relative boundary. The same approach works for convex functions: if a proper convex function f is not lower semicontinuous (i.e., its epigraph is convex and nonempty but is not closed), we can "correct" the function by replacing it with a new function whose epigraph is the closure of $\text{epi}\{f\}$. To justify this approach, of course we should be sure that the closure of the epigraph of a convex function is also an epigraph of such a function. This indeed is the case, and to see this, it suffices to note that a set G in $\mathbf{R}^{n+1}$ is the epigraph of a function taking values in $\mathbf{R} \cup \{+\infty\}$ if and only if the intersection of G with every vertical line $\{x = \text{const}, t \in \mathbf{R}\}$ is either empty or a closed ray of the form $\{x = \text{const}, t \geq \bar{t} > -\infty\}$. Now, it is absolutely evident that if $G = \text{cl}(\text{epi}\{f\})$ then the intersection of G with a vertical line is empty, or is a closed ray, or is the entire line (the last case can indeed take place – look at the closure of the epigraph of the function equal to $-\frac{1}{x}$ for $x > 0$ and $+\infty$ for $x \leq 0$). We see that in order to justify our idea of the "proper correction" of a convex function we need to prove that if f is convex then the last of the indicated three cases, i.e., that the intersection of $\text{cl}(\text{epi}\{f\})$ with a vertical line is the entire line, never occurs. However, we know from Proposition 10.11 that every convex function f is bounded from below on every compact set. Thus, $\text{cl}(\text{epi}\{f\})$ indeed cannot contain an entire vertical line. Therefore, we conclude that the closure of the epigraph of a convex function f is the epigraph of a certain function called the *closure* of f [notation: $\text{cl } f$] defined as:

$$\text{cl}(\text{epi}\{f\}) = \text{epi}\{\text{cl } f\}.$$

Of course, the function $\text{cl } f$ is convex (its epigraph is convex as it is $\text{cl}(\text{epi}\{f\})$ and $\text{epi}\{f\}$ itself is convex). Moreover, since the epigraph of $\text{cl } f$ is closed, $\text{cl } f$ is lsc. And of course we have the following immediate observation.

Observation 12.6 *The closure of an lsc convex function f is f itself.*

Proof Indeed, when f is convex, $\text{epi}\{f\}$ is convex, and when f is lsc, $\text{epi}\{f\}$ is closed, by Proposition 12.2. Hence, under the premise of this observation, $\text{epi}\{f\}$ is convex and closed and thus, by definition of $\text{cl } f$, is the same as $\text{epi}\{\text{cl } f\}$, implying that $f = \text{cl } f$. ∎

The following statement gives an instructive alternative description of $\text{cl } f$ in terms of f.

Proposition 12.7 *Let $f : \mathbf{R}^n \to \mathbf{R} \cup \{+\infty\}$ be a convex function. Then its closure $\text{cl } f$ satisfies the following:*

(i) *The affine minorants of $\text{cl } f$ are exactly the same as the affine minorants of f, and $\text{cl } f$ is the supremum of these minorants, i.e., for all $x \in \mathbf{R}^n$ we have*

$$\text{cl } f(x) = \sup_{\phi} \{\phi(x) : \phi \text{ is an affine minorant of } f\}. \tag{12.7}$$

Also, for every $x \in \text{rint}(\text{Dom}(\text{cl } f)) = \text{rint}(\text{Dom } f)$, we can replace sup on the right-hand side of (12.7) with max.

Moreover,

$$\begin{aligned}(a) \quad & f(x) \geq \text{cl } f(x), \quad \forall x \in \mathbf{R}^n, \\ (b) \quad & f(x) = \text{cl } f(x), \quad \forall x \in \text{rint}(\text{Dom } f), \\ (c) \quad & f(x) = \text{cl } f(x), \quad \forall x \notin \text{cl}(\text{Dom } f).\end{aligned} \quad (12.8)$$

Thus, the "correction" $f \mapsto \text{cl } f$ may vary f only at the points from $\text{rbd}(\text{Dom } f)$, *implying that*

$$\text{Dom } f \subseteq \text{Dom}(\text{cl } f) \subseteq \text{cl}(\text{Dom } f),$$

hence $\text{rint}(\text{Dom } f) = \text{rint}(\text{Dom}(\text{cl } f))$.

In addition, $\text{cl } f$ *is the supremum of all convex lower semicontinuous minorants of f.*

(ii) *For all $x \in \mathbf{R}^n$, we have*

$$\text{cl } f(x) = \lim_{r \to +0} \inf_{x': \|x'-x\|_2 \leq r} f(x').$$

Proof There is nothing to prove when $f \equiv +\infty$. In this case $\text{cl } f \equiv +\infty$ as well, $\text{Dom } f = \text{Dom cl } f = \emptyset$, and all claims are trivially satisfied. Thus, assume from now on that f is proper.

(i): We will prove this part in several simple steps.

By construction, $\text{epi}\{\text{cl } f\} = \text{cl}(\text{epi}\{f\}) \supseteq \text{epi}\{f\}$, which implies (12.8a). Also, as $\text{cl}(\text{epi}\{f\}) \subseteq [\text{cl}(\text{Dom } f)] \times \mathbf{R}$, we arrive at (12.8c).

Note that from (12.8a) we deduce that every affine minorant of $\text{cl } f$ is an affine minorant of f as well. Moreover, the reverse is also true. Indeed, let $g(x)$ be an affine minorant of f. Then, we clearly have $\text{epi}\{g\} \supseteq \text{epi}\{f\}$, and as $\text{epi}\{g\}$ is closed, we also get $\text{epi}\{g\} \supseteq \text{cl}(\text{epi}\{f\}) = \text{epi}\{\text{cl } f\}$. Note that the relation $\text{epi}\{g\} \supseteq \text{epi}\{\text{cl } f\}$ is simply the same as saying that g is an affine minorant of $\text{cl } f$. Thus, the affine minorants of f and of $\text{cl } f$ indeed are the same. Then, as $\text{cl } f$ is lsc and proper (since $\text{cl } f \leq f$ and f is proper), by applying Proposition 12.4 to $\text{cl } f$ and also applying Proposition 12.5 to f, we deduce (12.8b) and (12.7).

Finally, if g is a convex lsc minorant of f then g is definitely proper, and thus by Proposition 12.4 it is the supremum of all its affine minorants. These minorants of g are affine minorants of f as well, and thus also affine minorants of $\text{cl } f$. The bottom line is that g is $\leq$ the supremum of all affine minorants of f, which, as we already know, is $\text{cl } f$. Thus, a convex lsc minorant of f is a minorant of $\text{cl } f$, implying that the supremum $\widetilde{f}$ of these lsc convex minorants of f satisfies $\widetilde{f} \leq \text{cl } f$. The latter inequality is in fact an equality since $\text{cl } f$ itself is a convex lsc minorant of f by (12.8a). This completes the proof of part (i).

To verify (ii) we need to prove the following two facts:

(ii.1) For every sequence $\{x^i\}$ such that as $i \to \infty$ we have $x^i \to \bar{x}$ and $f(x^i) \to s$ (where s may be a finite number or infinity), we have $s \geq \text{cl } f(\bar{x})$.

(ii.2) For every $\bar{x}$, there exists a sequence $x^i \to \bar{x}$ such that $\lim_{i \to \infty} f(x^i) = \text{cl } f(\bar{x})$.

To prove (ii.1), note that under the premise of this claim we have $s \neq -\infty$ since f is below bounded on bounded subsets of $\mathbf{R}^n$ (Proposition 10.11). There is nothing to verify when $s = +\infty$. So, suppose $s \in \mathbf{R}$. Then, the point $[\bar{x}; s]$ is in $\text{cl}(\text{epi}\{f\}) = \text{epi}\{\text{cl } f\}$, and thus $\text{cl } f(\bar{x}) \leq s$, as claimed.

To prove (ii.2), note that the claim is trivially true when $\operatorname{cl} f(\bar{x}) = +\infty$. Indeed, in this case $f(\bar{x}) = +\infty$ as well, owing to (12.8a), and for all $i = 1, 2, \ldots$ we can take $x^i = \bar{x}$. Now, consider a point $\bar{x}$ such that $\operatorname{cl} f(\bar{x}) < \infty$. Then we have $[\bar{x}; \operatorname{cl} f(\bar{x})] \in \operatorname{epi}\{\operatorname{cl} f\} = \operatorname{cl}(\operatorname{epi}\{f\})$. Thus, there exists a sequence $[x^i; t_i] \in \operatorname{epi}\{f\}$ such that $[x^i; t_i] \to [\bar{x}; \operatorname{cl} f(\bar{x})]$ as $i \to \infty$. Passing to a subsequence, we can assume that the sequence $f(x^i)$ have a limit, finite or infinite, as $i \to \infty$. Hence, $\lim_{i \to \infty} x^i = \bar{x}$ and $\lim_{i \to \infty} f(x^i) = \liminf_{i \to \infty} f(x^i) \leq \lim_{i \to \infty} t_i = \operatorname{cl} f(\bar{x})$. Recall also that from (ii.1) we have $\lim_{i \to \infty} f(x^i) \geq \operatorname{cl} f(\bar{x})$, so we conclude that $\lim_{i \to \infty} f(x^i) = \operatorname{cl} f(\bar{x})$ as desired. ∎

12.2 Subgradients

Let $f : \mathbf{R}^n \to \mathbf{R} \cup \{+\infty\}$ be a convex function, and let $x \in \operatorname{Dom} f$. Recall from our discussion in the preceding section that f may admit an affine minorant $d^\top x - a$ which coincides with f at x, i.e.,

$$f(y) \geq d^\top y - a, \quad \forall y \in \mathbf{R}^n, \quad \text{and} \quad f(x) = d^\top x - a.$$

The equality relation above is equivalent to $a = d^\top x - f(x)$, and substituting this representation of a into the first inequality, we get

$$f(y) \geq f(x) + d^\top(y - x), \quad \forall y \in \mathbf{R}^n. \tag{12.9}$$

Thus, if f admits an affine minorant which is exact at x then there exists $d \in \mathbf{R}^n$, which gives rise to the inequality (12.9). In fact the reverse is also true: if d is such that (12.9) holds then the right-hand side of (12.9), regarded as a function of y, is an affine minorant of f which coincides with f at x.

Now note that (12.9) expresses a specific property of a vector d and leads to the following very important definition, which generalizes the notion of gradient for smooth convex functions to nonsmooth convex functions.

Definition 12.8 *[Subgradient of a convex function] Let $f : \mathbf{R}^n \to \mathbf{R} \cup \{+\infty\}$ be a convex function. Given $\bar{x} \in \mathbf{R}^n$, a vector $d \in \mathbf{R}^n$ is called a* subgradient *of f at $\bar{x}$ if d is the slope of an affine minorant of f which is exact at $\bar{x}$. That is, d is a subgradient of f at $\bar{x}$ if d satisfies*

$$f(y) \geq f(\bar{x}) + d^\top(y - \bar{x}), \quad \forall y \in \mathbf{R}^n.$$

The set of all subgradients of f at a point $\bar{x}$ is called the subdifferential *of f at $\bar{x}$ [notation: $\partial f(\bar{x})$].*

Figure 12.2 illustrates the notions of subgradient and subdifferential.

Let us make the following useful observation:

Let $f : \mathbf{R}^n \to \mathbf{R} \cup \{+\infty\}$ be a convex function, let $x, y \in \operatorname{Dom} f$, $g \in \partial f(x)$, and $h \in \partial f(y)$. Then

$$h^\top[y - x] \geq f(y) - f(x) \geq g^\top[y - x] \text{ and } [h - g]^\top[y - x] \geq 0.$$

Indeed, the first two inequalities are readily given by the definition of a subgradient, and the third inequality is an immediate consequence of the first two of them.

The subgradients of convex functions play an important role in the theory and numerical methods of convex optimization – they are quite reasonable surrogates of gradients in the

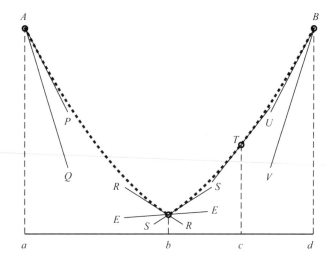

Figure 12.2 Subdifferentials of a univariate convex function. Dotted curve, convex function with domain $[a, d]$; point a, at the boundary point a of the domain, the subgradients of f are the slopes of lines like $\overline{AP}$ and $\overline{AQ}$, and $\partial f(a) = \{g : g \leq \text{slope}(\overline{AP})\}$; point b, at point b, the subgradients are the slopes of lines like $\overline{RR}$, $\overline{EE}$, $\overline{SS}$, and $\partial f(b) = \{g : \text{slope}(\overline{RR}) \leq g \leq \text{slope}(\overline{SS})\}$; point c, just one subgradient – the slope of the tangent line, taken at point T, to the graph of f; point d, similar to a,
$$\partial f(d) = \{g : g \geq \text{slope}(\overline{UB})\}.$$

cases when the latter do not exist. Let us present a simple and instructive illustration of this. Recall that Theorem 11.2 states that

> A necessary and sufficient condition for a convex function $f : \mathbf{R}^n \to \mathbf{R} \cup \{+\infty\}$ to attain its minimum at a point $x^* \in \text{int}(\text{Dom } f)$, where f is differentiable, is that $\nabla f(x^*) = 0$.

The "nonsmooth" version of this statement is as follows.

Proposition 12.9 *[Necessary and sufficient optimality condition for nonsmooth convex functions] A necessary and sufficient condition for a convex function $f : \mathbf{R}^n \to \mathbf{R} \cup \{+\infty\}$ to attain its minimum at a point $x^* \in \text{Dom } f$ is the inclusion $0 \in \partial f(x^*)$.*

Proof Given $x^* \in \text{Dom } f$, by the definition of the subdifferential, $0 \in \partial f(x^*)$ if and only if the vector 0 is a subgradient of f at x^*, which (by the definition of the subgradient) holds if and only if
$$f(y) \geq f(x^*) + 0^\top (y - x^*), \quad \forall y \in \mathbf{R}^n,$$
which simply states that x^* is a minimizer of f. ∎

In fact, if a convex function f is differentiable at $\bar{x} \in \text{int}(\text{Dom } f)$ then $\partial f(\bar{x})$ is the singleton $\{\nabla f(\bar{x})\}$ (see Proposition 12.10 below). Thus, Proposition 12.9 is indeed an extension of Theorem 11.2 to the case when f is possibly non-differentiable.

Looking at the (absolutely trivial!) proof of Proposition 12.9, one can ask: how can such a trivial necessary and sufficient optimality condition be useful? Why is it more informative than the tautological necessary and sufficient optimality condition "$f(x^*) \leq f(x)$ for all x"? Well, the usefulness of the Fermat optimality condition stems not from the condition *per se*, but from our knowledge on gradients, including (but not reduced to) our ability to compute gradients, to the extent given by standard calculus. Similarly, the usefulness of Proposition 12.9 stems from our knowledge on subgradients, including the ability, to the extent given by the calculus of subgradients, to compute subdifferentials of convex functions. Let us become acquainted with the basics of this calculus.

Here is a summary of the most elementary properties of subgradients.

Proposition 12.10 *Let $f : \mathbf{R}^n \to \mathbf{R} \cup \{+\infty\}$ be a convex function. Then:*

(i) *$\partial f(x)$ is a closed convex set for any $x \in \mathrm{Dom}\, f$, $\partial f(x) \neq \emptyset$ whenever $x \in \mathrm{rint}\,(\mathrm{Dom}\, f)$ (this is the most important fact about subgradients), and $\partial f(x)$ is bounded whenever $x \in \mathrm{int}(\mathrm{Dom}\, f)$.*
(ii) *If $x \in \mathrm{int}(\mathrm{Dom}\, f)$ and f is differentiable at x then $\nabla f(x)$ is the only subgradient of f at x: $\partial f(x) = \{\nabla f(x)\}$.*
(iii) *The "closedness of the subdifferential mapping:" Let $g_i \in \partial f(x_i)$ and $[x_i; g_i] \to [x; g]$ as $i \to \infty$. Assume also that $x \in \mathrm{Dom}\, f$ and f is continuous at x. Then $g \in \partial f(x)$.*
(iv) *Let Y be a nonempty convex compact subset of $\mathrm{int}(\mathrm{Dom}\, f)$. Then the set $G = \{[x; g] \in \mathbf{R}^n \times \mathbf{R}^n : x \in Y, g \in \partial f(x)\}$ is compact.*

Proof (i): The closedness and convexity of $\partial f(x)$ are evident from the definition of the subdifferential, as (12.9) is an infinite system of nonstrict linear inequalities, indexed by $y \in \mathrm{Dom}\, f$, on the variable d.

When $x \in \mathrm{rint}\,(\mathrm{Dom}\, f)$, Proposition 12.5 provides us with an affine function which underestimates f everywhere and coincides with f at x. The slope of this affine function is clearly a subgradient of f at x, and thus $\partial f(x) \neq \emptyset$.

The boundedness of $\partial f(x)$ when $x \in \mathrm{int}(\mathrm{Dom}\, f)$ is an immediate consequence of part (iv), to be proved soon.

(ii): Suppose $x \in \mathrm{int}(\mathrm{Dom}\, f)$ and f is differentiable at x. Then, by the gradient inequality we have $\nabla f(x) \in \partial f(x)$. Let us prove that, in this case, $\nabla f(x)$ is the only subgradient of f at x. Consider any $d \in \partial f(x)$. Then, by the definition of a subgradient, we have

$$f(y) - f(x) \geq d^\top (y - x), \quad \forall y \in \mathbf{R}^n.$$

Now, consider any fixed direction $h \in \mathbf{R}^n$ and any real number $t > 0$. By substituting $y = x + th$ into the preceding inequality and then dividing both sides of the resulting inequality by t, we obtain

$$\frac{f(x + th) - f(x)}{t} \geq d^\top h.$$

Taking the limit of both sides of this inequality as $t \to +0$, we get

$$h^\top \nabla f(x) \geq h^\top d.$$

Since h is an arbitrary direction, this inequality is valid for all $h \in \mathbf{R}^n$, which is possible if and only if $d = \nabla f(x)$.

(iii): Under the premise of this part, for every $y \in \mathbf{R}^n$ and for all $i = 1, 2, \ldots$, we have

$$f(y) \geq f(x_i) + g_i^\top (y - x_i).$$

Passing to the limit as $i \to \infty$, we get

$$f(y) \geq f(x) + g^\top (y - x), \quad \forall y \in \mathbf{R}^n,$$

and thus $g \in \partial f(x)$.

(iv): Let us start with the following observation:

> Let $f : \mathbf{R}^n \to \mathbf{R} \cup \{+\infty\}$ be convex, and consider $\bar{x} \in \mathrm{int}(\mathrm{Dom}\, f)$. Suppose f is Lipschitz continuous, with constant L with respect to $\|\cdot\|_2$, in a neighborhood V of $\bar{x}$, that is, $|f(x) - f(y)| \leq L\|x - y\|_2$ for all $x, y \in V$. Then
>
> $$\|g\|_2 \leq L, \quad \forall g \in \partial f(\bar{x}).$$

Indeed, when $g \in \partial f(x)$, by the subgradient inequality and Lipschitz continuity of f we have $g^\top (y - x) \leq f(y) - f(x) \leq L\|x - y\|_2$ for all y close to x, that is, $g^\top h \leq L\|h\|_2$ for all $h \in \mathbf{R}^n$, implying that $\|g\|_2 = \max_h \{g^\top h : \|h\|_2 \leq 1\} \leq L$.

Now, under the premise of (iv), we can find a compact set $Y' \subset \mathrm{int}(\mathrm{Dom}\, f)$ such that $Y \subset \mathrm{int}\, Y'$. By Theorem 10.9, f is Lipschitz continuous, with a certain constant L, on Y', implying by the preceding observation that $\|g\|_2 \leq L$ for all $g \in \partial f(x)$ with $x \in Y$. Then, as Y is bounded, the set $G = \{[x; g] \in \mathbf{R}^n \times \mathbf{R}^n : x \in Y, g \in \partial f(x)\}$ is bounded. Moreover, this set is closed by (iii) (recall that Y is compact and f is continuous on Y), so that G is compact. ∎

Proposition 12.10 sheds light on why subgradients are good surrogates of gradients: at a point where the gradient exists, the gradient itself is the unique subgradient; but, in contrast with the gradient, a subgradient exists basically everywhere (and certainly in the relative interior of the domain of the function).

Let us examine a simple function and its subgradients.

Example 12.1 Consider the function $f : \mathbf{R} \to \mathbf{R}$ given by

$$f(x) = |x| = \max\{x, -x\}.$$

This function f is, of course, convex (as the maximum of two linear forms x and $-x$). Whenever $x \neq 0$, f is differentiable at x with derivative $+1$ for $x > 0$ and -1 for $x < 0$. At the point $x = 0$, f is not differentiable; nevertheless, it must have subgradients at this point (since $0 \in \mathrm{int}(\mathrm{Dom}\, f)$). And indeed, it can immediately be seen that the subgradients of f at $x = 0$ are exactly the real numbers from the segment $[-1, 1]$. Thus,

$$\partial |x| = \begin{cases} \{-1\}, & \text{if } x < 0, \\ [-1, 1], & \text{if } x = 0, \\ \{+1\}, & \text{if } x > 0. \end{cases}$$

◇

It is important to note that at the points on the relative boundary of the domain of a convex function, even a "good" one, we may not have any subgradients. That is, it is possible to have $\partial f(x) = \varnothing$ for a convex function f at a point $x \in \mathrm{rbd}(\mathrm{Dom}\, f)$. We give an example of this next.

Example 12.2 Consider the function

$$f(x) = \begin{cases} -\sqrt{x}, & \text{if } x \geq 0, \\ +\infty, & \text{if } x < 0. \end{cases}$$

The convexity of this function follows from the convexity of its domain and Example 9.2. Consider the point $[0; f(0)] \in \text{rbd}(\text{epi}\{f\})$. It is clear that at the point $[0; f(0)]$ there is no non-vertical supporting line to the set $\text{epi}\{f\}$ and, consequently, there is no affine minorant of the function which is exact at $x = 0$. ◇

A significant – and important – part of Convex Analysis deals with *subgradient calculus*, which is the set of rules for computing subgradients of "composite" functions, such as sums, superpositions, maxima, etc., given the subgradients of the operands. These rules extend the standard Calculus rules to nonsmooth convex functions, and they are quite nice and useful. Here, we list several "self-evident" versions of these rules.

1. *Subdifferential of nonnegative weighted sum:* Let f, g be convex functions on $\mathbf{R}^n$. Consider $x \in \text{Dom } f \cap \text{Dom } g$ and $\lambda, \mu \in \mathbf{R}_+$. If $d \in \partial f(x)$ and $e \in \partial g(x)$ then $\lambda d + \mu e \in \partial[\lambda f + \mu g](x)$.

2. *Subdifferential of pointwise supremum:* Let $\{f_\alpha(\cdot)\}_{\alpha \in \mathcal{A}}$ be a family of convex functions on $\mathbf{R}^n$. Consider the convex function $\overline{f}(x) := \sup_{\alpha \in \mathcal{A}} f_\alpha(x)$ and $\bar{x} \in \text{Dom } \overline{f}$. Suppose that $\bar{\alpha}$ is such that $\overline{f}(\bar{x}) = f_{\bar{\alpha}}(\bar{x})$. Then, for any $\bar{d} \in \partial f_{\bar{\alpha}}(\bar{x})$, we have $\bar{d} \in \partial \overline{f}(\bar{x})$. Here is the justification: $\overline{f}(x) \geq f_{\bar{\alpha}}(x) \geq f_{\bar{\alpha}}(\bar{x}) + \bar{d}^\top(x - \bar{x})$ for all $x \in \mathbf{R}^n$, that is, $\overline{f}(x) \geq f_{\bar{\alpha}}(\bar{x}) + \bar{d}^\top(x - \bar{x})$ for all x, and this inequality becomes equality at $x = \bar{x}$, as $\overline{f}(\bar{x}) = f_{\bar{\alpha}}(\bar{x})$.

3. *Subdifferential of convex monotone/affine superposition [chain rule]:*
Let $f_1(x), \ldots, f_K(x)$ be convex functions on $\mathbf{R}^n$, and let F be a convex function on $\mathbf{R}^K$. Suppose for some $0 \leq k \leq K$ the functions $f_1, \ldots, f_k$ are affine, and also $F(y)$ is nondecreasing in $y_{k+1}, \ldots, y_K$. Recall that the superposition

$$g(x) := \begin{cases} F(f_1(x), \ldots, f_K(x)), & \text{if } f_k(x) < +\infty \text{ for all } k \leq K, \\ +\infty, & \text{otherwise,} \end{cases}$$

is a convex function of x. Consider $\bar{x} \in \bigcap_{i=1}^K \text{Dom}(f_i)$ such that

$$\bar{y} := [f_1(\bar{x}); \ldots; f_K(\bar{x})] \in \text{Dom } F.$$

Let $g^i \in \partial f_i(\bar{x})$ for all $i \leq K$ and $e = [e_1; \ldots; e_K] \in \partial F(\bar{y})$. Then the vector $\bar{d} := \sum_{i=1}^K e_i g^i$ satisfies $\bar{d} \in \partial[F(f_1, \ldots, f_K)](\bar{x})$.
The justification of this is as follows. Let $h \in \mathbf{R}^n$ be an arbitrary direction, and consider $x = \bar{x} + h$ and $y = \bar{y} + [(g^1)^\top h; (g^2)^\top h; \ldots; (g^K)^\top h]$. Then, for all $i \leq K$, we have

$$y_i = \bar{y}_i + (g^i)^\top h = f_i(\bar{x}) + (g^i)^\top(x - \bar{x}).$$

Moreover, for $i \leq k$, as $f_1, \ldots, f_k$ are affine we have $y_i = f_i(x)$; and, for $i > k$, as $g^i \in \partial f_i(\bar{x})$ we have $f_i(x) \geq y_i$. Consequently, using the partial monotonicity of F, we conclude that $F(f_1(x), \ldots, f_K(x)) \geq F(y)$. Now, $e \in \partial F(\bar{y})$, which implies the first inequality in the following chain:

$$F(y) \geq F(\bar{y}) + e^\top (y - \bar{y}) = F(f_1(\bar{x}), \ldots, f_K(\bar{x})) + \left(\sum_i e_i g^i\right)^\top h$$

$$= F(f_1(\bar{x}), \ldots, f_K(\bar{x})) + \bar{d}^\top (x - \bar{x}).$$

Indeed, the first equality follows from $y_i = \bar{y}_i + (g^i)^\top h$ for all i and the second equality from $x = \bar{x} + h$. Finally, this inequality combines with $F(f_1(x), \ldots, f_K(x)) \geq F(y)$ to imply that

$$F(f_1(x), \ldots, f_K(x)) \geq F(f_1(\bar{x}), \ldots, f_K(\bar{x})) + \bar{d}^\top (x - \bar{x}),$$

as desired.

Advanced versions of these subgradient calculus rules, under appropriate assumptions, describe how the entire subdifferentials of the resulting functions are obtained from those of the operands; the related considerations, however, are beyond our scope.

We close this section by providing an illustration of the second rule.

Example 12.3 In this example, we will examine the subgradients of the spectral norm of a symmetric matrix. For $X \in \mathbf{S}^n$, its spectral norm $\|X\|$ is given by

$$\|X\| := \max_{e \in \mathbf{R}^n} \left\{ |e^\top X e| : \|e\|_2 \leq 1 \right\}.$$

From this definition, it is clear that $\|X\|$ is a convex function of X.

Given X, we can compute a maximizer of the optimization problem defining $\|X\|$ by finding an eigenvalue–eigenvector pair (λ_X, e_*) of X such that λ_X is the largest in magnitude of the eigenvalues of X. Let e_X be the unit length normalization of e_*. Thus,

$$\|X\| = \max_{e: \|e\|_2 \leq 1} |e^\top X e| = \max_{e: \|e\|_2 \leq 1} |\mathrm{Tr}(X[ee^\top])| = |\mathrm{Tr}(X[e_X e_X^\top])|$$

$$= \mathrm{Tr}(X[\mathrm{sign}(\lambda_X) e_X e_X^\top]),$$

where the third equality follows from the choice of e_X, and the last equality is due to $\mathrm{Tr}(X[e_X e_X^\top]) = \lambda_X$. Then, using item 2 in our subgradient calculus rules, we conclude that the symmetric matrix $E(X) := \mathrm{sign}(\lambda_X) e_X e_X^\top$ is a subgradient of the function $\|\cdot\|$ at X. That is,

$$\|Y\| \geq \mathrm{Tr}(Y E(X)) = \|X\| + \mathrm{Tr}(E(X)(Y - X)), \qquad \forall Y \in \mathbf{S}^n,$$

where the equality holds due to the choice of $E(X)$ guaranteeing $\|X\| = \mathrm{Tr}(X E(X))$. To see that this is indeed a subgradient inequality, recall that the inner product on $\mathbf{S}^n$ is the Frobenius inner product $\langle A, B \rangle = \mathrm{Tr}(AB)$.

Let us close this example by discussing the smoothness properties of $\|X\|$. Recall that every norm $\|y\|$ in $\mathbf{R}^m$ is nonsmooth at the origin $y = 0$. However, $\|X\|$ is a nonsmooth function of X even at points other than $X = 0$. In fact, $\|\cdot\|$ is continuously differentiable in a neighborhood of every point $X \in \mathbf{S}^n$ where the maximum magnitude eigenvalue is unique and is of multiplicity 1, but it can lose smoothness at other points. ◇

12.3 Subdifferentials and Directional Derivatives of Convex Functions

Let f be a convex function on $\mathbf{R}^n$ and consider $x \in \mathrm{int}(\mathrm{Dom}\, f)$ (in fact the construction to follow admits an immediate generalization for the case when $x \in \mathrm{rint}\,(\mathrm{Dom}\, f)$ as well). Consider any direction $h \in \mathbf{R}^n$ and the univariate function

$$\phi(t) := f(x + th)$$

associated with h. Note that $\phi(t)$ is a real-valued convex function of t in a neighborhood of 0; thus for all small enough positive s and t we have

$$\frac{\phi(0) - \phi(-s)}{s} \leq \frac{\phi(t) - \phi(0)}{t}.$$

Moreover, the right-hand side of this inequality is nondecreasing in t, hence it implies the existence of a *directional derivative of f taken at x along the direction h*, i.e., the quantity

$$Df(x)[h] = \lim_{t \to +0} \frac{f(x + th) - f(x)}{t}.$$

As a function of h, the function $Df(x)[h]$ is clearly positively homogeneous of degree 1, i.e.,

$$Df(x)[\lambda h] = \lambda Df(x)[h], \quad \forall \lambda \geq 0.$$

In addition, $Df(x)[h]$ is a convex function of h. To see this, let $r > 0$ be such that the Euclidean ball B centered at x and of radius r is contained in $\mathrm{Dom}\, f$. Then the functions

$$f_s(h) := \frac{f(x + sh) - f(x)}{s}$$

are convex over the domain $h \in B$ as long as $0 < s \leq 1$. Moreover, as $s \to +0$, on B these functions converge pointwise to $Df(x)[h]$, which clearly implies the convexity of $Df(x)[h]$ as a function of $h \in B$. Finally, since, as a function of h, $Df(x)[h]$ is positively homogeneous of degree 1, its convexity on B clearly implies its convexity on the entire $\mathbf{R}^n$. Note that the convexity of f implies that

$$f(x+h) \geq f(x) + Df(x)[h], \quad \forall (x,h : x \in \mathrm{int}(\mathrm{Dom}\, f), x+h \in \mathrm{Dom}\, f). \quad (12.10)$$

We have arrived at the following result.

Lemma 12.11 *Let f be a convex function on $\mathbf{R}^n$. For any $x \in \mathrm{int}(\mathrm{Dom}\, f)$, the subdifferential $\partial f(x)$ of f at x is exactly the same as the subdifferential of $Df(x)[\cdot]$ at the origin.*

Proof Suppose d is a subgradient of f at x, and thus $f(x+th) - f(x) \geq t d^\top h$ holds for all $h \in \mathbf{R}^n$ and all small enough $t > 0$. This then implies that $Df(x)[h] \geq d^\top h$, that is, d is a subgradient of $Df(x)[h]$ at $h = 0$. For the reverse direction, suppose that d is a subgradient of $Df(x)[\cdot]$ at $h = 0$. Then we have $Df(x)[h] \geq d^\top h$ for all h, implying, by (12.10), that $f(x+h) \geq f(x) + d^\top h$ whenever $x+h \in \mathrm{Dom}\, f$. ∎

Our goal is to demonstrate that when $x \in \mathrm{int}(\mathrm{Dom}\, f)$, the subdifferential $\partial f(x)$ is large enough to fully define $Df(x)[\cdot]$:

Theorem 12.12 *Let f be a convex function on $\mathbf{R}^n$ and $x \in \mathrm{int}(\mathrm{Dom}\, f)$. Then*

$$Df(x)[h] = \max_{d} \left\{ d^\top h : d \in \partial f(x) \right\}. \quad (12.11)$$

Note that for a convex function f at $x \in \text{int}(\text{Dom } f)$, by Proposition 12.10 we know that $\partial f(x)$ is compact and thus in (12.11) the use of max as opposed to sup is justified. We are about to prove Theorem 12.12 using the fundamental *Hahn–Banach Theorem* (the finite-dimensional version) given below.

Theorem 12.13 *[Hahn–Banach Theorem, finite-dimensional version] Let $D(\cdot)$ be a real-valued convex positively homogeneous function, of degree 1, on $\mathbf{R}^n$, let $e \in \mathbf{R}^n$, and let E be a linear subspace of $\mathbf{R}^n$ such that*

$$e^\top z \leq D(z), \quad \forall z \in E.$$

Then, there exists $e' \in \mathbf{R}^n$ such that $[e']^\top z \equiv e^\top z$ for all $z \in E$ and $[e']^\top z \leq D(z)$ for all $z \in \mathbf{R}^n$. In other words, a linear functional, defined on a linear subspace of $\mathbf{R}^n$ and majorized on this subspace by $D(\cdot)$, can be extended from the subspace to a linear functional on the entire space in such a way that the extension is majorized by $D(\cdot)$ everywhere.

Proof of Theorem 12.12 In this proof we will show that Theorem 12.13 implies Theorem 12.12.

Consider any $d \in \partial f(x)$. Then, by Lemma 12.11 we have d is in the subdifferential $Df(x)[\cdot]$ at the origin, i.e., $Df(x)[h] \geq d^\top h$. Thus, we conclude

$$Df(x)[h] \geq \max_d \left\{ d^\top h : d \in \partial f(x) \right\}.$$

To prove the opposite inequality, let us fix $h \in \mathbf{R}^n$ and verify that $g := Df(x)[h] \leq \max_d \{ d^\top h : d \in \partial f(x) \}$. There is nothing to prove when $h = 0$, so let $h \neq 0$. Setting $\phi(t) := Df(x)[th]$, we get a convex (since, as we already know, $Df(x)[\cdot]$ is convex) univariate function such that $\phi(t) = gt$ for $t \geq 0$. Then this, together with the convexity of ϕ, implies that $\phi(t) \geq gt$ for all $t \in \mathbf{R}$. By applying the Hahn–Banach Theorem to the function $D(z) := Df(x)[z]$ (we already know that this function satisfies the premise of the Hahn–Banach Theorem), to $E := \mathbf{R}(h)$, and the linear form $e^\top (th) = gt$, $t \in \mathbf{R}$, on E, we conclude that there exists $e \in \mathbf{R}^n$ such that $e^\top u \leq Df(x)[u]$ for all $u \in \mathbf{R}^n$ and $e^\top h = g = Df(x)[h]$. By the first relation, e is a subgradient of $Df(x)[\cdot]$ at the origin and thus, by Lemma 12.11, $e \in \partial f(x)$, so that

$$\max_{d \in \partial f(x)} d^\top h \geq e^\top h = Df(x)[h].$$

Thus, the right-hand side of (12.11) is greater than or equal to the left-hand side. The opposite inequality has already been proved, so that (12.11) is an equality. ∎

Remark 12.14 The reasoning in the proof of Theorem 12.12 implies the following fact:

Let f be a convex function. Consider any $x \in \text{int}(\text{Dom } f)$ and any affine plane M such that $x \in M$. Then, "every subgradient, taken at x, of the restriction $f|_M$ of f to M can be obtained from a subgradient of f." That is, if e is such that $f(y) \geq f(x) + e^\top (y - x)$ for all $y \in M$ then there exists $e' \in \partial f(x)$ such that $e^\top (y - x) = (e')^\top (y - x)$ for all $y \in M$.

For completeness we also present a proof of the finite-dimensional version of the Hahn–Banach Theorem (Theorem 12.13).

12.3 Subdifferentials and Directional Derivatives of Convex Functions

Proof of Theorem 12.13 We are given a linear functional defined on a linear subspace E of $\mathbf{R}^n$ and this linear functional is majorized on this subspace by $D(\cdot)$; we want to prove that this linear functional can be extended from E to a linear functional on the entire space such that it is majorized by $D(\cdot)$ everywhere. To this end, it clearly suffices to prove this fact when E is of dimension $n-1$ (as, given this fact for E satisfying $\dim(E) = n-1$, we can build the desired extension in the general case by iterating extensions increasing the dimension by 1). Thus, suppose $\mathbf{R}^n = \mathbf{R}(g) + E$, where $g \notin E$. Note that in order to specify the desired vector e' all we need is to determine what the value of $\alpha := (e')^\top g$ should be, since then e' will be uniquely defined by the relation

$$(e')^\top (\lambda g + h) = \lambda \alpha + e^\top h, \quad \forall (\lambda \in \mathbf{R}, h \in E).$$

Therefore, we wish to find $\alpha \in \mathbf{R}$ such that

$$\lambda \alpha + e^\top h \leq D(\lambda g + h), \quad \forall (\lambda \in \mathbf{R}, h \in E).$$

Note that when $\lambda = 0$, the preceding inequality is automatically satisfied owing to the premise of the theorem on e. Moreover, as D is positively homogeneous of degree 1 and E is a linear subspace, all we need is to ensure that the preceding inequality is valid when $\lambda = \pm 1$, that is, to ensure that

$$\alpha \leq D(g+h) - e^\top h, \quad \forall h \in E \quad (a)$$
$$\alpha \geq -D(-g+h) + e^\top h, \quad \forall h \in E. \quad (b)$$

Now, to justify that an $\alpha \in \mathbf{R}$ satisfying the relations (a) and (b) above indeed exists, we need to show that every possible value of the right-hand side in (a) is greater than or equal to every possible value of the right-hand side in (b), that is, that $D(g+h) - e^\top h \geq -D(-g+h') + e^\top h'$ whenever $h, h' \in E$. By rearranging the terms, we thus need to show that

$$e^\top [h + h'] \leq D(h+g) + D(-g+h'), \quad \forall h, h' \in E. \quad (12.12)$$

Now, note that

$$\begin{aligned} D(h+g) + D(-g+h') &= 2 \left(\frac{1}{2} D(h+g) + \frac{1}{2} D(-g+h') \right) \\ &\geq 2 D \left(\frac{1}{2}(h+g) + \frac{1}{2}(-g+h') \right) \\ &= D(h+h') \\ &\geq e^\top [h+h'], \end{aligned}$$

where the first inequality follows from the convexity of the function D, the equalities hold because D is positively homogeneous of degree 1, and the last inequality is due to the facts that $h + h' \in E$ and that $e^\top z$ is majorized by $D(z)$ on E. Hence, (12.12) is proved. ∎

Remark 12.15 The advantage of the preceding proof of the Hahn–Banach Theorem in the finite-dimensional case is that it straightforwardly combines with what is called *transfinite induction* to yield the Hahn–Banach Theorem in the case when $\mathbf{R}^n$ is replaced with an arbitrary, perhaps infinite-dimensional, linear space and an extension of the linear functional from a linear subspace on the entire space which preserves majorization by a given convex and positively homogeneous function, of degree 1, on the space.

In the finite-dimensional case, alternatively we can prove the Hahn–Banach Theorem, via the Separation Theorem as follows: without loss of generality we can assume that $E \neq \mathbf{R}^n$. Define the sets $T := \{[x;t] : x \in \mathbf{R}^n, t \geq D(x)\}$ and $S := \{[h;t] : h \in E, t = e^\top h\}$. Thus, we get two nonempty convex sets with non-intersecting relative interiors (as $E \neq \mathbf{R}^n$ and $D(h)$ majorizes $e^\top h$ on E). Then, by the Separation Theorem there exists a nontrivial ($r \neq 0$) linear functional $r^\top[x;t] \equiv d^\top x + at$ which separates S and T, i.e., $\inf_{y \in T} r^\top y \geq \sup_{y \in S} r^\top y$. Moreover, since S is a linear subspace, we deduce that $\sup_{y \in S} r^\top y$ is either 0 or $+\infty$. Also, as $T \neq \emptyset$, we conclude that $+\infty > \inf_{y \in T} r^\top y \geq \sup_{y \in S} r^\top y$, and thus we must have $\sup_{y \in S} r^\top y = 0$. In addition, we claim that $a > 0$. Indeed,

$$0 = \sup_{y \in S} r^\top y \leq \inf_{y \in T} r^\top y = \inf_{x \in \mathbf{R}^n, t \in \mathbf{R}} \left\{ d^\top x + at : t \geq D(x) \right\}.$$

As the right-hand side value must be strictly greater than $-\infty$, we see that $a \geq 0$. Also, if $a = 0$ were to hold then from $r \neq 0$ we would have $d \neq 0$. Moreover, when $a = 0$ and $d \neq 0$, since $D(\cdot)$ is a finite-valued function we have $\inf_{x \in \mathbf{R}^n, t \in \mathbf{R}} \{d^\top x + at : t \geq D(x)\} = -\infty$. But, this is in contradiction to the infimum being bounded below by 0. Now that $a > 0$, by multiplying r by a^{-1} we get a separator of the form $r = [-e'; 1]$. Then

$$0 = \sup_{y \in S} r^\top y = \sup_{[h;t] \in S} r^\top [h;t] = \sup_{[h;t] \in S} \{(-e')^\top h + t\} = \sup_{h \in E, t = e^\top h} \{(-e')^\top h + t\}$$
$$= \sup_{h \in E} \{(-e')^\top h + e^\top h\},$$

and so we conclude that $(e')^\top h = e^\top h$ holds for all $h \in E$. Note also that the relation $0 = \sup_{[h;t] \in S} r^\top[h;t] \leq \inf_{[x;t] \in T} r^\top[x;t]$ is just $(e')^\top x \leq D(x)$ for all $x \in \mathbf{R}^n$. ◇

The Hahn–Banach Theorem is extremely important in its own right, and our way of proving Theorem 12.12 was motivated by the desire to acquaint the reader with the Hahn–Banach Theorem. If the justification of Theorem 12.12 had been our only goal, we could have achieved this goal in a much broader setting and at a cheaper cost; see the solution to Exercise IV.29, part D.4.

13

★ Legendre Transform

13.1 Legendre Transform: Definition and Examples

Let $f : \mathbf{R}^n \to \mathbf{R} \cup \{+\infty\}$ be a proper convex function. We know that f is "basically" the supremum of all its affine minorants. In fact, this is exactly the case when f is lower semicontinuous in addition to being convex and proper; otherwise (i.e., if it is not lower semicontinuous) the corresponding equality holds everywhere except, perhaps, at some points from rbd(Dom f). Now, let us look into the question of when an affine function $d^\top x - a$ is an affine minorant of f. This is the case if and only if we have

$$f(x) \geq d^\top x - a, \qquad \forall x \in \mathbf{R}^n,$$

which holds if and only if we have

$$a \geq d^\top x - f(x), \qquad \forall x \in \mathbf{R}^n.$$

Thus, we see that if the slope d of an affine function $d^\top x - a$ is fixed, then, in order for the function to be a minorant of f, it needs to satisfy

$$a \geq \sup_{x \in \mathbf{R}^n} \left\{ d^\top x - f(x) \right\}.$$

The supremum on the right-hand side of this inequality is a certain function of d, and we arrive at the following important definition.

Definition 13.1 *[Legendre transform (Fenchel dual) of a convex function]* Given a convex function $f : \mathbf{R}^n \to \mathbf{R} \cup \{+\infty\}$, its Legendre transform *(also called the* Fenchel conjugate *or* Fenchel dual*)* [notation: f^*] is the function

$$f^*(d) := \sup_{x \in \mathbf{R}^n} \left\{ d^\top x - f(x) \right\} : \mathbf{R}^n \to \mathbf{R} \cup \{+\infty\}.$$

Geometrically, the Legendre transform answers the following question: given a linear function with slope d, i.e., given the hyperplane $t = d^\top x$ in $\mathbf{R}^{n+1}$, what is the minimal "shift down" of this hyperplane to place it below the graph of f?

The definition of the Legendre transform immediately leads to a simple yet useful observation.

Fact 13.2 *Given a proper convex function $f : \mathbf{R}^n \to \mathbf{R} \cup \{+\infty\}$, its Legendre transform f^* is a proper convex lower semicontinuous function.*

Let us see some examples of simple functions and their Legendre transforms.

Example 13.1

1. Given $a \in \mathbf{R}$, consider the constant function
$$f(x) \equiv a.$$

Its Legendre transform is given by
$$f^*(d) = \sup_{x \in \mathbf{R}^n} \left\{ d^\top x - f(x) \right\} = \sup_{x \in \mathbf{R}^n} \left\{ d^\top x - a \right\} = \begin{cases} -a, & \text{if } d = 0, \\ +\infty, & \text{otherwise.} \end{cases}$$

2. Consider the affine function
$$f(x) = c^\top x + a, \quad \forall x \in \mathbf{R}^n.$$

Its Legendre transform is given by
$$f^*(d) = \sup_{x \in \mathbf{R}^n} \left\{ d^\top x - f(x) \right\} = \sup_{x \in \mathbf{R}^n} \left\{ d^\top x - (c^\top x + a) \right\} = \begin{cases} -a, & \text{if } d = c, \\ +\infty, & \text{otherwise.} \end{cases}$$

3. Consider the strictly convex quadratic function
$$f(x) = \frac{1}{2} x^\top A x,$$

where $A \in \mathbf{S}^n$ is positive definite. Its Legendre transform is given by
$$f^*(d) = \sup_{x \in \mathbf{R}^n} \left\{ d^\top x - f(x) \right\} = \sup_{x \in \mathbf{R}^n} \left\{ d^\top x - \left(\frac{1}{2} x^\top A x \right) \right\} = \frac{1}{2} d^\top A^{-1} d,$$

where the final equality is seen to hold by examining the first-order necessary and sufficient optimality condition (for the maximization-type objective) of differentiable concave functions.

4. Consider the function $f : \mathbf{R} \to \mathbf{R}$ given by $f(x) = |x|^p/p$, where $p \in (1, \infty)$. Then, using the first-order optimality conditions, we see that the Legendre transform of f is given by
$$f^*(d) = \sup_{x \in \mathbf{R}} \left\{ dx - \frac{|x|^p}{p} \right\} = \frac{|d|^q}{q},$$

where q satisfies $\frac{1}{p} + \frac{1}{q} = 1$.

5. Suppose f is a proper convex function and the function g is defined to be $g(x) = f(x - a)$. Then, the Legendre transform of g satisfies
$$g^*(d) = \sup_{x \in \mathbf{R}^n} \left\{ d^\top x - g(x) \right\} = \sup_{x \in \mathbf{R}^n} \left\{ d^\top x - f(x - a) \right\}$$
$$= \sup_{x' \in \mathbf{R}^n} \left\{ d^\top (a + x') - f(x') \right\} = d^\top a + f^*(d).$$

13.2 Legendre Transform: Main Properties

Given a function f, we define its *biconjugate* [notation: f^{**}] as the Legendre transform of the function f^*, that is,

$$f^{**} := (f^*)^*.$$

The most elementary (and the most fundamental) fact about the Legendre transform is its involutive property ("symmetry"), which we discuss next. In particular, this symmetry property of the Legendre transform gives us an alternative representation of f in terms of its affine minorants.

Proposition 13.3 *Let f be a proper convex function. Then its biconjugate f^{**} is exactly the closure of f, i.e.,*

$$f^{**} = \operatorname{cl} f.$$

In particular, when f is proper, convex, and also lower semicontinuous, *then it is precisely the Legendre transform of its Legendre transform, and thus*

$$f(x) = f^{**}(x) = \sup_d \left\{ x^\top d - f^*(d) \right\}.$$

Proof First, by Fact 13.2 f^* is a proper lsc convex function, so that f^{**}, once again by Fact 13.2, is a proper lsc convex function as well. Next, by definition,

$$f^{**}(x) = (f^*)^*(x) = \sup_{d \in \mathbf{R}^n} \left\{ x^\top d - f^*(d) \right\} = \sup_{d \in \mathbf{R}^n, a \geq f^*(d)} \left\{ d^\top x - a \right\}.$$

Now, recall from the origin of the Legendre transform that $a \geq f^*(d)$ if and only if the affine function $d^\top x - a$ is a minorant of f. Thus, $\sup_{d \in \mathbf{R}^n, a \geq f^*(d)} \left\{ d^\top x - a \right\}$ is exactly the supremum of all affine minorants of f, and this supremum, by Proposition 12.7 is simply the closure of f. Finally, when f is proper convex and lsc, $f = \operatorname{cl} f$ by Observation 12.6, that is $f^{**} = \operatorname{cl} f$ is the same as $f^{**} = f$. ∎

The Legendre transform is a very powerful descriptive tool, i.e., it is a "global" transformation, so that *local* properties of f^* correspond to *global* properties of f. Below we give a number of important consequences of Legendre transform highlighting this.

Let f be a proper convex lsc function.

A. By Proposition 13.3, the Legendre transform $f^*(d) = \sup_x \left\{ x^\top d - f(x) \right\}$ is a proper convex lsc function and $f(x) = \sup_d \left\{ x^\top d - f^*(d) \right\}$. Since $f^*(d) \geq d^\top x - f(x)$ for all x, we have

$$x^\top d \leq f(x) + f^*(d), \quad \forall x, d \in \mathbf{R}^n. \tag{13.1}$$

Moreover, the inequality in (13.1) becomes an equality if and only if $x \in \operatorname{Dom} f$ and $d \in \partial f(x)$, which is the same as saying if and only if $d \in \operatorname{Dom} f^*$ and $x \in \partial f^*(d)$.

Proof All we need to do is to justify the "moreover" part of the claim. Let $x, d \in \mathbf{R}^n$, and let us prove that $d^\top x = f(x) + f^*(d)$ if and only if $d \in \partial f(x)$. In one direction: when $d \in \partial f(x)$, we have $x \in \operatorname{Dom} f$ and $f(z) \geq f(x) + d^\top(z - x)$ for all $z \in \mathbf{R}^n$, so

$$f^*(d) = \sup_{z \in \mathbf{R}^n} \left\{ d^\top z - f(z) \right\} \leq \sup_{z \in \mathbf{R}^n} \left\{ d^\top z - f(x) - d^\top(z - x) \right\} = d^\top x - f(x).$$

Thus, $f^*(d) \leq d^\top x - f(x)$; since by (13.1) strict inequality here is impossible, we conclude that when $d \in \partial f(x)$, the inequality in (13.1) becomes an equality. In the opposite direction: let d, x be such that the inequality in (13.1) is an equality, and let us prove that $d \in \partial f(x)$. Indeed, in this case $x \in \text{Dom } f$ and $d^\top x - f(x) = f^*(d) \geq d^\top z - f(z)$ for all $z \in \mathbf{R}^n$ (recall the definition of the Legendre transform $f^*(d)$), that is, $f(z) \geq f(x) + d^\top(z - x)$ for all z, implying that $d \in \partial f(x)$.

We have seen that when f is proper convex and lsc, then $d \in \partial f(x)$ if and only if $d^\top x = f(x) + f^*(d)$. Hence, when f is proper convex and lsc, by Fact 13.2 and Proposition 13.3, we can swap the roles of f and f^*, so that the inequality in (13.1) is an equality if and only if $x \in \partial f^*(d)$. ∎

B. We always have $\inf_x f(x) = -f^*(0)$. Thus, f is bounded from below if and only if $0 \in \text{Dom } f^*$. Moreover, f attains its minimum if and only if $\partial f^*(0) \neq \varnothing$, in which case $\text{Argmin } f = \partial f^*(0)$.

Proof By definition of the Legendre transform we have $f^*(0) = -\inf_x f(x)$. Thus, f is below bounded if and only if $0 \in \text{Dom } f^*$. To see the rest of the claim, suppose now that $0 \in \text{Dom } f^*$ and so $f^*(0) \in \mathbf{R}$. As $\inf_x f(x) = -f^*(0)$ implies $f(x) + f^*(0) \geq 0$ for all $x \in \mathbf{R}^n$, we deduce that the equality $f(x) + f^*(0) = 0$ holds exactly for x's that are the minimizers of f. Also, by part **A**, we conclude that when $0 \in \text{Dom } f^*$, the inequality in (13.1) with $d = 0$ holds as an equality if and only if $x \in \partial f^*(0)$. Combining the last two statements, we conclude that when f is below bounded (or, equivalently, when $0 \in \text{Dom } f^*$), the set of minimizers of f is exactly $\partial f^*(0)$. ∎

C. By part **A**, we have (i) $\bar{d} \in \partial f(\bar{x})$ if and only if (ii) $\bar{x} \in \partial f^*(\bar{d})$, and moreover both (i) and (ii) hold simultaneously if and only if (iii) with $x = \bar{x}, d = \bar{d}$, the inequality in (13.1) holds as an equality.

C'. Here is a nice special case of **C**: *Let f be a proper convex lsc function on $\mathbf{R}^n$. Suppose that the domains of f and f^* are open, and that these functions are continuously differentiable and strictly convex on their domains. Then, the map $x \mapsto \nabla f(x) :$ Dom $f \to \mathbf{R}^n$ is a one-to-one mapping of Dom f onto Dom f^*, and the inverse of this map is given by $y \mapsto \nabla f^*(y) :$ Dom $f^* \to \mathbf{R}^n$.*

Proof First, we claim that under the given premise, $\nabla f(x)$ is an embedding of Dom f into $\mathbf{R}^n$, i.e., $\nabla f(x) = \nabla f(x')$ implies $x = x'$. Assume for contradiction that this is not the case. Consider the function $g(u) := f(u) - u^\top \nabla f(x)$. This function is strictly convex, since f is so. However, when $\nabla f(x) = \nabla f(x')$, both x and x' are minimizers of $g(u)$, which is not possible as $g(u)$ is strictly convex and, by Theorem 11.1, its minimizer, if any exists, must be unique. Thus, $x \mapsto \nabla f(x)$ is an embedding of Dom f into $\mathbf{R}^n$. Next, when $x \in \text{Dom } f$ and $d = \nabla f(x)$, we have $d \in \partial f(x)$, so that $d \in \text{Dom } f^*$ and $x \in \partial f^*(d)$ by **C**. As Dom f^* is open and f^* is continuously differentiable at its domain, the relation $x \in \partial f^*(d)$ is the same as $x = \nabla f^*(d)$. Thus, $x \mapsto \nabla f(x)$ is an embedding of Dom f into Dom f^* and its left inverse is the mapping $d \mapsto \nabla f^*(d) :$ Dom $f^* \to \mathbf{R}^n$. Recalling that f is proper lsc convex, then so is f^* (Fact 13.2), and $(f^*)^* = f$ (Proposition 13.3). Since the rest of our assumptions are symmetric with respect to f, f^* as well, the previous reasoning as applied to f^* in the role of f demonstrates that the mapping $d \mapsto \nabla f^*(d)$ is an embedding of Dom f^* into Dom f with left inverse $x \mapsto \nabla f(x)$. Taken together, these observations yield **C′**. ∎

D. Dom $f^* = \mathbf{R}^n$ if and only if $f(x)$ "grows at infinity faster than $\|x\|_2$," that is, if and only if the function $F(s) := \inf_{x:\|x\|_2 \leq s} f(x)/\|x\|_2$ blows up to ∞ as $s \to \infty$.

Proof Suppose $F(s) \to \infty$ as $s \to \infty$. Consider any fixed $d \in \mathbf{R}^n$. Then, there exists $\bar{s}$ such that $f(x) \geq \|d\|_2 \|x\|_2$ whenever $\|x\|_2 \geq \bar{s}$. Thus, whenever $\|x\|_2 \geq \bar{s}$ we have $d^\top x - f(x) \leq \|d\|_2 \|x\|_2 - f(x) \leq 0$. Also, as a convex function f is below bounded on the ball $\|x\|_2 \leq \bar{s}$ (in fact on any bounded set), the function $d^\top x - f(x)$ of x is bounded from above and we conclude that $d \in \text{Dom } f^*$. As this conclusion holds for any d, we conclude that Dom $f^* = \mathbf{R}^n$. Now, to see the reverse direction, suppose that $F(s)$ does not blow up to ∞ as $s \to \infty$. Then, we can find a sequence $\{x^i\}$ and $L \in \mathbf{R}$ such that $\|x^i\|_2 \to \infty$ as $i \to \infty$ and $f(x^i) \leq L\|x^i\|_2$ for all i. Passing to a subsequence, we can assume that as $i \to \infty$ we have $x^i/\|x^i\|_2 \to \xi$, where $\|\xi\|_2 = 1$. Now, select $d := 2L\xi$; then for all large enough i we have $f(x^i) \leq L\|x^i\|_2 \leq \frac{2}{3} d^\top x^i$. In addition, $d^\top x^i \to \infty$ as $i \to \infty$. Consequently, $d^\top x^i - f(x^i) \geq d^\top x^i - L\|x^i\|_2 \geq \frac{1}{3} d^\top x^i$ for large enough i. Hence, $d^\top x^i - f(x^i) \to +\infty$ as $i \to \infty$, that is, $d \notin \text{Dom } f^*$. ∎

The bottom line is that by investigating the Legendre transform of a convex function, we get a lot of "global" information on the function. This being said, a detailed investigation of the properties of the Legendre transform is beyond our scope. We close this chapter by listing several simple yet important consequences of the Legendre transform.

13.3 Young, Hölder, and Moment Inequalities

The Legendre transform leads to several important inequalities. Recall that from the definition of the Legendre transformation we have

$$f(x) + f^*(d) \geq x^\top d, \qquad \forall x, d \in \mathbf{R}^n.$$

We will see that specific choices of f in this inequality lead to several well-known inequalities.

13.3.1 Young's Inequality

Young's inequality reads as follows:

Let p and q be positive real numbers such that $\frac{1}{p} + \frac{1}{q} = 1$. Then,

$$xd \leq \frac{|x|^p}{p} + \frac{|d|^q}{q}, \qquad \forall x, d \in \mathbf{R}.$$

Proof Recall from Example 13.1 that the Legendre transform of the function $|x|^p/p$ is $|d|^q/q$. ∎

13.3.2 Hölder's Inequality

The admittedly simple-looking Young's inequality gives rise to the very nice and useful Hölder's inequality.

Let $1 \leq p \leq \infty$ and let $q = \frac{p}{p-1}$ (where $1/0 = +\infty$), so that $\frac{1}{p} + \frac{1}{q} = 1$. Then,

$$\sum_{i=1}^{n} |x_i y_i| \leq \|x\|_p \|y\|_q, \qquad \forall x, y \in \mathbf{R}^n. \tag{13.2}$$

Proof When $p = \infty$, we have $q = 1$ and (13.2) becomes the obvious relation

$$\sum_{i=1}^{n} |x_i y_i| \leq \left(\max_i \{|x_i|\} \right) \left(\sum_{i=1}^{n} |y_i| \right), \qquad \forall x, y \in \mathbf{R}^n.$$

By symmetry, we also see that when $p = 1$ we have $q = \infty$ and (13.2) is evident. Now, let $1 < p < \infty$, so that also $1 < q < \infty$. In this case we should prove that

$$\sum_{i=1}^{n} |x_i y_i| \leq \left(\sum_{i=1}^{n} |x_i|^p \right)^{1/p} \left(\sum_{i=1}^{n} |y_i|^q \right)^{1/q}.$$

When $x = 0$ or $y = 0$, this inequality is evident. So, we assume that $x \neq 0$ and $y \neq 0$. As both sides of this inequality are positively homogeneous of degree 1 with respect to x, and similarly with respect to y, without loss of generality we can assume that $\|x\|_p = \|y\|_q = 1$. Now, under this normalization, we should prove that $\sum_{i=1}^{n} |x_i y_i| \leq 1$. Recall that by Young's inequality we get $|x_i y_i| \leq \frac{|x_i|^p}{p} + \frac{|y_i|^q}{q}$ for all i, and so

$$\sum_{i=1}^{n} |x_i y_i| \leq \sum_{i=1}^{n} \left(\frac{|x_i|^p}{p} + \frac{|y_i|^q}{q} \right) = \frac{1}{p} \|x\|_p^p + \frac{1}{q} \|y\|_q^q = \frac{1}{p} + \frac{1}{q} = 1,$$

as desired. ∎

Note that for $p, q \in [1, \infty]$ such that $\frac{1}{p} + \frac{1}{q} = 1$, Hölder's inequality gives us

$$|x^\top y| \leq \|x\|_p \|y\|_q. \tag{13.3}$$

When $p = q = 2$, this is the well-known Cauchy–Schwarz inequality. Moreover, for every $p \in [1, \infty]$ the inequality (13.3) is *tight* in the sense that for every x there exists y with $\|y\|_q = 1$ such that

$$x^\top y = \|x\|_p \quad (= \|x\|_p \|y\|_q \text{ as } \|y\|_q = 1).$$

To justify this claim, note that when $x = 0$ we can select any y with $\|y\|_q = 1$. When $x \neq 0$ and $p < \infty$ we can set y to

$$y_i := \|x\|_p^{1-p} |x_i|^{p-1} \text{sign}(x_i), \qquad \forall i = 1, \ldots, n,$$

where we set $0^{p-1} = 0$ when $p = 1$. Finally, when $p = \infty$, that is, $q = 1$, we can find an index i_* of the entry of x that is largest in magnitude and set

$$y_i = \begin{cases} \text{sign}(x_{i_*}), & \text{if } i = i_*, \\ 0, & \text{if } i \neq i_*. \end{cases}$$

These observations together lead us to an extremely important, although simple, fact:

$$\|x\|_p = \max_y \left\{ y^\top x : \|y\|_q \leq 1 \right\}, \quad \text{where } \frac{1}{p} + \frac{1}{q} = 1. \quad (13.4)$$

From this, we deduce, in particular, that $\|x\|_p$ is convex (as an upper bound of a family of linear functions). Hence, by its convexity we deduce that for any x', x'' we have

$$\|x' + x''\|_p = 2 \left\| \frac{1}{2} x' + \frac{1}{2} x'' \right\|_p \leq 2 \left(\|x'\|_p / 2 + \|x''\|_p / 2 \right) = \|x'\|_p + \|x''\|_p,$$

which is just the triangle inequality. Thus, $\|x\|_p$ satisfies the triangle inequality; it clearly possesses the other two characteristic properties of a norm, namely positivity and homogeneity, as well. Consequently, $\|\cdot\|_p$ is a norm – a fact that we have announced twice already and proved (see Example 10.1).

13.3.3 Moment Inequality

A useful application of Hölder's inequality gives us another well-known inequality, as follows. The moment inequality reads:

For any $0 \neq a \in \mathbf{R}^n$, the function

$$f(\pi) := \ln(\|a\|_{1/\pi}) : [0, 1] \to \mathbf{R}$$

is a convex function of $\pi \in [0, 1]$. That is, by letting $p = 1/\pi$, and so $1 \leq p \leq \infty$, we have the following inequality:

$$\begin{array}{c} 1 \leq r < s < \infty, \ p \in [r, s] \implies \|a\|_p \leq \|a\|_r^\lambda \|a\|_s^{1-\lambda} \\ \lambda = \frac{r(s-p)}{p(s-r)} \quad \left(\iff \lambda \in [0, 1] \text{ and } \lambda \frac{1}{r} + (1-\lambda) \frac{1}{s} = \frac{1}{p} \right). \end{array} \quad (13.5)$$

Proof Let $\rho, \sigma \in (0, 1]$ be such that $\rho \neq \sigma$. Consider any $\lambda \in (0, 1)$ and set $\pi := \lambda \rho + (1 - \lambda) \sigma$. By defining $\theta := \lambda \rho / \pi$, we get $0 < \theta < 1$ and $1 - \theta = (1 - \lambda) \sigma / \pi$. Let $r := 1/\rho$, $s := 1/\sigma$, and $p := 1/\pi$, so we have $p = \theta r + (1 - \theta) s$. Then,

$$\sum_{i=1}^n |a_i|^p = \sum_{i=1}^n |a_i|^{\theta r} |a_i|^{(1-\theta)s} \leq \left(\sum_i^n |a_i|^r \right)^\theta \left(\sum_{i=1}^n |a_i|^s \right)^{1-\theta},$$

where the inequality follows from Hölder's inequality. Raising both sides of the resulting inequality to the power $1/p$ we arrive at

$$\|a\|_p \leq \|a\|_r^{r\theta/p} \|a\|_s^{s(1-\theta)/p} = \|a\|_{1/\rho}^\lambda \|a\|_{1/\sigma}^{1-\lambda}.$$

Recalling that $p = \theta r + (1 - \theta) s$ and $\ln(\cdot)$ is a monotone increasing function, we conclude that for any $\lambda \in (0, 1)$ the function $f(\pi) = \ln(\|a\|_{1/\pi})$ satisfies the inequality

$$f(\lambda \rho + (1 - \lambda)\sigma) \leq \lambda f(\rho) + (1 - \lambda) f(\sigma), \quad \forall (\rho, \sigma : \rho, \sigma \in (0, 1], \rho \neq \sigma).$$

Since this function is continuous on [0, 1], it is convex, as claimed. ∎

We close this section by discussing the dual norm of a norm. Let $\|\cdot\|$ be a norm on $\mathbf{R}^n$. We define its *dual* (a.k.a. *conjugate*) norm as the function

$$\|d\|_* := \sup_x \left\{ d^\top x : \|x\| \leq 1 \right\}.$$

As its name implies one can indeed show that this function $\|d\|_*$ is a norm.

Fact 13.4 *Let $\|\cdot\|$ be a norm on $\mathbf{R}^n$. Then, its dual norm $\|\cdot\|_*$ is indeed a norm. Moreover, the norm dual to $\|\cdot\|_*$ is the original norm $\|\cdot\|$, and the unit balls of conjugate to each other norms are polars of each other.*

For example, when $p \in [1, \infty]$, (13.4) says that the norm conjugate to $\|\cdot\|_p$ is $\|\cdot\|_q$ where $\frac{1}{p} + \frac{1}{q} = 1$.

We also have the following characterization of the Legendre transform of norms.

Fact 13.5 *Let $f(x) = \|x\|$ be a norm on $\mathbf{R}^n$. Then,*

$$f^*(d) = \begin{cases} 0, & \text{if } \|d\|_* \leq 1, \\ +\infty, & \text{otherwise.} \end{cases}$$

That is, the Legendre transform of $\|\cdot\|$ is the characteristic function of the unit ball of the conjugate norm.

13.4 Proofs of Facts

Fact 13.2 *Given a proper convex function $f : \mathbf{R}^n \to \mathbf{R} \cup \{+\infty\}$, its Legendre transform f^* is a proper convex lower semicontinuous function.*

Proof This fact follows immediately from the definition of the Legendre transform f^*: indeed, we lose nothing when replacing $\sup_{x \in \mathbf{R}^n} \{d^\top x - f(x)\}$ with $\sup_{x \in \text{Dom } f} \{d^\top x - f(x)\}$, so that the Legendre transform is the supremum of a nonempty (as f is proper) family of affine functions and as such is convex and lower semicontinuous. Since this supremum is finite at least at one point (namely, at every d which is the slope of an affine minorant of f and we know that such a minorant exists), f^* is a proper convex lsc function, as claimed. ∎

Fact 13.4 *Let $\|\cdot\|$ be a norm on $\mathbf{R}^n$. Then, its dual norm $\|\cdot\|_*$ is indeed a norm. Moreover, the norm dual to $\|\cdot\|_*$ is the original norm $\|\cdot\|$, and the unit balls of conjugate to each other norms are polars of each other.*

Proof From the definition of $\|\cdot\|_*$ it follows immediately that $\|\cdot\|_*$ satisfies all three conditions specifying a norm. To justify that the norm dual to $\|\cdot\|_*$ is $\|\cdot\|$, note that the unit ball of the dual norm is, by the definition of this norm, the polar of the unit ball of $\|\cdot\|$, and the latter set, like the unit ball of any norm, is closed, convex, and contains the origin. As a result, the unit balls of $\|\cdot\|$, $\|\cdot\|_*$ are polars of each other (Proposition 6.42), and the norm dual to its dual is the original norm – its unit ball is the polar of the unit ball of $\|\cdot\|_*$. ∎

Fact 13.5 *Let $f(x) = \|x\|$ be a norm on $\mathbf{R}^n$. Then,*

$$f^*(d) = \begin{cases} 0, & \text{if } \|d\|_* \leq 1, \\ +\infty, & \text{otherwise.} \end{cases}$$

That is, the Legendre transform of $\|\cdot\|$ is the characteristic function of the unit ball of the conjugate norm.

Proof Consider any fixed $d \in \mathbf{R}^n$. By the definition of the Legendre transform we have

$$f^*(d) = \sup_{x \in \mathbf{R}^n} \left\{ d^\top x - f(x) \right\} = \sup_{x \in \mathbf{R}^n} \left\{ d^\top x - \|x\| \right\}.$$

Now, consider the function $g_d(x) := d^\top x - \|x\|$, so that $f^*(d) = \sup_{x \in \mathbf{R}^n} g_d(x)$. The function $g_d(x)$ is positively homogeneous, of degree 1, in x, so that its supremum over the entire space is either 0 (this is the case when the function is nonpositive everywhere) or $+\infty$. By the same homogeneity, the function $g_d(x)$ is nonpositive everywhere if and only if it is nonpositive when $\|x\| = 1$, that is, if and only if $d^\top x \leq 1$ whenever $\|x\| = 1$, or, which is the same, when $d^\top x \leq 1$ whenever $\|x\| \leq 1$. The bottom line is that $f^*(d) = \sup_x g_d(x)$ is either 0 or $+\infty$, with the first option taking place if and only if $d^\top x \leq 1$ whenever $\|x\| \leq 1$, that is, if and only if $\|d\|_* \leq 1$. ∎

14

★ Functions of Eigenvalues of Symmetric Matrices

One might think that the calculus of convexity-preserving operations presented in section 10.1 does not look particularly deep. On the other hand, these "simple" rules are extremely useful and allow us to detect convexity and offer nice characterizations for a particular class of functions of symmetric matrices, which we will examine in this chapter.

Let $X \in \mathbf{S}^n$ be an $n \times n$ symmetric matrix, and let $\lambda(X)$ denote the vector of eigenvalues of X taken with their multiplicities and arranged in non-ascending order; see section D.1.1.C. In this chapter, we present a really deep result stating that whenever $f : \mathbf{R}^n \to \mathbf{R} \cup \{+\infty\}$ is convex and permutation symmetric, the function $F(X) := f(\lambda(X))$ is a convex function of the matrix $X \in \mathbf{S}^n$.

Recall that a function $f : \mathbf{R}^n \to \mathbf{R} \cup \{+\infty\}$ is called *permutation symmetric* if

$$f(x) = f(Px) \text{ for every } n \times n \text{ permutation matrix } P.$$

That is, a function is permutation symmetric if and only if its value remains unchanged when we permute the coordinates in its argument.

We start with the following observation.

Lemma 14.1 *Let $f : \mathbf{R}^n \to \mathbf{R} \cup \{+\infty\}$ be a convex permutation symmetric function. Then, for any $x \in \text{Dom } f$ and an $n \times n$ doubly stochastic matrix Π, we have $f(\Pi x) \leq f(x)$.*

Proof Recall that by Birkhoff's Theorem (Theorem 7.7), Π is a convex combination of permutation matrices P^i, i.e., there exist convex combination weights $\lambda_i \geq 0$ and permutation matrices P^i such that $\Pi = \sum_i \lambda_i P^i$ and $\sum_i \lambda_i = 1$. Then, as f is convex, we have

$$f(\Pi x) = f\left(\sum_i \lambda_i P^i x\right) \leq \sum_i \lambda_i f(P^i x) = \sum_i \lambda_i f(x) = f(x),$$

where the inequality follows from the convexity of f and the second equality is due to the fact that f is permutation symmetric. ∎

Our developments will also rely on the following fundamental fact.

Lemma 14.2 *For any $X \in \mathbf{S}^n$, the diagonal $\text{Dg}\{X\}$ of the matrix X is the image of the vector $\lambda(X)$ of the eigenvalues of X under multiplication by a doubly stochastic matrix. That is, there exists an $n \times n$ doubly stochastic matrix Π such that*

$$\text{Dg}\{X\} = \Pi \lambda(X).$$

Proof Consider the spectral decomposition of X, i.e.,

$$X = U^\top \operatorname{Diag}\{\lambda_1(X), \ldots, \lambda_n(X)\} U$$

where $U = [u_{ij}]_{i,j=1}^n$ is an orthogonal matrix. Define the matrix $\Pi := [u_{ji}^2]_{i,j=1}^n$. As U is an orthogonal matrix, we have that Π is indeed doubly stochastic (verify this yourself!). Moreover, denoting the ith standard basic orth by e_i, we get

$$\begin{aligned}
X_{ii} &= e_i^\top X e_i = e_i^\top (U^\top \operatorname{Diag}\{\lambda_1(X), \ldots, \lambda_n(X)\} U) e_i \\
&= \operatorname{Tr}(e_i^\top (U^\top \operatorname{Diag}\{\lambda_1(X), \ldots, \lambda_n(X)\} U) e_i) \\
&= \operatorname{Tr}(U e_i e_i^\top U^\top \operatorname{Diag}\{\lambda_1(X), \ldots, \lambda_n(X)\}) \\
&= \sum_{j=1}^n u_{ji}^2 \lambda_j(X) = [\Pi \lambda(X)]_i.
\end{aligned}$$
∎

Let us denote by $\mathcal{O}_n$ the set of all $n \times n$ orthogonal matrices. Lemmas 14.1 and 14.2 together give us the following very useful relation.

Proposition 14.3 *Let $f : \mathbf{R}^n \to \mathbf{R} \cup \{+\infty\}$ be a convex permutation symmetric function. Then, for every $X \in \mathbf{S}^n$, we have*

$$f(\lambda(X)) \geq f(\operatorname{Dg}\{X\}).$$

Furthermore, we also have

$$f(\lambda(X)) = \max_{V \in \mathcal{O}_n} f(\operatorname{Dg}\{V^\top X V\}). \tag{14.1}$$

In particular, the function $F(X) := f(\lambda(X))$ is a convex function of X.

Proof The first claim follows immediately from Lemmas 14.1 and 14.2.

To see the second claim, consider any $V \in \mathcal{O}_n$. Note that the matrix $V^\top X V$ has the same eigenvalues as X. Then, as f is a convex and permutation symmetric function, applying the first claim of this proposition to the matrix $V^\top X V$, we conclude that

$$f(\operatorname{Dg}\{V^\top X V\}) \leq f(\lambda(V^\top X V)) = f(\lambda(X)).$$

Taking the supremum over $V \in \mathcal{O}_n$ of both sides of this relation gives us

$$f(\lambda(X)) \geq \sup_{V \in \mathcal{O}_n} f(\operatorname{Dg}\{V^\top X V\}).$$

Note that for a properly chosen $V \in \mathcal{O}_n$ we have $\operatorname{Dg}\{V^\top X V\} = \lambda(X)$. Thus, the preceding inequality holds as an equality and the right-hand side supremum is achieved.

For the final claim, note that for any $V \in \mathcal{O}_n$, we have that the function $F_V(X) := f(\operatorname{Dg}\{V^\top X V\})$ is convex in X (as it is the composition of a convex function f and an affine map $X \mapsto \operatorname{Dg}\{V^\top X V\}$. Then, the final claim of the proposition follows from the second claim as $F(X)$ is the pointwise supremum of convex functions $\{F_V(X)\}_{V \in \mathcal{O}_n}$. ∎

Given a symmetric $n \times n$ matrix X, as a corollary of Proposition 14.3, we arrive at the following immediate relations between eigenvalues and diagonal entries of X.

1. For all $p \geq 1$, we have $\sum_{i=1}^n |X_{ii}|^p \leq \sum_{i=1}^n |\lambda_i(X)|^p$.
   (Consider the function $f(x) = \sum_{i=1}^n |x_i|^p$.)

2. Whenever X is positive semidefinite, we have $\prod_{i=1}^n X_{ii} \geq \text{Det}(X)$.
   (Consider $f(x) = -\sum_{i=1}^n \ln(x_i)$ over the domain where $x_i > 0$ for all i.)

3. Define the function $s_k(x) : \mathbf{R}^n \to \mathbf{R}$ to be the sum of the k largest entries of x (i.e., the sum of the first k entries in the vector obtained from x by writing down the coordinates of x in non-ascending order). Then, the function $S_k(X) := s_k(\lambda(X))$ is convex, and

$$s_k(\text{Dg}\{X\}) \leq S_k(X). \tag{14.2}$$

(Recall from Example 10.3 that $s_k(x)$ is a convex permutation symmetric function.)

Remark 14.4 Let us examine the convexity status of eigenvalues of symmetric matrices. Completely analogously to our discussion in Remark 10.6 for the vector case, we have by Proposition 14.3 that the largest eigenvalue $\lambda_1(X)$ of a symmetric $n \times n$ matrix X is a convex function of X. Therefore, the smallest eigenvalue $\lambda_n(X) = -\lambda_1(-X)$ is concave in X. On the other hand, the "intermediate" eigenvalues $\lambda_k(X)$, $1 < k < n$, of X, are neither convex nor concave functions of X. What *is* convex in X is the sum $S_k(X)$ of the $k \leq n$ largest eigenvalues of X, which we have just seen. This clearly implies that the sum of the k smallest eigenvalues of X is concave in X. The sum of the magnitudes (absolute values) of the k largest in magnitude eigenvalues of X is convex in X, since the function

$$\bar{s}_k(x) = s_k([|x_1|; \ldots; |x_n|]) = \max_{\substack{i_1,\ldots,i_k \in \{1,2,\ldots,n\}:\\ i_1 < i_2 < \ldots < i_k}} \{|x_{i_1}| + \cdots + |x_{i_k}|\}$$

is permutation symmetric and convex on $\mathbf{R}^n$. ◇

Example 14.1 Proposition 14.3 implies the convexity of the following functions of $X \in \mathbf{S}^n$:

- $F(X) = -\text{Det}^q(X)$ in the domain $X \succeq 0$, whenever $0 < q \leq \frac{1}{n}$.
  (Consider $f(x_1, \ldots, x_n) = -(x_1 \ldots x_n)^q : \mathbf{R}^n_+ \to \mathbf{R}$.)
- $F(X) = -\ln \text{Det}(X)$ in the domain $X \succ 0$.
  (Consider $f(x_1, \ldots, x_n) = -\sum_{i=1}^n \ln x_i : \text{int}(\mathbf{R}^n_+) \to \mathbf{R}$.)
- $F(X) = \text{Det}^{-q}(X)$ in the domain $X \succ 0$, whenever $q \in \mathbf{R}$ is positive.
  (Consider $f(x_1, \ldots, x_n) = (x_1 \ldots x_n)^{-q} : \text{int}(\mathbf{R}^n_+) \to \mathbf{R}$.)
- $\|X\|_p = \left(\sum_{i=1}^n |\lambda_i(X)|^p\right)^{1/p}$, where $p \geq 1$.
  (Consider $f(x_1, \ldots, x_n) = \|x\|_p$.)
- $\|X_+\|_p = \left(\sum_{i=1}^m (\max[\lambda_i(X), 0])^p\right)^{1/p}$, where $p \geq 1$.
  (Consider $f(x_1, \ldots, x_n) = \|x_+\|_p$, where $x_+ := [\max\{0, x_1\}; \ldots; \max\{0, x_n\}]$.) ◇

We say that a set $Q \in \mathbf{R}^n$ is permutation symmetric if for all $x \in Q$ and for all $n \times n$ permutation matrices P, we have $Px \in Q$ as well. Let us also mention the following useful corollary of Proposition 14.3.

★ *Functions of Eigenvalues of Symmetric Matrices*

Corollary 14.5 *Let Q be a nonempty, closed, convex, and permutation symmetric set in $\mathbf{R}^n$. Then, the set*

$$\mathcal{Q} := \{X \in \mathbf{S}^n : \lambda(X) \in Q\}$$

is closed and convex.

Proof From Fact D.21 we have that $\lambda(X)$ is continuous in X, thus $\mathcal{Q}$ is closed.
To prove that $\mathcal{Q}$ is convex, consider the function

$$f(x) := \min_{y \in Q} \|x - y\|_2.$$

Note that $\|x - y\|_2$ is a convex function of x and y over the convex domain $\{[x; y] \in \mathbf{R}^n \times \mathbf{R}^n : y \in Q\}$, and, as convexity is preserved by partial minimization, $f(x)$ is a convex real-valued function. The permutation symmetry of Q and $\|\cdot\|_2$ clearly implies the permutation symmetry of f. Then, by Proposition 14.3 the function $F(X) := f(\lambda(X))$ is a convex function of X. From the definition of the function f, we have $z \in Q$ if and only if $f(z) = 0$, which holds if and only if $f(z) \leq 0$. Thus,

$$\mathcal{Q} = \{X \in \mathbf{S}^n : f(\lambda(X)) \leq 0\} = \{X \in \mathbf{S}^n : F(X) \leq 0\},$$

that is, $\mathcal{Q}$ is a sublevel set of a convex function $F(X)$ of X, and is hence convex. ∎

Consider a univariate real-valued function g defined on some set $\text{Dom}\, g \subseteq \mathbf{R}$. In section D.1.5 we associate with the function $g(\cdot)$ the matrix-valued map $X \mapsto g(X) : \mathbf{S}^n \to \mathbf{S}^n$ as follows: the domain of this map is composed of all matrices $X \in \mathbf{S}^n$ with the spectrum $\sigma(X)$ (the subset of $\mathbf{R}$ composed of all eigenvalues of X) contained in $\text{Dom}\, g$, and for such a matrix X, we set

$$g(X) := U \,\text{Diag}\{g(\lambda_1), \ldots, g(\lambda_n)\} U^\top,$$

where $X = U \,\text{Diag}\{\lambda_1, \ldots, \lambda_n\} U^\top$ is an eigenvalue decomposition of X.

The following fact is quite important:

Fact 14.6 *Let $g : \mathbf{R} \to \mathbf{R} \cup \{+\infty\}$ be a convex function. Then, the function*

$$F(X) := \begin{cases} \text{Tr}(g(X)), & \text{if } \sigma(X) \subseteq \text{Dom}\, g, \\ +\infty, & \text{otherwise,} \end{cases} \quad : \mathbf{S}^n \to \mathbf{R} \cup \{+\infty\}$$

is convex.

We close this chapter with the following very useful fact from majorization.

Fact 14.7 *For $x \in \mathbf{R}^n$, let $x_\uparrow$ and $x_\downarrow$ be the vectors obtained by reordering the entries of x in non-decreasing and non-increasing orders, respectively. For example, $[1; 3; 2; 1]_\uparrow = [1; 1; 2; 3]$ and $[1; 3; 2; 1]_\downarrow = [3; 2; 1; 1]$.*

(i) For every $x, y \in \mathbf{R}^n$ and every $n \times n$ doubly stochastic matrix P, we have

$$[x_\uparrow]^\top y_\uparrow \geq x^\top P y \geq [x_\uparrow]^\top y_\downarrow.$$

As a result,

(ii) [Trace inequality] For every $A, B \in \mathbf{S}^n$, we have

$$\lambda^\top(A)\lambda(B) \geq \mathrm{Tr}(AB) \geq \lambda^\top(A)[\lambda(B)]_\uparrow.$$

14.1 Proofs of Facts

Fact 14.6 *Let $g : \mathbf{R} \to \mathbf{R} \cup \{+\infty\}$ be a convex function. Then, the function*

$$F(X) := \begin{cases} \mathrm{Tr}(g(X)), & \text{if } \sigma(X) \subseteq \mathrm{Dom}\, g, \\ +\infty, & \text{otherwise,} \end{cases} \quad : \mathbf{S}^n \to \mathbf{R} \cup \{+\infty\}$$

is convex.

Proof Define $\widehat{g}(x) := \sum_{i=1}^n g(x_i)$, so that $\widehat{g}$ is a permutation symmetric and convex function (since g is convex) on $\mathbf{R}^n$. Then, by Proposition 14.3 the function $\widehat{F}(x) := \widehat{g}(\lambda(X))$ is a convex function of $X \in \mathbf{S}^n$. Now, $\mathrm{Dom}\,\widehat{F} = \{X \in \mathbf{S}^n : \sigma(X) \subseteq \mathrm{Dom}\, g\}$, so $\mathrm{Dom}\,\widehat{F} = \mathrm{Dom}\, F$. Consider any $X \in \mathrm{Dom}\,\widehat{F}$ along with its eigenvalue decomposition given by $X = U \mathrm{Diag}\{\lambda(X)\} U^\top$. By the definition of $g(X)$, we have $g(X) = U \mathrm{Diag}\{g(\lambda_1(X)), \ldots, g(\lambda_n(X))\} U^\top$ and $F(X) = \mathrm{Tr}(g(X)) = \sum_{i=1}^n g(\lambda_i(X)) = \widehat{g}(\lambda(X)) = \widehat{F}(X)$. We conclude that F is just the convex function $\widehat{F}$. ∎

Fact 14.7 *For $x \in \mathbf{R}^n$, let $x_\uparrow$ and $x_\downarrow$ be the vectors obtained by reordering the entries of x in non-decreasing and non-increasing order, respectively. For example, $[1; 3; 2; 1]_\uparrow = [1; 1; 2; 3]$ and $[1; 3; 2; 1]_\downarrow = [3; 2; 1; 1]$.*

(i) For every $x, y \in \mathbf{R}^n$ and every $n \times n$ doubly stochastic matrix P, we have

$$[x_\uparrow]^\top y_\uparrow \geq x^\top P y \geq [x_\uparrow]^\top y_\downarrow.$$

As a result,

(ii) [Trace inequality] For every $A, B \in \mathbf{S}^n$, we have

$$\lambda^\top(A)\lambda(B) \geq \mathrm{Tr}(AB) \geq \lambda^\top(A)[\lambda(B)]_\uparrow.$$

Proof (i): First, we claim that for all $x, y \in \mathbf{R}^n$ we have

$$x_\downarrow^\top y_\downarrow \geq x^\top y \geq x_\downarrow^\top y_\uparrow. \qquad (*)$$

Indeed, by continuity, it suffices to verify this relation when all entries of x, like all entries in y, are distinct from each other. In such a case, observe that the inequalities to be proved remain intact when we simultaneously reorder, in the same order, entries in x and in y, so that we can assume without loss of generality that $x_1 \geq x_2 \geq \cdots \geq x_n$. Taking into account that $ac + bd - [ad + bc] = [a-b][c-d]$, we see that if $i < j$ and $\overline{y}$ is obtained from y by swapping its ith and jth entries, we have $x^\top \overline{y} \leq x^\top y$ when $y_i > y_j$ and $x^\top \overline{y} \geq x^\top y$ otherwise. Thus, the minimum (the maximum) of the inner products $x^\top z$ over the set of vectors z obtained by reordering entries in y is achieved when $z = y_\uparrow$ (respectively, $z = y_\downarrow$), as claimed.

In the situation of (i), by Birkhoff's Theorem, Py is a convex combination of vectors obtained from y by reordering entries, and so the relation in (i) is immediately implied by $(*)$.

(ii): Let $A = U \mathrm{Diag}\{\lambda(A)\} U^\top$ be the eigenvalue decomposition of A. Then,

$$\mathrm{Tr}(AB) = \mathrm{Tr}(U \mathrm{Diag}\{\lambda(A)\} U^\top B) = \mathrm{Tr}(\mathrm{Diag}\{\lambda(A)\}(U^\top BU)) = (\lambda(A))^\top \mu,$$

where μ is the diagonal of the matrix $U^\top BU$, i.e., $\mu = \mathrm{Dg}\{U^\top BU\}$. Then, by Lemma 14.2 we deduce that there exists a doubly stochastic matrix P such that

$$\mu = \mathrm{Dg}\{U^\top BU\} = P\lambda(U^\top BU) = P\lambda(B).$$

Thus, $\mathrm{Tr}(AB) = (\lambda(A))^\top \mu = (\lambda(A))^\top P\lambda(B)$. The desired inequality then follows from applying part (i) to the vectors $x := \lambda(A)$ and $y := \lambda(B)$. ∎

15

Exercises for Part III

15.1 Around Convex Functions

Exercise III.1 Which of the following functions are convex on the indicated domains:

- $f(x) \equiv 1$ on $\mathbf{R}$
- $f(x) = x$ on $\mathbf{R}$
- $f(x) = |x|$ on $\mathbf{R}$
- $f(x) = -|x|$ on $\mathbf{R}$
- $f(x) = -|x|$ on $\mathbf{R}_+ = \{x \in \mathbf{R} : x \geq 0\}$
- $f(x) = |2x - 3|$ on $\mathbf{R}$
- $f(x) = |2x^2 - 3|$ on $\mathbf{R}$
- $\exp\{x\}$ on $\mathbf{R}$
- $\exp\{x^2\}$ on $\mathbf{R}$
- $\exp\{-x^2\}$ on $\mathbf{R}$
- $\exp\{-x^2\}$ on $\{x \in \mathbf{R} : x \geq 100\}$
- $\ln(x)$ on $\{x \in \mathbf{R} : x > 0\}$
- $-\ln(x)$ on $\{x \in \mathbf{R} : x > 0\}$

Exercise III.2 [TrYs]

1. Prove the following fact:
   For every $C_i \in \mathbf{S}_+^m$, $i \leq I$, satisfying $\sum_{i \in I} C_i = I_m$ and for every $\lambda_i \in \mathbf{R}$, we have
   $$\mathrm{Tr}\left(\left(\sum_{i \in I}\lambda_i C_i\right)^2\right) \leq \mathrm{Tr}\left(\sum_{i \in I}\lambda_i^2 C_i\right).$$

2. Recall from Example 10.4 in section 10.2 that for $a_i \geq 0$, $\sum_i a_i > 0$, the function $\ln(\sum_i a_i \exp(\lambda_i))$ is a convex function of λ. Prove the following matrix analogy of this fact:
   For every $A_i \in \mathbf{S}_+^m$, $1 \leq i \leq I$ such that $\sum_i A_i \succ 0$, the function
   $$f(\lambda) = \ln \mathrm{Det}\left(\sum_i \exp(\lambda_i) A_i\right) : \mathbf{R}^I \to \mathbf{R}$$
   is convex.

3. Let A_i, $i \leq I$, be as in item 2. Is it true that the function
   $$g(x) = \ln \mathrm{Det}(\sum_i x_i^{-1} A_i) : \{x \in \mathbf{R}^I : x > 0\} \to \mathbf{R}$$
   is convex?

4. Let $B_i, i \leq I$, be $m_i \times n$ matrices such that $\sum_i B_i^\top B_i \succ 0$, and let
$$\Lambda = \{\lambda := (\lambda_1, \ldots, \lambda_I) : \lambda_i \in \mathbf{S}^{m_i}, \lambda_i \succ 0, i \leq I\}.$$
Prove that the function
$$h(\lambda) = \ln \operatorname{Det}\left(\sum_i B_i^\top \lambda_i^{-1} B_i\right) : \Lambda \to \mathbf{R}$$
is convex.

5. Let $B_i, i \leq I$, and Λ be as in the previous item. Prove that the matrix-valued function
$$F(\lambda) = \left[\sum_i B_i^\top \lambda_i^{-1} B_i\right]^{-1} : \Lambda \to \operatorname{int} \mathbf{S}_+^n$$
is $\succeq$-concave, that is, the $\succeq$-hypograph
$$\{(\lambda, Y) : \lambda \in \Lambda, Y \preceq F(\lambda)\}$$
of the function is convex.

Exercise III.3 [EdYs] A function f defined on a convex set Q is called *log-convex* on Q if it takes real positive values on Q and the function $\ln f$ is convex on Q. Prove that

1. a function that is log-convex on Q is convex on Q;
2. the sum (more generally, linear combination with positive coefficients) of two log-convex functions on Q also is log-convex on the set.

Exercise III.4 [EdYs] [Law of diminishing marginal returns] Consider the optimization problem
$$\operatorname{Opt}(r) = \max_x \{f(x) : G(x) \leq r \text{ and } x \in X\} \qquad (P[r])$$
where $X \subseteq \mathbf{R}^n$ is a nonempty convex set, $f(\cdot) : X \to \mathbf{R}$ is concave, and $G(x) = [g_1(x); \ldots; g_m(x)] : X \to \mathbf{R}^m$ is a vector-valued function with convex components, and let $\mathcal{R}$ be the set of those r for which $(P[r])$ is feasible. Prove that

1. $\mathcal{R}$ is a convex set with nonempty interior and this set is monotone, meaning that when $r \in \mathcal{R}$ and $r' \geq r$, one has $r' \in \mathcal{R}$.
2. The function $\operatorname{Opt}(r) : \mathcal{R} \to \mathbf{R} \cup \{+\infty\}$ satisfies the concavity inequality:
$$\forall (r, r' \in \mathcal{R}, \lambda \in [0,1]) : \operatorname{Opt}(\lambda r + (1-\lambda)r') \geq \lambda \operatorname{Opt}(r) + (1-\lambda)\operatorname{Opt}(r'). \qquad (!)$$
3. If $\operatorname{Opt}(r)$ is finite at some point $\bar{r} \in \operatorname{int} \mathcal{R}$, then $\operatorname{Opt}(r)$ is real-valued everywhere on $\mathcal{R}$. Moreover, when $X = \mathbf{R}^n$, and f and the components of G are affine, so that $(P[r])$ is an LP program, we can replace in the above claim the inclusion $r \in \operatorname{int} \mathcal{R}$ with the inclusion $r \in \mathcal{R}$: in the LP case, the function $\operatorname{Opt}(r)$ is either identically $+\infty$ everywhere on $\mathcal{R}$ or is real-valued at every point of $\mathcal{R}$.

Comment. Think about the problem $(P[r])$ as a problem where r is the vector of resources you create, and $f(\cdot)$ is your profit, so that the problem is to maximize your profit given your resources and "technological constraints" $x \in X$. Now let $\bar{r} \in \mathcal{R}$ and let e be a nonnegative vector, and let us look what happens when you select your vector of resources on the ray

$R = \bar{r} + \mathbf{R}_+e$, assuming that Opt($r$) on this ray is real-valued. When restricted to this ray, your best profit becomes a function $\phi(t)$ of the nonnegative variable t:

$$\phi(t) = \text{Opt}(\bar{r} + te).$$

Since $e \geq 0$, this function is nondecreasing, as it should be: the larger t, the more resources you have, and the larger is your profit. Not so nice news is that $\phi(t)$ is concave in t, meaning that the slope of this function does not increase as t grows. In other words, if it costs you $1 to pass from resources $\bar{x} + te$ to resources $\bar{x} + (t+1)e$, the return $\phi(t+1) - \phi(t)$ on one extra dollar of your investment goes down (or at least does not go up) as t grows. This is called *the law of diminishing marginal returns*.

Exercise III.5 [TrYs] [Follow-up to Exercise III.4] There are n goods j with per-unit prices $c_j > 0$, per-unit utilities $v_j > 0$, and maximum available amounts $\bar{x}_j$, $j \leq n$. Given a budget $R \geq 0$, you want to decide on the amounts x_j of goods to be purchased to maximize the total utility of the purchased goods, while respecting the budget and availability constraints. Pose the problem as, and verify that, the optimal value Opt(R) is a piecewise linear function of R. What are the breakpoints of this function? What are the slopes between breakpoints?

Exercise III.6 [TrYs] Let $\beta \in \mathbf{R}^n$ be such that $\beta_1 \geq \beta_2 \geq \cdots \geq \beta_n$. For $x \in \mathbf{R}^n$, let $x_{(k)}$ be the kth largest entry in x. Consider the function

$$f(x) = \sum_k \beta_k x_{(k)} = [\beta_1 - \beta_2]s_1(x) + [\beta_2 - \beta_3]s_2(x) + \cdots + [\beta_{n-1} - \beta_n]s_{n-1}(x) + \beta_n s_n(x),$$

where, as always, $s_k(x) = \sum_{i=1}^k x_{(i)}$. As we know from Exercise I.29, the functions $s_k(x)$, $k < n$, are polyhedrally representable:

$$t \geq s_k(x) \iff \exists z \geq 0, s : x_i \leq z_i + s, i \leq n, \sum_i z_i + ks \leq t,$$

and $s_n(x)$ is just linear:

$$s_n(x) = \sum_i x_i.$$

As a result, f admits the polyhedral representation

$$t \geq f(x) \iff \exists Z = [z_{ik}] \in \mathbf{R}^{n \times (n-1)}, s_k, t_k, k < n : \begin{cases} \forall (i \leq n, k < n) : z_{ik} \geq 0, x_i \leq z_{ik} + s_k, \\ \forall k < n : t_k \geq \sum_i z_{ik} + ks_k, \\ t \geq \sum_{k=1}^{n-1}[\beta_k - \beta_{k+1}]t_k + \beta_n \sum_{i=1}^n x_i. \end{cases}$$

This polyhedral representation has $2n^2 - n$ linear inequalities and $n^2 + n - 2$ extra variables. Now here goes the exercise:

1. Find an alternative polyhedral representation of f with $n^2 + 1$ linear inequalities and $2n$ extra variables.

2. [Computational study] Generate at random an orthogonal $n \times n$ matrix U and a vector β with nonincreasing entries and solve numerically the problem

$$\min_x \left\{ f(x) := \sum_k \beta_k x_{(k)} : \|Ux\|_\infty \leq 1 \right\},$$

utilizing the above polyhedral representations of f. For $n = 8, 16, 32, \ldots, 1024$, compare the running times corresponding to the two representations in question.

Exercise III.7 [EdYs] Let $a \in \mathbf{R}^n$ be a nonzero vector, and let $f(\rho) = \ln(\|a\|_{1/\rho})$, $\rho \in [0, 1]$. The moment inequality, see section 13.3.3, states that f is convex. Prove that the function is also nonincreasing and Lipschitz continuous, with Lipschitz constant $\ln n$, or, which is the same, that

$$1 \leq p \leq p' \leq \infty \implies \|a\|_p \geq \|a\|_{p'} \geq n^{1/p' - 1/p} \|a\|_p.$$

Exercise III.8 [TrYs] This exercise demonstrates the power of the symmetry principle. Consider the situation as follows: you are given noisy observations

$$\omega = Ax + \xi, \quad A = \text{Diag}\{\alpha_i, i \leq n\}$$

of an unknown signal x that is known to belong to the unit ball $\mathbf{B} = \{x \in \mathbf{R}^n : \|x\|_2 \leq 1\}$; here $\alpha_i > 0$ are given, and ξ is the standard (zero mean, unit covariance) Gaussian observation noise. Your goal is to recover from this observation the vector $y = Bx$, where $B = \text{Diag}\{\beta_i, i \leq n\}$ is given. You intend to recover y by the *linear estimate*

$$\widehat{y}_H(\omega) = H\omega,$$

where H is an $n \times n$ matrix that you are allowed to choose. For example, selecting $H = BA^{-1} = \text{Diag}\{\beta_i \alpha_i^{-1}\}$, you get an *unbiased* estimate:

$$\mathbf{E}\{\widehat{y}_H(Ax + \xi) - y\} = 0.$$

Let us quantify the quality of a candidate linear estimate $\widehat{y}_H$

– at a particular signal $x \in \mathbf{B}$ by the quantity

$$\text{Err}_x(H) = \sqrt{\mathbf{E}\{\|\widehat{y}_H(Ax + \xi) - Bx\|_2^2\}},$$

so that $\text{Err}_x^2(H)$ is the expected squared $\|\cdot\|_2$-distance between the estimate and the estimated quantity,

– on the entire set $\mathbf{B}$ of possible signals – by *risk* $\text{Risk}[H] = \max_{x \in \mathbf{B}} \text{Err}_x(H)$.

1. Find closed form expressions for $\text{Err}_x(H)$ and $\text{Risk}(H)$.
2. Formulate the problem of finding the linear estimate with minimal risk as the problem of minimizing a convex function and prove that the problem is solvable and admits an optimal solution H^* which is diagonal: $H^* = \text{Diag}\{\eta_i, i \leq n\}$.
3. Reduce the problem presented in item 2 to the problem of minimizing an easy-to-compute convex univariate function. Consider the case when $\beta_i = i^{-1}$ and $\alpha_i = [\sigma i^2]^{-1}$, $1 \leq i \leq n$, set $n = 10,000$, and fill in the following table:

σ	1.0	0.1	0.01	0.001	0.0001	0.00001	0.000001
Risk[H^*]							
Risk[BA^{-1}]							

where H^* is the minimum-risk linear estimate as yielded by the solution to the univariate problem you end up with, and Risk[BA^{-1}] is the risk of the unbiased linear estimate.

You should see from your numerical results that the minimal risk of linear estimation is much smaller than the risk of the unbiased linear estimate. Explain on a qualitative level why allowing for bias reduces the risk.

Exercise III.9 [EdYs] [1] Given sets of d-dimensional tentative nodes ($d = 2$ or $d = 3$) and of tentative bars of a TTD problem satisfying assumption $\mathfrak{R}$, let $\mathcal{V} = \mathbf{R}^M$ be the space of virtual displacements of the nodes, N be the number of tentative bars, and $W > 0$ be the allowed total bar volume; see Exercise I.16. Next let $\mathcal{C}(t, f) : \mathbf{R}_+^N \times \mathcal{V} \to \mathbf{R} \cup \{+\infty\}$ be the compliance of truss $t \geq 0$ w.r.t. load f (we identify trusses with the corresponding vectors t of bar volumes). Prove that

1. $\mathcal{C}(t, f)$ is a convex lsc function, positively homogeneous of homogeneity degree 1, of $[t; f]$ with $\mathbf{R}_{++}^N \times \mathcal{V} \subseteq \operatorname{Dom} \mathcal{C}$, where $\mathbf{R}_{++}^N = \operatorname{int} \mathbf{R}_+^N = \{t \in \mathbf{R}^n : t > 0\}$. This function is positively homogeneous, with degree -1, in t when f is fixed, and is positively homogeneous, of degree 2, in f when t is fixed. Besides this, $\mathcal{C}(t, f)$ is nonincreasing in $t \geq 0$: if $0 \leq t' \leq t$, then $\mathcal{C}(t, f) \leq \mathcal{C}(t', f)$ for every f.

2. The function $\operatorname{Opt}(W, f) = \inf_t \{\mathcal{C}(t, f) : t \geq 0, \sum_i t_i = W\}$ — the optimal value in the TTD problem (5.2) — with W restricted to reside in $\mathbf{R}_{++} = \{W > 0\}$ is a convex continuous function with domain $\mathbf{R}_{++} \times \mathcal{V}$. This function is positively homogeneous, of degree -1, in $W > 0$ and homogeneous, of homogeneity degree 2, in f:

$$\forall (\lambda > 0, \mu \geq 0): \operatorname{Opt}(\lambda W, \mu f) = \lambda^{-1} \mu^2 \operatorname{Opt}(W, f), \quad \forall (W, f) \in \mathbf{R}_{++} \times \mathcal{V}.$$

Moreover, the infimum in $\inf_t \{\mathcal{C}(t, f) : t \geq 0, \sum_i t_i = W\}$ is achieved whenever $W > 0$.

3. When on a certain bridge there is just one car, of unit weight, the compliance of the bridge does not exceed 1 whatever is the position of the car. How large could the compliance of the bridge be when there are 100 cars of total weight 70 on it?

To formulate the next two tasks, let us associate with a free node p the set $\mathcal{F}^p$ of all single-force loads stemming from forces g of magnitude $\|g\|_2$ not exceeding 1 and acting at node p. For a set S of free nodes, $\mathcal{F}^S$ is the set of all loads with nonzero forces acting solely at the nodes from S and with the sum of $\|\cdot\|_2$-magnitudes of the forces not exceeding 1, so that

$$\mathcal{F}^S = \operatorname{Conv}(\cup_{p \in S} \mathcal{F}^p)$$

(why?).

4. Let $S = \{p_1, \ldots, p_K\}$ be a K-element collection of free nodes from the nodal set. Assume that for every node p from S and every load $f \in \mathcal{F}^p$ there exists a truss of

---

[1] The preceding exercises in the TTD series are I.16 and I.18.

a given total weight W such that its compliance w.r.t. f does not exceed 1. Which, if any, of the following statements is true?

(i) For every load $f \in \mathcal{F}^S$, there exists a truss of total volume W with compliance w.r.t. f not exceeding 1.

(ii) There exists a truss of total volume W with compliance w.r.t. every load from $\mathcal{F}^S$ not exceeding 1.

(iii) For properly selected γ depending solely on d, there exists a truss of total volume $\gamma K W$ with compliance w.r.t. every load from $\mathcal{F}^S$ not exceeding 1.

★5. Prove the following statement:

In the situation of item 4 above, let $\gamma = 4$ when $d = 2$ and $\gamma = 7$ when $d = 3$. For every $k \leq K$ there exists a truss $\widehat{t}^k$ of total volume γW such that the compliance of t w.r.t. every load from $\mathcal{F}^{p_k}$ does not exceed 1. As a result, there exists a truss $\widetilde{t}$ of total volume $\gamma K W$ with compliance w.r.t. every load from $\mathcal{F}^S$ not exceeding 1.

15.2 Around Support, Characteristic, and Minkowski Functions

Exercise III.10 [EdYs] [Characteristic and support functions of convex sets] Let $X \subset \mathbf{R}^n$ be a nonempty convex set. The *characteristic* (a.k.a. *indicator*) *function* of X is, by definition, the function

$$\chi_X(x) = \begin{cases} 0, & x \in X, \\ +\infty, & x \notin X. \end{cases}$$

As can immediately be seen, this function is convex and proper. The Legendre transform of this function is called the *support function* $\phi_X(x)$ of X:

$$\phi_X(x) = \sup_u [x^\top u - \chi_X(u)] = \sup_{u \in X} x^\top u.$$

1. Prove that χ_X is lower semicontinuous (lsc) if and only if X is closed, and that the support functions of X and cl X are the same.

In the remaining part of this exercise, we will be interested in the properties of support functions and, in view of item 1, it makes sense to assume from now on that X, as well as being nonempty and convex, is also closed.

Prove the following facts:

2. $\phi_X(\cdot)$ is a proper lsc convex function which is positively homogeneous of degree 1:

$$\forall (x \in \mathrm{Dom}\,\phi_x, \lambda \geq 0) : \phi_X(\lambda x) = \lambda \phi_X(x).$$

In particular, the domain of ϕ_X is a cone. Demonstrate by example that this cone is not necessarily closed (look at the support function of the closed convex set $\{[v; w] \in \mathbf{R}^2 : v > 0, w \leq \ln v\}$).

3. Vice versa, every proper convex lsc function ϕ which is positively homogeneous of degree 1,

$$(x \in \mathrm{Dom}\,f, \lambda \geq 0) \Longrightarrow \phi(\lambda x) = \lambda \phi(x),$$

is the support function of a nonempty closed convex set; specifically, its subdifferential $\partial \phi(0)$ taken at the origin. In particular, $\phi_X(\cdot)$ "remembers" X: if X, Y are nonempty closed convex sets, then $\phi_X(\cdot) \equiv \phi_Y(\cdot)$ if and only if $X = Y$.

4. Let X, Y be two nonempty closed convex sets. Then $\phi_X(\cdot) \geq \phi_Y(\cdot)$ if and only if $Y \subseteq X$.
5. Dom $\phi_X = \mathbf{R}^n$ if and only if X is bounded.
6. Let X be the unit ball of some norm $\|\cdot\|$. Then ϕ_X is nothing other than the norm $\|\cdot\|_*$ conjugate to $\|\cdot\|$. In particular, when $p \in [1, \infty]$ and $X = \{x \in \mathbf{R}^n : \|x\|_p \leq 1\}$, we have $\phi_X(x) \equiv \|x\|_q$, $\frac{1}{q} + \frac{1}{p} = 1$.
7. Let $x \mapsto Ax + b : \mathbf{R}^n \to \mathbf{R}^m$ be an affine mapping, and let $Y = AX + b = \{Ax + b : x \in X\}$. Then

$$\phi_Y(v) = \phi_X(A^\top v) + b^\top v.$$

Exercise III.11 [EdYs] [Minkowski functions of convex sets] The goal of this exercise is to acquaint the reader with an important special family of convex functions – the Minkowski functions of convex sets.

Consider a proper *nonnegative* lower semicontinuous function $f : \mathbf{R}^n \to \mathbf{R} \cup \{+\infty\}$ which is *positively homogeneous of degree* 1, meaning that

$$x \in \text{Dom } f, t \geq 0 \implies tx \in \text{Dom } f \text{ and } f(tx) = tf(x).$$

Note that from the latter property of f and its properness it follows that $0 \in \text{Dom } f$ and $f(0) = 0$.

We can associate with f its *basic sublevel set*

$$X = \{x \in \mathbf{R}^n : f(x) \leq 1\}.$$

Note that X "remembers" f, specifically

$$\forall t > 0 : f(x) \leq t \iff f(t^{-1}x) \leq 1 \iff t^{-1}x \in X,$$

whence also

$$\forall x \in \mathbf{R}^n : f(x) = \inf\{t : t > 0, t^{-1}x \in X\}, \quad (15.1)$$
$$\inf\{t : t > 0, t \in \emptyset\} = +\infty \text{ by definition.}$$

Note that the basic sublevel set of our function f cannot be arbitrary: it is convex and closed (since f is convex lsc) and contains the origin (since $f(0) = 0$).

Now, given a closed convex set $X \subset \mathbf{R}^n$ containing the origin, we can associate with it a function $f : \mathbf{R}^n \to \mathbf{R} \cup \{+\infty\}$ by construction from (15.1), specifically, as

$$f(x) = \inf\{t : t > 0, t^{-1}x \in X\}. \quad (15.2)$$

This function is called the *Minkowski function* (M.f.) of X.

Here goes your first task:

1. Prove that when $X \subset \mathbf{R}^n$ is convex, closed, bounded, and contains the origin, the function f given by (15.2) is a proper, nonnegative, convex lsc function that is positively homogeneous of degree 1, and X is the basic sublevel set of f. Moreover, f is nothing other than the support function ϕ_{X_*} of the polar X_* of X.

15.2 Around Support, Characteristic, and Minkowski Functions

Your next tasks are as follows:

2. What are the Minkowski functions of
   - the singleton $\{0\}$?
   - a linear subspace?
   - a closed cone $\mathbf{K}$?
   - the unit ball of a norm $\|\cdot\|$?
3. Prove that the Minkowski functions f_X, f_Y of closed convex sets X, Y containing the origin are linked by the relation $f_X \geq f_Y$ if and only if $X \subseteq Y$.
4. When does the Minkowski function of a set X (convex, closed, bounded, and containing the origin) not take value $+\infty$?
5. What is the set of zeros of the Minkowski function of a set X (convex, closed, bounded, and containing the origin)?
6. What is the Minkowski function of the intersection $\cap_{k \leq K} X_k$ of closed convex sets containing the origin?

Exercise III.12 [EdYs]

1. Recall that the closed conic transform
$$\overline{\mathrm{ConeT}}(X) = \mathrm{cl}\left\{[x;t] \in \mathbf{R}^n \times \mathbf{R} : t > 0,\ x/t \in X\right\}$$
of a nonempty convex set $X \subset \mathbf{R}^n$ (see section 1.5) is a closed cone such that
$$\mathrm{cl}(X) = \{x : [x;1] \in \overline{\mathrm{ConeT}}(X)\}.$$
What is the cone dual to $\overline{\mathrm{ConeT}}(X)$?

2. Let $X \subset \mathbf{R}^n$ be a nonempty closed convex set and $X^+ = \overline{\mathrm{ConeT}}(X)$. Prove that
$$X_t^+ := \{x : [x;t] \in X^+\} = \begin{cases} tX, & t > 0\ (a) \\ \mathrm{Rec}(X), & t = 0\ (b) \\ \varnothing, & t < 0\ (c). \end{cases}$$

3. Let $X_1, \ldots, X_K$ be closed convex sets in $\mathbf{R}^n$ with nonempty intersection X. Prove that
$$\overline{\mathrm{ConeT}}(X) = \bigcap_k \overline{\mathrm{ConeT}}(X_k).$$

4. Let $X = \bigcap_{k \leq K} X_k$, where $X_1, \ldots, X_K$ are closed convex sets in $\mathbf{R}^n$ such that $X_K \cap \mathrm{int}\, X_1 \cap \mathrm{int}\, X_2 \cdots \cap \mathrm{int}\, X_{K-1} \neq \varnothing$. Prove that $\phi_X(y) \leq a$ if and only if there exist y_k, $k \leq K$, such that
$$y = \sum_k y_k \text{ and } \sum_k \phi_{X_k}(y_k) \leq a. \tag{$*$}$$

   In words: *In the situation in question, the supremum of a linear form on $\bigcap_k X_k$ does not exceed some a if and only if the form can be decomposed into the sum of K forms with the sum of their suprema over the respective sets X_k not exceeding a.*

5. Prove the following polyhedral version of the claim in item 4:
   Let $X_k = \{x \in \mathbf{R}^n : A_k x \leq b_k\}$, $k \leq K$, be polyhedral sets with nonempty intersection X. A linear form does not exceed some $a \in \mathbf{R}$ everywhere on X if and only if the form

can be decomposed into the sum of K linear forms with the sum of their maxima on the respective sets X_k not exceeding a.

Exercise III.13 [TrYs] Let $X \subset \mathbf{R}^n$ be a nonempty polyhedral set given by the polyhedral representation

$$X = \{x : \exists u : Ax + Bu \leq r\}.$$

Build a polyhedral representation of the epigraph of the support function of X. For a non-polyhedral extension, see Exercise IV.36.

Exercise III.14 [TrYs] Compute in closed analytical form the support functions of the following sets:

1. The ellipsoid $\{x \in \mathbf{R}^n : (x-c)^\top C(x-c) \leq 1\}$ with $C \succ 0$.
2. The probabilistic simplex $\{x \in \mathbf{R}_+^n : \sum_i x_i = 1\}$.
3. The nonnegative part of the unit $\|\cdot\|_p$-ball: $X = \{x \in \mathbf{R}_+^n : \|x\|_p \leq 1\}$, $p \in [1, \infty]$.
4. The positive semidefinite part of the unit $\|\cdot\|_{p,\mathrm{Sh}}$ norm: $X = \{x \in \mathbf{S}_+^n : \|x\|_{p,\mathrm{Sh}} \leq 1\}$.

15.3 Around Subdifferentials

Exercise III.15 [EdYs] Let f be a convex function and $\bar{x} \in \mathrm{Dom}\, f \subseteq \mathbf{R}^n$. Prove that the property of $g \in \mathbf{R}^n$ of being a subgradient of f at $\bar{x}$ is local: the inequality

$$f(x) \geq f(\bar{x}) + g^\top (x - \bar{x}) \qquad (*)$$

holds true for all $x \in \mathbf{R}^n$ if and only if it holds true for all x in a neighborhood of $\bar{x}$.

Exercise III.16 [EdYs] [Subdifferentials of norms] Let $\|\cdot\|$ be a norm on $\mathbf{R}^n$, and let $\|\cdot\|_*$ be its conjugate (see Fact 13.4). Prove that

1. The subdifferential of $\|\cdot\|$ taken at the origin is the unit ball B_* of $\|\cdot\|_*$, or, which is the same, the polar

$$\{u : u^\top x \leq 1 \,\forall (x : \|x\| \leq 1)\}$$

of the unit ball B of the norm $\|\cdot\|$.
2. When $x \neq 0$, the subdifferential of $\|\cdot\|$ taken at x is the set $\{u \in B_* : u^\top x = \|x\|\}$. In particular, the subdifferential of $\|\cdot\|$ remains intact when replacing x with tx, $t > 0$, and is reflected with respect to the origin when x is replaced with tx, $t < 0$.

Exercise III.17 [EdYs] [Shatten norms] Let $p \in [1, \infty]$. The space $\mathbf{S}^n$ of symmetric $n \times n$ matrices can be equipped with *Shatten p-norms* – matrix analogies of the standard $\|\cdot\|_p$-norms on $\mathbf{R}^n$. Specifically, the Shatten p-norm $\|\cdot\|_{p,\mathrm{Sh}}$ of a symmetric matrix X is defined as

$$\|X\|_{p,\mathrm{Sh}} = \|\lambda(X)\|_p,$$

where $\lambda(X)$, as always, is the vector of eigenvalues of X.

1. Prove that Shatten norms are indeed norms, and that the norm conjugate to $\|\cdot\|_{p,\mathrm{Sh}}$ is $\|\cdot\|_{q,\mathrm{Sh}}$, $\frac{1}{p} + \frac{1}{q} = 1$:

$$\|X\|_{q,\mathrm{Sh}} = \max_Y \{\mathrm{Tr}(XY) : \|Y\|_{p,\mathrm{Sh}} \leq 1\}. \qquad (15.3)$$

2. Verify that $\|\cdot\|_{2,\mathrm{Sh}}$ is nothing other than the Frobenius norm of X, and $\|X\|_{\infty,\mathrm{Sh}}$ is the same as the spectral norm of X.

Exercise III.18 [EdYs] [Chain rule for subdifferentials] Let $Y \subseteq \mathbf{R}^m$ and $X \subseteq \mathbf{R}^n$ be nonempty convex sets, $\bar{y} \in Y$, $\bar{x} \in X$, let $f(\cdot) : Y \to \mathbf{R}$ be a convex function, and let $A(\cdot) : X \to Y$ with $A(\bar{x}) = \bar{y}$. Further, let $\mathbf{K}$ be a closed cone in $\mathbf{R}^n$. A function f is called **K**-*monotone* on Y, if for $y, y' \in Y$ such that $y' - y \in \mathbf{K}$ it holds that $f(y') \geq f(y)$, and A is called **K**-*convex* on X if for all $x, x' \in X$ and $\lambda \in [0,1]$ it holds that $\lambda A(x) + (1-\lambda)A(x') - A(\lambda x + (1-\lambda)x') \in \mathbf{K}$.[2]

Prove that

1. A is **K**-convex on X if and only if for every $\phi \in \mathbf{K}_*$ the real-valued function $\phi^\top A(x)$ is convex on X.
2. Let A be **K**-convex on X and differentiable at $\bar{x}$. Let $A'(\bar{x})$ denote the Jacobian of A at $\bar{x}$. Prove that
$$\forall x \in X : A(x) - [A(\bar{x}) + A'(\bar{x})[x - \bar{x}]] \in \mathbf{K}. \qquad (*)$$
3. Let f be **K**-monotone on Y and let A be **K**-convex on X. Prove that the function $f \circ A(x) = f(A(x))$ is real-valued and is also convex on X.
4. Let f be **K**-monotone on Y. Prove that $\partial f(\bar{y}) \subseteq \mathbf{K}_*$ provided $\bar{y} \in \operatorname{int} Y$.
5. [Chain rule] Let $\bar{y} \in \operatorname{int} Y$, $\bar{x} \in \operatorname{int} X$; let f be **K**-monotone on Y, and let A be **K**-convex on X and differentiable at $\bar{x}$. Prove that
$$\partial f \circ A(\bar{x}) = [A'(\bar{x})]^\top \partial f(\bar{y}) = \{[A'(\bar{x})]^\top g : g \in \partial f(\bar{y})\}. \qquad (!)$$

Exercise III.19 [EdYs] Recall that the sum $S_k(X)$ of the $k \leq n$ largest eigenvalues of $X \in \mathbf{S}^n$ is a convex function of X, see Remark 14.4. Point out *a* subgradient of $S_k(\cdot)$ at a point $\bar{X} \in \mathbf{S}^n$. As a special case, find a subgradient of the maximal eigenvalue $\lambda_{\max}(X)$ of $X \in \mathbf{S}^n$ treated as a function of X.

15.4 Around the Legendre Transform

Exercise III.20 [TrYs] Compute Legendre transforms of the following univariate functions:

1. $f(x) = -\ln x$, $\operatorname{Dom} f = (0, \infty)$.
2. $f(x) = e^x$, $\operatorname{Dom} f = \mathbf{R}$.
3. $f(x) = x \ln x$, $\operatorname{Dom} f = [0, \infty)$ ($0 \ln 0 = 0$ by definition).
4. $f(x) = x^p/p$, $\operatorname{Dom} f = [0, \infty)$; here $p > 1$.

Exercise III.21 [TrYs] Compute Legendre transforms of the following functions:

- [log-barrier for nonnegative orthant $\mathbf{R}^n_+$] $f(x) = -\sum_{i=1}^n \ln x_i : \operatorname{int} \mathbf{R}^n_+ \to \mathbf{R}$
- [log-det barrier for semidefinite cone $\mathbf{S}^n_+$] $f(x) = -\ln \operatorname{Det}(x) : \operatorname{int} \mathbf{S}^n_+ \to \mathbf{R}$ (start by proving the convexity of f).

---

[2] We shall study cone-monotonicity and cone-convexity in more detail in Part IV.

Exercise III.22 [EdYs] [Computing Legendre transform of the log-barrier $-\ln(x_n^2 - x_1^2 - \cdots - x_{n-1}^2)$ for Lorentz cone] Consider the optimization problem

$$\max_{x,t} \left\{ \xi^\top x + \tau t + \ln(t^2 - x^\top x) : (t,x) \in X = \{(t,x) : t > \sqrt{x^\top x}\} \right\}$$

where $\xi \in \mathbf{R}^n$, $\tau \in \mathbf{R}$ are parameters. Is the problem convex?[3] For what values of ξ, τ this problem is solvable? What is the optimal value? Is it convex in the parameters?

Exercise III.23 [EdYs] Consider the optimization problem

$$\max_{x,y} \{f(x,y) = ax + by + \ln(\ln y - x) + \ln(y) : (x,y) \in X = \{(x,y) : y > \exp\{x\}\}\},$$

where $a, b \in \mathbf{R}$ are parameters. Is the problem convex? For what values of a, b this problem is solvable? What is the optimal value? Is it convex in the parameters?

Exercise III.24 [TrYs] Compute Legendre transforms of the following functions:

- ["geometric mean"] $f(x) = -\prod_{i \leq n} x_i^{\pi_i} : \mathbf{R}_+^n \to \mathbf{R}$, where the $\pi_i > 0$ sum up to 1 and $n > 1$.
- ["inverse geometric mean"] $f(x) = \prod_{i \leq n} x_i^{-\pi_i} : \text{int } \mathbf{R}_+^n \to \mathbf{R}$, where $\pi_i > 0$.

Exercise III.25 [EdYs] Prove the following version of the results of section 13.2: Suppose that $f : \mathbf{R}^n \to \mathbf{R} \cup \{+\infty\}$ is a proper convex lsc function with an open domain G, f is twice continuously differentiable, with positive definite Hessian, on G (thus, f is strongly convex with respect to the Euclidean norm on G). Also, assume that the following holds:

(!) Whenever y is such that the function $y^\top x - f(x)$, as a function of x, is bounded from above, the function achieves its maximum over x.

Let f^* be the Legendre transform of f, and let G_* be the domain of f^*. Prove that G_* is an open convex set, f^* is twice continuously differentiable, with a positive definite Hessian, on G_*, and the mappings $x \mapsto \nabla f(x)$ and $y \mapsto \nabla f^*(y)$ are inverse to each other one-to-one correspondences between G and G_*.
Hint: Use Implicit Function Theorem (Theorem IV.21.5).

Exercise III.26 [EdYs] The goal of this exercise is to investigate the relation between the smoothness of a convex function and the strong convexity of its Legendre transform. Let us start with the following preliminary task:

1. Let $\|\cdot\|$ be a norm on $\mathbf{R}^n$. Recall that the unit ball of the conjugate norm $\|y\|_* = \max_x\{y^\top x : \|x\| \leq 1\}$ is the polar of the unit ball of $\|\cdot\|$ norm, and the norm conjugate to $\|\cdot\|_*$ is the norm $\|\cdot\|$ itself. Prove that the functions $\frac{1}{2}\|x\|^2$ and $\frac{1}{2}\|d\|_*^2$ are Legendre transforms of each other.

The subsequent tasks need certain preamble.
Smooth convex functions. Let f be a convex function, let $\|\cdot\|$ be a norm on $\mathbf{R}^n$, and let L be a nonnegative real. We say that f is $(L, \|\cdot\|)$-*smooth*, if Dom $f = \mathbf{R}^n$ and

---

[3] A *maximization* problem with objective $f(\cdot)$ and certain constraints and domain is called convex if the equivalent minimization problem with the objective $(-f)$ and the original constraints and domain is convex.

15.4 Around the Legendre Transform

$$f(z) \leq f(x) + e^\top[z-x] + \frac{L}{2}\|z-x\|^2, \qquad \forall (x, z \in \mathbf{R}^n, e \in \partial f(x)).$$

It is easily seen that *a convex function* $f : \mathbf{R}^n \to \mathbf{R}$ *is* $(L, \|\cdot\|)$*-smooth if and only if it is continuously differentiable, and the mapping* $x \mapsto \nabla f(x)$ *is Lipschitz continuous, with constant L, from the norm* $\|\cdot\|$ *to the norm* $\|\cdot\|_*$:

$$\|\nabla f(x) - \nabla f(y)\|_* \leq L\|x-y\|, \qquad \forall x, y,$$

same as *if and only if f is continuously differentiable and*

$$[x-y]^\top [\nabla f(x) - \nabla f(y)] \leq L\|x-y\|^2, \qquad \forall x, y.$$

A twice continuously differentiable function $f : \mathbf{R}^n \to \mathbf{R}$ is $(L, \|\cdot\|)$-smooth if and only if

$$0 \leq \frac{d^2}{dt^2}\Big|_{t=0} f(x+th) \leq L\|h\|^2, \qquad \forall x, h.$$

Example: Convex quadratic function $f(x) = \frac{1}{2}x^\top Q x - q^\top x + c$, where $Q \succeq 0$, is $(L, \|\cdot\|_2)$-smooth whenever the eigenvalues of Q are upper-bounded by L.

Strongly convex functions. Let $g : \mathbf{R}^n \to \mathbf{R} \cup \{+\infty\}$ be a proper convex lsc function, let $\|\cdot\|_*$ be the conjugate of a norm $\|\cdot\|$, and let L be a positive real. We say that g is $(L, \|\cdot\|_*)$-*strongly convex*, if for every $\bar{y} \in \text{Dom}\, g$ the following holds

$$g(y) \geq g(\bar{y}) + [y-\bar{y}]^\top e + \frac{1}{2L}\|y-\bar{y}\|_*^2, \qquad \forall (y \in \mathbf{R}^n, e \in \partial g(\bar{y})).$$

From this inequality it follows that *a proper convex lsc function g is $(L, \|\cdot\|_*)$-strongly convex if and only if*

$$[e'-e]^\top[y'-y] \geq \frac{1}{L}\|y'-y\|_*^2, \qquad \forall (y, y' \in \text{Dom}\, g, e \in \partial g(y), e' \in \partial g(y')).$$

A twice continuously differentiable on rint(Dom g) convex lsc function g is $(L, \|\cdot\|_*)$-strongly convex if and only if $\frac{d^2}{dt^2}\Big|_{t=0} g(y+th) \geq L^{-1}\|h\|_*^2$ for all $y \in \text{rint}(\text{Dom}\, g)$ and all h from the linear subspace parallel to Aff(Dom g).

Example: Convex quadratic function $f(x) = \frac{1}{2}x^\top Q x - q^\top x + c : \mathbf{R}^n \to \mathbf{R}$ is $(L, \|\cdot\|_*)$-strongly convex if and only if $Q \succ 0$ and all eigenvalues of Q are lower-bounded by L^{-1}.

Note: When f is convex quadratic form with the matrix of the quadratic part equal to $Q \succ 0$, the Legendre transform f^* of f is convex quadratic form with the matrix of the quadratic part equal to Q^{-1}.

From the examples above, *a quadratic form with positive definite matrix of the quadratic part, the form is $(L, \|\cdot\|_2)$-smooth if and only if the Legendre transform f^* of f is $(L, \|\cdot\|_2)$-strongly convex.*

The point of the exercise is to justify the following far-reaching extension of the latter observation:

> The Legendre transform f^* of an $(L, \|\cdot\|)$-smooth convex function $f : \mathbf{R}^n \to \mathbf{R}$ is $(L, \|\cdot\|_*)$-strongly convex. Vice versa, if the Legendre transform f^* of a convex function $f : \mathbf{R}^n \to \mathbf{R} \cup \{\infty\}$ is $(L, \|\cdot\|_*)$-strongly convex, then f is $(L, \|\cdot\|)$-smooth.

2. Justify the above claim.

Your concluding task is as follows:

3. Verify that the function $f(x) = \ln(\sum_{i=1}^n e^{x_i})$ is $(1, \|\cdot\|_\infty)$-smooth, compute its Legendre transform f^*, and make conclusions about strong convexity of f^* (the latter plays an important role in the design of *proximal first-order algorithms* for the minimization of convex functions over the probabilistic simplex).

15.5 Miscellaneous Exercises

Exercise III.27 [EdYs] [Multi-factor Hölder inequality] Given positive reals $q_1, \ldots, q_n$ and $p \in [1, \infty)$, we define the weighted p-norm of a vector $x \in \mathbf{R}^n$ as

$$|x|_p = \left(\sum_{j=1}^n q_j |x_j|^p\right)^{1/p}.$$

This clearly is a norm which becomes the standard norm $\|\cdot\|_p$ when $q_j = 1$, $j \leq n$. Like $\|x\|_p$, the quantity $|x|_p$ has a limit, namely, $\|x\|_\infty$, as $p \to \infty$, and we define $|\cdot|_\infty$ as this limit.

Now let p_i, $i \leq k$, be positive reals such that

$$\sum_{i=1}^k \frac{1}{p_i} = 1.$$

1. Prove that for nonnegative reals $a_1, \ldots, a_k$ one has

$$a_1 a_2 \cdots a_k \leq \frac{a_1^{p_1}}{p_1} + \cdots + \frac{a_k^{p_k}}{p_k}$$

or, equivalently (set $b_i = a_i^{p_i}$)

$$\forall b \geq 0 : b_1^{1/p_1} b_2^{1/p_2} \cdots b_k^{1/p_k} \leq \frac{b_1}{p_1} + \frac{b_2}{p_2} + \cdots + \frac{b_k}{p_k}.$$

Note: the special case $p_i = k$, $i \leq k$, of this inequality is the inequality between the geometric and arithmetic means.

2. Let $x^1, \ldots, x^k \in \mathbf{R}^n$, and let $x^1 x^2 \cdots x^k$ be the entrywise product of $x^1, \ldots, x^k$:

$$[x^1 x^2 \cdots x^k]_j = x_j^1 x_j^2 \cdots x_j^k, \quad 1 \leq j \leq n.$$

Prove that

$$|x^1 x^2 \cdots x^k|_1 \leq \sum_{i=1}^k \frac{|x^i|_{p_i}^{p_i}}{p_i}. \tag{$*$}$$

3. Prove the *multi-factor Hölder inequality*: for vectors $x^i \in \mathbf{R}^n$, $i \leq k$, one has

$$|x^1 x^2 \cdots x^k|_1 \leq |x^1|_{p_1} |x^2|_{p_2} \cdots |x^k|_{p_k} \tag{\#}$$

Note: (#) was stated for positive *reals* $p_1, \ldots, p_k$ with $\sum_i 1/p_i = 1$. It can immediately be seen that (#) remains true when $p_i = \infty$ for some i (and, of course, $1/p_i$ is set to 0 for these i).

Further note: (#) is the general form of Hölder inequality, which in the main text was proved for $k = 2$ and $|\cdot|_{p_i} = \|\cdot\|_{p_i}$. Needless to say, this inequality extends to the case when the x_i are functions $x_i(\omega)$ on a space with measure $q(\cdot)$ and the finite sums $\sum_{j=1}^n q_j f_j$ are replaced with integrals, resulting in

$$\int \left| \prod_{i=1}^k x_i(\omega) \right| q(d\omega) \leq \prod_{i=1}^k \left[\int |x_i(\omega)|^{p_i} q(d\omega) \right]^{1/p_i},$$

provided some measurability conditions are satisfied. In this textbook, however, we do not touch on infinite-dimensional spaces of functions and the related norms.

Exercise III.28 [EdYs] [Muirhead's inequality] For any $u \in \mathbf{R}^n$ and $z \in \mathbf{R}_{++}^n := \{z \in \mathbf{R}^n : z > 0\}$ define

$$f_z(u) = \frac{1}{n!} \sum_\sigma z_{\sigma(1)}^{u_1} \cdots z_{\sigma(n)}^{u_n},$$

where the sum is over all permutations σ of $\{1, \ldots, n\}$. Show that if P is a doubly stochastic $n \times n$ matrix, then

$$f_z(Pu) \leq f_z(u), \quad \forall (u \in \mathbf{R}^n, z \in \mathbf{R}_{++}^n).$$

Exercise III.29 [EdYs] Prove that a convex lsc function f with polyhedral domain is continuous on its domain. Does the conclusion remain true when either of the assumptions that (a) convex f is lsc, or (b) Dom f is polyhedral is lifted?

Exercise III.30 [TrYs] Let $a_1, \ldots, a_n > 0$, $\alpha, \beta > 0$. Solve the optimization problem

$$\min_x \left\{ \sum_{i=1}^n \frac{a_i}{x_i^\alpha} : x > 0, \sum_i x_i^\beta \leq 1 \right\}.$$

Exercise III.31 [TrYs] [Computational study] Consider the following situation: there are K radar units with the kth unit capable of locating targets within the ellipsoid $E_k = \{x \in \mathbf{R}^n : (x - c_k)^\top C_k (x - c_k) \leq 1\}$ ($C_k \succ 0$); the position of the target as measured by the kth unit is

$$y_k = x + \sigma_k \zeta_k,$$

where x is the actual position of the target, and ζ_k is the standard (zero mean, unit covariance) Gaussian observation noise; the ζ_k are independent across k. Given measurements $y_1, \ldots, y_K$ of target's location x known to belong to the "common field of view" $E = \cap_k E_k$ of the radar units, which we assume to possess a nonempty interior, we want to estimate a given linear form $e^\top x$ of x by using the linear estimate

$$\widehat{x} = \sum_k h_k^\top y_k + h.$$

We are interested in finding the estimate (e.g., the parameters $h_1, \ldots, h_K, h$) minimizing the risk

$$\text{Risk2} = \max_{x \in E} \sqrt{\mathbf{E}\left\{\left[e^\top x - \sum_k h_k^\top [x + \sigma_k \zeta_k] - h\right]^2\right\}}.$$

1. Pose the problem as a convex optimization program.
2. Process the problem numerically (See Figure 15.1) and look at the results.
   Recommended setup:
   - $K = 3, n = 2, [c_1, c_2, c_3] = \begin{bmatrix} 1.000 & -0.500 & -0.500 \\ 0 & 0.866 & -0.866 \end{bmatrix}$,

   $C_1 = \begin{bmatrix} 0.2500 & 0 \\ 0 & 1.5000 \end{bmatrix}, C_2 = \begin{bmatrix} 1.1875 & 0.5413 \\ 0.5413 & 0.5625 \end{bmatrix}, C_3 = \begin{bmatrix} 1.1875 & -0.5413 \\ -0.5413 & 0.5625 \end{bmatrix}$

   - $\sigma_1 = 0.1, \sigma_2 = 0.2, \sigma_3 = 0.3$
   - $e = [1; 1]/\sqrt{2}$.

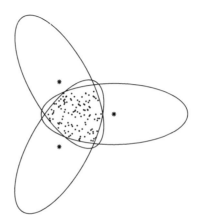

Figure 15.1 Three radar units (stars) and their common field of view (dotted).

Exercise III.32 [EdYs] For any $k \leq m$ and $X \in \mathbf{S}^m$, recall that $S_k(X)$ denotes the sum of the k largest eigenvalues of the matrix X. Given $X \in \mathbf{S}^m$, define $R[X] := \{V^\top X V : V \in \mathcal{O}_m\}$ where $\mathcal{O}_m = \{V \in \mathbf{R}^{m \times m} : VV^\top = I_m\}$ is the set of all $m \times m$ orthogonal matrices. Prove that, for any two symmetric matrices $X, Y \in \mathbf{S}^m$, we have

$Y \in \text{Conv}(R[X])$ if and only if $S_k(Y) \leq S_k(X)$ for all $k < m$ and $\text{Tr}(Y) = \text{Tr}(X)$.

Part IV

Convex Programming, Lagrange Duality, Saddle Points

16

Convex Programming Problems and Convex Theorem of the Alternative

16.1 Mathematical Programming and Convex Programming Problems

For the reader's convenience, we start by reproducing the basic optimization terminology presented in section 4.5.1.

A (constrained) mathematical programming problem has the following form:

$$(P) \quad \min_x \left\{ f(x) : \begin{array}{l} x \in X, \\ g(x) \equiv [g_1(x); \ldots; g_m(x)] \leq 0, \\ h(x) \equiv [h_1(x); \ldots; h_k(x)] = 0 \end{array} \right\}, \quad (16.1)$$

where

- [domain] $X \subseteq \mathbf{R}^n$ is called the *domain* of the problem.
- [objective] f is called the *objective* (function) of the problem,
- [constraints] g_i, $i = 1, \ldots, m$, are called the (functional) *inequality constraints*, and h_j, $j = 1, \ldots, k$, are called the *equality constraints*.[1]

We always assume that $X \neq \emptyset$ and that the objective and the constraints are well defined on X. Moreover, we typically skip indicating X when $X = \mathbf{R}^n$. Thus, *in the sequel, unless the domain is explicitly present in the formulation, it is the entire space $\mathbf{R}^n$.*

We use the following standard terminology related to (16.1):

- [feasible solution] A point $x \in \mathbf{R}^n$ is called a *feasible solution* to (16.1), if $x \in X$, $g_i(x) \leq 0$, $i = 1, \ldots, m$, and $h_j(x) = 0$, $j = 1, \ldots, k$, i.e., if x satisfies all restrictions imposed by the formulation of the problem.
  - [feasible set] The set of all feasible solutions is called the *feasible set* of the problem.
  - [feasible problem] A problem with a nonempty feasible set (i.e., one which admits feasible solutions) is called *feasible* (or *consistent*).
  - [active constraint] An inequality constraint $g_i(\cdot) \leq 0$ is called *active at a given feasible solution x*, if this constraint is satisfied at the point x as an equality rather than strict inequality, i.e., if

$$g_i(x) = 0.$$

Each equality constraint $h_j(x) = 0$ by definition is active at every feasible solution x.

---

[1] Rigorously speaking, the constraints are not the *functions* g_i, h_j, but the *relations* $g_i(x) \leq 0, h_j(x) = 0$. We will use the word "constraints" in both of these senses, and it will always be clear what is meant. For example, we will say that "x satisfies the constraints" to refer to the relations, and we will say that "the constraints are differentiable" to refer to the underlying functions.

235

- [optimal value] The *optimal value* of the problem refers to the quantity

$$f^* := \begin{cases} \inf_x \{f(x) : x \in X, g(x) \leq 0, h(x) = 0\}, & \text{if the problem is feasible,} \\ +\infty, & \text{if the problem is infeasible.} \end{cases}$$

  – [below boundedness] The problem is called *below bounded*, if its optimal value is $> -\infty$, i.e., if the objective is bounded from below on the feasible set.
- [optimal solution] A point $x \in \mathbf{R}^n$ is called an *optimal solution* to (16.1) if x is feasible and $f(x) \leq f(x')$ for any other feasible solution x', i.e., if

$$x \in \text{Argmin}\left\{f(x') : x' \in X, g(x') \leq 0, h(x') = 0\right\}.$$

  – [solvable problem] A problem is called *solvable*, if it admits optimal solutions.
  – [optimal set] The set of all optimal solutions to a problem is called its *optimal set*.

The terminology above is for minimization problems; for maximization modifications to it see section 4.5.1. Recall that the *solvability* of problem (P) is a strictly stronger statement than its *feasibility and below boundedness*; see Remark 4.6.

16.1.1 Convex Programming Problem

A mathematical programming problem (P) is called *convex* (or a convex programming problem), if

- X is a *convex* subset of $\mathbf{R}^n$,
- $f, g_1, \ldots, g_m$ are *real-valued convex* functions on X, and
- there are no equality constraints at all.

Note that instead of saying that there are no equality constraints, we could say that there *are* constraints of this type but only *linear* (affine) ones; this latter case can be immediately reduced to the case without equality constraints by replacing $\mathbf{R}^n$ with the affine subspace given by the (linear) equality constraints.

16.2 Convex Theorem of the Alternative

The simplest case of a convex problem is, of course, a linear programming problem – one where $X = \mathbf{R}^n$ and the objective and all the constraints are linear. The main descriptive components of LP are LP duality and optimality conditions; our primary goal in this and forthcoming chapters is to extend duality and optimality conditions from linear to convex programming.

The origins of our developments are based on the following simple observation: the fact that a point x^* is an optimal solution can be expressed in terms of the feasibility or infeasibility of certain systems of constraints. These systems in our current setup of convex optimization problems are given by

$$x \in X, \ f(x) \leq c, \ g_j(x) \leq 0, \ j = 1, \ldots, m, \tag{16.2}$$

and

$$x \in X, \ f(x) < c, \ g_j(x) \leq 0, \ j = 1, \ldots, m; \tag{16.3}$$

16.2 Convex Theorem of the Alternative

here c is a parameter. The optimality of x^* for the problem means precisely that for appropriately chosen c (this choice, of course, is $c = f(x^*)$) the first of these systems is feasible and x^* is its feasible solution, while the second system is infeasible. Next, in the case of LP, we converted the "negative" part of this simple observation – the claim that (16.3) is infeasible – into a positive statement, using the General Theorem of the Alternative (Theorem 4.3), and this gave us the LP Duality Theorem (Theorem 4.10).

We will follow the same approach for convex optimization problems. To this end, we need a "convex analogy" to the Theorem of the Alternative – something like the latter statement, but for the case when the inequalities in question are given by convex functions rather than linear functions (and, besides, we now have to handle a "convex inclusion" $x \in X$).

Indeed, it is easy to *guess* the result we need. How did we come to the formulation of the Theorem of the Alternative? The main question, basically, boiled down to how to express in an affirmative manner the fact that a system of linear inequalities has no solutions. To this end, we observed that if we can combine, in a linear fashion, the inequalities of the system and get an obviously false inequality such as $0 \le -1$, then the system is infeasible. Note that this condition is nothing other than a certain affirmative statement with respect to the weights with which we are combining the original inequalities.

Now, the scheme of the above reasoning has nothing tied to the linearity (and even convexity) of the inequalities in question. Indeed, consider an *arbitrary* system of constraints of the type (16.3):

$$\begin{aligned} f(x) &< c \\ g_j(x) &\le 0, \quad j = 1, \ldots, m \\ x &\in X. \end{aligned} \tag{I}$$

Here, all we assume is that X is a nonempty subset in $\mathbf{R}^n$ and $f, g_1, \ldots, g_m$ are real-valued functions on X. Then, it is absolutely evident that

If there exist nonnegative weights $\lambda_1, \ldots, \lambda_m$ such that the inequality

$$f(x) + \sum_{j=1}^{m} \lambda_j g_j(x) < c \tag{16.4}$$

has no solutions in X, then (I) *also has no solutions.*

Indeed, a solution to (I) is clearly a solution to (16.4) – the latter inequality is just a combination of the inequalities from (I) with the weights 1 (for the first inequality) and λ_j (for the remaining inequalities).

Now, what does it mean for (16.4) to have no solutions in the domain X? A necessary and sufficient condition for this is that the infimum of the left-hand side of (16.4) over the domain $x \in X$ is greater than or equal to c. Thus, we arrive at the following evident result.

Proposition 16.1 *[Sufficient condition for the infeasibility of* (I)*] Consider a system* (I) *with arbitrary data and assume that the system*

$$\begin{aligned} \inf_{x \in X} \left[f(x) + \sum_{j=1}^{m} \lambda_j g_j(x) \right] &\ge c \\ \lambda_j &\ge 0, \; j = 1, \ldots, m \end{aligned} \tag{II}$$

with unknowns $\lambda_1, \ldots, \lambda_m$ has a solution. Then, (I) *is infeasible.*

Let us stress that Proposition 16.1 is completely general; it does not require any assumptions (not even convexity) on the entities involved.

That said, Proposition 16.1, unfortunately, is not particularly helpful: the actual power of the Theorem of the Alternative (and the key fact utilized in the proof of the Linear Programming Duality Theorem) is not the *sufficiency* of the condition of the proposition for the infeasibility of (I) but the *necessity* of this condition. Justification of the necessity of the condition in question has nothing to do with the evident reasoning that established its sufficiency. In the linear case ($X = \mathbf{R}^n$, $f, g_1, \ldots, g_m$ are linear), we established the necessity via the Homogeneous Farkas' Lemma. We will next prove the necessity of the condition for the *convex* case. At this step, we already need some additional, although minor, assumptions; and in the general nonconvex case the sufficient condition stated in Proposition 16.1 is simply *not* necessary for the infeasibility of (I). This, of course, is very bad, yet expected, news – this is the reason why there are difficult optimization problems that we do not know how to solve efficiently.

The "preface" just presented outlines our action plan. Let us carry out our plan by formally defining the aforementioned minor regularity assumptions.

Definition 16.2 *[Slater condition] Let $X \subseteq \mathbf{R}^n$ and let $g_1, \ldots, g_m$ be real-valued functions on X. We say that these functions satisfy the* Slater condition *on X, if there exists a strictly feasible solution x, that is, $x \in \text{rint } X$ such that $g_j(x) < 0$, $j = 1, \ldots, m$.*

We say that an inequality constrained problem

$$\min_x \{f(x) : g_j(x) \leq 0, \ j = 1, \ldots, m, \ x \in X\} \quad \text{(IC)}$$
$$[\text{where } f, g_1, \ldots, g_m \text{ are real-valued functions on } X]$$

satisfies the Slater condition (synonym: is strictly feasible), if $g_1, \ldots, g_m$ satisfy this condition on X.

In the case where some constraints are linear, we rely on a slightly relaxed regularity condition.

Definition 16.3 *[Relaxed Slater condition] Let $X \subseteq \mathbf{R}^n$, and let $g_1, \ldots, g_m$ be real-valued functions on X. We say that $g_1, \ldots, g_m$ satisfy the* relaxed Slater condition *on X, if there exists $x \in \text{rint } X$ such that $g_j(x) \leq 0$ for all $1 \leq j \leq m$, and $g_j(x) < 0$ for all j with non-affine g_j.*

An inequality-constrained problem (IC) *is said to satisfy the relaxed Slater condition (synonym: is essentially strictly feasible), if $g_1, \ldots, g_m$ satisfy this condition on X.*

A system of equality and inequality constraints

$$g_j(x) \leq 0, \ j = 1, \ldots, m, \ h_i(x) = 0, \ i = 1, \ldots, k, \ x \in X \quad \text{(C)}$$
$$[\text{where } g_1, \ldots, g_m, h_1, \ldots, h_k \text{ are real-valued functions on } X]$$

is said to satisfy the relaxed Slater condition (synonym: is essentially strictly feasible), if all h_i are affine functions, and there exists an essentially strictly feasible solution, *that is, a feasible solution $x \in \text{rint } X$ where all inequality constraints $g_j(x) \leq 0$ with non-affine g_j are satisfied as strict inequalities. An optimization problem of minimizing a (real-valued on X) objective f under constraints* (C) *is called* essentially strictly feasible, *if the system of constraints is so.*

Note: (C) is essentially strictly feasible if and only if the equivalent inequality reformulation

$$g_j(x) \leq 0, \ j = 1, \ldots, m, \ \pm h_i(x) \leq 0, \ i = 1, \ldots, k, \ x \in X,$$

of (C) is essentially strictly feasible.

Clearly, the validity of the Slater condition implies the validity of the relaxed Slater condition (why?). We are about to establish the following fundamental fact.

Theorem 16.4 *[Convex Theorem of the Alternative] Let $X \subseteq \mathbf{R}^n$ be convex, let $f, g_1, \ldots, g_m$ be real-valued convex functions on X, and let $g_1, \ldots, g_m$ satisfy the relaxed Slater condition on X. Then, system (I) is feasible if and only if system (II) is infeasible.*

Theorem 16.4 is a special case of Theorem 16.13, to be formulated and proved in the next section.

16.3 ★ Convex Theorem of the Alternative – Cone-Constrained Form

We will now present and prove a form of Theorem 16.4 that will be stronger than the original form. To this end, we need a few definitions and concepts related to cones.

Definition 16.5 *[Regular cone] A cone $\mathbf{K} \subset \mathbf{R}^n$ is called a* regular cone *if $\mathbf{K}$ is closed, convex, full-dimensional (i.e., possesses a nonempty interior), and is pointed (i.e., $\mathbf{K} \cap (-\mathbf{K}) = \{0\}$).*

In our developments, we will frequently examine the dual cones as well. Therefore, we introduce the following elementary fact on the regularity of dual cones.

Fact 16.6 *(i) A cone $\mathbf{K} \subseteq \mathbf{R}^n$ is regular if and only if its dual cone $\mathbf{K}_* = \{y \in \mathbf{R}^n : y^\top x \geq 0, \ \forall x \in \mathbf{K}\}$ is regular.*
(ii) Given regular cones $\mathbf{K}_1, \ldots, \mathbf{K}_m$, their direct product $\mathbf{K}_1 \times \cdots \times \mathbf{K}_m$ is also regular.

There are a number of "magic cones" that are regular and play a crucial role in convex optimization. In particular, many convex optimization problems in practice can be posed as optimization problems involving domains expressed using these cones as the basic building blocks.

Fact 16.7 *The following cones (see the examples discussed in section 1.2.4) are regular:*

(i) A nonnegative ray, $\mathbf{R}_+$.
(ii) A Lorentz (a.k.a., second-order, or ice-cream) cone,
$$\mathbf{L}^n = \{x \in \mathbf{R}^n : x_n \geq \sqrt{x_1^2 + \cdots + x_{n-1}^2}\} \ (\mathbf{L}^1 := \mathbf{R}_+).$$
(iii) A positive semidefinite cone, $\mathbf{S}_+^n = \{X \in \mathbf{S}^n : a^\top X a \geq 0, \ \forall a \in \mathbf{R}^n\}$.

Our developments will be based on an important concept, which we introduce now.

Definition 16.8 *[Cone-convexity] Let $\mathbf{K} \subset \mathbf{R}^\nu$ be a regular cone. A map $h(\cdot) : \mathrm{Dom}\, h \to \mathbf{R}^\nu$ is called $\mathbf{K}$-convex if $\mathrm{Dom}\, h$ is a convex set in some $\mathbf{R}^n$ and for every $x, y \in \mathrm{Dom}\, h$ and $\lambda \in [0, 1]$ we have*

$$\lambda h(x) + (1-\lambda)h(y) - h(\lambda x + (1-\lambda)y) \in \mathbf{K}. \tag{16.5}$$

Note that in the simplest case, that of $\mathbf{K} = \mathbf{R}_+^\nu$ (the nonnegative orthant is a regular cone!), the $\mathbf{K}$-convexity of a map h means exactly that the components of h are convex functions with common domain.

An instructive example of a "genuine cone-convex" function is as follows:

Lemma 16.9 *Let $\mathbf{K} = \mathbf{S}_+^m$, and consider $h : \mathbf{R}^{m \times n} \to \mathbf{S}^m$ defined as $h(x) = xx^\top$. Then h is $\mathbf{K}$-convex.*

Proof Indeed, for any $x, y \in \mathbf{R}^{m \times n}$ and $\lambda \in [0, 1]$, we have

$$(\lambda x + (1-\lambda)y)(\lambda x + (1-\lambda)y)^\top = \lambda x x^\top + (1-\lambda) y y^\top - \lambda(1-\lambda)(x-y)(x-y)^\top.$$

Therefore,

$$\lambda h(x) + (1-\lambda)h(y) - h(\lambda x + (1-\lambda)y) = \underbrace{\lambda(1-\lambda)}_{\geq 0} \underbrace{(x-y)(x-y)^\top}_{\geq 0} \succeq 0. \qquad \blacksquare$$

See Chapter 20 for other instructive examples of $\mathbf{K}$-convex functions and their "calculus."

Indeed, $\mathbf{K}$-convexity can be expressed in terms of the usual convexity owing to the following immediate observation.

Fact 16.10 *Let $\mathbf{K}$ be a regular cone in $\mathbf{R}^\nu$, $Z \subseteq \mathbf{R}^n$ be a nonempty convex set, and $h : Z \to \mathbf{R}^\nu$ be a mapping with $\mathrm{Dom}\, h = Z$. Then, h is $\mathbf{K}$-convex if and only if for all $\mu \in \mathbf{K}_*$ (where $\mathbf{K}_*$ is the cone dual to $\mathbf{K}$) we have that the real-valued functions $\mu^\top h(\cdot)$ are convex on Z.*

Given a regular cone $\mathbf{K} \subset \mathbf{R}^\nu$, we can associate with it the $\mathbf{K}$-*inequality between vectors of* $\mathbf{R}^\nu$: we say that $a \in \mathbf{R}^\nu$ is $\mathbf{K}$-greater than or equal to $b \in \mathbf{R}^\nu$ (notation: $a \geq_\mathbf{K} b$, or, equivalently, $b \leq_\mathbf{K} a$) when $a - b \in \mathbf{K}$:

$$a \geq_\mathbf{K} b \quad \iff \quad b \leq_\mathbf{K} a \quad \iff \quad a - b \in \mathbf{K}.$$

For example, when $\mathbf{K} = \mathbf{R}_+^\nu$ is a nonnegative orthant, $\geq_\mathbf{K}$ is the standard coordinate-wise vector inequality $\geq$. That is, $a \geq b$ means that every entry of a is greater than or equal to, in the standard arithmetic sense, the corresponding entry in b. A $\mathbf{K}$-vector inequality possesses all the algebraic properties of $\geq$.

Fact 16.11 *Let $\mathbf{K} \subset \mathbf{R}^\nu$ be a regular cone. Then, any $\mathbf{K}$-inequality $a \geq_\mathbf{K} b$ satisfies all the following properties:*

(i) *It is a partial order on $\mathbf{R}^\nu$, i.e., the relation $a \geq_\mathbf{K} b$ is*
  - *reflexive: $a \geq_\mathbf{K} a$ for all a;*
  - *anti-symmetric: $a \geq_\mathbf{K} b$ and $b \geq_\mathbf{K} a$ if and only if $a = b$;*
  - *transitive: if $a \geq_\mathbf{K} b$ and $b \geq_\mathbf{K} c$, then $a \geq_\mathbf{K} c$.*

(ii) *It is compatible with linear operations, i.e., $\geq_\mathbf{K}$-inequalities can be*
  - *summed: if $a \geq_\mathbf{K} b$ and $c \geq_\mathbf{K} d$, then $a + c \geq_\mathbf{K} b + d$;*
  - *multiplied by nonnegative reals: if $a \geq_\mathbf{K} b$ and λ is a nonnegative real, then $\lambda a \geq_\mathbf{K} \lambda b$.*

16.3 ★ Convex Theorem of the Alternative – Cone-Constrained Form

(iii) It is compatible with convergence, i.e., one can "pass to the sidewise limits" (in which we independently take the limits of both sides of a family of inequalities) in the $\geq_{\mathbf{K}}$-inequality:
- if $a_t \geq_{\mathbf{K}} b_t$, $t = 1, 2, \ldots$, and $a_t \to a$ and $b_t \to b$ as $t \to \infty$, then $a \geq_{\mathbf{K}} b$.

(iv) It gives rise to a strict version $>_{\mathbf{K}}$ of the $\geq_{\mathbf{K}}$-inequality $a >_{\mathbf{K}} b$ (equivalently: $b <_{\mathbf{K}} a$), meaning that $a - b \in \mathrm{int}\,\mathbf{K}$. The strict $\mathbf{K}$-inequality possesses the basic properties of the coordinate-wise $>$, specifically,
- $>_{\mathbf{K}}$ is stable: if $a >_{\mathbf{K}} b$ and a', b' are close enough to a, b respectively, then $a' >_{\mathbf{K}} b'$;
- if $a >_{\mathbf{K}} b$, λ is a positive real, and $c \geq_{\mathbf{K}} d$, then $\lambda a >_{\mathbf{K}} \lambda b$ and $a + c >_{\mathbf{K}} b + d$.

In summary, the arithmetics of $\geq_{\mathbf{K}}$ and $>_{\mathbf{K}}$ inequalities is completely similar to that of the usual $\geq$ and $>$ inequalities. Verification of the claims made in Fact 16.11 is immediate and is left to the reader.

In the standard approach to nonlinear convex optimization, a mathematical programming problem that is convex has the following form:

$$\min_{x \in X}\{f(x) : g(x) := [g_1(x); \ldots; g_m(x)] \leq 0,\ [h_1(x); \ldots; h_k(x)] = 0\},$$

where X is a convex set, $f(x), g_i(x) : X \to \mathbf{R}$, $1 \leq i \leq m$, are convex functions, and $h_j(x)$, $1 \leq j \leq k$, are affine functions. In this form, the nonlinearity "sits" in the functions g_i and/or the non-polyhedrality of the set X. Since the 1990s, it has been realized that, along with this form, it is extremely convenient to consider the *conic form* where the nonlinearity "sits" in the inequality relations $\leq$. That is, the usual coordinate-wise $\leq$ is replaced with $\leq_{\mathbf{K}}$, where $\mathbf{K}$ is a regular cone. The resulting *convex program in cone-constrained form* reads

$$\min_{x \in X}\left\{f(x) : \overline{g}(x) := Ax - b \leq 0,\ \underbrace{\widehat{g}(x) \leq_{\mathbf{K}} 0}_{\iff \widehat{g}(x) \in -\mathbf{K}}\right\}, \tag{16.6}$$

where $X \subseteq \mathbf{R}^n$ is a convex set, $f : X \to \mathbf{R}$ is a convex function, $\mathbf{K} \subset \mathbf{R}^\nu$ is a regular cone, $A \in \mathbf{R}^{k \times n}$, and $\widehat{g} : X \to \mathbf{R}^\nu$ is $\mathbf{K}$-convex. Note that the $\mathbf{K}$-convexity of $\widehat{g}$ in our new notation is simply equivalent to the requirement

$$\widehat{g}(\lambda x + (1 - \lambda)y)) \leq_{\mathbf{K}} \lambda \widehat{g}(x) + (1 - \lambda)\widehat{g}(y), \qquad \forall x, y \in X,\ \forall \lambda \in [0, 1].$$

Indeed, when $\mathbf{K}$ is the nonnegative orthant, (16.6) recovers the mathematical programming form (16.1) of a convex problem.

It turns out that with the "cone-constrained approach" to convex programming, we lose nothing when restricting ourselves to $X = \mathbf{R}^n$, linear $f(x)$, and affine $\widehat{g}(x)$; this specific version of (16.6) is called the "conic problem" (to be considered in more detail later). That said, it makes sense to speak about a "less extreme" form of a convex program, specifically, the one presented in (16.6); we call problems of this form "convex problems in cone-constrained form," reserving the words "conic problems" for problems (16.6) with $X = \mathbf{R}^n$, linear f, and affine $\widehat{g}$; see section 18.4.

The developments in section 16.2 can be naturally extended to the cone-constrained case as follows. Let $X \subseteq \mathbf{R}^n$ be a nonempty convex set, $\mathbf{K} \subset \mathbf{R}^\nu$ be a regular cone, f be a real-valued function on X, and $\widehat{g}(\cdot)$ be a mapping from X into $\mathbf{R}^\nu$. Instead of the feasibility or infeasibility of system (I) we can speak about the feasibility or infeasibility of a system of constraints,

$$\begin{aligned} f(x) &< c \\ \overline{g}(x) := Ax - b &\leq 0 \\ \widehat{g}(x) &\leq_{\mathbf{K}} 0 \quad [\iff \widehat{g}(x) \in -\mathbf{K}] \\ x &\in X \end{aligned} \quad \text{(ConI)}$$

in variables x, i.e., this is the cone-constrained analogy to the inequality-constrained system (I). We call system (ConI) *convex*, if, in addition to the already assumed convexity of X, the function f is convex on X, and the map $\widehat{g}$ is $\mathbf{K}$-convex on X.

Denoting by $\mathbf{K}_*$ the cone dual to $\mathbf{K}$, a *sufficient* condition for the infeasibility of (ConI) is the feasibility of the following system of constraints,

$$\begin{aligned} \inf_{x \in X} \left[f(x) + \overline{\lambda}^\top \overline{g}(x) + \widehat{\lambda}^\top \widehat{g}(x) \right] &\geq c \\ \overline{\lambda} &\geq 0 \\ \widehat{\lambda} &\geq_{\mathbf{K}_*} 0 \quad [\iff \widehat{\lambda} \in \mathbf{K}_*] \end{aligned} \quad \text{(ConII)}$$

in variables $\lambda = (\overline{\lambda}, \widehat{\lambda})$. Indeed, given a feasible solution $\overline{\lambda}, \widehat{\lambda}$ to (ConII) and "aggregating" the constraints in (ConI) with weights $1, \overline{\lambda}, \widehat{\lambda}$ (i.e., taking sidewise inner products of constraints in (ConI) based on these aggregation weights and summing the results), we arrive at the inequality

$$f(x) + \overline{\lambda}^\top \overline{g}(x) + \widehat{\lambda}^\top \widehat{g}(x) < c$$

which, since $\overline{\lambda} \geq 0$ and $\widehat{\lambda} \in \mathbf{K}_*$, is a consequence of (ConI) – must be satisfied at every feasible solution to (ConI). On the other hand, the aggregated inequality contradicts the first constraint in (ConII), and we conclude that under the circumstances (ConI) is infeasible.

The "cone-constrained" version of the Slater/relaxed Slater condition is as follows:

Definition 16.12 *[Cone-constrained Slater/relaxed Slater condition] We say that the system of constraints*

$$\overline{g}(x) := Ax - b \leq 0, \quad \widehat{g}(x) \leq_{\mathbf{K}} 0 \quad \text{(S)}$$

in variables x satisfies

(i) *the Slater condition on X, if there exists $\bar{x} \in \mathrm{rint}\, X$ such that $\overline{g}(\bar{x}) < 0$ and $\widehat{g}(\bar{x}) <_{\mathbf{K}} 0$ (i.e., $\widehat{g}(\bar{x}) \in -\mathrm{int}\, \mathbf{K}$),*
(ii) *the relaxed Slater condition on X, if there exists $\bar{x} \in \mathrm{rint}\, X$ such that $\overline{g}(\bar{x}) \leq 0$ and $\widehat{g}(\bar{x}) <_{\mathbf{K}} 0$.*

We say that an optimization problem in cone-constrained form, i.e., a problem of the form

$$\min_{x \in X} \{ f(x) : \overline{g}(x) := Ax - b \leq 0, \widehat{g}(x) \leq_{\mathbf{K}} 0 \} \quad \text{(ConeIC)}$$

satisfies the Slater/relaxed Slater condition, if the system of its constraints satisfies this condition on X.

Note that in the case $\mathbf{K} = \mathbf{R}_+^\nu$, (ConI) and (ConII) become, respectively, (I) and (II), and the cone-constrained versions of the Slater/relaxed Slater condition become the usual ones.

16.3 ★ Convex Theorem of the Alternative – Cone-Constrained Form

We are now ready to state the *Convex Theorem of the Alternative in cone-constrained form*, dealing with a convex cone-constrained system (ConI), and obtain Theorem 16.4 as a particular case of this result.

Theorem 16.13 *[Convex Theorem of the Alternative in cone-constrained form] Let $\mathbf{K} \subset \mathbf{R}^\nu$ be a regular cone, let $X \subseteq \mathbf{R}^n$ be nonempty and convex, let f be a real-valued convex function on X, $\overline{g}(x) = Ax - b$ be affine, and $\widehat{g}(x) : X \to \mathbf{R}^\nu$ be $\mathbf{K}$-convex. Suppose that system (S) satisfies the cone-constrained relaxed Slater condition on X. Then, the system (ConI) is feasible if and only if the system (ConII) is infeasible.*

Note. In some cases (ConI) may have no affine (i.e., polyhedral) part $\overline{g}(x) := Ax - b \leq 0$ and/or no "general part" $\widehat{g}(x) \leq_{\mathbf{K}} 0$; the absence of one or both of these parts leads to self-evident modifications in (ConII). To unify our forthcoming considerations, it is convenient to assume that both these parts are present. This assumption is "for free": it can immediately be seen that in our context, in the absence of one or both g-constraints in (ConI) we lose nothing when adding an artificial affine part $\overline{g}(x) := 0^\top x - 1 \leq 0$ instead of the missing affine part, and/or an artificial general part $\widehat{g}(x) := 0^\top x - 1 \leq_{\mathbf{K}} 0$ with $\mathbf{K} = \mathbf{R}_+$, instead of the missing general part. Thus, we lose nothing when assuming from the very beginning that both polyhedral and general parts are present.

It can be seen that the Convex Theorem of the Alternative (Theorem 16.4) is the special case of Theorem 16.13 when $\mathbf{K}$ is a nonnegative orthant.

In the proof of Theorem 16.13, we will use Lemma 16.14, a generalization of the Inhomogeneous Farkas' Lemma. We will present the proof of this result after the proof of Theorem 16.13.

Lemma 16.14 *Let $X \subseteq \mathbf{R}^n$ be a convex set with $0 \in \operatorname{int} X$. Let $g(x) = Ax + a : \mathbf{R}^n \to \mathbf{R}^k$ and $d^\top x + \delta : \mathbf{R}^n \to \mathbf{R}$ be affine functions such that $g(0) \leq 0$ and*

$$x \in X, \; g(x) \leq 0 \implies d^\top x + \delta \leq 0.$$

Then, there exists a vector $\mu \geq 0$ such that

$$d^\top x + \delta \leq \mu^\top g(x), \; \forall x \in X.$$

Proof of Theorem 16.13 The first part of the statement – "if (ConII) has a solution, then (ConI) has no solutions" – has been already verified. What we need is to prove the reverse statement. Thus, let us assume that (ConI) has no solutions, and let us prove that then (ConII) has a solution.

(1) Without loss of generality we may assume that X is full-dimensional: $\operatorname{rint} X = \operatorname{int} X$ (indeed, otherwise we can replace our "universe" $\mathbf{R}^n$ with the affine span of X). Besides this, if needed, by shifting f by a constant, we can assume that $c = 0$. Thus, we are in the case where

$$\left. \begin{array}{rl} f(x) & < \; 0, \\ \overline{g}(x) := Ax - b & \leq \; 0, \\ \widehat{g}(x) & \leq_{\mathbf{K}} 0, \quad [\iff \widehat{g}(x) \in -\mathbf{K}] \\ x & \in \; X; \end{array} \right\} \qquad \text{(ConI)}$$

$$\left.\begin{array}{rl} \inf\limits_{x \in X}\left[f(x) + \overline{\lambda}^\top \overline{g}(x) + \widehat{\lambda}^\top \widehat{g}(x)\right] &\geq 0, \\ \overline{\lambda} &\geq 0, \\ \widehat{\lambda} \geq_{\mathbf{K}_*} 0 \; [&\Longleftrightarrow \; \widehat{\lambda} \in \mathbf{K}_*]. \end{array}\right\} \quad \text{(ConII)}$$

Moreover, because the system satisfies the cone-constrained version of the relaxed Slater condition on X, there exists some $\bar{x} \in \operatorname{rint} X = \operatorname{int} X$ such that $\overline{g}(\bar{x}) \leq 0$ and $\widehat{g}(\bar{x}) \in -\operatorname{int} \mathbf{K}$. If necessary, by shifting X (that is, passing from variables x to variables $x - \bar{x}$; this clearly does not affect the statement we need to prove) we can assume that $\bar{x}$ is just the origin, so that we have

$$0 \in \operatorname{int} X, \quad \overline{g}(0) \leq 0, \quad \widehat{g}(0) <_{\mathbf{K}} 0. \tag{16.7}$$

Recall that we are in the situation when (ConI) is infeasible, that is, the optimization problem

$$\operatorname{Opt}(P) := \min_x \{f(x) : x \in X, \; \overline{g}(x) \leq 0, \; \widehat{g}(x) \leq_{\mathbf{K}} 0\} \tag{P}$$

satisfies $\operatorname{Opt}(P) \geq 0$, and our goal is to show that (ConII) is feasible.

(2) Define

$$Y := \{x \in X : \overline{g}(x) \leq 0\} = \{x \in X : Ax - b \leq 0\},$$

along with

$$S := \left\{t = [t_0; t_1] \in \mathbf{R} \times \mathbf{R}^\nu : t_0 < 0, \; t_1 \leq_{\mathbf{K}} 0\right\},$$
$$T := \left\{t = [t_0; t_1] \in \mathbf{R} \times \mathbf{R}^\nu : \exists x \in Y \text{ s.t. } f(x) \leq t_0, \; \widehat{g}(x) \leq_{\mathbf{K}} t_1\right\},$$

so that both sets S and T are nonempty (as (P) is feasible by (16.7)) and convex (since X, Y, and f are convex, and $\widehat{g}(x)$ is $\mathbf{K}$-convex on X). Moreover, S and T do not intersect since $\operatorname{Opt}(P) \geq 0$. Then, by the Separation Theorem (Theorem 6.3) S and T can be separated by an appropriately chosen linear form α. Thus,

$$\sup_{t \in S} \alpha^\top t \leq \inf_{t \in T} \alpha^\top t \tag{16.8}$$

and the linear form $\alpha^\top t$ is nonconstant on $S \cup T$, implying that $\alpha \neq 0$. Denote $\alpha = [\alpha_0; \alpha_1]$ with $\alpha_0 \in \mathbf{R}$ and $\alpha_1 \in \mathbf{R}^\nu$. We claim that $\alpha_0 \geq 0$ and $\alpha_1 \in \mathbf{K}_*$. Suppose this is not the case; then either $\alpha_0 < 0$ or $\alpha_1 \notin \mathbf{K}_*$ (i.e., there exists $\bar{\tau} \in \mathbf{K}$ such that $\alpha_1^\top \bar{\tau} < 0$) or both. But, in either of these cases we would have $\sup_{t \in S} \alpha^\top t = +\infty$ (look at what happens with $\alpha^\top t$ on the ray $\{t = [t_0; t_1] : t_0 \leq -1, t_1 = 0\} \subset S$ when $\alpha_0 < 0$, and on the ray $\{[t_0 = -1, t_1 = -s\bar{\tau}] : s \geq 0\} \subset S$ when $\alpha_1 \notin \mathbf{K}_*$ and $\bar{\tau} \in \mathbf{K}$ is such that $\alpha_1^\top \bar{\tau} < 0$) and this would contradict (16.8) as T is nonempty. Then, taking into account that $\alpha_0 \geq 0$ and $\alpha_1 \in \mathbf{K}_*$, we have $\sup_{t \in S} \alpha^\top t = 0$, and (16.8) reads

$$0 \leq \inf_{t \in T} \alpha^\top t. \tag{16.9}$$

(3) We now claim that $\alpha_0 > 0$. Note that the point $\bar{t} := [\bar{t}_0; \bar{t}_1]$ with the components $\bar{t}_0 := f(0)$ and $\bar{t}_1 := \widehat{g}(0)$ belongs to T (since $0 \in Y$ by (16.7)), thus by (16.9) it holds that $\alpha_0 f(0) + \alpha_1^\top \widehat{g}(0) \geq 0$. Assume for contradiction that $\alpha_0 = 0$. Then, we deduce $\alpha_1^\top \widehat{g}(0) \geq 0$, which due to $-\widehat{g}(0) \in \operatorname{int} \mathbf{K}$ (see (16.7)) and $\alpha_1 \in \mathbf{K}_*$ is possible only when $\alpha_1 = 0$; see Fact 6.28(iii). Thus, we conclude that $\alpha_1 = 0$ in addition to $\alpha_0 = 0$, which is impossible, since $\alpha \neq 0$.

Given that we have proved $\alpha_0 > 0$, in the sequel we set $\bar{\alpha}_1 := \alpha_1/\alpha_0$ and
$$h(x) := f(x) + \bar{\alpha}_1^\top \widehat{g}(x).$$

Note that $h(\cdot)$ is convex on X due to the convexity of X and f, the **K**-convexity of $\widehat{g}$, and the inclusion $\bar{\alpha}_1 \in \mathbf{K}_*$; see Fact 16.10.

Observe that (16.9) remains valid when we replace α with $\bar{\alpha} := \alpha/\alpha_0$. Moreover, when $x \in Y$, we have $[f(x); \widehat{g}(x)] \in T$ and thus from (16.9) and the definition of $h(x)$ we deduce that
$$x \in Y \implies h(x) \geq 0. \tag{16.10}$$

(4.i) Consider the convex sets
$$Q := \left\{[x; \tau] \in \mathbf{R}^n \times \mathbf{R} : x \in X, \overline{g}(x) := Ax - b \leq 0, \tau < 0\right\},$$
$$W := \left\{[x; \tau] \in \mathbf{R}^n \times \mathbf{R} : x \in X, \tau \geq h(x)\right\}.$$

These sets are clearly nonempty and do not intersect (since the x-component of a point from $Q \cap W$ would satisfy the premise but violate the conclusion in (16.10)). By the Separation Theorem, there exists $[e; \beta] \neq 0$ such that
$$\sup_{[x;\tau] \in Q} \left\{e^\top x + \beta\tau\right\} \leq \inf_{[x;\tau] \in W} \left\{e^\top x + \beta\tau\right\}.$$

As W is nonempty, we have $\inf_{[x;\tau] \in W}[e^\top x + \beta\tau] < +\infty$. Then, taking into account the definition of Q, we deduce that $\beta \geq 0$ (since otherwise the left-hand side in the preceding inequality would be $+\infty$). With $\beta \geq 0$ in mind and considering the definitions of Q and W, the preceding inequality reads
$$\sup_x \left\{e^\top x : x \in X, \overline{g}(x) \leq 0\right\} \leq \inf_x \left\{e^\top x + \beta h(x) : x \in X\right\}. \tag{16.11}$$

Let us define $a := \sup_x \left\{e^\top x : x \in X, \overline{g}(x) \leq 0\right\}$. Note that in (16.11), $\sup_x$ and $\inf_x$ are taken over nonempty sets, implying that $a \in \mathbf{R}$.

(4.ii) Recall that we saw that $\beta \geq 0$ in (16.11). We claim that in fact $\beta > 0$. Assume for contradiction that $\beta = 0$. Then, using the definition of a and $\beta = 0$, (16.11) implies
$$[e^\top x - a \leq 0, \forall(x \in X : \overline{g}(x) \leq 0)] \quad \text{and} \quad [e^\top x \geq a, \forall x \in X]. \tag{16.12}$$

Taking into account that $0 \in X$ and $\overline{g}(0) \leq 0$ by (16.7), the first relation in (16.12) says that $a \geq 0$. Then, as $a \geq 0$, from the second relation in (16.12) we deduce that $e^\top x \geq 0$ for $x \in X$. As $0 \in \text{int } X$, there exists a small enough $\varepsilon > 0$ such that $(0 - \varepsilon e) \in X$. Thus, from $e^\top(-\varepsilon e) \geq 0$, we deduce that $e = 0$. But, $e = 0$ is impossible, since we are in the case when $[e; \beta] \neq 0$ and $\beta = 0$.

(4.iii) Thus, in (16.11) we must have $\beta > 0$. Then, by replacing e with $\beta^{-1} e$, we can assume that (16.11) holds true with $\beta = 1$. Once again recalling that $a = \sup_x \left\{e^\top x : x \in X, \overline{g}(x) \leq 0\right\}$, the inequality (16.11) becomes
$$a \leq h(x) + e^\top x, \quad \forall x \in X. \tag{16.13}$$

By the definition of a, we also have
$$d^\top x + \delta := e^\top x - a \leq 0, \quad \forall(x \in X : \overline{g}(x) \leq 0),$$

so that the data X, $g(x) := \overline{g}(x)$ augmented with the affine function $d^\top x + \delta$ just defined satisfy the premise in Lemma 16.14 (recall that (16.7) holds true). Applying Lemma 16.14, we conclude that there exists $\mu \geq 0$ such that

$$e^\top x - a \leq \mu^\top \overline{g}(x), \quad \forall x \in X.$$

Combining this relation and (16.13), we conclude that

$$h(x) + \mu^\top \overline{g}(x) \geq 0, \quad \forall x \in X. \tag{16.14}$$

Recalling that $h(x) = f(x) + \bar{\alpha}_1^\top \widehat{g}(x)$ with $\bar{\alpha}_1 \in \mathbf{K}_*$ and setting $\overline{\lambda} = \mu$, $\widehat{\lambda} = \bar{\alpha}_1$, we get $\overline{\lambda} \geq 0$, $\widehat{\lambda} \in \mathbf{K}_*$ while by (16.14) it holds that

$$f(x) + \overline{\lambda}^\top \overline{g}(x) + \widehat{\lambda}^\top \widehat{g}(x) \geq 0, \quad \forall x \in X,$$

meaning that $\overline{\lambda}, \widehat{\lambda}$ solve (ConII) (recall that we are in the case $c = 0$). ∎

Proof of Lemma 16.14 Recall that $g(x) = Ax + a$. Let us define the following cones:

$$M_1 := \mathrm{cl}\left\{[x; t] \in \mathbf{R}^n \times \mathbf{R} : t > 0,\ x/t \in X\right\},$$
$$M_2 := \left\{y = [x; t] \in \mathbf{R}^n \times \mathbf{R} : Ax + ta \leq 0\right\}.$$

The cone M_2 is polyhedral, and the cone M_1 is closed (in fact it is the closed conic transform $\overline{\mathrm{ConeT}(X)}$ of X, see section 1.5) with a nonempty interior (since the point $[0; \ldots; 0; 1] \in \mathbf{R}^{n+1}$ belongs to $\mathrm{int}\, M_1$ owing to $0 \in \mathrm{int}\, X$). Moreover, $(\mathrm{int}\, M_1) \cap M_2 \neq \emptyset$ as the point $[0; \ldots; 0; 1] \in \mathrm{int}\, M_1$ belongs to M_2, since $g(0) \leq 0$.

Let $f := [d; \delta] \in \mathbf{R}^n \times \mathbf{R}$. We claim that the linear form $f^\top [x; t] = d^\top x + t\delta$ is nonpositive on $M_1 \cap M_2$. Indeed, consider any $y := [z; t] \in M_1 \cap M_2$, and define $y_s = [z_s; t_s] = (1-s)y + s[0; \ldots; 0; 1]$. Then, for $s = 0$ we have $y_s = y$ and it is in $M_1 \cap M_2$, and when $s = 1$ we have $y_s = [0; \ldots; 0; 1] \in M_1 \cap M_2$. As $M_1 \cap M_2$ is convex, we observe that $y_s \in M_1 \cap M_2$ for $0 \leq s \leq 1$. Now, consider the case when $0 < s \leq 1$. Then, we have $t_s > 0$ (since $s > 0$ and also $t \geq 0$ since $y \in M_1$), and $y_s \in M_1$ implies that $w_s := z_s/t_s \in \mathrm{cl}\, X$ (why?), while $y_s \in M_2$ along with $t_s > 0$ implies that $g(w_s) \leq 0$. Since $0 \in \mathrm{int}\, X$, $w_s \in \mathrm{cl}\, X$ implies that $\theta w_s \in X$ for all $\theta \in [0, 1)$, while $g(0) \leq 0$ along with $g(w_s) \leq 0$ implies that $g(\theta w_s) \leq 0$ for all $\theta \in [0, 1)$. Then, by invoking the premise of the lemma, we conclude that

$$d^\top (\theta w_s) + \delta \leq 0, \quad \forall \theta \in [0, 1).$$

Hence, whenever $0 < s \leq 1$, we have $d^\top w_s + \delta \leq 0$, or, equivalently, $f^\top y_s \leq 0$. As $s \to +0$, we have $f^\top y_s \to f^\top y = f^\top [z; t]$, implying that $f^\top [z; t] \leq 0$, as claimed.

Therefore, we have shown that $f^\top [x; t] = d^\top x + t\delta \leq 0$ for every $[x; t] \in (M_1 \cap M_2)$. That is, $(-f) \in (M_1 \cap M_2)_*$, where, as usual, M_* is the cone dual to the cone M. To summarize, we have that M_1, M_2 are closed cones such that $(\mathrm{int}\, M_1) \cap M_2 \neq \emptyset$ and $(-f) \in (M_1 \cap M_2)_*$. Applying to M_1, M_2 the Dubovitski–Milutin Lemma (Proposition 6.32), we conclude that $(M_1 \cap M_2)_* = (M_1)_* + (M_2)_*$. Since $(-f) \in (M_1 \cap M_2)_*$, there exist $\psi \in (M_1)_*$ and $\phi \in (M_2)_*$ such that $f = [d; \delta] = -\phi - \psi$. The inclusion $\phi \in (M_2)_*$ means that the homogeneous linear inequality $\phi^\top y \geq 0$ in variables $y \in \mathbf{R}^{n+1}$ is a consequence of the system of homogeneous linear inequalities given by $[A, a] y \leq 0$. Hence, by the Homogeneous Farkas' Lemma (Lemma 4.1), $-\phi$ is a conic combination of the transposes of the rows of the matrix $[A, a]$, so that $\phi^\top [x; 1] = -\mu^\top g(x)$ for some nonnegative μ and all $x \in \mathbf{R}^n$. Thus, for all $x \in \mathbf{R}^n$, we deduce that

$$d^\top x + \delta = [d; \delta]^\top [x; 1] = f^\top [x; 1] = -\phi^\top [x; 1] - \psi^\top [x; 1] = \mu^\top g(x) - \psi^\top [x; 1].$$

Finally, note that $[x; 1] \in M_1$ whenever $x \in X$. Then, as $\psi \in (M_1)_*$, we have $\psi^\top [x; 1] \geq 0$ for all $x \in X$. Thus, for all $x \in X$, we have $0 \leq \psi^\top [x; 1] = \mu^\top g(x) - (d^\top x + \delta)$ and so μ satisfies precisely the requirements stated in the lemma. ∎

To complete the story about the Convex Theorem of the Alternative, let us present an example which demonstrates that the relaxed Slater condition is crucial for the validity of Theorem 16.13.

Example 16.1 Consider the following special case of (ConI):

$$f(x) \equiv x < 0, \quad \overline{g}(x) \equiv 0 \leq 0, \quad \widehat{g}(x) \equiv x^2 \leq 0 \quad \text{(ConI)}$$

(here the embedding space is $\mathbf{R}$, $X = \mathbf{R}$, $c = 0$, and $\mathbf{K} = \mathbf{R}_+$, that is, this is just a system of scalar convex constraints). Here, the system (ConII) is the system of constraints

$$\inf_{x \in \mathbf{R}} \left[x + \overline{\lambda} \cdot 0 + \widehat{\lambda} x^2 \right] \geq 0, \quad \overline{\lambda} \geq 0, \quad \widehat{\lambda} \geq 0 \quad \text{(ConII)}$$

on variables $\overline{\lambda}, \widehat{\lambda} \in \mathbf{R}$.

The system (ConI) is clearly infeasible. The system (ConII) is infeasible as well – it can immediately be seen that whenever $\overline{\lambda}$ and $\widehat{\lambda}$ are nonnegative, the quantity $x + \overline{\lambda} \cdot 0 + \widehat{\lambda} x^2$ is negative for all small-in-magnitude $x < 0$, that is, the first inequality in (ConII) is incompatible with the remaining two inequalities of the system.

Note that in this example the only missing component of the premise in Theorem 16.13 is the relaxed Slater condition. Let us now examine what happens when we replace the constraint $\widehat{g}(x) \equiv x^2 \leq 0$ with $\widehat{g}(x) \equiv x^2 - 2\epsilon x \leq 0$, where $\epsilon > 0$. In this case, (ConI) remains infeasible, and we gain the validity of the relaxed (relaxed, not plain!) Slater condition. Then, as all the conditions of the Convex Theorem of the Alternative are now met, we deduce that (ConII), which now reads

$$\inf_x \left[x + \overline{\lambda} \cdot 0 + \widehat{\lambda}(x^2 - 2\epsilon x) \right] \geq 0, \quad \overline{\lambda} \geq 0, \quad \widehat{\lambda} \geq 0,$$

must be feasible. In fact, $\overline{\lambda} = 0, \widehat{\lambda} = 1/(2\epsilon)$ is a feasible solution to (ConII). ◇

16.4 Proofs of Facts

Fact 16.6

(i) A cone $\mathbf{K} \subseteq \mathbf{R}^n$ is regular if and only if its dual cone $\mathbf{K}_* = \{y \in \mathbf{R}^n : y^\top x \geq 0, \forall x \in \mathbf{K}\}$ is regular.
(ii) Given regular cones $\mathbf{K}_1, \ldots, \mathbf{K}_m$, their direct product $\mathbf{K}_1 \times \cdots \times \mathbf{K}_m$ is also regular.

Proof (i): This is an immediate consequence of Fact 6.28.
(ii): This is evident. ∎

Fact 16.7 *The following cones* (see the examples discussed in section 1.2.4) *are regular:*

(i) *A nonnegative ray,* $\mathbf{R}_+$.
(ii) *The Lorentz (a.k.a., second-order, or ice-cream) cone,* $\mathbf{L}^n = \{x \in \mathbf{R}^n : x_n \geq \sqrt{x_1^2 + \cdots + x_{n-1}^2}\}$ ($\mathbf{L}^1 := \mathbf{R}_+$).
(iii) *A positive semidefinite cone,* $\mathbf{S}_+^n = \{X \in \mathbf{S}^n : a^\top X a \geq 0, \forall a \in \mathbf{R}^n\}$.

Proof The regularity of the above cones is self-evident. ∎

Fact 16.10 *Let* $\mathbf{K}$ *be a regular cone in* $\mathbf{R}^\nu$, $Z \subseteq \mathbf{R}^n$ *be a nonempty convex set and* $h : Z \to \mathbf{R}^\nu$ *be a mapping with* $\mathrm{Dom}\, h = Z$. *Then, h is $\mathbf{K}$-convex if and only if for all* $\mu \in \mathbf{K}_*$ *(where $\mathbf{K}_*$ is the cone dual to $\mathbf{K}$) we have that the real-valued functions $\mu^\top h(\cdot)$ are convex on Z.*

Proof Since $\mathbf{K}$ is a closed cone, it is the cone dual to $\mathbf{K}_*$; that is, (16.5) holds for all $x, y \in Z$ and $\lambda \in [0, 1]$ if and only if

$$\mu^\top h(\lambda x + (1 - \lambda)y) \leq \lambda \mu^\top h(x) + (1 - \lambda)\mu^\top h(y), \quad \forall (x, y \in Z, \lambda \in [0, 1], \mu \in \mathbf{K}_*).$$

∎

17

Lagrange Function and Lagrange Duality

17.1 Lagrange Function

The Convex Theorem of the Alternative brings to our attention the function

$$\underline{L}(\lambda) := \inf_{x \in X} \left[f(x) + \sum_{j=1}^{m} \lambda_j g_j(x) \right], \tag{17.1}$$

and the related aggregate function

$$L(x, \lambda) := f(x) + \sum_{j=1}^{m} \lambda_j g_j(x) \tag{17.2}$$

from which $\underline{L}(\lambda)$ originates. The aggregate function in (17.2) is called the *Lagrange* (or *Lagrangian*) *function* of the inequality-constrained optimization program

$$\min \{ f(x) : g_j(x) \leq 0, \ j = 1, \ldots, m, \ x \in X \} \\ \left[\text{where } f, g_1, \ldots, g_m \text{ are real-valued functions on } X \right]. \tag{IC}$$

The Lagrange function $L(x, \lambda)$ of an optimization problem is a very important entity as most optimality conditions are expressed in terms of this function. Let us start with translating our developments from section 16.2 to the language of the Lagrange function.

17.2 Convex Programming Duality Theorem

We start by developing the duality theorem for convex optimization problems.

Theorem 17.1 *Consider an arbitrary inequality-constrained optimization program*

$$\min \{ f(x) : g_j(x) \leq 0, \ j = 1, \ldots, m, \ x \in X \} \\ \left[\text{where } f, g_1, \ldots, g_m \text{ are real-valued functions on } X \right] \tag{IC}$$

along with its Lagrange function

$$L(x, \lambda) := f(x) + \sum_{i=1}^{m} \lambda_i g_i(x) : X \times \mathbf{R}_+^m \to \mathbf{R}.$$

Then,

(i) [Weak Duality] For every $\lambda \geq 0$, the infimum of the Lagrange function in $x \in X$, that is,

$$\underline{L}(\lambda) := \inf_{x \in X} L(x, \lambda),$$

is a lower bound on the optimal value of (IC), *so that the optimal value of the optimization problem*

$$\sup_{\lambda \geq 0} \underline{L}(\lambda) \tag{IC*}$$

also is a lower bound for the optimal value in (IC).

(ii) [Convex Duality Theorem] *If the problem* (IC)
- *is convex,*
- *is below bounded, and*
- *satisfies the relaxed Slater condition,*

then the optimal value of (IC*) *is attained and is equal to the optimal value in* (IC).

Proof Let c^* be the optimal value of (IC).

(i): This part is just Proposition 16.1 (why?). It makes sense, however, to repeat here the corresponding reasoning:

Consider any $\lambda \in \mathbf{R}_+^m$. Then, by definition of the Lagrange function, for any x that is feasible for (IC) we have

$$L(x, \lambda) = f(x) + \sum_{j=1}^m \lambda_j g_j(x) \leq f(x),$$

where the inequality follows from the facts that $\lambda \in \mathbf{R}_+^m$ and that the feasibility of x for (IC) implies that $g_j(x) \leq 0$ for all $j = 1, \ldots, m$. Then, we immediately arrive at

$$\underline{L}(\lambda) = \inf_{x \in X} L(x, \lambda) \leq \inf_{x \in X: g(x) \leq 0} L(x, \lambda) \leq \inf_{x \in X: g(x) \leq 0} f(x) = c^*,$$

as desired.

(ii): This part is an immediate consequence of the Convex Theorem of the Alternative. Note that the system

$$f(x) < c^*, \quad g_j(x) \leq 0, \ j = 1, \ldots, m,$$

has no solutions in X, and by Theorem 16.4, the system (II) associated with $c = c^*$ has a solution, i.e., there exists $\lambda^* \geq 0$ such that $\underline{L}(\lambda^*) \geq c^*$. But, we know from part (i) that strict inequality is impossible here and, moreover, that $\underline{L}(\lambda) \leq c^*$ for every $\lambda \geq 0$. Thus, $\underline{L}(\lambda^*) = c^*$ and λ^* is a maximizer of $\underline{L}$ over $\lambda \geq 0$. ∎

17.3 Lagrange Duality and Saddle Points

Theorem 17.1 establishes a certain connection between two optimization problems, i.e., the "primal" problem

$$\begin{array}{l} \min_x \{f(x) : g_j(x) \leq 0, \ j = 1, \ldots, m, \ x \in X\} \\ \left[\text{where } f, g_1, \ldots, g_m \text{ are real-valued functions on } X\right], \end{array} \tag{IC}$$

and its *Lagrange dual problem*

$$\begin{array}{l} \max_\lambda \{\underline{L}(\lambda) : \lambda \in \mathbf{R}_+^m\} \\ \left[\text{where } \underline{L}(\lambda) = \inf_{x \in X} L(x, \lambda) \text{ and } L(x, \lambda) = f(x) + \sum_{i=1}^m \lambda_i g_i(x)\right]. \end{array} \tag{IC*}$$

17.3 Lagrange Duality and Saddle Points

Here, the variables λ of the dual problem are called the *Lagrange multipliers* of the primal problem. Theorem 17.1 states that the optimal value of the dual problem is at most that of the primal problem, and under some favorable circumstances (i.e., when the primal problem is convex, below bounded, and satisfies the relaxed Slater condition) the optimal values in this pair of problems are equal to each other.

In our formulation there may seem to be some asymmetry between the primal and the dual problems. In fact, both these problems are related to the Lagrange function in a quite symmetric way. Indeed, consider the problem

$$\min_{x \in X} \overline{L}(x), \quad \text{where } \overline{L}(x) := \sup_{\lambda \geq 0} L(x, \lambda).$$

By definition of the Lagrange function $L(x, \lambda)$, the function $\overline{L}(x)$ is clearly $+\infty$ at every point $x \in X$ which is not feasible for (IC) and is $f(x)$ on the feasible set of (IC), so that this problem is equivalent to (IC). We see that both the primal and the dual problems originate from the Lagrange function: in the primal problem, we *minimize* over X the result of the *maximization* of $L(x, \lambda)$ in $\lambda \geq 0$, i.e., the primal problem is

$$\min_{x \in X} \sup_{\lambda \in \mathbf{R}_+^m} L(x, \lambda),$$

and in the dual program we *maximize* over $\lambda \geq 0$ the result of the *minimization* of $L(x, \lambda)$ in $x \in X$, i.e., the dual problem is

$$\max_{\lambda \in \mathbf{R}_+^m} \inf_{x \in X} L(x, \lambda).$$

This is a particular (and the most important) example of a *two-person zero-sum game*, which we will explore later.

We have seen that under certain convexity and regularity assumptions the optimal values in (IC) and (IC*) are equal to each. There is also another way of stating when these optimal values are equal – this is always the case when the Lagrange function possesses a *saddle point*, i.e., there exists a pair $x^* \in X, \lambda^* \geq 0$ such that at the pair the function $L(x, \lambda)$ attains its minimum as a function of $x \in X$ and attains its maximum as a function of $\lambda \geq 0$:

$$L(x, \lambda^*) \geq L(x^*, \lambda^*) \geq L(x^*, \lambda), \quad \forall x \in X, \forall \lambda \geq 0.$$

This then leads to the following easily demonstrable fact (prove it from scratch or look at Theorem 22.2).

Proposition 17.2 *The primal–dual pair of solutions $(x^*, \lambda^*) \in X \times \mathbf{R}_+^k$ is a saddle point of the Lagrange function L of* (IC) *if and only if x^* is an optimal solution to* (IC), *λ^* is an optimal solution to* (IC*), *and the optimal values in the indicated problems are equal to each other.*

18

★ Convex Programming in Cone-Constrained Form

The results from sections 17.1 and 17.2 relating to convex optimization problems in the standard MP format admit instructive extensions to the case of convex problems in cone-constrained form. We next present these extensions.

18.1 Convex Problem in Cone-Constrained Form

The convex problem in cone-constrained form is an optimization problem of the form

$$\text{Opt}(P) = \min_{x \in X} \{f(x) : \overline{g}(x) \leq 0, \widehat{g}(x) \leq_K 0\}, \tag{P}$$

where $X \subseteq \mathbf{R}^n$ is a nonempty convex set, $f : X \to \mathbf{R}$ is a convex function, $\overline{g}(x) := Ax - b$ is an affine function from $\mathbf{R}^n$ to $\mathbf{R}^k$, $\mathbf{K} \subset \mathbf{R}^\nu$ is a regular cone, and $\widehat{g}(\cdot) : X \to \mathbf{R}^\nu$ is **K**-convex.

Example 18.1 The positive semidefinite cone $\mathbf{S}^n_+$ and the notation $A \succeq B$, $B \preceq A$, $A \succ B$, $B \prec A$ for the associated non-strict and strict conic inequalities are introduced in appendix section D.2.2. As we know from Fact 16.7 and Example 6.10, the cone $\mathbf{S}^n_+$ is regular and self-dual. Recall from Lemma 16.9 that the function from $\mathbf{S}^n$ to $\mathbf{S}^n$ given by $\widehat{g}(x) = xx^\top = xx = x^2$ is $\succeq$-convex. As a result, the problem

$$\begin{aligned}\text{Opt}(P) &= \min_{x=(t,y)\in\mathbf{R}\times\mathbf{S}^n} \{t : \text{Tr}(y) \leq t, \ y^2 \preceq B\} \\ &= \min_{x=(t,y)\in\mathbf{R}\times\mathbf{S}^n} \{t : \langle y, I_n \rangle - t \leq 0, \ y^2 - B \preceq 0\},\end{aligned} \tag{18.1}$$

where B is a positive definite matrix and $\langle \cdot, \cdot \rangle$ is the Frobenius inner product, is a convex program in cone-constrained form. ◇

18.2 Cone-Constrained Lagrange Function

Consider (P) with a convex objective f, a convex domain X, an affine map $\overline{g}(\cdot) : \mathbf{R}^n \to \mathbf{R}^k$, a regular cone $\mathbf{K} \subset \mathbf{R}^\nu$, and a **K**-convex function $\widehat{g}(\cdot) : X \to \mathbf{R}^\nu$. Let $\Lambda := \mathbf{R}^k_+ \times \mathbf{K}_*$, where $\mathbf{K}_*$ is the cone dual to $\mathbf{K}$, and consider $\lambda := [\overline{\lambda}; \widehat{\lambda}] \in \Lambda$. Then, the *cone-constrained Lagrange function* of (P) is defined as

$$L(x; \lambda) := f(x) + \overline{\lambda}^\top \overline{g}(x) + \widehat{\lambda}^\top \widehat{g}(x) : X \times \Lambda \to \mathbf{R}.$$

18.3 Convex Programming Duality Theorem in Cone-Constrained Form

By construction, for any $\lambda \in \Lambda$, we conclude that $L(x; \lambda)$ as a function of x underestimates $f(x)$ everywhere on the feasible domain of (P).

Example 18.2 (continued from Example 18.1) We see that the cone-constrained Lagrange function of (18.1) is given by

$$L(t, y; \overline{\lambda}, \widehat{\lambda}) = t + \overline{\lambda}[\mathrm{Tr}(y) - t] + \mathrm{Tr}(\widehat{\lambda}(y^2 - B)) : [\mathbf{R} \times \mathbf{S}^n] \times [\mathbf{R}_+ \times \mathbf{S}^n_+] \to \mathbf{R}.$$

$\diamond$

The *cone-constrained Lagrange dual* of (P) is the optimization problem

$$\mathrm{Opt}(D) := \max \left\{ \underline{L}(\lambda) : \lambda \in \Lambda \right\} \quad [\text{where } \underline{L}(\lambda) := \inf_{x \in X} L(x; \lambda)], \quad (D)$$

and where $\Lambda := \mathbf{R}^k_+ \times \mathbf{K}_*$. From $L(x; \lambda) \leq f(x)$ for all $x \in X$ and $\lambda \in \Lambda$, we clearly extract

$$\mathrm{Opt}(D) \leq \mathrm{Opt}(P). \quad [\text{Weak Duality}]$$

Note that Weak Duality is independent of any assumptions of convexity on f and X and on the **K**-convexity of $\widehat{g}$.

Example 18.3 (Continued from Example 18.1) It can immediately be seen (check it, or look at the solution to Exercise IV.22) that for the cone-constrained Lagrange function of (18.1) we have $\mathrm{Dom}(\underline{L}) = \{[\overline{\lambda}; \widehat{\lambda}] : \overline{\lambda} = 1, \widehat{\lambda} \succ 0\}$ and

$$\underline{L}(\overline{\lambda}, \widehat{\lambda}) = \begin{cases} -\frac{1}{4}\mathrm{Tr}(\widehat{\lambda}^{-1}) - \mathrm{Tr}(\widehat{\lambda} B), & \text{if } [\overline{\lambda}, \widehat{\lambda}] \in \mathrm{Dom}(\underline{L}), \\ -\infty, & \text{otherwise.} \end{cases} \quad (18.2)$$

Then, the cone-constrained Lagrange dual of (18.1) is the problem

$$\mathrm{Opt}(D) = \max_{\widehat{\lambda} \succ 0, \overline{\lambda} \geq 0} \left\{ -\frac{1}{4}\mathrm{Tr}(\widehat{\lambda}^{-1}) - \mathrm{Tr}(\widehat{\lambda} B) : \overline{\lambda} = 1 \right\}$$

$$= \max_{\widehat{\lambda} \succ 0} \left\{ -\frac{1}{4}\mathrm{Tr}(\widehat{\lambda}^{-1}) - \mathrm{Tr}(\widehat{\lambda} B) \right\}. \quad (18.3)$$

18.3 Convex Programming Duality Theorem in Cone-Constrained Form

For cone-constrained problems, we have the following strong duality theorem.

Theorem 18.1 *[Convex programming Duality Theorem in cone-constrained form] Consider the convex cone-constrained problem (P), that is, X is convex, f is real-valued and convex on X, and $\widehat{g}(\cdot)$ is well defined and **K**-convex on X. Assume that the problem is below bounded and satisfies the relaxed Slater condition. Then, (D) is solvable and $\mathrm{Opt}(P) = \mathrm{Opt}(D)$.*

Note that part (ii) of Theorem 17.1, the only nontrivial part of the theorem, is just the special case of Theorem 18.1 for which **K** is a nonnegative orthant.

Proof of Theorem 18.1 This proof is immediate. Under the premise of the theorem, $c := \mathrm{Opt}(P)$ is a real, and the system of constraints (ConI) associated with this c has no solutions.

The relaxed Slater condition along with the Convex Theorem of the Alternative in cone-constrained form (Theorem 16.13) imply the feasibility of (ConII), i.e., the existence of $\lambda_* = [\bar{\lambda}_*; \hat{\lambda}_*] \in \Lambda$ such that

$$\underline{L}(\lambda_*) = \inf_{x \in X} \left\{ f(x) + \bar{\lambda}_*^\top \bar{g}(x) + \hat{\lambda}_*^\top \hat{g}(x) \right\} \geq c = \mathrm{Opt}(P).$$

Thus, we deduce that (D) has a feasible solution with objective value $\geq \mathrm{Opt}(P)$, by Weak Duality, this value is exactly $\mathrm{Opt}(P)$, the solution in question is optimal for (D), and $\mathrm{Opt}(P) = \mathrm{Opt}(D)$. ∎

Example 18.4 (continued from Example 18.1) Problem (18.1) is clearly below bounded and satisfies the Slater condition (since $B \succ 0$). By Theorem 18.1 the dual problem (18.3) is solvable and has the same optimal value as (18.1). The solution for the (convex!) dual problem (18.3) can be found by applying the Fermat rule, which states that at a local minimizer of a differentiable function, the gradient should vanish. To this end, note also that for a positive definite $n \times n$ matrix y and $h \in \mathbf{S}^n$ it holds that

$$\frac{d}{dt}\Big|_{t=0}(y + th)^{-1} = -y^{-1}hy^{-1}$$

(why?). Then, the Fermat rule says that the optimal solution to (18.3) is

$$\bar{\lambda}_* = 1, \quad \hat{\lambda}_* = \frac{1}{2} B^{-1/2}$$

and $\mathrm{Opt}(P) = \mathrm{Opt}(D) = -\mathrm{Tr}(B^{1/2})$. ◇

18.3.1 "Subgradient Interpretation" of Lagrange Multipliers

Consider any $\Delta := [\bar{\delta}; \hat{\delta}] \in \mathbf{R}^k \times \mathbf{R}^v$, and define the maps $\bar{g}_\Delta(x) := Ax - b - \bar{\delta} = \bar{g}(x) - \bar{\delta}$ and $\hat{g}_\Delta(x) := \hat{g}(x) - \hat{\delta}$ along with the parametric family of convex cone-constrained problems (P_Δ) given by

$$\mathrm{Opt}(P_\Delta) := \min_{x \in X} \left\{ f(x) : \bar{g}_\Delta(x) \leq 0, \hat{g}_\Delta(x) \leq_K 0 \right\}$$
$$= \min_{x \in X} \left\{ f(x) : \bar{g}(x) - \bar{\delta} \leq 0, \hat{g}(x) - \hat{\delta} \leq_K 0 \right\}.$$

Notice that (P) is part of this family, as it is precisely (P_0).

The cone-constrained Lagrange duals of these problems (P_Δ) also form a parametric family. Specifically, by setting $\lambda = [\bar{\lambda}; \hat{\lambda}]$ and $\Lambda = \mathbf{R}_+^k \times \mathbf{K}_*$, we arrive at the cone-constrained Lagrange function of (P_Δ) as

$$L_\Delta(x, \lambda) := f(x) + \bar{\lambda}^\top (\bar{g}(x) - \bar{\delta}) + \hat{\lambda}^\top (\hat{g}(x) - \hat{\delta}),$$

and the resulting dual family of problems (D_Δ), given by

$$\mathrm{Opt}(D_\Delta) := \max_{\lambda \in \Lambda} \left\{ \underline{L}_\Delta(\lambda) \right\},$$

where

$$\underline{L}_\Delta(\lambda) := \inf_{x \in X} \left\{ L_\Delta(x, \lambda) \right\} = \inf_{x \in X} \left\{ f(x) + \bar{\lambda}^\top (\bar{g}(x) - \bar{\delta}) + \hat{\lambda}^\top (\hat{g}(x) - \hat{\delta}) \right\}.$$

Since $L_\Delta(x,\lambda) = L_0(x,\lambda) - \bar{\lambda}^\top \bar{\delta} - \widehat{\lambda}^\top \widehat{\delta}$, we deduce $\underline{L}_\Delta(\lambda) = \underline{L}_0(\lambda) - \bar{\lambda}^\top \bar{\delta} - \widehat{\lambda}^\top \widehat{\delta}$, where by definition $\underline{L}_0(\lambda) = \inf_{x \in X} \left\{ f(x) + \bar{\lambda}^\top \bar{g}(x) + \widehat{\lambda}^\top \widehat{g}(x) \right\}$. Thus,

$$\mathrm{Opt}(D_\Delta) = \max_{\lambda \in \Lambda} \left\{ \underline{L}_\Delta(\lambda) \right\} = \max_{\lambda \in \Lambda} \left\{ \underline{L}_0(\lambda) - \bar{\lambda}^\top \bar{\delta} - \widehat{\lambda}^\top \widehat{\delta} \right\}.$$

We have the following nice and instructive fact, which provides further insights into the optimum value sensitivity of these parametric families of problems.

Fact 18.2 *Consider the parametric family (P_Δ) of convex cone-constrained problems along with the family (D_Δ) of their cone-constrained Lagrange duals. Then,*

(i) *If $\underline{L}_0(\mu) > -\infty$ for some $\mu = [\bar{\mu}; \widehat{\mu}] \in \Lambda$ then the primal optimal value $\mathrm{Opt}(P_\Delta)$ takes values in $\mathbf{R} \cup \{+\infty\}$ and is a convex function of Δ.*
(ii) *If (D_0) is solvable with optimal solution $\lambda_* = [\bar{\lambda}_*; \widehat{\lambda}_*]$ and $\mathrm{Opt}(D_0) = \mathrm{Opt}(P_0)$ then $-\lambda_*$ is a subgradient of $\mathrm{Opt}(P_\Delta)$ at the point $\Delta = 0$, i.e.,*

$$\mathrm{Opt}(P_\Delta) \geq \mathrm{Opt}(P_0) - \bar{\lambda}_*^\top \bar{\delta} - \widehat{\lambda}_*^\top \widehat{\delta}, \qquad \forall (\Delta = [\bar{\delta}; \widehat{\delta}]).$$

The premises in (i) and (ii) hold when (P_0) satisfies the relaxed Slater condition and is below bounded.

Example 18.5 (continued from Example 18.1) Problem (18.1) can be embedded into the parametric family of problems

$$\mathrm{Opt}(P_R) := \min_{x=(t,y) \in \mathbf{R} \times \mathbf{S}^n} \left\{ t : \mathrm{Tr}(y) \leq t, \ y^2 \preceq B + R \right\} \qquad (P[R])$$

with R varying through $\mathbf{S}^n$. Taking into account all we have established so far for this problem and also considering Fact 18.2, we arrive at

$$\mathrm{Tr}((B+R)^{1/2}) = -\mathrm{Opt}(P_R) \leq \mathrm{Tr}(B^{1/2}) + \frac{1}{2}\mathrm{Tr}(B^{-1/2}R), \qquad \forall (R \succ -B). \quad (18.4)$$

Note that this is just the gradient inequality for the concave function (see Fact 14.6) $\mathrm{Tr}(X^{1/2}) : \mathbf{S}^n_+ \to \mathbf{R}$; see Fact D.24. ◇

18.4 Conic Programming and Conic Duality Theorem

In this section, we will consider a special case of convex problem in cone-constrained form, namely *conic programs* in today's optimization terminology. In conic programs, $f(x)$ is linear, X is the entire space, and $\widehat{g}(x) = Px - p$ is affine. Note that affine mapping is $\mathbf{K}$-convex for any cone $\mathbf{K}$. Thus, the conic problem automatically satisfies the convexity restrictions from the cone-constrained Convex Programming Duality Theorem (Theorem 18.1) and is an optimization problem of the form

$$\mathrm{Opt}(P) = \min_{x \in \mathbf{R}^n} \left\{ c^\top x : Ax - b \leq 0, \ Px - p \leq_\mathbf{K} 0 \right\}, \quad (18.5)$$

where $\mathbf{K}$ is a regular cone in certain $\mathbf{R}^\nu$.

The simplest example of a conic problem is an LP problem, where $\mathbf{K}$ is a nonnegative orthant. Another instructive example is the conic reformulation of the convex quadratic quadratically constrained problem. This example relies on the following useful observation.

Fact 18.3 Consider the convex quadratic constraint $x^\top A^\top A x \leq b^\top x + c$, where $A \in \mathbf{R}^{d \times n}$, $b \in \mathbf{R}^n$, and $c \in \mathbf{R}$. This constraint can be equivalently rewritten as a conic constraint involving the Lorentz cone, i.e.,

$$x^\top A^\top A x \leq b^\top x + c$$
$$\iff [2Ax;\ b^\top x + c - 1;\ b^\top x + c + 1] \in \mathbf{L}^{d+2}$$
$$\iff 4x^\top A^\top A x + (b^\top x + c - 1)^2 \leq (b^\top x + c + 1)^2 \text{ and } b^\top x + c + 1 \geq 0.$$

Fact 18.3 immediately leads to the following useful result.

Fact 18.4 Given $A_j \in \mathbf{R}^{d_j \times n}$, $b_j \in \mathbf{R}^n$, and $c_j \in \mathbf{R}$, for $0 \leq j \leq m$, the convex quadratic quadratically constrained optimization problem

$$\min_x \left\{ x^\top A_0^\top A_0 x + b_0^\top x + c_0 :\ x^\top A_j^\top A_j x + b_j^\top x + c_j \leq 0,\ 1 \leq j \leq m \right\}$$

is equivalent to the conic problem on the product of $m+1$ Lorentz cones; specifically, it admits the conic formulation given by

$$\min_{x,t} \left\{ t :\ \begin{array}{c} A[x;t] + b := [\alpha_0(x,t); \alpha_1(x); \ldots; \alpha_m(x)] \\ \in \mathbf{L}^{d_0+2} \times \mathbf{L}^{d_1+2} \times \cdots \times \mathbf{L}^{d_m+2} \end{array} \right\},$$

where $\alpha_0(x,t) := [2A_0 x;\ t - b_0^\top x - c_0 - 1;\ t - b_0^\top x - c_0 + 1]$ and $\alpha_j(x) := [2A_j x;\ -b_j^\top x - c_j - 1;\ -b_j^\top x - c_j + 1]$ for all $1 \leq j \leq m$.

The cone-constrained Lagrange dual problem of problem (18.5) reads

$$\max_{\bar{\lambda},\hat{\lambda}} \left[\inf_x \left\{ c^\top x + \bar{\lambda}^\top (Ax - b) + \hat{\lambda}^\top (Px - p) :\ \bar{\lambda} \geq 0,\ \hat{\lambda} \in \mathbf{K}_* \right\} \right],$$

which, as it can be seen, is just the problem

$$\mathrm{Opt}(D) = \max_{\bar{\lambda},\hat{\lambda}} \left\{ -b^\top \bar{\lambda} - p^\top \hat{\lambda} :\ A^\top \bar{\lambda} + P^\top \hat{\lambda} + c = 0,\ \bar{\lambda} \geq 0,\ \hat{\lambda} \in \mathbf{K}_* \right\}. \quad (D)$$

Recall that by Fact 16.6(i) the cone dual to a regular cone also is a regular cone. As a result, the problem (D), called the *conic dual* of the conic problem (18.5), also is a conic problem. An immediate computation (utilizing the fact that $(\mathbf{K}_*)_* = \mathbf{K}$ for every regular cone $\mathbf{K}$) shows that *conic duality is symmetric*.

Fact 18.5 *Conic duality is symmetric, i.e., the conic dual to conic problem (D) is (equivalent to) conic problem (P), i.e., (18.5).*

In view of primal–dual symmetry, the Convex Duality Theorem in cone-constrained form (Theorem 18.1) in the conic programming case takes the following nice form.

Theorem 18.6 *[Conic Duality Theorem] Consider a primal–dual pair of conic problems*

$$\mathrm{Opt}(P) := \min_{x \in \mathbf{R}^n} \left\{ c^\top x :\ Ax - b \leq 0,\ Px - x \leq_{\mathbf{K}} 0 \right\}, \quad (P)$$

$$\mathrm{Opt}(D) := \max_{\bar{\lambda},\hat{\lambda}} \left\{ -b^\top \bar{\lambda} - p^\top \hat{\lambda} :\ A^\top \bar{\lambda} + P^\top \hat{\lambda} + c = 0,\ \bar{\lambda} \geq 0,\ \hat{\lambda} \in \mathbf{K}_* \right\}. \quad (D)$$

Then, we always have $\text{Opt}(D) \leq \text{Opt}(P)$. Moreover, if one problem in the pair is bounded and satisfies the relaxed Slater condition, then the other problem in the pair is solvable, and $\text{Opt}(P) = \text{Opt}(D)$. Finally, if both of the problems satisfy the relaxed Slater condition, then both are solvable with equal optimal values.

Proof This proof is immediate. Weak duality has already been verified. To verify the second claim, note that by primal–dual symmetry we can assume that the bounded problem satisfying the relaxed Slater condition is (P). But, then the claim in question is given by Theorem 18.1. Finally, if both problems satisfy the relaxed Slater condition (and in particular are feasible), by Weak Duality, both are bounded, and therefore solvable with equal optimal values by the preceding claim. ∎

Application example: $\mathcal{S}$-Lemma The $\mathcal{S}$-Lemma is an extremely useful fact that has applications in optimization, engineering, and control.

Lemma 18.7 *[$\mathcal{S}$-Lemma] Let $A, B \in \mathbf{S}^n$ be such that*

$$\exists \bar{x} : \quad \bar{x}^\top A \bar{x} > 0. \tag{18.6}$$

Then, the implication

$$x^\top A x \geq 0 \quad \Longrightarrow \quad x^\top B x \geq 0 \tag{18.7}$$

holds if and only if

$$\exists \lambda \geq 0: \quad B \succeq \lambda A. \tag{18.8}$$

Note that the $\mathcal{S}$-Lemma is a statement of the same flavor as the Homogeneous Farkas' Lemma: the latter states that a homogeneous linear inequality $b^\top x \geq 0$ is a consequence of a system of homogeneous linear inequalities $a_i^\top x \geq 0$, $1 \leq i \leq k$, if and only if the target inequality can be obtained from the inequalities of the system by taking a weighted sum with nonnegative weights; we could add to "taking a weighted sum" "and adding an identically true homogeneous linear inequality." The simple reason for this is that there exists only one inequality of the latter type, $0^\top x \geq 0$. Similarly, the $\mathcal{S}$-Lemma says that (whenever (18.6) holds) homogeneous quadratic inequality $x^\top B x \geq 0$ is a consequence of a (single-inequality) system of homogeneous quadratic inequalities $x^\top A x \geq 0$ if and only if the target inequality can be obtained by taking a weighted sum, with nonnegative weights, of the inequalities of the system (that is, by taking a nonnegative multiple of the inequality $x^\top A x \geq 0$) and adding an identically true homogeneous quadratic inequality (there are plenty of them, these are inequalities $x^\top C x \geq 0$ with $C \succeq 0$).

Note that the possibility that the target inequality can be obtained by summing up, with nonnegative weights, inequalities from a certain system and adding an identically true inequality is clearly a *sufficient* condition for the target inequality to be a consequence of the system. The actual power of the Homogeneous Farkas' Lemma and the $\mathcal{S}$-Lemma is in the fact that this evident sufficient condition is also necessary for the conclusion in question to be valid (in the case of linear inequalities, whenever the system is finite; in the case of the $\mathcal{S}$-Lemma, when the system is a single-inequality one and (18.6) holds). The fact that, in the quadratic case, to guarantee necessity the system should be a single-inequality system, however unpleasant, is a must. In fact, a straightforward "quadratic version" of the

Homogeneous Farkas' Lemma fails, in general, to be true even when there are just two quadratic inequalities in the system. This being said, even that poor, as compared to its linear inequalities analogy, $\mathcal{S}$-Lemma is extremely useful.

In preparation for the $\mathcal{S}$-Lemma, we will first prove the following weaker statement.

Lemma 18.8 *Let $A, B \in \mathbf{S}^n$. Suppose $\exists \bar{x}$ satisfying $\bar{x}^\top A \bar{x} > 0$. Then, the implication*

$$\{X \succeq 0, \ \mathrm{Tr}(AX) \geq 0\} \implies \mathrm{Tr}(BX) \geq 0 \qquad (18.9)$$

holds if and only if $B \succeq \lambda A$ for some $\lambda \geq 0$.

Proof The "if" part of this lemma is evident. To prove the "only if" part, consider the conic problem

$$\mathrm{Opt}(P) = \min_X \{\mathrm{Tr}(BX) : \mathrm{Tr}(AX) \geq 0, \ X \succeq 0\} \qquad (P)$$

along with its conic dual, which is given by

$$\mathrm{Opt}(D) = \max_{\lambda, Y}\{0 \cdot \lambda + \mathrm{Tr}(0_{n \times n} Y) : \lambda \geq 0, \ Y \succeq 0, \ Y + \lambda A = B\} \qquad (D)$$

(derive the conic dual of (P) yourself by utilizing the fact that $\mathbf{S}^n_+$ is self-dual). Note that from the premise of (18.6), we deduce that, for large enough nonnegative t, the solution $\bar{X} := I_n + t \bar{x} \bar{x}^\top$ will ensure that the Slater condition holds true. Moreover, under the premise of (18.9), $\mathrm{Opt}(P)$ is bounded from below by 0 as well. Then, by the Conic Duality Theorem, the dual (D) is solvable, implying that $B \succeq \lambda A$ for some $\lambda \geq 0$, as required in this lemma and completing the proof. ∎

Note that (18.7) is just (18.9) with X restricted to be of rank ≤ 1. Indeed, $X \succeq 0$ is of rank ≤ 1 if and only if $X = xx^\top$ for some vector x, and in this case $\mathrm{Tr}(PX) = x^\top P x$ for every symmetric P of appropriate size. We are now ready to complete the proof of $\mathcal{S}$-Lemma.

Proof of Lemma 18.7 ($\mathcal{S}$-Lemma) The "if" part is evident. To prove the "only if" part, assume that implication (18.7) holds true, and let us verify that $B \succeq \lambda A$ for some $\lambda \geq 0$. By Lemma 18.8, all we need to do is to show that the validity implication (18.7) implies the validity of implication (18.9). Thus, assume that $x^\top A x \geq 0$ does imply that $x^\top B x \geq 0$, and let $X \succeq 0$ be such that $\mathrm{Tr}(AX) \geq 0$; all we need is to prove that in this case $\mathrm{Tr}(BX) \geq 0$ holds as well. To this end, let $X^{1/2} A X^{1/2} = U \mathrm{Diag}\{\mu\} U^\top$ be the eigenvalue decomposition of $X^{1/2} A X^{1/2}$. By defining $\bar{\mu} := \mathrm{Tr}(\mathrm{Diag}\{\mu\})$ and using the relation $X^{1/2} A X^{1/2} = U \mathrm{Diag}\{\mu\} U^\top$, we arrive at

$$\bar{\mu} = \mathrm{Tr}(\mathrm{Diag}\{\mu\}) = \mathrm{Tr}(U^\top X^{1/2} A X^{1/2} U) = \mathrm{Tr}(X^{1/2} A X^{1/2}) = \mathrm{Tr}(AX) \geq 0.$$

Now, consider an n-dimensional Rademacher random vector ζ, i.e., an n-dimensional vector with entries which, independently of each other, take values ± 1 with probabilities $1/2$. By setting $\xi := X^{1/2} U \zeta$, we get

$$\xi^\top A \xi = \zeta^\top (U^\top X^{1/2} A X^{1/2} U) \zeta = \zeta^\top \mathrm{Diag}\{\mu\} \zeta$$
$$= \mathrm{Tr}(\mathrm{Diag}\{\mu\} \zeta \zeta^\top) = \mathrm{Tr}(\mathrm{Diag}\{\mu\}) = \bar{\mu} \geq 0.$$

Thus, $\xi^\top A \xi \geq 0$ for all realizations of ξ. Recalling that we are in the case when (18.7) holds, we conclude that $\xi^\top B \xi \geq 0$ for all realizations of ξ, or, which is the same,

18.4 Conic Programming and Conic Duality Theorem

$\zeta^\top(U^\top X^{1/2}BX^{1/2}U)\zeta \geq 0$ for all realizations of ζ. Passing to expectations and recalling that ζ is a Rademacher random vector, we get

$$0 \leq \mathbf{E}_\zeta\left[\zeta^\top(U^\top X^{1/2}BX^{1/2}U)\zeta\right] = \mathrm{Tr}(U^\top X^{1/2}BX^{1/2}U)$$
$$= \mathrm{Tr}(X^{1/2}BX^{1/2}) = \mathrm{Tr}(BX),$$

that is, $\mathrm{Tr}(BX) \geq 0$, so that (18.9) does hold true. ∎

Inhomogeneous $\mathcal{S}$-Lemma. The $\mathcal{S}$-Lemma provides a necessary and sufficient condition for a homogeneous quadratic inequality $x^\top Bx \geq 0$ to be a consequence of a strictly feasible homogeneous quadratic inequality $x^\top Ax \geq 0$. What about the inhomogeneous case? When is an inhomogeneous quadratic inequality

$$x^\top Bx + 2b^\top x + \beta \geq 0 \quad (B)$$

a consequence of a strictly feasible inhomogeneous quadratic inequality

$$x^\top Ax + 2a^\top x + \alpha \geq 0? \quad (A)$$

The answer is easy to guess. The implication $(A) \implies (B)$ is just the implication

$$\forall(t \neq 0, x): \ x^\top Ax + 2ta^\top x + \alpha t^2 \geq 0 \implies x^\top Bx + 2tb^\top x + \beta t^2 \geq 0 \quad (*)$$

(substitute x/t instead of x into (A) and look what happens to (B)). We understand the situation when a slightly stronger implication,

$$\forall(t, x): \ x^\top Ax + 2ta^\top x + \alpha t^2 \geq 0 \implies x^\top Bx + 2tb^\top x + \beta t^2 \geq 0, \quad (**)$$

holds true: the homogeneous inequality in the premise of $(**)$ is strictly feasible along with (A), so that, by the homogeneous $\mathcal{S}$-Lemma, $(**)$ holds true if and only if

$$\exists \lambda \geq 0: \ \begin{bmatrix} B & b \\ b^\top & \beta \end{bmatrix} \succeq \lambda \begin{bmatrix} A & a \\ a^\top & \alpha \end{bmatrix}. \quad (18.10)$$

Thus, *if* we knew that in the case of a strictly feasible (A) the validity of implication $(*)$ is the same as the validity of implication $(**)$, we could be sure that the first of these implications holds if and only if (18.10) holds. The above "if" indeed is true.

Lemma 18.9 *[Inhomogeneous $\mathcal{S}$-Lemma] Let $A, B \in \mathbf{S}^n$. Suppose there exists $\bar{x}$ such that $\bar{x}^\top A\bar{x} + 2a^\top \bar{x} + \alpha > 0$. Then, the implication $(A) \implies (B)$ holds if and only if (18.10) holds.*

Proof Suppose the premise holds, i.e., $\bar{x}^\top A\bar{x} + 2a^\top \bar{x} + \alpha > 0$ for some $\bar{x}$. From the discussion preceding this lemma all we need to verify is that the validity of $(*)$ is exactly the same as the validity of $(**)$. Clearly, the validity of $(**)$ implies the validity of $(*)$, so our task boils down to demonstrating that, under the premise of the lemma, the validity of $(*)$ implies the validity of $(**)$. Thus, assume that $(*)$ is valid, and let us prove that $(**)$ is valid as well. All we need to prove is that $y^\top Ay \geq 0$ implies $y^\top By \geq 0$. Thus, assume that y is such that $y^\top Ay \geq 0$, and let us prove that $y^\top By \geq 0$ as well. Define $x_t := t\bar{x} + (1-t)y$, and consider the univariate quadratic functions $q_a(t) := x_t^\top Ax_t + 2ta^\top x_t + \alpha t^2$, $q_b(t) := x_t^\top Bx_t + 2tb^\top x_t + \beta t^2$. We have so far seen that

(i) for all $t \neq 0$, $q_a(t) \geq 0 \implies q_b(t) \geq 0$,
(ii) $q_a(1) > 0$ and $q_a(0) \geq 0$,

and we would like to show that $q_b(0) \geq 0$. Note that q_a and q_b are linear or quadratic functions of t and thus they are continuous in t. Now, consider the following cases (the reader might like to sketch q_a in these cases):

- If $q_a(0) > 0$, by the continuity of q_a we have $q_a(t) > 0$ for all small enough (in magnitude) nonzero t, and so in such a case, by (i) we also get $q_b(t) \geq 0$ for all small enough in magnitude nonzero t, implying, by continuity, that $q_b(0) \geq 0$.
- If $q_a(0) = 0$, the reasoning goes as follows. When t varies from 0 to 1, the linear or quadratic function $q_a(t)$ varies from 0 to something positive. It follows that
  - either $q_a(t) \geq 0$, $0 \leq t \leq 1$, implying by (i) that $q_b(t) \geq 0$ for $t \in (0, 1]$, and so $q_b(0) \geq 0$ holds by continuity of $q_b(t)$ at $t = 0$,
  - or $q_a(\bar{t}) < 0$ holds for some $\bar{t} \in (0, 1)$. Assuming that this is the case, the linear or quadratic function $q_a(t)$ is zero at $t = 0$, negative somewhere on $(0, 1)$, and positive at $t = 1$. Therefore, q_a is a quadratic, not linear, function of t which has exactly one root in the interval $(0, 1)$. Let this root in $(0, 1)$ be t_1. Recall that the other root of the quadratic function q_a is $t = 0$; thus we must have $q_a(t) = c(t - 0)(t - t_1)$ for some $c \in \mathbf{R}$. From $t_1 < 1$ and $q_a(1) > 0$ it follows that $c > 0$; this, in turn, combines with $t_1 > 0$ to imply that $q_a(t) > 0$ when $t < 0$. By (i) it follows that $q_b(t) \geq 0$ for $t < 0$, whence by continuity $q_b(0) \geq 0$. ∎

As an important consequence of the Inhomogeneous $\mathcal{S}$-Lemma we arrive at the following observation, which states that the semidefinite programming relaxation of the quadratically constrained quadratic program with a single strictly feasible quadratic constraint is exact.

Corollary 18.10 *Let $A, B \in \mathbf{S}^n$. Suppose there exists $\bar{x}$ such that $\bar{x}^\top A \bar{x} + 2a^\top \bar{x} + \alpha > 0$. Then,*

$$\beta^* := \inf_x \left\{ x^\top B x + 2b^\top x : x^\top A x + 2a^\top x + \alpha \geq 0 \right\}$$
$$= \max_{\beta, \lambda} \left\{ \beta : \lambda \geq 0, \begin{bmatrix} B - \lambda A & b - \lambda a \\ b^\top - \lambda a^\top & -\lambda \alpha - \beta \end{bmatrix} \succeq 0 \right\}.$$

(Here, as always, the optimal value of an infeasible maximization problem is $-\infty$.)

Proof Define the quadratic functions $q(x) := x^\top A x + 2a^\top x + \alpha$ and $q_\beta(x) := x^\top B x + 2b^\top x - \beta$. Note that β^* is the supremum of all β's satisfying $\beta \leq \inf_x \{ x^\top B x + 2b^\top x : x^\top A x + 2a^\top x + \alpha \geq 0 \}$, that is, those β for which the implication

$$q(x) \geq 0 \implies q_\beta(x) \geq 0$$

holds true. By the premise of the corollary, there exists $\bar{x}$ satisfying $q(\bar{x}) > 0$, so that by the Inhomogeneous $\mathcal{S}$-lemma (see Lemma 18.9) β^* is the supremum of those β which can be augmented by appropriate λ to yield feasible solutions to the maximization problem in the corollary's formulation, or, which is the same, β^* is the optimal value in the latter problem. ∎

Example 18.6 (Continued from Example 18.1) Invoking the Schur Complement Lemma (Proposition D.33), we can rewrite (18.1) equivalently as the conic problem

$$\text{Opt}(P) = \min_{t \in \mathbf{R}, y \in \mathbf{S}^n} \left\{ t : \text{Tr}(y) \leq t, \begin{bmatrix} B & y \\ y & I_n \end{bmatrix} \succeq 0 \right\}. \tag{18.11}$$

The conic dual of (18.11) can be obtained as follows: we equip the scalar inequality $t - \text{Tr}(y) \geq 0$ with a Lagrange multiplier $\bar{\lambda} \geq 0$ and the inequality $\begin{bmatrix} B & y \\ y & I_n \end{bmatrix} \succeq 0$ with a Lagrange multiplier $\begin{bmatrix} U & V \\ V^\top & W \end{bmatrix} \succeq 0$ (recall that the semidefinite cone is self-dual, so that legitimate Lagrange multipliers for $\succeq$-constraints are $\succeq$-nonnegative), and sum the termwise inner products of the constraints of (18.11), thus arriving at the aggregated inequality

$$\bar{\lambda}(t - \text{Tr}(y)) + 2\text{Tr}(yV^\top) \geq -\text{Tr}(BU) - \text{Tr}(W),$$

which by construction is a consequence of the constraints in (18.11). We then impose on the Lagrange multipliers, in addition to the above conic constraints, the restriction that the left-hand side in the aggregated inequality is, in $t \in \mathbf{R}$, $y \in \mathbf{S}^n$, identically equal to the objective in (18.11), that is, the restrictions

$$\bar{\lambda} = 1, \quad V + V^\top = I_n.$$

Then, the dual problem is given by

$$\text{Opt}(D) = \max_{\substack{\bar{\lambda} \in \mathbf{R}, \\ U, W \in \mathbf{S}^n, V \in \mathbf{R}^{n \times n}}} \left\{ -(\text{Tr}(BU) + \text{Tr}(W)) : \begin{array}{l} \bar{\lambda} = 1, \\ V + V^\top = I_n, \\ \begin{bmatrix} U & V \\ V^\top & W \end{bmatrix} \succeq 0 \end{array} \right\}. \tag{18.12}$$

Thus, the dual is precisely the problem of maximizing under the outlined restriction the right-hand side of the aggregated inequality.

Since problem (18.11) satisfies the Slater condition (as $B \succ 0$) and is below bounded (why?), the dual problem is solvable and $\text{Opt}(P) = \text{Opt}(D)$. Moreover, the dual problem also satisfies the relaxed Slater condition (why?), so that both the primal and the dual problems are solvable. ◇

18.5 Proofs of Facts

Fact 18.2 *Consider the parametric family (P_Δ) of convex cone-constrained problems along with the family (D_Δ) of their cone-constrained Lagrange duals. Then,*

(i) *If $\underline{L}_0(\mu) > -\infty$ for some $\mu = [\bar{\mu}; \widehat{\mu}] \in \Lambda$ then the primal optimal value $\text{Opt}(P_\Delta)$ takes values in $\mathbf{R} \cup \{+\infty\}$ and is a convex function of Δ.*
(ii) *If (D_0) is solvable with optimal solution $\lambda_* = [\bar{\lambda}_*; \widehat{\lambda}_*]$ and $\text{Opt}(D_0) = \text{Opt}(P_0)$ then $-\lambda_*$ is a subgradient of $\text{Opt}(P_\Delta)$ at the point $\Delta = 0$, i.e.,*

$$\text{Opt}(P_\Delta) \geq \text{Opt}(P_0) - \bar{\lambda}_*^\top \bar{\delta} - \widehat{\lambda}_*^\top \widehat{\delta}, \quad \forall (\Delta = [\bar{\delta}; \widehat{\delta}]).$$

The premises in (i) *and* (ii) *hold when (P_0) satisfies the relaxed Slater condition and is below bounded.*

Proof   Under the premise of (i), by Weak Duality we have for all Δ

$$\mathrm{Opt}(P_\Delta) \geq \underline{L}_\Delta(\mu) = \underline{L}_0(\mu) - \overline{\mu}^\top \overline{\delta} - \widehat{\mu}^\top \widehat{\delta}. \tag{18.13}$$

That is, for all Δ, $\mathrm{Opt}(P_\Delta) > -\infty$ (as $\underline{L}_0(\mu) > -\infty$). To prove that in this case $\mathrm{Opt}(P_\Delta)$ is convex in Δ, it suffices to verify that when $\Delta', \Delta'' \in \mathrm{Dom}\,\mathrm{Opt}(P_\cdot)$ and $\theta \in (0, 1)$, we have $\mathrm{Opt}(P_{\theta \Delta' + (1-\theta)\Delta''}) \leq \theta \, \mathrm{Opt}(P_{\Delta'}) + (1 - \theta)\mathrm{Opt}(P_{\Delta''})$. Let us denote $\Delta' = [\overline{\delta}'; \widehat{\delta}']$ and $\Delta'' = [\overline{\delta}''; \widehat{\delta}'']$. Then, by the definitions of $\mathrm{Opt}(P_{\Delta'})$ and $\mathrm{Opt}(P_{\Delta''})$, given $\epsilon > 0$, we can find $x'_\epsilon, x''_\epsilon \in X$ such that

$$f(x'_\epsilon) \leq \mathrm{Opt}(P_{\Delta'}) + \epsilon, \quad \overline{g}(x'_\epsilon) \leq \overline{\delta}', \quad \widehat{g}(x'_\epsilon) \leq_{\mathbf{K}} \widehat{\delta}',$$
$$f(x''_\epsilon) \leq \mathrm{Opt}(P_{\Delta''}) + \epsilon, \quad \overline{g}(x''_\epsilon) \leq \overline{\delta}'', \quad \widehat{g}(x''_\epsilon) \leq_{\mathbf{K}} \widehat{\delta}''.$$

Hence, by the convexity of f and $\overline{g}$ and the **K**-convexity of $\widehat{g}$, the point $x_\epsilon := \theta x'_\epsilon + (1-\theta) x''_\epsilon$ satisfies

$$f(x_\epsilon) \leq (\theta\,\mathrm{Opt}(P_{\Delta'}) + (1 - \theta)\mathrm{Opt}(P_{\Delta''})) + \epsilon, \quad \overline{g}(x_\epsilon) \leq \theta \overline{\delta}' + (1-\theta)\overline{\delta}'',$$
$$\widehat{g}(x_\epsilon) \leq_{\mathbf{K}} \theta \widehat{\delta}' + (1-\theta)\widehat{\delta}'',$$

implying that $\mathrm{Opt}(P_{\theta\Delta'+(1-\theta)\Delta''}) \leq (\theta\,\mathrm{Opt}(P_{\Delta'}) + (1 - \theta)\mathrm{Opt}(P_{\Delta''})) + \epsilon$. Since $\epsilon > 0$ is arbitrary, we arrive at the desired relation $\mathrm{Opt}(P_{\theta\Delta'+(1-\theta)\Delta''}) \leq \theta\,\mathrm{Opt}(P_{\Delta'}) + (1 - \theta)\mathrm{Opt}(P_{\Delta''})$. This justifies part (i).

To justify (ii), note that, under the premise of this claim, the premise of (i) is satisfied by taking $\mu := \lambda_*$. Then, (18.13) holds true for this μ as well. Moreover, as $\underline{L}_\Delta(\lambda_*) = \mathrm{Opt}(P_0)$, we see that (18.13) with $\mu = \lambda_*$ says that $-\lambda_*$ is a subgradient of $\mathrm{Opt}(P_\cdot)$ at the origin.

Finally, when (P_0) is below bounded and satisfies the relaxed Slater condition, the premise of (ii) (and therefore of (i) as well) holds true by Theorem 18.1. ∎

Fact 18.3   *Consider the convex quadratic constraint $x^\top A^\top A x \leq b^\top x + c$, where $A \in \mathbf{R}^{d \times n}$, $b \in \mathbf{R}^n$, and $c \in \mathbf{R}$. This constraint can be equivalently rewritten as a conic constraint involving a Lorentz cone, i.e.,*

$$x^\top A^\top A x \leq b^\top x + c$$
$$\iff [2Ax;\, b^\top x + c - 1;\, b^\top x + c + 1] \in \mathbf{L}^{d+2}$$
$$\iff 4x^\top A^\top A x + (b^\top x + c - 1)^2 \leq (b^\top x + c + 1)^2 \text{ and } b^\top x + c + 1 \geq 0.$$

Proof   Note that $b^\top x + c = \frac{(b^\top x+c+1)^2 - (b^\top x+c-1)^2}{4}$. Thus, if x is feasible for the quadratic constraint (and, in particular, $b^\top x + c \geq 0$, implying that $b^\top x + c + 1 > 0$), x is feasible for the conic constraint, and clearly vice versa. ∎

Fact 18.5   *Consider the conic problem given in (18.5),*

$$\mathrm{Opt}(P) = \min_{x \in \mathbf{R}^n}\left\{c^\top x : Ax - b \leq 0,\ Px - p \leq_{\mathbf{K}} 0\right\}, \tag{P}$$

along with its conic dual problem

$$\mathrm{Opt}(D) = \max_{\overline{\lambda},\widehat{\lambda}}\left\{-b^\top \overline{\lambda} - p^\top \widehat{\lambda} : A^\top \overline{\lambda} + P^\top \widehat{\lambda} + c = 0,\ \overline{\lambda} \geq 0,\ \widehat{\lambda} \in \mathbf{K}_*\right\}. \tag{D}$$

18.5 Proofs of Facts

Conic duality is symmetric, i.e., the conic dual to conic problem (D) is (equivalent to) conic problem (P), i.e., (18.5).

Proof In order to apply our recipe for building the conic dual to a conic problem, let us rewrite (D) in the minimization form with ≤ 0-type affine constraints and a single conic constraint, i.e.,

$$-\text{Opt}(D) = \min_{\bar{\lambda},\hat{\lambda}} \left\{ b^\top\bar{\lambda} + p^\top\hat{\lambda} : \begin{array}{c} A^\top\bar{\lambda} + P^\top\hat{\lambda} + c \leq 0 \\ -A^\top\bar{\lambda} - P^\top\hat{\lambda} - c \leq 0, -\hat{\lambda} \in -\mathbf{K}_* \\ -\bar{\lambda} \leq 0 \end{array} \right\}. \quad (D)$$

The dual to this problem, in view of $[\mathbf{K}_*]_* = \mathbf{K}$, reads

$$\max_{u,v,w,y} \left\{ c^\top[u-v] : b + A[u-v] - w = 0, \ P[u-v] + p - y = 0, \ u \geq 0, \ v \geq 0, \ w \geq 0, \ y \in \mathbf{K} \right\}.$$

By setting $x := v - u$ and eliminating y and w, the latter problem becomes

$$\max_{x} \left\{ -c^\top x : Ax - b \leq 0, \ Px - p \leq_{\mathbf{K}} 0 \right\},$$

which is just (P). ∎

19

Optimality Conditions in Convex Programming

Using our results on convex optimization duality, we next derive optimality conditions for convex programs.

19.1 Saddle Point Form of Optimality Conditions

Theorem 19.1 *[Saddle point formulation of optimality conditions in convex programming]*
Consider the optimization problem

$$\min_x \{f(x): g_j(x) \leq 0, \ j = 1, \ldots, m, \ x \in X\} \quad \text{(IC)}$$
$$\left[\text{where } f, g_1, \ldots, g_m \text{ are real-valued functions on } X\right],$$

along with its Lagrange dual problem

$$\max_\lambda \{\underline{L}(\lambda): \lambda \in \mathbf{R}^m_+\}, \quad \text{(IC*)}$$
$$\left[\text{where } \underline{L}(\lambda) = \inf_{x \in X} L(x, \lambda) \text{ and } L(x, \lambda) = f(x) + \sum_{i=1}^m \lambda_i g_i(x)\right].$$

Let $x^* \in X$. Then,

(i) A sufficient condition for x^* to be an optimal solution to (IC) is the existence of a vector of Lagrange multipliers $\lambda^* \geq 0$ such that (x^*, λ^*) is a saddle point of the Lagrange function $L(x, \lambda)$, i.e., a point where $L(x, \lambda)$ attains its minimum as a function of $x \in X$ and attains its maximum as a function of $\lambda \geq 0$:

$$L(x, \lambda^*) \geq L(x^*, \lambda^*) \geq L(x^*, \lambda) \quad \forall x \in X, \ \forall \lambda \geq 0. \quad (19.1)$$

(ii) Furthermore, if the problem (IC) is convex and satisfies the relaxed Slater condition, then the above condition is necessary for the optimality of x^*: if x^* is optimal for (IC), then there exists $\lambda^* \geq 0$ such that (x^*, λ^*) is a saddle point of the Lagrange function.

Proof (i): Assume that for a given $x^* \in X$ there exists $\lambda^* \geq 0$ such that (19.1) is satisfied, and let us prove that then x^* is optimal for (IC). First, we claim that x^* is feasible. Assume for contradiction that $g_j(x^*) > 0$ for some j. Then, of course, $\sup_{\lambda \geq 0} L(x^*, \lambda) = +\infty$ (look what happens when all λ's, except λ_j, are fixed and $\lambda_j \to +\infty$). But, $\sup_{\lambda \geq 0} L(x^*, \lambda) = +\infty$ is forbidden by the second inequality in (19.1). Thus, x^* must be feasible to (IC).

Since x^* is feasible, $\sup_{\lambda \geq 0} L(x^*, \lambda) = f(x^*)$, and we conclude from the second inequality in (19.1) that $L(x^*, \lambda^*) = \sup_{\lambda \geq 0} L(x^*, \lambda) = f(x^*)$. Finally, let us examine the first inequality in (19.1). This relation now reads

19.1 Saddle Point Form of Optimality Conditions

$$f(x) + \sum_{j=1}^{m} \lambda_j^* g_j(x) \geq f(x^*), \quad \forall x \in X.$$

Recall that for any x that is feasible for (IC), we have $g_j(x) \leq 0$ for all j. Together with $\lambda^* \geq 0$, we then deduce that for any x feasible for (IC), we have $f(x) \geq f(x) + \sum_{j=1}^{m} \lambda_j^* g_j(x)$. But then the above relation immediately implies that x^* is optimal.

(ii): Assume that (IC) is a convex program, x^* is its optimal solution, and the problem satisfies the relaxed Slater condition. We will prove that then there exists $\lambda^* \geq 0$ such that (x^*, λ^*) is a saddle point of the Lagrange function, i.e., that (19.1) is satisfied. As we know from the Convex Programming Duality Theorem (Theorem 17.1.ii), the dual problem (IC*) has a solution $\lambda^* \geq 0$ and the optimal value of the dual problem is equal to the optimal value of the primal problem, i.e., to $f(x^*)$:

$$f(x^*) = \underline{L}(\lambda^*) \equiv \inf_{x \in X} L(x, \lambda^*). \tag{19.2}$$

Then, we immediately conclude that

$$\lambda_j^* > 0 \implies g_j(x^*) = 0$$

(this is called *complementary slackness*: positive Lagrange multipliers can be associated only with active constraints, i.e., the ones satisfied as equalities). Indeed, from (19.2) it certainly follows that

$$f(x^*) \leq L(x^*, \lambda^*) = f(x^*) + \sum_{j=1}^{m} \lambda_j^* g_j(x^*);$$

the terms in the summation expression on the right-hand side are nonpositive (since x^* is feasible for (IC) and $\lambda^* \geq 0$), and the sum itself is nonnegative due to our inequality. Note that this is possible if and only if all the terms in the summation expression are zero, and this is precisely complementary slackness.

From complementary slackness we immediately conclude that $f(x^*) = L(x^*, \lambda^*)$, so that (19.2) results in

$$L(x^*, \lambda^*) = f(x^*) = \inf_{x \in X} L(x, \lambda^*).$$

On the other hand, since x^* is feasible for (IC), from the definition of the Lagrangian function we deduce that $L(x^*, \lambda) \leq f(x^*)$ whenever $\lambda \geq 0$. Combining our observations, we conclude that

$$L(x^*, \lambda) \leq L(x^*, \lambda^*) \leq L(x, \lambda^*)$$

for all $x \in X$ and all $\lambda \geq 0$. ∎

Note that Theorem 19.1(i) is valid for an arbitrary inequality-constrained optimization problem, not necessarily a convex one. However, in the nonconvex case the *sufficient* condition for optimality given by Theorem 19.1(i) is extremely far from being necessary and is "almost never" satisfied. In contrast with this, in the convex case the condition in question is not only sufficient, but also "nearly necessary" – it certainly is necessary when (IC) is a convex program satisfying the relaxed Slater condition.

Theorem 19.1 presents the saddle point form of the optimality conditions for convex problems in standard mathematical programming form (that is, with the constraints represented by scalar convex inequalities). Similar results can be obtained for convex cone-constrained problems as follows.

Theorem 19.2 *[Saddle point formulation of optimality conditions in Convex Cone-constrained Programming] Consider a convex cone-constrained problem*

$$\mathrm{Opt}(P) = \min_{x \in X} \{f(x) : \overline{g}(x) := Ax - b \leq 0, \widehat{g}(x) \leq_{\mathbf{K}} 0\} \quad (P)$$

(X is convex, $f : X \to \mathbf{R}$ is convex, $\overline{g}(\cdot) : \mathbf{R}^n \to \mathbf{R}^k$ is affine, $\mathbf{K}$ is a regular cone, and $\widehat{g}$ is $\mathbf{K}$-convex on X), along with its Cone-constrained Lagrange dual problem

$$\mathrm{Opt}(D) = \max_{\lambda := [\overline{\lambda}; \widehat{\lambda}]} \left\{ \overline{L}(\lambda) := \inf_{x \in X} \left[f(x) + \overline{\lambda}^\top \overline{g}(x) + \widehat{\lambda}^\top \widehat{g}(x) \right] : \overline{\lambda} \geq 0, \widehat{\lambda} \in \mathbf{K}_* \right\}. \quad (D)$$

Suppose that (P) is bounded and satisfies the relaxed Slater condition. Then, a point $x^* \in X$ is an optimal solution to (P) if and only if x^* can be augmented by properly selected $\lambda^* \in \Lambda := \mathbf{R}_+^k \times [\mathbf{K}_*]$ to be a saddle point of the cone-constrained Lagrange function

$$L(x; [\overline{\lambda}; \widehat{\lambda}]) := f(x) + \overline{\lambda}^\top \overline{g}(x) + \widehat{\lambda}^\top \widehat{g}(x)$$

on $X \times \Lambda$.

Proof The proof basically repeats that of Theorem 19.1. In one direction: assume that $x^* \in X$ can be augmented by $\lambda^* = [\overline{\lambda}^*; \widehat{\lambda}^*] \in \Lambda$ to form a saddle point of $L(x; \lambda)$ on $X \times \Lambda$, and let us prove that x^* is an optimal solution to (P). Observe, first, that from

$$L(x^*; \lambda^*) = \sup_{\lambda \in \Lambda} L(x^*; [\overline{\lambda}; \widehat{\lambda}]) = f(x^*) + \sup_{\lambda \in \Lambda} \left[\overline{\lambda}^\top \overline{g}(x^*) + \widehat{\lambda}^\top \widehat{g}(x^*) \right] \quad (19.3)$$

it follows that the linear form $\widehat{\lambda}^\top \widehat{g}(x^*)$ of $\widehat{\lambda}$ is bounded from above on the cone $\mathbf{K}_*$, implying that $-\widehat{g}(x^*) \in [\mathbf{K}_*]_* = \mathbf{K}$. Similarly, (19.3) says that the linear form $\overline{\lambda}^\top \overline{g}(x^*)$ of $\overline{\lambda}$ is bounded from above on the cone $\mathbf{R}_+^k$, implying that $-\overline{g}(x^*)$ belongs to the dual of this cone, that is, to $\mathbf{R}_+^k$. Thus, x^* is feasible for (P). As x^* is feasible for (P), the right-hand side of the second equality in (19.3) is $f(x^*)$, and thus (19.3) says that $L(x^*; \lambda^*) = f(x_*)$. With this in mind, the relation $L(x; \lambda^*) \geq L(x^*; \lambda^*)$ (which is satisfied for all $x \in X$, since (x^*, λ^*) is a saddle point of L) reads $L(x; \lambda^*) \geq f(x^*)$. This combines with the relation $f(x) \geq L(x; \lambda^*)$ (which, due to $\lambda^* \in \Lambda$, holds true for all x feasible for (P)) to imply that $\mathrm{Opt}(P) \geq f(x^*)$. The bottom line is that x^* is a feasible solution to (P) satisfying $\mathrm{Opt}(P) \geq f(x^*)$; thus, x^* is an optimal solution to (P), as claimed.

In the opposite direction: let x^* be an optimal solution to (P), and let us verify that x^* is the first component of a saddle point of $L(x; \lambda)$ on $X \times \Lambda$. Indeed, (P) is a convex essentially strictly feasible cone-constrained problem; being solvable, it is below bounded. Applying Convex Programming Duality Theorem in cone-constrained form (Theorem 18.1), the dual problem (D) is solvable with optimal value $\mathrm{Opt}(D) = \mathrm{Opt}(P)$. Denoting by $\lambda^* = [\overline{\lambda}^*; \widehat{\lambda}^*]$ an optimal solution to (D), we have

$$f(x^*) = \mathrm{Opt}(P) = \mathrm{Opt}(D) = \overline{L}(\lambda^*) = \inf_{x \in X} L(x; \lambda^*), \quad (19.4)$$

whence $f(x^*) \leq L(x^*; \lambda^*) = f(x^*) + [\overline{\lambda}^*]^\top \overline{g}(x^*) + [\widehat{\lambda}^*]^\top \widehat{g}(x^*)$, that is, $[\overline{\lambda}^*]^\top \overline{g}(x^*) + [\widehat{\lambda}^*]^\top \widehat{g}(x^*) \geq 0$. Both terms in the latter sum are nonpositive (as x^* is feasible for (P) and $\lambda^* \in \Lambda$), while their sum is nonnegative, so that $[\overline{\lambda}^*]^\top \overline{g}(x^*) = 0$ and $[\widehat{\lambda}^*]^\top \widehat{g}(x^*) = 0$. We conclude that the inequality $f(x^*) \leq L(x^*; \lambda^*)$ is in fact equality, so that (19.4) reads $\inf_{x \in X} L(x; \lambda^*) = L(x^*; \lambda^*)$. Next, $L(x^*; \lambda) \leq f(x^*)$ for $\lambda \in \Lambda$ owing to the feasibility of x^* for (P), which combines with the already proved equality $L(x^*; \lambda^*) = f(x^*)$ to imply that $\sup_{\lambda \in \Lambda} L(x^*; \lambda) = L(x^*; \lambda^*)$. Thus, (x^*, λ^*) is the desired saddle point of L. ∎

19.2 Karush–Kuhn–Tucker Form of Optimality Conditions

Theorem 19.1 provides, basically, the strongest known optimality conditions for a convex programming problem. These conditions, however, are "implicit" – they are expressed in terms of the saddle point of the Lagrange function, and it is unclear how to verify whether a given solution is a saddle point of the Lagrange function. Fortunately, the proof of Theorem 19.1 yields more or less explicit optimality conditions.

Recall that the normal cone $N_X(x)$ of a set $X \subseteq \mathbf{R}^n$ at a point $x \in X$ as defined by (11.5) is

$$N_X(x) = \left\{ h \in \mathbf{R}^n : h^\top (x' - x) \leq 0, \forall x' \in X \right\}.$$

Now let us define the notion of the *Karush–Kuhn–Tucker point* of an inequality-constrained optimization problem.

Definition 19.3 *[Karush–Kuhn–Tucker point of inequality-constrained mathematical programming problem] Consider the optimization problem*

$$\min_x \left\{ f(x) : g_j(x) \leq 0, j = 1, \ldots, m, x \in X \right\} \quad \text{(IC)}$$
$$\left[\text{where } f, g_1, \ldots, g_m \text{ are real-valued functions on } X \right].$$

A point $x^* \in \mathbf{R}^n$ is called a **Karush–Kuhn–Tucker (KKT) point** *of* (IC) *if x^* is feasible, f, $g_1, \ldots, g_m$ are differentiable at x^*, and there exist nonnegative Lagrange multipliers λ_j^*, $j = 1, \ldots, m$, such that*

$$\lambda_j^* g_j(x^*) = 0, \ j = 1, \ldots, m \ \text{[complementary slackness]} \quad (19.5)$$

and

$$\nabla f(x^*) + \sum_{j=1}^m \lambda_j^* \nabla g_j(x^*) \in -N_X(x^*) \ \text{[KKT equation]} \quad (19.6)$$

where $N_X(x^*)$ is the normal cone of X at x^*.

We are now ready to state more explicit optimality conditions for convex programs, on the basis of KKT points.

Theorem 19.4 *[Karush–Kuhn–Tucker optimality conditions in convex programming] Let* (IC) *be a convex program (i.e., X is nonempty and convex, and $f, g_1, \ldots, g_m$ are convex on X). Let $x^* \in X$, and let the functions $f, g_1, \ldots, g_m$ be differentiable at x^*. Then,*

(i) [*Sufficiency*] If x^* is a KKT point of (IC) then x^* is an optimal solution to the problem.

(ii) [*Necessity and sufficiency*] If, in addition to the premise, the relaxed Slater condition holds, x^* is an optimal solution to (IC) if and only if x^* is a KKT point of the problem.

Proof (i): Suppose x^* is a KKT point of problem (IC), and let us prove that x^* is an optimal solution to (IC). By Theorem 19.1, it suffices to demonstrate that augmenting x^* by a properly selected $\lambda \geq 0$, we get a saddle point (x^*, λ) of the Lagrange function on $X \times \mathbf{R}_+^m$. Let λ^* be the Lagrange multiplier associated with x^* according to the definition of a KKT point. We claim that (x^*, λ^*) is a saddle point of the Lagrange function. Indeed, complementary slackness says that $L(x^*, \lambda^*) = f(x^*)$, while owing to the feasibility of x^* we have $\sup_{\lambda \geq 0} \sum_{j=1}^n \lambda_j g_j(x^*) = 0$. Taken together, these observations say that $L(x^*, \lambda^*) = \sup_{\lambda \geq 0} L(x^*, \lambda)$. Moreover, the function $\phi(x) := L(x, \lambda^*) : X \to \mathbf{R}$ is convex and differentiable at x^*, and by the KKT equation we have $\nabla \phi(x^*) \in -N_X(x^*)$. Invoking Proposition 11.3, we conclude that x^* minimizes ϕ on X, that is, $L(x, \lambda^*) \geq L(x^*, \lambda^*)$ for all $x \in X$. Thus, (x^*, λ^*) is a saddle point of the Lagrange function.

(ii): In view of (i), all we need to do to prove (ii) is to demonstrate that if x^* is an optimal solution to (IC), (IC) is a convex program that satisfies the relaxed Slater condition, and the objective and constraints of (IC) are differentiable at x^*, then x^* is a KKT point. Indeed, let x^* and (IC) satisfy the above "if." By Theorem 19.1(ii), x^* can be augmented by some $\lambda^* \geq 0$ to yield a saddle point (x^*, λ^*) of $L(x, \lambda)$ on $X \times \mathbf{R}_+^m$. Then, the saddle point inequalities (19.1) give us

$$f(x^*) + \sum_{j=1}^m \lambda_j^* g_j(x^*) = L(x^*, \lambda^*) \geq \sup_{\lambda \geq 0} L(x^*, \lambda) = f(x^*) + \sup_{\lambda \geq 0} \sum_{j=1}^m \lambda_j g_j(x^*). \quad (19.7)$$

Moreover, as x^* is feasible to (IC), we have $g_j(x^*) \leq 0$ for all j, whence

$$\sup_{\lambda \geq 0} \sum_j \lambda_i g_j(x^*) = 0.$$

Therefore (19.7) implies that $\sum_j \lambda_j^* g_j(x^*) \geq 0$. This inequality, in view of $\lambda_j^* \geq 0$ and $g_j(x^*) \leq 0$ for all j, implies that $\lambda_j^* g_j(x^*) = 0$ for all j, i.e., the complementary slackness condition (19.5) holds. The relation $L(x, \lambda^*) \geq L(x^*, \lambda^*)$ for all $x \in X$ implies that the function $\phi(x) := L(x, \lambda^*)$ attains its minimum on X at $x = x^*$. Note also that $\phi(x)$ is convex on X and differentiable at x^*; thus, by Proposition 11.3, we deduce that the KKT equation (19.6) also holds. ∎

Note that the optimality conditions stated in Theorem 11.2 and Proposition 11.3 are a particular case of Theorem 19.4 corresponding to $m = 0$.

Remark 19.5 A standard special case of Theorem 19.4 that is worth discussing explicitly occurs when x^* is in the (relative) interior of X.

When $x^* \in \mathrm{int}\, X$, we have $N_X(x^*) = \{0\}$, so that (19.6) reads

$$\nabla f(x^*) + \sum_{j=1}^m \lambda_j^* \nabla g_j(x^*) = 0.$$

When $x^* \in \text{rint } X$, $N_X(x^*)$ is the orthogonal complement to the linear subspace $\mathcal{L}$ to which $\text{Aff}(X)$ is parallel, so that (19.6) reads

$$\left(\nabla f(x^*) + \sum_{j=1}^m \lambda_j^* \nabla g_j(x^*)\right) \text{ is orthogonal to } \mathcal{L} := \text{Lin}(X - x_*).$$

◇

19.3 Cone-Constrained KKT Optimality Conditions

The cone-constrained version of the notion of a KKT point is defined as follows:

Definition 19.6 *[Karush–Kuhn–Tucker point of a cone-constrained problem] Consider the cone-constrained optimization problem*

$$\begin{bmatrix} \min\{f(x) : \overline{g}(x) := Ax - b \leq 0, \, \widehat{g}(x) \leq_{\mathbf{K}} 0, \, x \in X\} \\ \text{where } X \subseteq \mathbf{R}^n, \, f : X \to \mathbf{R}, \, \overline{g} : \mathbf{R}^n \to \mathbf{R}^k, \\ \widehat{g} : X \to \mathbf{R}^\nu, \, \mathbf{K} \subset \mathbf{R}^\nu \text{ is a regular cone} \end{bmatrix}. \quad (\text{ConeC})$$

A point $x^* \in \mathbf{R}^n$ is called a Karush–Kuhn–Tucker (KKT) point of (ConeC), if x^* is feasible, f and $\widehat{g}$ are differentiable at x^*, and there exist Lagrange multipliers $\overline{\lambda} = [\overline{\lambda}_1; \ldots; \overline{\lambda}_k] \geq 0$ and $\widehat{\lambda} \in \mathbf{K}_*$ such that

$$\overline{\lambda}_j [\overline{g}(x^*)]_j = 0, \, \forall j \leq k, \text{ and } \widehat{\lambda}^\top \widehat{g}(x^*) = 0 \quad \text{[complementary slackness]} \quad (19.8)$$

$$\text{and } \nabla_x \left[f(x) + \overline{\lambda}^\top \overline{g}(x) + \widehat{\lambda}^\top \widehat{g}(x) \right]\bigg|_{x=x^*} \in -N_X(x^*) \quad \text{[KKT equation]} \quad (19.9)$$

where $N_X(x^*)$ is the normal cone of X at x^*; see (11.5).

On the basis of this definition, the cone-constrained version of Theorem 19.4 is as follows.

Theorem 19.7 *[Karush–Kuhn–Tucker optimality conditions in cone-constrained convex programming] Consider a convex cone-constrained problem*

$$\text{Opt}(P) = \min_{x \in X} \{f(x) : \overline{g}(x) := Ax - b \leq 0, \, \widehat{g}(x) \leq_{\mathbf{K}} 0\} \quad (P)$$

(X is convex, $f : X \to \mathbf{R}$ is convex, $A \in \mathbf{R}^{k \times n}$, $\mathbf{K}$ is a regular cone, and $\widehat{g}$ is $\mathbf{K}$-convex on X). Suppose $x^* \in X$ is a feasible solution to the problem, and let f and $\widehat{g}$ be differentiable at x^*.

(i) *If x^* is a KKT point (as defined by Definition 19.6) of (P), then x^* is an optimal solution to (P).*

(ii) *If x^* is optimal solution to (P) and, if addition to the above premise, (P) satisfies the cone-constrained relaxed Slater condition, then x^* is a KKT point, as defined by Definition 19.6, of (P).*

The proof of this theorem follows verbatim from the proof of Theorem 19.4, with Theorem 19.2 in the role of Theorem 19.1.

Application: Optimal value in parametric convex cone-constrained problem. What follows is a far-reaching extension of the subgradient interpretation of Lagrange multipliers presented in section 18.3.1. Consider a parametric family of convex cone-constrained problems defined by a parameter $p \in P$:

$$\text{Opt}(p) := \min_{x \in X} \{f(x, p) : g(x, p) \leq_M 0\}, \qquad (\text{P}[p])$$

where

(a) $X \subseteq \mathbf{R}^n$ and $P \subseteq \mathbf{R}^\mu$ are nonempty convex sets.
(b) $\mathbf{M}$ is a regular cone in some $\mathbf{R}^\nu$.
(c) $f : X \times P \to \mathbf{R}$ is convex, and $g : X \times P \to \mathbf{R}^\nu$ is $\mathbf{M}$-convex.[1]

Next, we make the following additional assumption:

(d) Suppose $\bar{x} \in X$ and $\bar{p} \in P$ are such that
   - $\bar{x}$ is a KKT point, as defined by Definition 19.6, of the convex cone-constrained problem (P[$\bar{p}$]) (and therefore, by Theorem 19.7, it is an optimal solution to the problem), and
   - $f(x, p)$ and $g(x, p)$ are differentiable at the point $[\bar{x}; \bar{p}]$, the derivatives being

$$Df([\bar{x}; \bar{p}])[[dx; dp]] =: F_x^\top dx + F_p^\top dp, \quad Dg([\bar{x}; \bar{p}])[[dx; dp]] =: G_x dx + G_p dp.$$

and let $\bar{\mu} \in \mathbf{M}_*$ be the Lagrange multiplier associated with $\bar{x}$ and (P[$\bar{p}$]), so that

$$\bar{\mu}^\top g(\bar{x}, \bar{p}) = 0 \text{ and } [x - \bar{x}]^\top [F_x + G_x^\top \bar{\mu}] \geq 0, \forall x \in X \qquad (19.10)$$

(cf. Definition 19.6).

Proposition 19.8 *Under assumptions (a)–(d), Opt($\cdot$) is a convex function on P taking values in $\mathbf{R} \cup \{+\infty\}$ and finite at $\bar{p}$, and the vector*

$$F_p + G_p^\top \bar{\mu}$$

is a subgradient of Opt($\cdot$) at $\bar{p}$:

$$\text{Opt}(p) \geq \text{Opt}(\bar{p}) + [p - \bar{p}]^\top [F_p + G_p^\top \bar{\mu}], \quad \forall p \in P. \qquad (19.11)$$

Proof Observe first that, as f is convex, by the gradient inequality we have

$$f(x, p) \geq f(\bar{x}, \bar{p}) + F_x^\top [x - \bar{x}] + F_p^\top [p - \bar{p}], \quad \forall (x \in X, p \in P). \qquad (19.12)$$

Also, from $\bar{\mu} \in \mathbf{M}_*$ and the $\mathbf{M}$-convexity of g, by Fact 16.10 it follows that the function $\bar{\mu}^\top g(x, p)$ is convex on $X \times P$, so that by gradient inequality we have

$$\bar{\mu}^\top g(x, p) \geq \underbrace{\bar{\mu}^\top g(\bar{x}, \bar{p})}_{=0 \text{ by (19.10)}} + \bar{\mu}^\top G_x [x - \bar{x}] + \bar{\mu}^\top G_p [p - \bar{p}], \quad \forall (x \in X, p \in P). \qquad (19.13)$$

---

[1] In what follows, splitting the constraint in a cone-constrained problem into a system of scalar linear inequalities and a conic inequality does not play any role, and, in order to save notation, (P[p]) uses a "single-constraint" format of the cone-constrained problem. Of course, the two-constraint format $\bar{g}(x) := Ax - b \leq 0$, $\widehat{g}(x) \leq_\mathbf{K} 0$ reduces to the single-constraint format by setting $g(x) = [\bar{g}(x); \widehat{g}(x)]$ and $\mathbf{M} = \mathbf{R}_+^k \times \mathbf{K}$.

Now let $p \in P$ and x be feasible for (P[p]). Then,

$$\begin{aligned}
f(x, p) &\geq f(x, p) + \overline{\mu}^\top g(x, p) \\
&\geq f(\overline{x}, \overline{p}) + F_x^\top [x - \overline{x}] + F_p^\top [p - \overline{p}] + \overline{\mu}^\top G_x [x - \overline{x}] + \overline{\mu}^\top G_p [p - \overline{p}] \\
&= \text{Opt}(\overline{p}) + (F_x + G_x^\top \overline{\mu})^\top [x - \overline{x}] + (F_p + G_p^\top \overline{\mu})^\top [p - \overline{p}] \\
&\geq \text{Opt}(\overline{p}) + (F_p + G_p^\top \overline{\mu})^\top [p - \overline{p}],
\end{aligned}$$

where the first inequality follows from $\overline{\mu} \in \mathbf{M}_*$ and $g(x, p) \leq_\mathbf{M} 0$, the second is due to (19.12) and (19.13), the equality holds by recalling that $\overline{x}$ is optimal for ($P[\overline{p}]$), and the last inequality is due to (19.10). As the resulting inequality holds true for all x feasible for (P[p]), it justifies (19.11).

To complete the proof, we need to verify the convexity of $\text{Opt}(\cdot)$. By the relation in (19.11), $\text{Opt}(p)$ for $p \in P$ is either a real or $+\infty$, as is required for a convex function. For any $p', p'' \in P \cap \text{Dom}(\text{Opt}(\cdot))$ and $\lambda \in [0, 1]$, we need to check that

$$\text{Opt}(\lambda p' + (1 - \lambda) p'') \leq \lambda \text{Opt}(p') + (1 - \lambda) \text{Opt}(p''). \tag{19.14}$$

This is immediate (cf. the proof of Fact 18.2): given $\epsilon > 0$, we can find $x', x'' \in X$ such that

$$g(x', p') \leq_\mathbf{M} 0, \quad g(x'', p'') \leq_\mathbf{M} 0, \quad f(x', p') \leq \text{Opt}(p') + \epsilon, \quad f(x'', p'') \leq \text{Opt}(p'') + \epsilon.$$

Setting $p := \lambda p' + (1 - \lambda) p''$, $x := \lambda x' + (1 - \lambda) x''$ and invoking the convexity of f and the **M**-convexity of g, we get

$$g(x, p) \leq_\mathbf{M} 0, \quad f(x) \leq [\lambda \text{Opt}(p') + (1 - \lambda) \text{Opt}(p'')] + \epsilon.$$

Finally, since $\epsilon > 0$ is arbitrary, we arrive at (19.14). ∎

Note that the result of section 18.3.1 is simply what Proposition 19.8 states in the case where f is independent of p, $\mathbf{M} = \mathbf{R}_+^k \times \mathbf{K}$, $p = [\overline{\delta}, \widehat{\delta}] \in \mathbf{R}^k \times \mathbf{R}^v$, and $g(x, p) = [\overline{g}(x) - \overline{\delta}; \widehat{g}(x) - \widehat{\delta}]$.

19.4 Optimality Conditions in Conic Programming

We continue by discussing the case of conic programming.

Theorem 19.9 *[Optimality conditions in conic programming] Consider a primal–dual pair of conic problems (cf. section 18.4)*

$$\text{Opt}(P) = \min_{x \in \mathbf{R}^n} \left\{ c^\top x : Ax - b \leq 0, \; Px - p \leq_\mathbf{K} 0 \right\}, \tag{P}$$

$$\text{Opt}(D) = \max_{\overline{\lambda}, \widehat{\lambda}} \left\{ -b^\top \overline{\lambda} - p^\top \widehat{\lambda} : A^\top \overline{\lambda} + P^\top \widehat{\lambda} + c = 0, \; \overline{\lambda} \geq 0, \; \widehat{\lambda} \in \mathbf{K}_* \right\}. \tag{D}$$

Suppose that both problems satisfy the relaxed Slater condition. Then, a pair of feasible solutions x_* to (P) and $\lambda_* := [\overline{\lambda}_*; \widehat{\lambda}_*]$ to (D) is optimal to the respective problems if and only if

$$\text{DualityGap}(x_*; \lambda_*) := c^\top x_* - [-b^\top \overline{\lambda}_* - p^\top \widehat{\lambda}_*] = 0 \qquad \text{[zero duality gap]}$$

which holds if and only if

$$\bar{\lambda}_*^\top [b - Ax_*] + \widehat{\lambda}_*^\top [p - Px_*] = 0. \qquad \text{[complementary slackness]}$$

Remark 19.10 Under the premise of Theorem 19.9, from the feasibility of x_* and λ_* for their respective problems it follows that $b - Ax_* \geq 0$, $p - Px_* \in \mathbf{K}$, $\bar{\lambda}_* \geq 0$, and $\widehat{\lambda}_* \in \mathbf{K}_*$. Therefore, complementary slackness (which says that the sum of two inner products, each being of a vector from a regular cone and a vector from the dual of this cone and as such automatically nonnegative) is zero is a very strong restriction. This comment is applicable to the relation $[\widehat{\lambda}_*]^\top \widehat{g}(x^*) = 0$ in (19.8). ◇

Proof of Theorem 19.9 By the Conic Duality Theorem (Theorem 18.6) we are in the case when $\mathrm{Opt}(P) = \mathrm{Opt}(D) \in \mathbf{R}$, and therefore, for any x and $\lambda := [\bar{\lambda}; \widehat{\lambda}]$, we have

$$\mathrm{DualityGap}(x; \lambda) = [c^\top x - \mathrm{Opt}(P)] + [\mathrm{Opt}(D) - (-b^\top \bar{\lambda} - p^\top \widehat{\lambda})].$$

Now, when x is feasible for (P), the *primal optimality gap* $c^\top x - \mathrm{Opt}(P)$ is nonnegative and is zero if and only if x is optimal for (P). Similarly, when $\lambda = [\bar{\lambda}; \widehat{\lambda}]$ is feasible for (D), the *dual optimality gap* $\mathrm{Opt}(D) - (-b^\top \bar{\lambda} - p^\top \widehat{\lambda})$ is nonnegative and is zero if and only if λ is optimal for (D). We conclude that whenever x is feasible for (P) and λ is feasible for (D), the duality gap $\mathrm{DualityGap}(x; \lambda)$ (which, as we have seen, is the sum of the corresponding optimality gaps) is nonnegative and is zero if and only if both these optimality gaps are zero, that is, if and only if x is optimal for (P) and λ is optimal for (D).

It remains to note that the complementary slackness condition is equivalent to the zero duality gap condition. To this end, note that since x_* and λ_* are feasible for their respective problems, we have

$$\begin{aligned}
\mathrm{DualityGap}(x_*; \lambda_*) &= c^\top x_* + b^\top \bar{\lambda}_* + p^\top \widehat{\lambda}_* \\
&= -[A^\top \bar{\lambda}_* + P^\top \widehat{\lambda}_*]^\top x_* + b^\top \bar{\lambda}_* + p^\top \widehat{\lambda}_* \\
&= \bar{\lambda}_*^\top [b - Ax_*] + \widehat{\lambda}_*^\top [p - Px_*].
\end{aligned}$$

Therefore, complementary slackness, for the solutions x_* and λ_* that are feasible for the respective problems, is exactly the same as a zero duality gap. ∎

Example 19.1 (continued from Example 18.1) Consider the primal–dual pair of conic problems (18.11) and (18.12). We claim that the primal solution $y = -B^{1/2}, t = -\mathrm{Tr}(B^{1/2})$ and the dual solution $\bar{\lambda} = 1, U = \frac{1}{2}B^{-1/2}, V = \frac{1}{2}I_n, W = \frac{1}{2}B^{1/2}$ are optimal for their respective problems. Indeed, it is immediately seen that these solutions are feasible for the respective problems (to check the feasibility of the dual solution, use the Schur Complement Lemma). Moreover, the objective value of the primal solution is equal to the objective value of the dual solution, and both these quantities are equal to $-\mathrm{Tr}(B^{1/2})$. Thus, the zero duality gap does indeed hold true. ◇

19.5 Duality in Linear and Convex Quadratic Programming

The fundamental role of the Lagrange function and Lagrange duality in optimization is clear already from the optimality conditions given by Theorem 19.1, but this role is not

19.5 Duality in Linear and Convex Quadratic Programming

restricted by this theorem only. There are several cases when we can explicitly write down the Lagrange dual, and whenever this is the case, we get a pair of optimization programs that are explicitly formulated and closely related to each other, the *primal–dual pair*. By analyzing this pair of problems simultaneously, we get more information about their properties (and have the possibility of solving the problems numerically in a more efficient way) than it is possible when we restrict ourselves to only one problem of the pair. The detailed investigation of duality in "well-structured" convex programming deals with cone-constrained Lagrange duality and conic problems. This being said, there are cases where in fact "plain" Lagrange duality is quite appropriate. Let us look at two of these particular cases.

19.5.1 Linear Programming Duality

Let us start with some general observations. Note that the Karush–Kuhn–Tucker condition under the assumption of Theorem 19.4, i.e., that the problem

$$\min_x \{f(x) : g(x) \leq 0, \ j = 1, \ldots, m, \ x \in X\} \tag{IC}$$

is convex, x^* is a feasible solution to the problem, and $f, g_1, \ldots, g_m$ are differentiable at x^*, is exactly the condition that $(x^*, \lambda^* := (\lambda_1^*, \ldots, \lambda_m^*))$ is a saddle point of the Lagrange function

$$L(x, \lambda) = f(x) + \sum_{j=1}^{m} \lambda_j g_j(x) \tag{19.15}$$

on $X \times \mathbf{R}_+^m$. The equalities (19.5) taken together with feasibility of x^* state that $L(x^*, \lambda)$ attains its maximum in $\lambda \geq 0$ at λ^*, and (19.6) states that when λ is fixed at λ^* the function $L(x, \lambda^*)$ attains its minimum in $x \in X$ at $x = x^*$.

Now consider the particular case of (IC) where $X = \mathbf{R}^n$ is the entire space, the objective f is convex and everywhere differentiable and the constraints $g_1, \ldots, g_m$ are *linear*. In this case the relaxed Slater condition holds whenever there is a feasible solution to (IC), and, when that is the case, Theorem 19.4 states that the KKT (Karush–Kuhn–Tucker) condition is necessary and sufficient for the optimality of x^*; as we just have explained, this is the same as saying that the necessary and sufficient condition of optimality for x^* is that x^*, along with certain $\lambda^* \geq 0$, forms a saddle point of the Lagrange function. Combining these observations with Proposition 17.2, we get the following simple result.

Proposition 19.11 *Let* (IC) *be a convex program with* $X = \mathbf{R}^n$, *everywhere differentiable objective* f, *and linear constraints* $g_1, \ldots, g_m$.

Then, x^* *is an optimal solution to* (IC) *if and only if there exists* $\lambda^* \geq 0$ *such that* (x^*, λ^*) *is a saddle point of the Lagrange function* (19.15) *(regarded as a function of* $x \in \mathbf{R}^n$ *and* $\lambda \geq 0$*). In particular,* (IC) *is solvable if and only if* L *has saddle points, and if that is the case then both* (IC) *and its Lagrange dual*

$$\max_{\lambda \geq 0} \left\{ \underline{L}(\lambda) := \inf_x L(x, \lambda) \right\} \tag{IC*}$$

are solvable with equal optimal objective values.

Let us look what Proposition 19.11 says in the linear programming case, i.e., when (IC) is the problem given by

$$\min_x \left\{ f(x) := c^T x : \ g_j(x) := b_j - a_j^T x \leq 0, \ j = 1, \ldots, m \right\}. \tag{P}$$

In order to get to the Lagrange dual, we form the Lagrange function of (IC):

$$L(x, \lambda) = f(x) + \sum_{j=1}^m \lambda_j g_j(x) = \left(c - \sum_{j=1}^m \lambda_j a_j \right)^T x + \sum_{j=1}^m \lambda_j b_j,$$

and minimize it in $x \in \mathbf{R}^n$; this will give us the dual objective. In our case the minimization in x is immediate: the minimal value is equal to $-\infty$, if $c - \sum_{j=1}^m \lambda_j a_j \neq 0$, and it is $\sum_{j=1}^m \lambda_j b_j$, otherwise. Hence, we see that the Lagrange dual is given by

$$\max_\lambda \left\{ b^T \lambda : \ \sum_{j=1}^m \lambda_j a_j = c, \ \lambda \geq 0 \right\}. \tag{D}$$

Therefore, the Lagrange dual problem is precisely the usual LP dual to (P), and Proposition 19.11 is one of the equivalent forms of the LP Duality Theorem (Theorem 4.10) which we already know.

19.5.2 Convex Quadratic Programming Duality

Now consider the case when the original problem is a linearly constrained convex quadratic program given by

$$\min_x \left\{ f(x) := \frac{1}{2} x^T Q x + c^T x : \ g_j(x) := b_j - a_j^T x \leq 0, \ j = 1, \ldots, m \right\}, \tag{P}$$

where the objective is a strictly convex quadratic form, so that the matrix $Q = Q^T$ is positive definite, i.e., $x^T Q x > 0$ whenever $x \neq 0$. It is convenient to rewrite the constraints in vector–matrix form using the notation

$$g(x) = b - Ax \leq 0, \text{ where } b := [b_1; \ldots; b_m], \ A := \begin{bmatrix} a_1^T \\ \vdots \\ a_m^T \end{bmatrix}.$$

In order to form the Lagrange dual to program (P), we write down the Lagrange function

$$L(x, \lambda) = f(x) + \sum_{j=1}^m \lambda_j g_j(x)$$

$$= \frac{1}{2} x^T Q x + c^T x + \lambda^T (b - Ax) = \frac{1}{2} x^T Q x - (A^T \lambda - c)^T x + b^T \lambda,$$

and minimize it in x. Since the function is convex and differentiable in x, the minimum, if exists, is given by the Fermat rule

$$\nabla_x L(x, \lambda) = 0,$$

19.5 Duality in Linear and Convex Quadratic Programming

which in our situation becomes

$$Qx = A^\top \lambda - c.$$

Since Q is positive definite, it is nonsingular, so that the Fermat equation has a unique solution which is the minimizer of $L(\cdot, \lambda)$. This solution is

$$x(\lambda) := Q^{-1}(A^\top \lambda - c).$$

Substituting the expression for $x(\lambda)$ into the expression for the Lagrange function, we get the dual objective

$$\underline{L}(\lambda) = -\frac{1}{2}(A^\top \lambda - c)^\top Q^{-1}(A^\top \lambda - c) + b^\top \lambda.$$

Thus, the dual problem is to maximize this objective over the nonnegative orthant. Let us rewrite this dual problem equivalently by introducing additional variables

$$t := -Q^{-1}(A^\top \lambda - c) \quad \Longrightarrow \quad (A^\top \lambda - c)^\top Q^{-1}(A^\top \lambda - c) = t^\top Qt.$$

With this substitution, the dual problem becomes

$$\max_{\lambda, t} \left\{ -\frac{1}{2} t^\top Qt + b^\top \lambda : \ A^\top \lambda + Qt = c, \ \lambda \geq 0 \right\}. \tag{D}$$

We see that the dual problem also turns out to be a linearly constrained convex quadratic program.

Remark 19.12 Note also that a feasible quadratic program in the form of (P) with a positive definite matrix Q is automatically solvable. This relies on the following simple general fact:

> Let (IC) be a feasible program with closed domain X, an objective and constraints that are continuous on X, and such that $f(x) \to \infty$ as $x \in X$ "goes to infinity" (i.e., $\|x\|_2 \to \infty$). Then (IC) is solvable.

In the case of our quadratic program (P), as Q is positive definite we have $f(x) \to \infty$ as $\|x\|_2 \to \infty$. Then, the solvability of (P) follows from the simple fact stated above. Readers are encouraged to prove this simple fact for themselves. ◇

On the basis of this remark, Proposition 19.11 leads to the following result.

Theorem 19.13 *[Duality Theorem in quadratic programming]*
Let (P) be a feasible quadratic program with positive definite symmetric matrix Q in the objective. Then, both (P) and (D) are solvable, and the optimal values in these problems are equal to each other.

A pair of primal and dual feasible solutions, say $(x; (\lambda, t))$, to these problems is optimal

(i) *if and only if the* zero duality gap *optimality condition holds, i.e., the primal objective value at x is equal to the dual objective value at (λ, t)*
or, equivalently,
(ii) *if and only if the following holds*

$$\lambda_i(Ax - b)_i = 0, \ i = 1, \ldots, m, \ \text{and} \ t = -x. \tag{19.16}$$

Proof (i): Proposition 19.11 implies that the optimal value in the minimization problem (P) is equal to the optimal value in the maximization problem (D). It follows that the value of the primal objective at any primal feasible solution is $\geq$ the value of the dual objective at any dual feasible solution, and equality is possible if and only if these values coincide with the optimal values in the problems, as claimed in (i).

(ii): Let $\Delta(x, (\lambda, t))$ be the difference between the primal objective value of the primal feasible solution x and the dual objective value of the dual feasible solution (λ, t):

$$\Delta(x, (\lambda, t)) := \left(c^T x + \frac{1}{2} x^T Q x\right) - \left(b^T \lambda - \frac{1}{2} t^T Q t\right)$$

$$= (A^T \lambda + Q t)^T x + \frac{1}{2} x^T Q x + \frac{1}{2} t^T Q t - b^T \lambda$$

$$= \lambda^T (A x - b) + \frac{1}{2} (x + t)^T Q (x + t),$$

where the second equation follows since $A^T \lambda + Q t = c$. Whenever x is primal feasible we have $Ax - b \geq 0$, and similarly the dual feasibility of (λ, t) implies that $\lambda \geq 0$. Since Q is positive definite as well, we then deduce that the first and second terms in the above representation of $\Delta(x, (\lambda, t))$ are nonnegative for every pair $(x; (\lambda, t))$ of primal and dual feasible solutions. Thus, for such a pair, $\Delta(x, (\lambda, t)) = 0$ holds if and only if $\lambda^T(Ax-b) = 0$ and $(x+t)^T Q(x+t) = 0$. The first of these equalities, due to $\lambda \geq 0$ and $Ax \geq b$, is equivalent to $\lambda_j (Ax - b)_j = 0$ for all $j = 1, \ldots, m$; the second equality, due to the positive definiteness of Q, is equivalent to $x + t = 0$. ∎

20
★ Cone-Convex Functions: Elementary Calculus and Examples

So far, when speaking about **K**-convex functions, we have assumed the cone K to be regular, that is, pointed, closed, and with a nonempty interior. In the considerations in this chapter, the nonemptiness of int **K** is of no importance, so that within this chapter we specify a **K**-convex mapping f, where **K** is a closed pointed cone in some embedding Euclidean space F, as a mapping f defined on a convex domain Dom $f \subseteq \mathbf{R}^n$ and taking values in F such that

$$f(\lambda x + (1-\lambda)y) \leq_{\mathbf{K}} \lambda f(x) + (1-\lambda)f(y), \quad \forall (x, y \in \text{Dom } f, \lambda \in [0, 1]),$$

where, as always, $a \leq_{\mathbf{K}} b$ means that $b - a \in \mathbf{K}$. In particular, we allow for $\mathbf{K} = \{0\}$, in which case, $a \leq_{\mathbf{K}} b$ is equivalent to $a = b$; in this extreme case, the **K**-convexity of f means that f is affine on Dom f.

Note that Fact 16.10, as is clear from its proof, remains true when the regularity of cone **K** given in its premise is relaxed to closedness and pointedness.

The calculus presented below resembles the usual calculus of real-valued monotone and convex functions, and almost all the claims to follow are nearly self-evident, and therefore not all of them are accompanied by proofs. The absence of a proof in the text means that providing the proof is an exercise for the reader.[1]

20.1 Epigraph Characterization of Cone-Convexity

Let $f(x) : \text{Dom } f \to \mathbf{R}^\nu$ be a mapping with convex domain Dom $f \subseteq \mathbf{R}^n$, and let $\mathbf{K} \subset \mathbf{R}^\nu$ be a closed pointed cone. The following claim is evident.

Fact 20.1 *A function f is **K**-convex if and only if its **K**-epigraph*

$$\text{epi}_{\mathbf{K}}(f) := \left\{ (x, y) \in \text{Dom } f \times \mathbf{R}^\nu : y \geq_{\mathbf{K}} f(x) \right\}$$

is convex.

Let us examine some **K**-convex functions. In what follows, we use $\succeq$-convexity as synonym of $\mathbf{S}_+^m$-convexity, with context-specified m.

Example 20.1 Consider the "convex matrix-valued quadratic function" of $x \in \mathbf{R}^{p \times q}$ given by

$$f(x) := [AxB][AxB]^\top + [CxD + D^\top x^\top C^\top] + H,$$

---
[1] This being said, all missing proofs can be found in the solutions to Exercise IV.29.

where $A, C \in \mathbf{R}^{m \times p}$, $B \in \mathbf{R}^{q \times n}$, $D \in \mathbf{R}^{q \times m}$, $H \in \mathbf{S}^m$. Note that $f : \mathbf{R}^{p \times q} \mapsto \mathbf{S}^m$. We claim that f is $\succeq$-convex. Indeed, by the Schur Complement Lemma we have

$$\mathrm{epi}_{\mathbf{S}_+^m}\{f\} = \left\{ (x,y) : \left[\begin{array}{c|c} y - (CxD + D^\top x^\top C^\top) - H & AxB \\ \hline B^\top x^\top A^\top & I_n \end{array} \right] \succeq 0 \right\}$$

and the right-hand side set is clearly convex (as it is the inverse image of $\mathbf{S}_+^{m+n}$ under an affine mapping). ◇

Example 20.2 Consider the matrix-valued fractional-quadratic function

$$f(u,v) := u^\top v^{-1} u$$

with domain $\mathrm{Dom}\, f = \{(u,v) : u \in \mathbf{R}^{p \times q},\ v \in \mathrm{int}\,\mathbf{S}_+^p\}$. We claim that f is $\succeq$-convex. Indeed, by the Schur Complement Lemma we have

$$\mathrm{epi}_{\mathbf{S}_+^q}\{f\} = \left\{ [(u,v,y) : \left[\begin{array}{c|c} y & u^\top \\ \hline u & v \end{array} \right] \succeq 0, v \succ 0 \right\},$$

and the right-hand side set is convex (by the same reason as in Example 20.1). ◇

20.2 Testing Cone-Convexity and Cone-Monotonicity

20.2.1 Cone-Monotonicity

Let us start with a new (for us) notion which will play an important role in the calculus of cone-convexity.

Definition 20.2 *[(**U**, **K**)-monotonicity] Let E and F be Euclidean spaces equipped with closed pointed cones* **U** *and* **K**, *Q be a nonempty convex subset of E, and $f(x) : Q \to F$ be a mapping. We say that this mapping is* (**U**, **K**)-*monotone on Q, if $f(x) \leq_\mathbf{K} f(x')$ holds for every $x, x' \in Q$ satisfying $x \leq_\mathbf{U} x'$.*

For example, when **U** and **K** are nonnegative orthants, e.g., $E = \mathbf{R}^n$ and $F = \mathbf{R}^m$, the (**U**, **K**)-monotonicity of F on Q means that whenever $x \leq x'$ and $x, x' \in Q$, one has $F(x) \leq F(x')$. An instructive extreme example is the one where $\mathbf{U} = \{0\}$; in this case, every mapping $f : Q \to F$ is (**U**, **K**)-monotone.

20.2.2 Differential Criteria for Cone-Convexity and Cone-Monotonicity

We next present differential characterizations of cone-convexity and cone-monotonicity. The following claim is nearly evident.

Proposition 20.3 *Let E, F be Euclidean spaces equipped with closed pointed cones* **U** *and* **K**, *let $\mathrm{Dom}\, f \subseteq E$ be a convex set with a nonempty interior, and let $f : \mathrm{Dom}\, f \to F$ be a mapping. Then,*

(i) *f is* **K**-*convex if and only if the scalar function $\langle g, f(x) \rangle : \mathrm{Dom}\, f \to \mathbf{R}$ is convex for every $g \in \mathbf{K}_*$. In particular, assuming that f is continuous on $\mathrm{Dom}\, f$ and twice differentiable on $\mathrm{int}(\mathrm{Dom}\, f)$, we deduce that f is* **K**-*convex if and only if*

20.2 Testing Cone-Convexity and Cone-Monotonicity

$$\left.\frac{d^2}{dt^2}\right|_{t=0} f(x+th) \geq_{\mathbf{K}} 0, \qquad \forall (x \in \text{int}(\text{Dom } f),\ h \in E).$$

(ii) Assuming f is continuous on $\text{Dom } f$ and differentiable on $\text{int}(\text{Dom } f)$, f is $(\mathbf{U}, \mathbf{K})$-monotone on $\text{Dom } f$ if and only if

$$\left.\frac{d}{dt}\right|_{t=0} f(x+th) \geq_{\mathbf{K}} 0, \qquad \forall (h \in \mathbf{U},\ x \in \text{int}(\text{Dom } f)).$$

Example 20.3 The function $f(x) = xx^\top : \mathbf{R}^{m \times n} \to \mathbf{S}^m$ is $\mathbf{S}^m_+$-convex. Indeed,

$$\left.\frac{d}{dt}\right|_{t=0} f(x+th)[h] = xh^\top + hx^\top, \quad \left.\frac{d^2}{dt^2}\right|_{t=0} f(x+th) = 2hh^\top \succeq 0.$$

We can arrive at the same conclusion by specifying appropriately the data in Example 20.1. It is worthwhile to compare our new tools with the basic verification of the $\succeq$-convexity of a similar matrix-valued function in the proof of Lemma 16.9. $\diamondsuit$

Example 20.4 The function $x \mapsto f(x) := x^{-1} : \text{int } \mathbf{S}^m_+ \to \text{int } \mathbf{S}^m_+$ is $(-\mathbf{S}^m_+, \mathbf{S}^m_+)$-monotone and $\mathbf{S}^m_+$-convex. Indeed, as is shown in Example C.8 in section C.1.6, for $x \in \text{Dom } f$ and $h \in \mathbf{S}^m$ it holds that

$$Df(x)[h] := \left.\frac{d}{dt}\right|_{t=0} f(x+th) = -x^{-1}hx^{-1}.$$

Thus, $Df(x)[h] \succeq 0$ whenever $h \in -\mathbf{S}^m_+$, which, by Proposition 20.3(ii), implies the desired monotonicity. From the above expression for $Df(x)[h]$ it follows that

$$\begin{aligned}
D^2 f(x)[h, h] &:= \left.\frac{d^2}{dt^2}\right|_{t=0} f(x+th) \\
&= \left.\frac{d}{dt}\right|_{t=0} \left(-(x+th)^{-1}h(x+th)^{-1}\right) \\
&= (x^{-1}hx^{-1})hx^{-1} + x^{-1}h(x^{-1}hx^{-1}) \\
&= 2x^{-1}hx^{-1}hx^{-1} \\
&= 2x^{-1/2}(x^{-1/2}hx^{-1/2})^2 x^{-1/2} \succeq 0.
\end{aligned}$$

Then, by Proposition 20.3(i) we deduce the $\mathbf{S}^m_+$-convexity of $f(x)$ as well.

One can easily construct numerical examples showing that the function $f(x) = x^{-2} : \text{int } \mathbf{S}^m_+ \to \mathbf{S}^m_+$ is neither $(-\mathbf{S}^m_+, \mathbf{S}^m_+)$-monotone nor $\mathbf{S}^m_+$-convex. $\diamondsuit$

Our next example is less trivial and relies on the following well-known result, which is important in its own right.

Theorem 20.4 *Let $f(s) : (a, b) \to \mathbf{R}$ be an analytic function on an interval of real axis. Then, the function $F(x) = f(x) : \text{Dom } F \to \mathbf{S}^m$ where $\text{Dom } F := \{x \in \mathbf{S}^m : \lambda_i(x) \in (a, b), 1 \leq i \leq m\}$ (see section D.1.5 for the precise definition of the map F) is infinitely many times differentiable on the open set Δ.*

The justification of this well-known fact requires tools (integral Cauchy formula) which go beyond the prerequisites in calculus that we take for granted in this book.

We are now ready for our next example.

Example 20.5 The function $f(x) = x^{1/2} : \mathbf{S}_+^m \to \mathbf{S}^m$ is $(\mathbf{S}_+^m, \mathbf{S}_+^m)$-monotone and $\mathbf{S}_+^m$-concave (the latter, of course, means that $-f(x)$ is $\mathbf{S}_+^m$-convex).

Here is the justification. By Proposition D.26, $f(x)$ is continuous on $\mathbf{S}_+^m$, so that it suffices to prove that the function possesses the desired monotonicity and concavity properties in the interior of $\mathbf{S}_+^m$. By Theorem 20.4, $f(x)$ is infinitely many times differentiable on int $\mathbf{S}_+^m$. Let us compute the derivative of $f(x)$. Given $x \succ 0$ and $h \in \mathbf{S}^m$ and setting $d := Df(x)[h]$, we have, by differentiating the identity $f^2(x) \equiv x$,

$$x^{1/2} \cdot d + d \cdot x^{1/2} = h. \tag{20.1}$$

Rewriting this linear equation in the variable $d \in \mathbf{S}^m$ in an orthonormal eigenbasis of x, we see that the equation has a unique solution. A solution (and therefore *the* solution) to the equation is given by

$$d = \int_0^\infty \exp\{-x^{1/2}t\} \, h \, \exp\{-x^{1/2}t\} \, dt. \tag{20.2}$$

Indeed, consider $u \in \mathbf{R}^{n \times n}$. Then, one has

$$\frac{d}{dt} \exp\{tu\} = \frac{d}{dt}\left[\sum_{i=0}^\infty \frac{1}{i!} t^i u^i\right] = u \sum_{i=0}^\infty \frac{1}{i!} t^i u^i = u \exp\{tu\} = \exp\{tu\} \, u$$

(for the definition of the exponent of a square matrix, see Remark D.25). Moreover, by integrating in t over the ray $\mathbf{R}_+$, the identity

$$\frac{d}{dt}\left[\exp\{-x^{1/2}t\} \, h \, \exp\{-x^{1/2}t\}\right]$$
$$= -x^{1/2} \exp\{-x^{1/2}t\} \, h \, \exp\{-x^{1/2}t\} - \exp\{-x^{1/2}t\} \, h \, \exp\{-x^{1/2}t\} \, x^{1/2},$$

we arrive at the result.[2]

From (20.2) we deduce that $d \succeq 0$ whenever $h \succeq 0$, and thus by Proposition 20.3(ii), we conclude that f is $(\mathbf{S}_+^m, \mathbf{S}_+^m)$-monotone. Now, differentiating in t the identity (20.1) with x replaced by $x + th$, that is, the identity

$$f(x+th) Df(x+th)[h] + Df(x+th)[h] f(x+th) \equiv h,$$

and setting $t = 0$, we get

$$0 = 2d^2 + x^{1/2} D^2 f(x)[h,h] + D^2 f(x)[h,h] x^{1/2},$$

---

[2] Note that when $x \succ 0$ and h commute (as is the case when $x, h \in \mathbf{S}^1$ and $x > 0$), (20.2) becomes

$$d = \int_0^\infty \exp\{-x^{1/2}t\} \, h \, \exp\{-x^{1/2}t\} \, dt = h \int_0^\infty \exp\{-2x^{1/2}t\} \, dt$$
$$= h \int_0^\infty \frac{d}{dt}\left[-[2x^{1/2}]^{-1} \exp\{-2x^{1/2}t\}\right] dt = [2x^{1/2}]^{-1} h,$$

in full accordance with the formula $D[\sqrt{s}][ds] = \frac{1}{2\sqrt{s}} ds$ for the derivative of the univariate function $\sqrt{s}$. Thus, (20.2) is a matrix version of the formula for the derivative of the usual square root; this was not readily predictable in advance.

whence, as above,

$$D^2 f(x)[h,h] = -2 \int_0^\infty \exp\{-tx^{1/2}\} d^2 \exp\{-tx^{1/2}\} dt,$$

so that $D^2 f(x)[h,h] \preceq 0$; thus $f(x) = x^{1/2}$ is $\mathbf{S}_+^m$-concave. ◇

20.3 Elementary Calculus of Cone-Convexity

We start with the elementary calculus of cone-convexity.

A. Let F be a Euclidean space and $\mathbf{K} \subset F$ be a closed pointed cone. Then,

- **A.1** An affine mapping $f(x) = Ax + b : \mathbf{R}^n \to F$ is $\mathbf{K}$-convex.
- **A.2** When $f_i(x)$, $i = 1, \ldots, m$, are $\mathbf{K}$-convex functions with common domain $D \subseteq E$, E being a finite-dimensional space, and $\lambda_i \geq 0$, the function $\sum_i \lambda_i f_i(x)$ with the domain D is $\mathbf{K}$-convex.
- **A.3** When $f(x) : \text{Dom } f \to F$ is $\mathbf{K}$-convex, and $x = Ay + b : G \to E$ is an affine mapping, E being the embedding space of $\text{Dom } f$, the function $g(y) := f(Ay + b)$ with domain $\text{Dom } g = \{y : Ay + b \in \text{Dom } f\}$ is $\mathbf{K}$-convex.
- **A.4** When $f(x) : \text{Dom } f \to \mathbf{R}^\nu$ is $\mathbf{U}$-convex, $\mathbf{U}$ being a closed pointed cone in $\mathbf{R}^\nu$, and $y = Az + b : \mathbf{R}^\nu \to F$ is an affine mapping such that $Au \in \mathbf{K}$ whenever $u \in \mathbf{U}$, the function $g(x) := Af(x) + b$, with $\text{Dom } g := \text{Dom } f$, is $\mathbf{K}$-convex.

We next present the "conic version" of the convex monotone superposition rules from section 10.1.

B. Let

- $\mathbf{K} \subset \mathbf{R}^\nu$ be a closed pointed cone,
- $F(y) : \text{Dom } F \to \mathbf{R}^\nu$ be a mapping with convex domain $\text{Dom } F \subseteq \mathbf{R}^{n_1} \times \cdots \times \mathbf{R}^{n_K}$, so that an argument $y = [y_1; \ldots; y_K]$ of F is a block vector with blocks y_k of dimension n_k, $1 \leq k \leq K$,
- $\mathbf{U}_k \subset \mathbf{R}^{n_k}$, $k \leq K$, be closed pointed cones,
- $f_k(x) : D \to \mathbf{R}^{n_k}$, $1 \leq k \leq K$, be mappings with common convex domain $D \subseteq \mathbf{R}^n$.

Assume that F is $\mathbf{K}$-convex and $(\mathbf{U}_1 \times \cdots \times \mathbf{U}_K, \mathbf{K})$-monotone, f_k are $\mathbf{U}_k$-convex, $k \leq K$, and

$$f(x) := [f_1(x); \ldots; f_K(x)] \in \text{Dom } F, \quad \forall x \in D.$$

Then, the function

$$G(x) := F(f(x)) : D \to \mathbf{R}^\nu$$

is $\mathbf{K}$-convex.

Remark 20.5 Note that in the preceding superposition rule **B**, some $\mathbf{U}_k$ may be trivial: $\mathbf{U}_k = \{0\}$. For these k, the $\mathbf{U}_k$-convexity of f_k is the same as f_k being an affine function, and $(\mathbf{U}_k, \mathbf{K})$-monotonicity of F in y_k holds true automatically. Thus, the above rule covers the usual convex monotone superposition rule from section 10.1, where the affinity of some of the inner functions f_k allowed us to lift the requirement for the outer function F to be monotone in the respective y_k. ◇

★ *Cone-Convex Functions: Elementary Calculus and Examples*

Let us illustrate these calculus rules on some examples.

Example 20.6 Suppose $f(x) : \mathbf{S}^m_+ \to \mathbf{S}^m_+$ and $g(x) : \mathbf{S}^m_+ \to \mathbf{S}^m_+$ are $(\mathbf{S}^m_+, \mathbf{S}^m_+)$-monotone and $\mathbf{S}^m_+$-concave; then, as an immediate corollary of **B**, so is the mapping $h(x) := f(g(x))$.

Therefore, recalling also Example 20.5, we conclude that for any positive integer k the function $x^{1/2^k} : \mathbf{S}^m_+ \to \mathbf{S}^m_+$ is $(\mathbf{S}^m_+, \mathbf{S}^m_+)$-monotone and $\mathbf{S}^m_+$-concave. Taking into account that

$$\lim_{\alpha \to +0} \frac{s^\alpha - 1}{\alpha} = \ln s, \quad s > 0.$$

we conclude that the matrix logarithm, i.e., the function

$$f(x) := \ln(x) : \text{int}\,\mathbf{S}^m_+ \to \mathbf{S}^m$$

is $(\mathbf{S}^m_+, \mathbf{S}^m_+)$-monotone and $\mathbf{S}^m_+$-concave.

Finally, in contrast with these, the matrix exponent $\exp\{x\} : \mathbf{S}^m \to \mathbf{S}^m$ possesses no monotonicity or convexity properties unless $m = 1$. ◇

Example 20.7 Let the function $f(x) : \text{Dom}\, f \to \text{int}\,\mathbf{S}^m_+$ be continuous on the convex domain $\text{Dom}\, f \subseteq \mathbf{R}^n$ with a nonempty interior, and assume that f is twice differentiable on $\text{int}\,\text{Dom}\, f$. When f is $\mathbf{S}^m_+$-concave, the function $g(x) := (f(x))^{-1}$ is $\mathbf{S}^m_+$-convex.

Indeed, for $x \in \text{Dom}\, f$ and $h \in \mathbf{R}^n$ we have

$$Dg(x)[h] = -g(x)Df(x)[h]g(x),$$
$$D^2 g(x)[h,h] = 2g(x)Df(x)[h]g(x)Df(x)[h]g(x) - g(x)D^2 f(x)[h,h]g(x)$$
$$= 2g^{1/2}(x)\left(g^{1/2}(x)Df(x)[h]g^{1/2}(x)\right)^2 g^{1/2}(x) - g(x)D^2 f(x)[h,h]g(x)$$
$$\succeq 0$$

(recall that $D^2 f(x)[h,h] \preceq 0$ by Proposition 20.3(i) as f is $\mathbf{S}^m_+$-concave). ◇

Example 20.8 All the following functions,

$$f(u,v,w,z) = \begin{bmatrix} u & v^\top \\ \hline v & w - z^{1/2} \end{bmatrix}, \text{ where}$$

$\text{Dom}\, f := \{(u,v,w,z) : u \in \mathbf{S}^m, v \in \mathbf{R}^{n\times m}, w \in \mathbf{S}^n, z \in \mathbf{S}^n, z \succeq 0\}$;

$$g(u,v,w,z) = \begin{bmatrix} u & v^\top \\ \hline v & w + z^{-1/2} \end{bmatrix}, \text{ where}$$

$\text{Dom}\, g := \{(u,v,w,z) : u \in \mathbf{S}^m, v \in \mathbf{R}^{n\times m}, w \in \mathbf{S}^n, z \in \mathbf{S}^n, z \succ 0\}$;

$$h(u,v,w,z) = \begin{bmatrix} u & v^\top \\ \hline v & w + z^{-1} \end{bmatrix}, \text{ where}$$

$\text{Dom}\, h := \{(u,v,w,z) : u \in \mathbf{S}^m, v \in \mathbf{R}^{n\times m}, w \in \mathbf{S}^n, z \in \mathbf{S}^n, z \succ 0\}$;

$$e(u,v,w,z) = \begin{bmatrix} u & v^\top \\ \hline v & w + z^2 \end{bmatrix}, \text{ where}$$

$\text{Dom}\, e := \{(u,v,w,z) : u \in \mathbf{S}^m, v \in \mathbf{R}^{n\times m}, w \in \mathbf{S}^n, z \in \mathbf{S}^n\}$,

are $\mathbf{S}^{m+n}_+$-convex.

20.3 Elementary Calculus of Cone-Convexity

To justify our claim, note that the function

$$F(y_1, y_2, y_3) := \begin{bmatrix} y_1 & y_2^\top \\ \hline y_2 & y_3 \end{bmatrix} : \mathbf{S}^m \times \mathbf{R}^{n \times m} \times \mathbf{S}^n \to \mathbf{S}^{m+n}$$

is affine and therefore $\mathbf{S}_+^{m+n}$-convex. By Proposition 20.3(ii), this function is $(\mathbf{S}_+^n, \mathbf{S}_+^{m+n})$-monotone in y_3. Applying **B** above with $\mathbf{U}_1 = \{0\} \subset \mathbf{S}^m$, $\mathbf{U}_2 = \{0\} \subset \mathbf{R}^{m \times n}$, $\mathbf{U}_3 = \mathbf{S}_+^n$ and

$$f_1(u, v, w, z) := u, \quad f_2(u, v, w, z) := v, \quad f_3(u, v, w, z) := w - z^{1/2}$$

(the latter function is $\mathbf{S}_+^n$-convex in its domain $\{(u, v, w, z) : z \succeq 0\}$ by Example 20.5), we conclude that f is $\mathbf{S}_+^{m+n}$-convex. Similar reasoning, with $f_3(u, v, w, z)$ replaced with

- $w + z^{-1/2}$ (this function is $\mathbf{S}_+^n$-convex in the domain $z \succ 0$ by $\succeq$-concavity of $z^{1/2}$, $z \succeq 0$ (Example 20.5) and Example 20.7), or

- $w + z^{-1}$ (this function is $\mathbf{S}_+^n$-convex in the domain $z \succ 0$ by Example 20.7), or

- $w + z^2$ (this function is $\mathbf{S}_+^n$-convex by Example 20.3),

justifies the announced $\mathbf{S}_+^{m+n}$-convexity of the functions g, h, e as well. ◇

21

★ Mathematical Programming Optimality Conditions

The goal of this chapter is to develop optimality conditions for a general-type mathematical programming problem

$$\min_x \left\{ f(x) : \underbrace{g_1(x) \leq 0, \ldots, g_m(x) \leq 0,}_{\text{inequality constraints}} \atop \underbrace{h_1(x) = 0, \ldots, h_k(x) = 0}_{\text{equality constraints}} \right\}, \quad (21.1)$$

where m, k are nonnegative integers and the objective f and the constraints g_j, h_i are real-valued functions, each well defined on its own subset of $\mathbf{R}^n$. This topic seemingly "goes beyond convexity;" however, related developments utilize convex analysis tools, and it would be unwise to skip it completely.

As in the case of convex programs, optimality conditions are aimed at answering the following question:

> Given a feasible solution x_* to (21.1), what are necessary and/or sufficient conditions for x_* to be an optimal solution to the problem?

The conditions we are looking for should be verifiable: given x_* and local information on the objective and the constraints (i.e., their values and derivatives taken at x_*) we should be able to check whether the conditions are or are not satisfied.

Beyond the convex case, no verifiable sufficient conditions for x_* to be *globally optimal* (i.e., $f(x_*) \leq f(x)$ for every feasible x) are known. The existing optimality conditions focus on the *local optimality* of x_*, defined as follows:

Definition 21.1 *[Local optimality] A feasible solution to (21.1) is called* locally optimal *if the objective and the constraints are well defined in a neighborhood of x_* and x_* has the best – the smallest – objective value among all feasible solutions that are close enough to it, that is, there exists $r > 0$ such that*

$$x \text{ is feasible for (21.1) and } \|x - x_*\|_2 \leq r \implies f(x_*) \leq f(x).$$

The classical MP optimality conditions that we are about to present are applicable only when x_* is a *regular* feasible solution to (21.1), with regularity defined as follows:

Definition 21.2 *[Regular solution] A vector $x_* \in \mathbf{R}^n$ is called a* regular solution *to problem (21.1), if*

- x_* *is a feasible solution to the problem,*

- the objective and the constraints are well defined and continuously differentiable in a neighborhood of x_*, and
- the gradients, taken at x_*, of the constraints active at x_* (i.e., the constraints which are satisfied at x_* as equalities) are linearly independent.

Geometrically and informally, the regularity of a solution x_* means that the set S that is cut from a small enough neighborhood of x_* by the equality versions of the constraints active at x_* is a smooth surface.

21.1 Formulating Optimality Conditions

21.1.1 Default Assumption

From now on, unless the opposite is explicitly stated,

> We assume that x_* is a regular solution to (21.1) and that the objective and the constraints are twice continuously differentiable in a neighborhood of x_*.

We will express our optimality conditions in terms of the Lagrange function of (21.1) given by

$$L(x; \lambda, \mu) := f(x) + \sum_{j=1}^{m} \lambda_j g_j(x) + \sum_{i=1}^{k} \mu_i h_i(x).$$

This function is well defined on the direct product of a neighborhood X of x_* and the entire space $\mathbf{R}_\lambda^m \times \mathbf{R}_\mu^k$ of *Lagrange multipliers* λ, μ and is twice continuously differentiable in $x \in X$ and linear in $[\lambda; \mu]$.

We are now ready to state our optimality conditions (first the necessary condition and then the sufficient condition) for general mathematical programming programs (MPs) under our *Default Assumption*.

Proposition 21.3 *[Necessary optimality condition for general MP] Suppose the Default Assumption holds and let x_* be a locally optimal solution to problem (21.1). Then,*

(i) *[first-order part]* x_* *is a KKT* (Karush–Kuhn–Tucker) *point of the problem, i.e., there exist $\lambda^* \in \mathbf{R}_+^m$ and $\mu^* \in \mathbf{R}^k$ satisfying*

$$\lambda_j^* g_j(x_*) = 0, \text{ for all } 1 \leq j \leq m, \qquad \text{[complementary slackness]}$$
$$\nabla_x\big|_{x=x_*} L(x; \lambda^*, \mu^*) = 0. \qquad \text{[KKT equation]}$$

(ii) *[second-order part] The second-order directional derivatives of $L(\cdot; \lambda^*, \mu^*)$ taken at x_* along the directions from the linear subspace*

$$T_\mathrm{n} := \left\{ d \in \mathbf{R}^n : \begin{array}{l} d \text{ is orthogonal to the gradients taken at } x_* \\ \text{of all constraints that are active at } x_* \end{array} \right\}$$

are nonnegative, i.e.,

$$d^\top \nabla_x^2\big|_{x=x_*} L(x; \lambda^*, \mu^*) d \geq 0, \quad \forall d \in T_\mathrm{n}.$$

Note that, under our Default Assumption, if x_* is a KKT point of (21.1), then the corresponding Lagrange multipliers λ^*, μ^* are uniquely defined by x_*. Indeed, the KKT equation says that $-\nabla f(x_*) = \sum_j \lambda_j^* \nabla g_j(x_*) + \sum_i \mu_i^* \nabla h_i(x_*)$. By complementary slackness, the Lagrange multipliers λ_j^* for inequality constraints that are non-active at x_* are zero, and the rest of the λ_j^* taken together with the μ_i^* form the coefficients in a representation of $-\nabla f(x_*)$ as a linear combination of *linearly independent* (since x_* is regular) gradients, taken at x_*, of the constraints that are active at x_*, and thus are uniquely defined.

Proposition 21.4 *[Sufficient optimality condition for general MP] Suppose the Default Assumption holds, and let x_* be such that*

(i) *[first-order part] x_* is a KKT point of (21.1) as defined in item (i) of Proposition 21.3, and*

(ii) *[second-order part] For the Lagrange multipliers λ^*, μ^* associated with the KKT point x_*, we have that the second-order directional derivatives of $L(\cdot; \lambda^*, \mu^*)$ taken at x_* along the nonzero directions from the linear subspace*

$$T_s := \left\{ d \in \mathbf{R}^n : \begin{array}{l} d \text{ is orthogonal to the gradients taken at } x_* \\ \text{of all equality constraints } h_i \text{ and all inequality} \\ \text{constraints } g_j \text{ corresponding to } \lambda_j^* > 0 \end{array} \right\}$$

are positive, i.e.,

$$d^\top \nabla_x^2 \big|_{x=x_*} L(x; \lambda^*, \mu^*) d > 0, \quad \forall d \in T_s \setminus \{0\}.$$

Then, x_* is a locally optimal solution to (21.1).

Note that, by complementary slackness, $\lambda_j^* > 0$ only when the constraint $g_j(x) \leq 0$ is active at x_*. As a result, the linear subspaces participating in the second-order parts of these two optimality conditions are embedded into one another: $T_n \subseteq T_s$, and in general this inclusion is strict. In fact, $T_n = T_s$ holds if and only if all inequality constraints g_j that are active at x_* are associated with positive, and not just nonnegative, Lagrange multipliers λ_j^*.

21.2 Justifying Optimality Conditions

Justifying optimality conditions: preliminary step. It may happen that some of the inequality constraints are non-active at x_*. It can immediately be seen that, in our context, eliminating these constraints from (21.1) changes nothing: on the one hand, x_* remains a regular solution to the resulting problem and is a *locally* optimal solution to the latter problem if and only if it is locally optimal to the former problem. On the other hand, the satisfiability statuses of the optimality conditions in both problems are exactly the same, since these conditions are expressed in terms of two entities:

(i) the functions of x obtained from the respective Lagrange functions by setting λ, μ to λ^*, μ^*; note that complementary slackness enforces these functions for both problems in question to be the same;

(ii) the linear subspaces T_n and T_s; these linear subspaces are defined in terms of the gradients taken at x_* of the constraints *active* at x_* and for both problems in question are the same.

21.2 Justifying Optimality Conditions

The bottom line is that *it suffices to verify the validity of our optimality conditions in the special case when all inequality constraints in (21.1) are active at x_**, and this is what we assume from now on.

21.2.1 Main Tool: Implicit Function Theorem

We are about to justify the optimality conditions by utilizing the following fundamental fact (which is one of the forms of the Implicit Function Theorem available in any graduate textbook on multivariate calculus):

Theorem 21.5 *Let $x_* \in \mathbf{R}^n$, and let $\phi_1, \ldots, \phi_p$ be real-valued functions that are well defined and $\kappa \geq 1$ times continuously differentiable in a neighborhood U of x_* and are normalized by the condition $\phi_\ell(x_*) = 0$, $\ell \leq p$. Assume that the gradients of $\phi_\ell(x)$ taken at x_*, $\ell \leq p$, are linearly independent. Then, there exists a neighborhood $X \subseteq U$ of x_*, a neighborhood Y of $y_* := 0 \in \mathbf{R}^n$, and a one-to-one mapping $y(x)$ of X onto Y which, along with its inverse $x(y)$ (this is a one-to-one mapping of Y onto X), possesses the following properties:*

(i) $y_* := y(x_*) = 0$ ($\Longleftrightarrow x(0) = x_*$);
(ii) $y(x)$ and $x(y)$ are κ times continuously differentiable on X, resp. on Y;
(iii) in y-variables, the functions $\phi_\ell(\cdot)$, $\ell \leq p$, become just the first p coordinates y_ℓ, $\ell \leq p$, of y:

$$\phi_\ell(x(y)) \equiv y_\ell, \; \forall (y \in Y, \ell \leq p) \quad (\Longleftrightarrow \phi_\ell(x) = y_\ell(x), \; \forall (x \in X, \ell \leq p).) \quad (21.2)$$

21.2.2 Strategy

Let us apply the Implicit Function Theorem to the $m + k$ functions given by

$$\phi_j(x) := g_j(x), \; j \leq m, \quad \phi_{m+i}(x) := h_i(x), \; i \leq k. \quad (21.3)$$

Since we are in the case when x_* is a regular solution and *all* constraints of the problem of interest are active at x_*, the resulting $p = m + k$ functions ϕ_ℓ satisfy the premise in the Implicit Function Theorem, with κ set to 2. Let X, Y, $x(\cdot)$, $y(\cdot)$ be the entities given by the theorem as applied to our ϕ_ℓ and x_*. Taking into account (21.2), substituting $x = x(y)$ and defining $\phi(y) := f(x(y))$, we convert the original problem (21.1) into the linearly constrained mathematical programming problem

$$\min_y \{\phi(y) : y_j \leq 0, \; j \leq m, \; y_{m+i} = 0, \; i \leq k\}. \quad (21.4)$$

Taking into account that $y(x)$ and $x(y)$ are twice continuously differentiable mappings that are inverse to each other, of neighborhoods X of x_* and Y of $y_* = y(x_*) = 0$ onto each other, the function $\phi(y)$ is twice continuously differentiable in a neighborhood of $y_* = 0$, and x_* is a locally optimal solution to (21.1) if and only if $y_* = 0$ is a locally optimal solution to (21.4). Moreover, by looking at the problem (21.4), we conclude that $y_* = 0$ is a regular solution to it.

Our course of action will be as follows:

A. We start with verifying that our optimality conditions as applied to problem (21.4) and its regular solution $y_* = 0$ are indeed valid.

B. We "transfer" these *valid* optimality conditions from problem (21.4) to the problem of interest, (21.1). Taking into account that x_* is locally optimal for the latter problem if and only if $y_* = 0$ is a locally optimal solution to the former problem, we end up with *valid* necessary and sufficient optimality conditions for the problem of interest. As we shall see, these valid conditions are *exactly* the conditions stated in Propositions 21.3 and 21.4; this will complete the justification of these two propositions.

Note that when implementing our strategy we should not bother about complementary slackness, since in both problems (21.1) and (21.4) all constraints are active at x_*, resp. y_*, making complementary slackness trivially true.

21.2.3 Justifying Optimality Conditions for (21.4)

We are about to verify the validity of Propositions 21.3 and 21.4 as applied to problem (21.4); verification is nearly immediate. Let

$$\widehat{L}(y; \lambda, \mu) := \phi(y) + \sum_{j=1}^{m} \lambda_j y_j + \sum_{i=1}^{k} \mu_i y_{m+i}$$

be the Lagrange function of (21.4). Here are the straightforward specifications of the optimality conditions stated in Propositions 21.3 and 21.4 as applied to the solution $y_* = 0$ of problem (21.4):

- **Claim N**: *The condition **N** is as follows:*

  N(i) [first-order part] $y_* = 0$ is a KKT point of (21.4), i.e., there exist $\lambda^* \in \mathbf{R}^m_+$ and $\mu^* \in \mathbf{R}^k$ such that $\nabla_y\big|_{y=0} \widehat{L}(y; \lambda^*, \mu^*) = 0$.

  N(ii) [second-order part] *The second-order directional derivatives of the function* $\widehat{L}(\cdot; \lambda^*, \mu^*)$, *or, which is the same under the circumstances, of the function* $\phi(y)$, *taken at the point* $y_* = 0$ *along any direction from the linear space*

  $$\overline{T}_n := \left\{ d \in \mathbf{R}^n : e_\ell^\top d = 0, \forall \ell \leq m+k \right\}$$

  ($e_1, \ldots, e_n$, as always, are the standard basic orths in $\mathbf{R}^n$) are nonnegative, i.e.,

  $$e_\ell^\top d = 0, \forall \ell \leq m+k \implies d^\top \nabla^2_y\big|_{y=0} \widehat{L}(y; \lambda^*, \mu^*) d = d^\top \nabla^2 \phi(0) d \geq 0.$$

  The condition **N** is necessary for $y_* = 0$ to be a locally optimal solution to (21.4).

- **Claim S**: *The condition **S** is as follows:*

  S(i) [first-order part] $y_* = 0$ is a KKT point of (21.4) (see item **N(i)** above),

  S(ii) [second-order part] *The second-order directional derivative of the function* $\widehat{L}(\cdot; \lambda^*, \mu^*)$ *(where* λ^*, μ^* *are the Lagrange multipliers associated with the KKT point* $y_* = 0$*), or, which is the same under the circumstances, of the function* $\phi(y)$, *taken at the point* $y_* = 0$ *along every nonzero direction from the linear subspace*

  $$\overline{T}_s := \left\{ d \in \mathbf{R}^n : e_\ell^\top d = 0, \forall (\ell \leq m : \lambda_\ell^* > 0),\ e_{m+\ell}^\top d = 0, \forall \ell \leq k \right\}$$

is positive, i.e.,

$$d \neq 0 \text{ and } e_\ell^\top d = 0, \forall (\ell \leq m : \lambda_j^* > 0) \text{ and } e_{m+\ell}^\top d = 0, \forall \ell \leq k$$
$$\implies d^\top \nabla_y^2\big|_{y=0} \widehat{L}(y; \lambda^*, \mu^*) d = d^\top \nabla^2 \phi(0) d > 0.$$

The condition **S** *is sufficient for* $y_* = 0$ *to be a locally optimal solution to* (21.4).

We are about to justify Claims N and S.

1. Observe that the feasible set of (21.4) is the polyhedral cone

$$\mathbf{F} := \left\{ d \in \mathbf{R}^n : e_\ell^\top d \leq 0, \forall \ell \leq m, \ e_{m+\ell}^\top d = 0, \forall \ell \leq k \right\},$$

so that a necessary condition for $y_* = 0$ to be a locally optimal solution to (21.4) is that, for every $d \in \mathbf{F}$, 0 is the local minimizer of the restriction of $\phi(\cdot)$ on the ray $\{td : t \geq 0\}$. Next, given a univariate function ψ that is well defined and twice continuously differentiable in a neighborhood of the origin, elementary calculus says that a necessary condition for 0 to be a local minimizer of the restriction of ψ onto the nonnegative ray is

$$\psi'(0) \geq 0 \text{ and } \psi''(0) \geq 0 \text{ when } \psi'(0) = 0.$$

Thus, we conclude that

C: *A necessary condition for* $y_* = 0$ *to be a locally optimal solution to* (21.4) *is that for, every direction* $d \in \mathbf{F}$,

(i) *the first-order directional derivative* $d^\top \nabla \phi(0)$ *of* ϕ *taken at the origin along the direction d is nonnegative, and*

(ii) *if the first-order directional derivative from* (i) *is zero, then the second-order directional derivative* $d^\top \nabla^2 \phi(0) d$ *of* ϕ *taken at the origin along the direction d is nonnegative.*

Now, **C**(i) is simply the fact that the homogeneous linear inequality $d^\top \nabla \phi(0) \geq 0$ in variables $d \in \mathbf{R}^n$ is a consequence of the following system of homogeneous linear inequalities in variables d:

$$-e_j^\top d \geq 0, \forall j \leq m, \quad \pm e_{m+i}^\top d \geq 0, \forall i \leq k.$$

By the Homogeneous Farkas' Lemma, this is the same as saying that $-\nabla \phi(0)$ is a linear combination, with nonnegative coefficients, of the vectors $e_j, j \leq m$, $\pm e_{m+i}, i \leq k$, or, which again is the same, that $-\nabla \phi(0)$ is a linear combination of the standard basic orths e_ℓ, $\ell \leq m+k$, with the first m coefficients nonnegative. Thus, **C**(i) is just the fact that

$$\nabla \phi(0) + \sum_{j=1}^m \lambda_j^* e_j + \sum_{i=1}^k \mu_i^* e_{m+i} = 0$$

for some nonnegative λ_j^*. The bottom line is that **C**(i) amounts to the fact that $y_* = 0$ is a KKT point of problem (21.4).

Let us define and examine the set

$$\mathbf{K} := \left\{ d \in \mathbf{F} : d^\top \nabla \phi(0) = 0 \right\}.$$

When C(i) holds true, by the representation of $\nabla\phi(0)$ in terms of $\lambda^* \in \mathbf{R}^m_+$, $\mu^* \in \mathbf{R}^k$ and also by definition of $\mathbf{F}$, we have

$$\mathbf{K} = \left\{ d \in \mathbf{F} : d^\top \nabla\phi(0) = 0 \right\}$$

$$= \left\{ d \in \mathbf{F} : d^\top \left(\sum_{j=1}^m \lambda_j^* e_j + \sum_{i=1}^k \mu_i^* e_{m+i} \right) = 0 \right\}$$

$$= \left\{ d \in \mathbf{R}^n : \begin{array}{l} e_j^\top d \leq 0, \forall j \leq m, \; e_{m+i}^\top d = 0, \forall i \leq k, \\ \sum_{j=1}^m \lambda_j^* d_j + \sum_{i=1}^k \mu_i^* d_{m+i} = 0 \end{array} \right\}$$

$$= \left\{ d \in \mathbf{R}^n : \begin{array}{l} e_j^\top d \leq 0, \forall j \leq m, \; e_{m+i}^\top d = 0, \forall i \leq k, \\ \sum_{j=1}^m \lambda_j^* d_j = 0 \end{array} \right\}$$

$$= \left\{ d \in \mathbf{R}^n : \begin{array}{l} e_j^\top d \leq 0, \forall (j \leq m : \lambda_j^* = 0) \\ e_j^\top d = 0, \forall (j \leq m : \lambda_j^* > 0) \\ e_{m+i}^\top d = 0, \forall i \leq k \end{array} \right\}.$$

Note that C(ii) requires the quadratic form $d^\top \nabla^2 \phi(0) d$ to be nonnegative on the cone $\mathbf{K}$. The bottom line is that we have justified the following:

Fact N: *The condition* **N***, that*

$y_* = 0$ is a KKT point of (21.4) such that the second-order directional derivatives $d^\top \nabla^2 \phi(0) d$ of ϕ (or, which is the same, of the Lagrange function $\widehat{L}(\cdot; \lambda^*, \mu^*)$, with λ^*, μ^* given by the KKT property of y_*) taken at $y_* = 0$ along all directions from cone $\mathbf{K}$ are nonnegative, is necessary for $y_* = 0$ to be a locally optimal solution to (21.4).

2. We have already outlined that a necessary condition for 0 to be a local minimizer of the restriction onto $\mathbf{R}_+$ of a univariate function ψ that is well defined and twice continuously differentiable in a neighborhood of 0 is "$\psi'(0) \geq 0$ and $\psi''(0) \geq 0$ when $\psi'(0) = 0$." In fact, a slightly strengthened version of this condition is a sufficient condition for the local optimality of 0 for ψ over $\mathbf{R}_+$ as well. Specifically, if for ψ it holds that $\psi'(0) \geq 0$ and $\psi''(0) > 0$ when $\psi'(0) = 0$, then 0 is a local minimizer of the restriction of ψ onto $\mathbf{R}_+$. This suggests the following *educated guess:*

Let all first-order directional derivatives $d^\top \nabla\phi(0)$ of the function ϕ taken at the origin along directions from the cone $\mathbf{F}$ be nonnegative, and the second-order directional derivatives $d^\top \nabla^2 \phi(0) d$ taken at the origin along all nonzero directions from $\mathbf{F}$ which are orthogonal to $\nabla\phi(0)$ be positive. Then $y_* = 0$ is a locally optimal solution to (21.4).

An absolutely straightforward verification demonstrates that our educated guess is correct. Modulo this verification (it requires only the fact that a continuous real-valued function on a nonempty closed and bounded set attains its minimum on the set; establishing this verification is left to the reader), we have arrived at the following:

Fact S: *The condition* **S***, that*

$y_* = 0$ is a KKT point of problem (21.4) such that the second-order directional derivatives $d^\top \nabla^2 \phi(0) d$ of ϕ (or, which is the same, of the Lagrange function $\widehat{L}(\cdot; \lambda^*, \mu^*)$, with λ^*, μ^* given by the KKT property of y_*) taken at $y_* = 0$ along all nonzero directions from cone **K** *are positive, is sufficient for $y_* = 0$ to be a locally optimal solution to (21.4).*

Note that the sufficient optimality condition **S*** is obtained from **N*** by strengthening the nonnegativity of the quadratic form $d^\top \nabla^2 \phi(0) d = d^\top \nabla_y^2|_{y=0} \widehat{L}(y; \lambda^*, \mu^*) d$ on the cone **K** to positivity of the form on the "nonzero part" **K**\\{0} of that cone.

3. So far, we have established conditions **N***, **S*** that are rather close to each other and are necessary, respectively sufficient, for $y_* = 0$ to be a locally optimal solution to (21.4). Unfortunately, these conditions are difficult to verify, since checking nonnegativity or positivity outside the origin of a quadratic form on a polyhedral cone, in contrast with checking the form's nonnegativity or positivity outside the origin on a linear subspace, is a computationally intractable task even when the cone is as simple as the nonnegative orthant. To overcome, to some extent, this difficulty, we

- modify the necessary optimality condition **N*** by replacing nonnegativity of the quadratic form $d^\top \nabla^2 \phi(0) d$ on the cone **K** with nonnegativity of the form on the largest linear subspace contained in **K**. This linear subspace, as is immediately seen, is $\overline{T}_n$, and the resulting "spoiled" necessary optimality condition is nothing but condition **N**;
- modify the sufficient optimality condition **S*** by replacing positivity of the quadratic form $d^\top \nabla^2 \phi(0) d$ on the "nonzero part" **K**\\{0} of the cone **K** with positivity of the form on the nonzero part of the smallest linear subspace containing **K**. This linear subspace, as is immediately seen, is $\overline{T}_s$, and the resulting "spoiled" sufficient optimality condition is nothing but condition **S**.

Claims **N** and **S** are justified.

21.2.4 Justifying Propositions 21.3 and 21.4

Let us start by summarizing some of our observations.

O.1 x_* is a locally optimal solution to (21.1) if and only if $y_* = y(x_*) = 0$ is a locally optimal solution to (21.4).

O.2 Setting

$$\phi_\ell(x) := e_\ell^\top y(x), \quad \forall \ell = 1, \ldots, n,$$

where e_ℓ are the standard basic orths in $\mathbf{R}^n$, we get functions that are twice continuously differentiable in a neighborhood X of x_* such that

$$\phi_\ell(x) = \begin{cases} g_\ell(x), & \text{if } \ell \leq m, \\ h_{\ell-m}(x), & \text{if } m < \ell \leq m+k, \end{cases} \quad \forall x \in X$$

(see (21.2), (21.3)).

O.3 The Lagrange functions $L(x; \lambda, \mu)$, $\widehat{L}(y; \lambda, \mu)$ of problems (21.1) and (21.4) are linked by the relation
$$L(x; \lambda, \mu) = \widehat{L}(y(x); \lambda, \mu), \quad \forall (x \in X, \lambda \in \mathbf{R}^m, \mu \in \mathbf{R}^k).$$

Let
$$J := [\nabla y_1(x_*), \ldots, \nabla y_n(x_*)]^\top$$

be the Jacobian taken at $x = x_*$ of the mapping $x \mapsto y(x)$; this $n \times n$ matrix is nonsingular, the inverse being the Jacobian taken at $y_* := y(x_*) = 0$ of the mapping $y \mapsto x(y)$. By the chain rule we have
$$\nabla_x\big|_{x=x_*} L(x; \lambda, \mu) = J^\top \nabla_y\big|_{y=y_*=0} \widehat{L}(y; \lambda, \mu), \quad \forall (\lambda \in \mathbf{R}^m, \mu \in \mathbf{R}^k).$$

Taking into account the nonsingularity of J and recalling that all constraints in (21.1) and (21.4) are active at x_*, y_*, respectively, we conclude that *x_* is a KKT point of (21.1) if and only if $y_* = 0$ is a KKT point of (21.4), and in this case the Lagrange multipliers λ^*, μ^* certifying the KKT properties of the points are the same for both points.*

O.4 By the same chain rule we have
$$\nabla \phi_\ell(x_*) = J^\top e_\ell, \quad \forall \ell \leq n,$$

implying that *a direction d is orthogonal to $\nabla \phi_\ell(x_*)$ if and only if the direction Jd is orthogonal to e_ℓ.*

O.5 Finally, by elementary calculus we have
$$\begin{aligned} \forall (d \in \mathbf{R}^n, \lambda, \mu): \\ d^\top \nabla_x^2\big|_{x=x_*} L(x; \lambda, \mu) d &= (Jd)^\top \nabla_y^2\big|_{y=y_*=0} \widehat{L}(y; \lambda, \mu)(Jd) \\ &\quad + \left(\nabla_y\big|_{y=y_*} \widehat{L}(y; \lambda, \mu)\right)^\top \frac{d^2}{dt^2}\bigg|_{t=0} y(x_* + td). \end{aligned} \quad (21.5)$$

Justifying Proposition 21.3

Let x_* be a locally optimal solution to (21.1), so that by **O.1** the point $y_* = 0$ is a locally optimal solution to (21.4). Due to the latter fact and the (already verified) Claim **N**, condition **N** is satisfied, whence, in particular, y_* is a KKT point of (21.4), the associated Lagrange multipliers being some $\lambda^* \geq 0, \mu^*$. Consequently, x_* is a KKT point of (21.1), see **O.3**, the associated Lagrange multipliers being the same λ^*, μ^*; we have justified the first-order part of the conclusion in Proposition 21.3. Next, by **O.4** we have $\overline{T}_n = JT_n$. Besides this, by **O.5** with λ, μ set to λ^*, μ^*, we have
$$d^\top \nabla_x^2\big|_{x=x_*} L(x; \lambda^*, \mu^*) d = (Jd)^\top \nabla_y^2\big|_{y=0} \widehat{L}(y; \lambda^*, \mu^*)(Jd) \quad (21.6)$$

for all d, since the last term on the right-hand side of (21.5) vanishes when $\lambda = \lambda^*, \mu = \mu^*$ – we already know that $y_* = 0$ is a KKT point of (21.4), the Lagrange multipliers being λ^*, μ^*. As we have just seen, when $d \in T_n$, we have $Jd \in \overline{T}_n$, so that the right-hand side in (21.6) is nonnegative by the second-order part of condition **N**. Thus, the second-order part of the conclusion in Proposition 21.3 holds as well. ∎

Justifying Proposition 21.4

Let the premise of Proposition 21.4 hold. Then, by the first-order part of this premise, x_* is a KKT point of (21.1), the associated Lagrange multipliers being some $\lambda^* \geq 0$ and μ^*. By **O.3**, $y_* = 0$ is a KKT point of problem (21.4), the Lagrange multipliers being the same λ^*, μ^*. Thus, the first-order part of condition **S** is satisfied. Next, the linear subspace T_s is cut off from $\mathbf{R}^n$ by the requirement that a direction from T_s should be orthogonal to the gradients, taken at x_*, of some of the functions ϕ_ℓ, and the linear subspace $\overline{T}_s$ is cut off from $\mathbf{R}^n$ by the requirement that a direction from $\overline{T}_s$ should be orthogonal to some of the vectors e_ℓ. The indices ℓ participating in these requirements are fully specified, via the same rules for both subspaces in question, by the vectors of Lagrange multipliers associated with the inequality constraints of the respective problems (21.1), (21.4). As has already been explained, these two vectors of Lagrange multipliers coincide with each other, which combines with **O.4** to imply that $\overline{T}_s = JT_s$. Invoking (21.5) with λ, μ set to λ^*, μ^* and taking into account that $y_* = 0$ is a KKT point of (21.4), the Lagrange multipliers being λ^*, μ^*, we conclude that (21.6) still holds true, so that the second-order part of the premise in Proposition 21.4 implies that the second-order part of condition **S** is satisfied. Thus, the whole of condition **S** is satisfied, so that $y_* = 0$ is a locally optimal solution to (21.4) due to the already verified Claim **S**. It remains to note that the local optimality of y_* as a solution to (21.4) implies, by **O.1**, the local optimality of x_* as a solution to (21.1). ∎

21.3 Concluding Remarks

A. We have obtained the claims in Propositions 21.3 and 21.4, by "translating" to the problem of interest, (21.1), Claims **N, S** stating optimality conditions established for problem (21.4). In turn, Claims **N, S** were "spoiled" versions of the "tight" necessary optimality condition **N*** and the "tight" sufficient optimality condition **S*** for (21.4). We could also translate to problem (21.1) the conditions **N***, **S*** as they are, thus arriving at "tight" necessary and sufficient conditions **N⁺, S⁺** for the local optimality of x_* in problem (21.1). As can immediately be seen, **N⁺, S⁺** are obtained from Propositions 21.3 and 21.4 as follows. Define

$$H := \nabla_x^2 \big|_{x=x_*} L(x; \lambda^*, \mu^*).$$

Then,

- in the second-order part of the claim of Proposition 21.3, the requirement $d^\top H d \geq 0$ for all directions d from the linear subspace T_n is strengthened to $d^\top H d \geq 0$ for all directions d from the following cone (which contains T_n):

$$\overline{\mathbf{K}} := \left\{ d \in \mathbf{R}^n : \begin{array}{l} d^\top \nabla h_i(x_*) = 0, \quad \forall i, \\ d^\top \nabla g_j(x_*) = 0, \text{ for all } j \text{ with } \lambda_j^* > 0, \\ d^\top \nabla g_j(x_*) \leq 0, \text{ for all } j \text{ with } g_j(x^*) = 0 \text{ and } \lambda_j^* = 0 \end{array} \right\},$$

- in the second-order part of the claim of Proposition 21.4, the requirement $d^\top H d > 0$ for all nonzero directions d from the linear subspace

$$T_s = \left\{ d : d^\top \nabla h_i(x_*) = 0, i \leq k, d^\top \nabla g_i(x_*) = 0, \forall (j \leq m: \lambda_j^* > 0) \right\}$$

is relaxed to $d^\top H d > 0$ for all nonzero directions d from the cone $\overline{\mathbf{K}}$ contained in T_s.

As we have explained, the rationale to "spoil" $\mathbf{N}^+, \mathbf{S}^+$ is the desire to end up with *efficiently verifiable* optimality conditions. Let us also examine the special cases where the tight conditions are verifiable as they stand. Here are these cases (below, x_* is a regular feasible solution to (21.1) which happens to be a KKT point of the problem, and λ^* is the associated vector of Lagrange multipliers for the inequality constraints):

- [nondegeneracy] All inequality constraints g_j active at x_* are associated with positive Lagrange multipliers λ_j^*.
  In this case, $\overline{\mathbf{K}} = T_\mathrm{n} = T_s$.
- Among the inequality constraints active at x_*, there is exactly one with zero Lagrange multiplier λ_j^*.
  In this case, $\overline{\mathbf{K}}$ is cut off from T_s by a single homogeneous linear inequality, so that the conditions for a homogeneous quadratic form to be nonnegative or positive outside the origin on the cone $\overline{\mathbf{K}}$ and on the subspace T_s are the same. Consequently, in the case in question $\mathbf{N}^+$ is obtained from Proposition 21.3 by replacing in the formulation of the second-order part the subspace T_n with a larger subspace T_s, and Proposition 21.4 states exactly the same as condition $\mathbf{S}^+$.
- Among the inequality constraints active at x_*, there are exactly two with zero Lagrange multipliers λ_j^*, say, the jth and j'th.
  In this case, the cone $\overline{\mathbf{K}}$ is cut off from the linear subspace T_s by two homogeneous linear constraints; thus
  $$\overline{\mathbf{K}} = \left\{ d \in T_s : a^\top d \leq 0,\ b^\top d \leq 0 \right\},$$
  where a and b are the orthoprojections of $\nabla g_j(x_*)$ and $\nabla g_{j'}(x_*)$ onto T_s. Note that a and b are linearly independent due to the regularity of x_* and the origin of T_s. As a result, a homogeneous quadratic form $d^\top H d$ is nonnegative or positive outside the origin on the cone $\overline{\mathbf{K}}$ if and only if it is so everywhere on the set
  $$\{d \in T_s : d^\top \underbrace{[ab^\top + ba^\top]}_{:=E} d \geq 0\},$$
  or, by invoking the $\mathcal{S}$-Lemma (Lemma 18.7), if and only if there exists $\theta \geq 0$ such that the quadratic form $d^\top [H - \theta E] d$ is nonnegative or positive outside the origin on the linear subspace T_s. The bottom line is that in the case in question the optimality conditions $\mathbf{N}^+$ and $\mathbf{S}^+$ are efficiently verifiable and may "outperform" the standard optimality conditions stated in Propositions 21.3 and 21.4. In order to see an example of when they may "outperform," look at what the conditions say in the case of the problem $\min_{x_1, x_2}\{f(x) : x_1 \leq 0, x_2 \leq 0\}$ when $x_* = 0$ and $\nabla f(0) = 0$.

B. Finally, we remark that on the closest inspection, the first-order part of Proposition 21.3 remains a necessary condition for local optimality in the case when we relax our Default Assumption by replacing twice continuous differentiability of the objective and the constraints in a neighborhood of a regular solution x_* with plain continuous differentiability; the validity of this modification is readily given by inspecting the relevant parts of the above derivations and the Implicit Function Theorem applied with $\kappa = 1$ rather than with $\kappa = 2$.

21.3 Concluding Remarks

The resulting "truncated" version of Proposition 21.3 is called the necessary First-Order-Optimality condition.

Aside from rare cases when the optimality conditions allow us to find an optimal solution in closed analytical form, their role in traditional mathematical programming is in "guiding" algorithms and justifying their convergence properties. Specifically, when the current iterate of an iterative algorithm does not satisfy the necessary optimality condition (e.g., the gradient of the smooth objective which we want to minimize over the entire space is nonzero), this condition usually does suggest a way to "improve" the iterate (say, in the example above, moving along the antigradient direction reduces the value of the objective, provided the step is chosen properly), and the algorithm utilizes this improvement and in this sense is guided by the optimality condition. However, in this textbook we do not touch on algorithms at all. Therefore, to illustrate the role of optimality conditions, we now consider the rare situation where these conditions allow the establishment of an important theoretical result, namely, the $\mathcal{S}$-Lemma (Lemma 18.7).

21.3.1 Illustration: $\mathcal{S}$-Lemma Revisited

The proof to follow is very much less elegant than the original one; the advantage of the new proof is its "bare hands" nature – it is what you can develop if you do not want to think much.

The only nontrivial part of the $\mathcal{S}$-Lemma is the claim that if a homogeneous quadratic inequality

$$x^\top B x \geq 0 \qquad (B)$$

is a consequence of a strictly feasible quadratic inequality

$$x^\top A x \geq 0, \qquad (A)$$

then $B - \theta A \succeq 0$ for properly selected $\theta \geq 0$. The proof of this claim via optimality conditions goes as follows. Define $f(x) := x^\top B x$ and $h(x) := x^\top A x - 1$ and consider the following equality-constrained optimization problem

$$\text{Opt} := \min_x \{f(x) : h(x) = 0\}. \qquad (P)$$

Since (A) is strictly feasible, this problem (P) is feasible, and since (B) is a consequence of (A), we have $\text{Opt} \geq 0$. *Assume for a moment that the problem is solvable*, and let x_* be its optimal solution. Note that x_* is a regular solution to our problem, since $\nabla h(x_*) = 2Ax_* \neq 0$ due to $x_*^\top A x_* = 1$. Applying the necessary optimality condition, we get a real number μ^* such that

$$\nabla f(x_*) + \mu^* \nabla h(x_*) = 0, \text{ and}$$
$$d^\top (\nabla^2 f(x_*) + \mu^* \nabla^2 h(x_*)) d \geq 0, \quad \forall (d : d^\top \nabla h(x_*) = 0).$$

That is, by defining $H := B + \mu^* A$ we arrive at

(a) $Hx_* = 0$ and (b) $d^\top H d \geq 0, \forall (d : d^\top A x_* = 0)$.

Thus, we conclude that $x_*^\top H x_* = x_*^\top (B + \mu^* A) x_* = 0$, that is, $\mathrm{Opt} + \mu^* = 0$, implying that $\mu^* \leq 0$ due to $\mathrm{Opt} \geq 0$. Next, for any $g \in \mathbf{R}^n$ and setting $\gamma := g^\top A x_*$ and $d := g - (g^\top A x_*) x_* = g - \gamma x_*$, we get

$$(c)\ d^\top A x_* = 0$$

due to $x_*^\top A x_* = 1$. Consequently, for all $g \in \mathbf{R}^n$ we have

$$g^\top H g = (d + \gamma x_*)^\top H (d + \gamma x_*)$$
$$= \underbrace{d^\top H d}_{\geq 0 \text{ by (c,b)}} + 2\gamma d^\top \underbrace{H x_*}_{=0 \text{ by (a)}} + \gamma^2 x_*^\top \underbrace{H x_*}_{=0} \geq 0,$$

that is, $H \succeq 0$. Thus, setting $\theta = -\mu^* \geq 0$, we get $B - \theta A \succeq 0$, as claimed.

It remains to get rid of the assumption that (P) is solvable. To this end let $\epsilon > 0$ and define $B_\epsilon := B + \epsilon I$. When x satisfies (A), we have $x^\top B_\epsilon x = x^\top B x + \epsilon x^\top x \geq \epsilon x^\top x$, so that replacing in (P) matrix B with B_ϵ, we get a feasible optimization problem with nonnegative optimal value and objective tending to ∞ along every sequence of feasible solutions with norms tending to ∞, implying that the problem is solvable. By the above reasoning applied to B_ϵ in the role of B, there exists $\theta_\epsilon \geq 0$ such that $B_\epsilon \succeq \theta_\epsilon A$. Let $\bar{x}$ be a once and forever fixed feasible solution to (P). Then, the feasibility of $\bar{x}$ (i.e., $\bar{x}^\top A \bar{x} = 1$) and $B_\epsilon \succeq \theta_\epsilon A$ together imply $\bar{x}^\top B_\epsilon \bar{x} \geq \theta_\epsilon \bar{x}^\top A \bar{x} = \theta_\epsilon$. By plugging in the definition of B_ϵ, we see that $\bar{x}^\top B \bar{x} + \epsilon \|\bar{x}\|_2^2 \geq \theta_\epsilon$, and so as $\epsilon \to +0$ the sequence θ_ϵ remains bounded. Thus, defining θ to be a limiting point of θ_ϵ as $\epsilon \to +0$, we get $\theta \geq 0$, and $B_\epsilon \succeq \theta_\epsilon A$ combines with $B_\epsilon \to B$ as $\epsilon \to +0$ to imply that $B \succeq \theta A$. ∎

22

Saddle Points

22.1 Definition and Game Theory Interpretation

When speaking about the "saddle point" formulation of optimality conditions in Convex Programming, we touched on the topic of saddle points, which is very interesting in its own right. Let us recall our situation and the definition of saddle points.

Definition 22.1 *[Saddle points] Let $X \subseteq \mathbf{R}^n$ and $\Lambda \subseteq \mathbf{R}^m$ be two nonempty sets, and let*

$$L(x,\lambda): X \times \Lambda \to \mathbf{R}$$

be a real-valued function of $x \in X$ and $\lambda \in \Lambda$. We say that a point $(x^*, \lambda^*) \in X \times \Lambda$ is a saddle point of L on $X \times \Lambda$ if L attains at this point its maximum in $\lambda \in \Lambda$ and its minimum in $x \in X$, i.e.,

$$L(x, \lambda^*) \geq L(x^*, \lambda^*) \geq L(x^*, \lambda), \quad \forall (x, \lambda) \in X \times \Lambda. \tag{22.1}$$

The notion of a saddle point admits a natural interpretation in *game terms*. Consider a *two-person zero-sum game* where player I chooses $x \in X$ and player II chooses $\lambda \in \Lambda$; after the players have made their decisions, player I pays to player II the sum $L(x, \lambda)$. Of course, I is interested in minimizing his payment, while II is interested in maximizing her income. What is the natural notion of the *equilibrium* in such a game, i.e., what are the choices (x, λ) of players I and II such that neither player has any interest in unilaterally changing his or her choice? It can immediately be seen that the equilibria are exactly the saddle points of the cost function L. Indeed, if the players' decisions (x, y) are chosen to be a saddle point (x^*, λ^*) satisfying (22.1), then player I is not interested in passing from x^* to another choice, given that II keeps her choice $\lambda = \lambda^*$ fixed: the first inequality in (22.1) shows that such a choice cannot decrease the payment of I. Similarly, player II is not interested in choosing something different from λ^*, given that I keeps his choice $x = x^*$, as such an action cannot increase the income of II. On the other hand, if the players' decisions (x, λ) do not correspond to a saddle point, then either player I can decrease his payment by passing from x to another choice, given that II keeps her choice at λ (this is the case when the first inequality in (22.1) is violated), or similarly for player II. Thus, we conclude that the equilibria are exactly the saddle points.

This game interpretation of the notion of a saddle point motivates deep insight into the structure of the set of saddle points. Consider the following two situations:

(A) player I makes his choice first, and player II makes her choice already knowing the choice of player I;

(B) vice versa, player II chooses first, and player I makes his choice already knowing the choice of player II.

In case (A) the reasoning of player I is as follows. If I choose some x, then player II will of course choose the λ which maximizes, for my x, my payment $L(x, \lambda)$, so that I will pay the sum

$$\overline{L}(x) := \sup_{\lambda \in \Lambda} L(x, \lambda).$$

Consequently, my policy should be to choose an x which minimizes my *loss function* $\overline{L}$, i.e., an x which solves the optimization problem

$$\mathrm{Opt}(P) = \inf_{x \in X} \overline{L}(x); \qquad (P)$$

with this policy my anticipated payment will be

$$\mathrm{Opt}(P) = \inf_{x \in X} \overline{L}(x) = \inf_{x \in X} \sup_{\lambda \in \Lambda} L(x, \lambda).$$

In case (B), similar reasoning on the part of player II results in her *profit function* being given by

$$\underline{L}(\lambda) := \inf_{x \in X} L(x, \lambda),$$

and thus player II's objective becomes to maximize in λ $\underline{L}(\lambda)$, resulting in the following optimization problem:

$$\mathrm{Opt}(D) = \sup_{\lambda \in \Lambda} \underline{L}(\lambda). \qquad (D)$$

On the basis of this policy, the anticipated profit of II is given by

$$\mathrm{Opt}(D) = \sup_{\lambda \in \Lambda} \underline{L}(\lambda) = \sup_{\lambda \in \Lambda} \inf_{x \in X} L(x, \lambda).$$

Note that these two sets of reasoning relate to two *different* games: one with the priority of player II (when making her decision, player II already knows the choice of player I), i.e., the case of (A); and the other with the similar priority of player I, i.e., the case of (B). Therefore, generally speaking, we should not expect that the anticipated loss of player I in (A) is equal to the anticipated profit of player II in (B). What can be guessed is that the anticipated loss of player I in (B) is *less than or equal to* the anticipated profit of player II in (A), since the conditions of game (B) are better for player I than those of game (A). Thus, we may guess that *independently of any structural property of the function* $L(x, \lambda)$, the following inequality holds:

$$\mathrm{Opt}(D) = \sup_{\lambda \in \Lambda} \inf_{x \in X} L(x, \lambda) \leq \mathrm{Opt}(P) = \inf_{x \in X} \sup_{\lambda \in \Lambda} L(x, \lambda). \qquad (22.2)$$

This inequality indeed is true, which is seen from the following reasoning:

$$\inf_{x \in X} L(x, \lambda) \leq L(y, \lambda), \qquad \forall y \in X,$$

$$\implies \sup_{\lambda \in \Lambda} \inf_{x \in X} L(x, \lambda) \leq \sup_{\lambda \in \Lambda} L(y, \lambda) =: \overline{L}(y), \qquad \forall y \in X.$$

Consequently, the quantity $\sup_{\lambda \in \Lambda} \inf_{x \in X} L(x, \lambda)$ is a lower bound for the function $\overline{L}(y)$ for all $y \in X$, and therefore it is a lower bound for the infimum of the latter function over $y \in X$, i.e., it is a lower bound for $\inf_{y \in X} \sup_{\lambda \in \Lambda} L(y, \lambda)$.

Now let us look at what happens when the game in question has a saddle point (x^*, λ^*), so that the following relations hold:

$$L(x, \lambda^*) \geq L(x^*, \lambda^*) \geq L(x^*, \lambda), \quad \forall (x, \lambda) \in X \times \Lambda. \tag{22.3}$$

We claim that if this is the case, then we have the following property:

(*) x^* is an optimal solution to (P), λ^* is an optimal solution to (D) and the optimal values in these two optimization problems are equal to each other (and are equal to the quantity $L(x^*, \lambda^*)$).

Indeed, from (22.3) it follows that

$$\underline{L}(\lambda^*) \geq L(x^*, \lambda^*) \geq \overline{L}(x^*),$$

hence, of course,

$$\sup_{\lambda \in \Lambda} \underline{L}(\lambda) \geq \underline{L}(\lambda^*) \geq L(x^*, \lambda^*) \geq \overline{L}(x^*) \geq \inf_{x \in X} \overline{L}(x).$$

On the other hand, from (22.2) we deduce that $\sup_{\lambda \in \Lambda} \underline{L}(\lambda) \leq \inf_{x \in X} \overline{L}(x)$, which is possible if and only if all the inequalities in the chain are equalities, which is exactly what is said in property (*).

Thus, if (x^*, λ^*) is a saddle point of L, then (*) holds. We are about to demonstrate that the inverse is also true.

Theorem 22.2 *[Structure of the set of saddle points] Let $L : X \times Y \to \mathbf{R}$ be a function. The function L has a saddle point if and only if the related optimization problems (P) and (D) are solvable with real value $\mathrm{Opt}(P) = \mathrm{Opt}(D)$. In such a case, the saddle points of L are exactly all the pairs (x^*, λ^*) where x^* is an optimal solution to (P) and λ^* is an optimal solution to (D), and the value of the cost function $L(\cdot, \cdot)$ at every one of these points is equal to the common optimal value in (P) and (D).*

Proof We have already established "half" the theorem: if there are saddle points of L, then their components are optimal solutions to (P), respectively (D), and the optimal objective values, $\mathrm{Opt}(P)$ and $\mathrm{Opt}(D)$, of these two problems are equal to each other and to the value of L at the saddle point in question.

To complete the proof, we should demonstrate that if x^* is an optimal solution to (P), λ^* is an optimal solution to (D), and the (real) optimal objective values of these two problems are equal to each other then (x^*, λ^*) is a saddle point of L. But this is also immediate, as for all $x \in X$, $\lambda \in \Lambda$ we have

$$L(x, \lambda^*) \geq \underline{L}(\lambda^*) = \overline{L}(x^*) \geq L(x^*, \lambda).$$

Here, the first inequality holds by the definition of $\underline{L}$, the equality holds by the assumption that the optimum objective values of the problems (P) and (D) are the same, and the last inequality follows from the definition of $\overline{L}$. Hence,

$$L(x, \lambda^*) \geq L(x^*, \lambda) \quad \forall x \in X, \lambda \in \Lambda.$$

By substituting $\lambda = \lambda^*$ into the right-hand side of this inequality, we get $L(x, \lambda^*) \geq L(x^*, \lambda^*)$, and also substituting $x = x^*$ into the right-hand side of our inequality we get $L(x^*, \lambda^*) \geq L(x^*, \lambda)$. Thus (x^*, λ^*) is indeed a saddle point of L. ∎

22.2 Existence of Saddle Points: Sion–Kakutani Theorem

It is easily seen that a "quite respectable" function L on a direct product-type convex domain may have no saddle points.

Example 22.1 Consider the function $L(x, \lambda) := (x - \lambda)^2$ on the unit square $[0, 1] \times [0, 1]$. Then, we have

$$\overline{L}(x) = \sup_{\lambda \in [0,1]} (x - \lambda)^2 = \max\{x^2, (x-1)^2\},$$
$$\underline{L}(\lambda) = \inf_{x \in [0,1]} (x - \lambda)^2 = 0, \ \lambda \in [0, 1],$$

so that the optimal objective value of (P) is $\frac{1}{4}$, and the optimal objective value of (D) is 0. Hence, Theorem 22.2 implies that L has no saddle points. ◇

On the other hand, there are generic cases when L does have a saddle point, for example, when

$$L(x, \lambda) = f(x) + \sum_{i=1}^{m} \lambda_i g_i(x) : X \times \mathbf{R}_+^m \to \mathbf{R}$$

is the Lagrange function of a *solvable* convex program satisfying the Slater condition. Setting $\Lambda = \mathbf{R}_+^m$, we get

$$\overline{L}(x) = \begin{cases} f(x), & \text{if } g_j(x) \leq 0, \ j \leq m \\ +\infty, & \text{otherwise} \end{cases} : X \to \mathbf{R} \cup \{+\infty\},$$

so that, in the notation from Theorem 22.2, the problem (P) stemming from L, Λ, X is (equivalent to) the problem

$$\min_x \{f(x) : g_j(x) \leq 0, \ j \leq m, \ x \in X\} \qquad (*)$$

underlying the Lagrange function in question, and problem (D) is the standard Lagrange dual of $(*)$, cf. section 17.3. Theorem 19.1 states that in this case saddle points exist, which, in particular, combines with Theorem 22.2 to imply that the value of L at a saddle point is equal to the optimal value of (P), that is, of $(*)$ – a fact that was not explicitly articulated in Theorem 19.1 and which we will use later.

Note that, in the case in question, $\Lambda = \mathbf{R}_+^m$ and X are nonempty and convex, and L is convex in x for every fixed $\lambda \in \Lambda$ and is linear (and therefore concave) in λ for every fixed $x \in X$. As we shall see in a while, these structural properties of L – convexity in x for every fixed λ and concavity in λ for every fixed x – take upon themselves the "main responsibility" for the fact that in this case saddle points exist. We state this formally as the following result, whose origin dates back to von Neumann.

22.3 Proof of Sion–Kakutani Theorem

Theorem 22.3 *[Existence of saddle points of a convex–concave function (Sion–Kakutani)] Let X and Λ be nonempty convex compact sets in $\mathbf{R}^n$ and $\mathbf{R}^m$, respectively, and let*

$$L(x, \lambda) : X \times \Lambda \to \mathbf{R}$$

be a continuous function which is convex in $x \in X$ for every fixed $\lambda \in \Lambda$ and is concave in $\lambda \in \Lambda$ for every fixed $x \in X$. Then, L has saddle points on $X \times \Lambda$.

In the proof of Theorem 22.3 we will use a basic fact from Real Analysis.

Fact 22.4 *Let X and Λ be nonempty compact sets in $\mathbf{R}^n$ and $\mathbf{R}^m$, respectively, and let $L(x, \lambda) : X \times \Lambda \to \mathbf{R}$ be a continuous function. Then, the problems (P) and (D) are solvable.*

In order to prove Theorem 22.3, we need one more critical ingredient, which is quite significant on its own right.

Lemma 22.5 *[Minimax Lemma] Let X be a nonempty convex compact set, and let $f_0, \ldots, f_N$ be a collection of $N+1$ convex and continuous functions on X. Then, there exist convex combination weights $\mu_i \in \mathbf{R}_+$, $i = 0, \ldots, N$ such that $\sum_{i=0}^{N} \mu_i = 1$ and*

$$\min_{x \in X} \max_{i=0,\ldots,N} f_i(x) = \min_{x \in X} \sum_{i=0}^{N} \mu_i f_i(x).$$

Remark 22.6 The minimum of *every* convex combination of a collection of *arbitrary* functions is less than or equal to the minimax of the collection. This evident fact can be also obtained from applying (22.2) to the function

$$M(x, \mu) := \sum_{i=0}^{N} \mu_i f_i(x)$$

on the direct product of X and the standard simplex

$$\Delta = \left\{ \mu \in \mathbf{R}^{N+1} : \mu_i \geq 0, \forall 0 \leq i \leq N, \sum_{i=0}^{N} \mu_i = 1 \right\}.$$

The interesting part of the Minimax Lemma states that if functions f_i are convex and continuous on a convex compact set X, then the indicated inequality is in fact equality. You can easily verify that this is simply the claim that the function M possesses a saddle point. Thus, the Minimax Lemma is in fact a particular case of the Sion–Kakutani Theorem (i.e., Theorem 22.3). ◇

We will first give a direct proof of this particular case of Theorem 22.3, stated in Lemma 22.5, and then use it to prove the general case given in Theorem 22.3.

22.3 Proof of Sion–Kakutani Theorem

22.3.1 Proof of Minimax Lemma

Consider the optimization program

$$\min_{t,x} \{ t : x \in X, f_i(x) - t \leq 0, \forall i = 0, \ldots, N \}. \tag{S}$$

This is clearly a convex problem with optimal objective value equal to

$$t^* := \min_{x \in X} \max_{i=0,\ldots,N} f_i(x).$$

Note that (t, x) is a feasible solution for (S) if and only if $x \in X$ and $t \geq \max_{i=0,\ldots,N} f_i(x)$. Problem (S) clearly satisfies the Slater condition and is solvable (since X is a compact set and f_i, $i = 0, \ldots, N$, are continuous on X; therefore their maximum is also continuous on X and thus attains its minimum on the compact set X). Let (t^*, x^*) be an optimal solution to problem (S). According to Theorem 19.1, there exists $\lambda^* \geq 0$ such that $((t^*, x^*), \lambda^*)$ is a saddle point of the corresponding Lagrange function

$$L(t, x; \lambda) = t + \sum_{i=0}^{N} \lambda_i (f_i(x) - t) = t \left(1 - \sum_{i=0}^{N} \lambda_i\right) + \sum_{i=0}^{N} \lambda_i f_i(x)$$

on $(t, x; \lambda) \in [\mathbf{R} \times X] \times \mathbf{R}_+^{N+1}$, and the value of this function at $((t^*, x^*), \lambda^*)$ is equal to t^*, i.e., the optimal objective value of (S), see the discussion preceding Theorem 22.3.

Now, since $L(t, x; \lambda^*)$ attains its minimum in (t, x) over the set $\mathbf{R} \times X$ at (t^*, x^*), we must have

$$\sum_{i=0}^{N} \lambda_i^* = 1;$$

otherwise the minimum of L in (t, x) would be $-\infty$ (look what happens when $\sum_i \lambda_i < 1$ and $t \to -\infty$, and what happens when $\sum_i \lambda_i > 1$ and $t \to +\infty$). Thus,

$$\left(\min_{x \in X} \max_{i=0,\ldots,N} f_i(x)\right) = t^* = \min_{t \in \mathbf{R}, x \in X} \left(t \cdot 0 + \sum_{i=0}^{N} \lambda_i^* f_i(x)\right),$$

so that

$$\min_{x \in X} \max_{i=0,\ldots,N} f_i(x) = \min_{x \in X} \sum_{i=0}^{N} \lambda_i^* f_i(x)$$

for some $\lambda_i^* \geq 0$, $\forall 0 \leq i \leq N$ satisfying $\sum_{i=0}^{N} \lambda_i^* = 1$, as desired. ∎

We are now ready to prove Theorem 22.3, the Sion–Kakutani Theorem.

22.3.2 From Minimax Lemma to the Proof of Sion–Kakutani Theorem

On the basis of Theorem 22.2, all we need to prove is that

(i) the optimization problems (P) and (D) are solvable, and
(ii) the optimal values in (P) and (D) are equal to each other.

In fact, (i) is valid independently of the convexity–concavity of L and the convexity of the sets X and Λ, and it is simply given by Fact 22.4.

The essence of the matter in the proof of Theorem 22.3 lies in (ii). In order to prove (ii), of course, we will heavily exploit the convexity–concavity of L and the convexity of the sets X and Λ. We will prove this in several steps and use the Minimax Lemma (i.e., Lemma 22.5) in the last step.

22.3 Proof of Sion–Kakutani Theorem

1. We need to prove that the optimal objective values of (P) and (D) (which, by (i), are well-defined real numbers) are equal to each other, i.e., the following relation holds:

$$\inf_{x \in X} \sup_{\lambda \in \Lambda} L(x, \lambda) = \sup_{\lambda \in \Lambda} \inf_{x \in X} L(x, \lambda).$$

We already know from (22.2) (which, by the way, makes no structural assumptions on L, X, or Λ) that

$$\inf_{x \in X} \sup_{\lambda \in \Lambda} L(x, \lambda) \geq \sup_{\lambda \in \Lambda} \inf_{x \in X} L(x, \lambda).$$

So, all we need to do is to prove the reverse inequality. Without loss of generality we can assume that $\mathrm{Opt}(D)$ is zero, i.e., $\mathrm{Opt}(D) = \sup_{\lambda \in \Lambda} \inf_{x \in X} L(x, \lambda) = 0$ (we can achieve this by shifting the function L by a constant quantity if needed). Then, all we need to prove is that $\mathrm{Opt}(P)$ cannot be positive, i.e., $\mathrm{Opt}(P) = \inf_{x \in X} \sup_{\lambda \in \Lambda} L(x, \lambda) \leq 0$.

2. What does that if $\mathrm{Opt}(D)$ is zero mean? When $\mathrm{Opt}(D) = 0$, then the function $\underline{L}(\lambda) = \inf_{x \in X} L(x, \lambda)$ is nonpositive for every $\lambda \in \Lambda$; equivalently, for every $\lambda \in \Lambda$, the function $L(x, \lambda)$, which is a convex and continuous function of $x \in X$, has a nonpositive minimal value over $x \in X$. Since X is compact, this minimal value is achieved, so that for any $\lambda \in \Lambda$ the set

$$X(\lambda) := \{x \in X : L(x, \lambda) \leq 0\}$$

is nonempty. Moreover, as X is convex and for every $\lambda \in \Lambda$ the function L is convex in $x \in X$, the set $X(\lambda)$ is convex (it is simply the sublevel set of a convex function, so Proposition 9.6 applies here). Note also that the set $X(\lambda)$ is closed since X is closed and $L(x, \lambda)$ is continuous in $x \in X$. Thus, if $\mathrm{Opt}(D) = 0$, then the set $X(\lambda)$ is a nonempty convex compact set for every $\lambda \in \Lambda$.

3. What does that if $\mathrm{Opt}(P) \leq 0$ mean? It means exactly that there is a point $x \in X$ where the function $\overline{L}$ is nonpositive, i.e., there exists a point $x \in X$ where $L(x, \lambda) \leq 0$ for all $\lambda \in \Lambda$. In other words, to prove that $\mathrm{Opt}(P) \leq 0$ is the same as proving that *the sets $X(\lambda)$, $\lambda \in \Lambda$, have a point in common*.

4. With the above observations we see that the situation is as follows: we are given a family of nonempty closed convex subsets $X(\lambda)$, $\lambda \in \Lambda$, of a compact set X, and we need to prove that these sets has a point in common. By Helly Theorem II, in order to prove that all $X(\lambda)$ have a point in common, it suffices to prove that every one of the $N + 1$ sets of this family, where $N := \dim(X)$, has a point in common. Let $X(\lambda_0), \ldots, X(\lambda_N)$ be $N + 1$ sets from our family; we want to prove that the sets have a point in common. To this end, by defining

$$f_i(x) := L(x, \lambda_i), \ i = 0, \ldots, N,$$

all we need to prove is that there exists a point x where all our functions are nonpositive, or, equivalently prove that the minimax of our collection of functions f_i for $i = 0, \ldots, N$, i.e., the quantity

$$\alpha := \min_{x \in X} \max_{i=0,\ldots,N} f_i(x),$$

is nonpositive.

Note that as L is convex and continuous in x, all the functions f_i are convex and continuous. Since X is compact and all f_i are convex and continuous, we can then apply the

Minimax Lemma (i.e., Lemma 22.5) and deduce that α is the minimum in $x \in X$ of a certain convex combination $\phi(x) := \sum_{i=0}^{N} \nu_i f_i(x)$ (where $\nu_i \geq 0, \sum_i \nu_i = 1$) of the functions $f_i(x)$. Thus, we arrive at

$$\phi(x) = \sum_{i=0}^{N} \nu_i f_i(x) = \sum_{i=0}^{N} \nu_i L(x, \lambda_i) \leq L\left(x, \sum_{i=0}^{N} \nu_i \lambda_i\right),$$

where the last inequality follows from the concavity of L in λ (this is the only – and crucial – point where we use this assumption). Then, we see that $\phi(\cdot)$ is majorized by $L(\cdot, \bar{\lambda})$ where $\bar{\lambda} := \sum_{i=0}^{N} \nu_i \lambda_i$ for some convex combination weights ν_i, so that $\bar{\lambda} \in \Lambda$ by the convexity of Λ. Thus, the minimum of ϕ in $x \in X$ – and we already know that this minimum is exactly α – is nonpositive (recall that the minimum of L in x is nonpositive for every $\lambda \in \Lambda$). ∎

22.4 Sion–Kakutani Theorem: A Refinement

The next theorem lifts the assumption of boundedness of X and Λ in Theorem 22.3 – now only one of these sets needs to be bounded – at the price of some weakening of the conclusion. In particular, when only one of the sets is bounded, we can no longer claim to be able to find the associated saddle points (as they may not exist); yet we can switch the order in which we take the minimum over x and the maximum over λ, i.e., the optimum objective values of the associated problems are still the same.

Theorem 22.7 *[Swapping* min *and* max *in convex–concave saddle point problem (Sion–Kakutani)] Let X and Λ be nonempty convex sets in $\mathbf{R}^n$ and $\mathbf{R}^m$, respectively, and suppose that X is compact. Let*

$$L(x, \lambda) : X \times \Lambda \to \mathbf{R}$$

be a continuous function which is convex in $x \in X$ for every fixed $\lambda \in \Lambda$ and is concave in $\lambda \in \Lambda$ for every fixed $x \in X$. Then,

$$\inf_{x \in X} \sup_{\lambda \in \Lambda} L(x, \lambda) = \sup_{\lambda \in \Lambda} \inf_{x \in X} L(x, \lambda). \tag{22.4}$$

Proof Once again, we already know from (22.2) (which, by the way, makes no structural assumptions on L, X, or Λ) that

$$\inf_{x \in X} \sup_{\lambda \in \Lambda} L(x, \lambda) \geq \sup_{\lambda \in \Lambda} \inf_{x \in X} L(x, \lambda).$$

Hence, there is nothing to prove when the right-hand side in (22.4) is $+\infty$. So, we assume that this is not the case. Since X is compact and L is continuous in $x \in X$, $\inf_{x \in X} L(x, \lambda) > -\infty$ for every $\lambda \in \Lambda$, so that the right hand side in (22.4) cannot be $-\infty$, either. Then, $\sup_{\lambda \in \Lambda} \inf_{x \in X} L(x, \lambda) \in (-\infty, \infty)$, and thus it must be a real number. By shifting the function L by a constant (if needed), without loss of generality we can assume that this real number is 0, i.e.,

$$\sup_{\lambda \in \Lambda} \inf_{x \in X} L(x, \lambda) = 0.$$

All we need to prove now is that the left-hand side in (22.4) is nonpositive. Assume, on the contrary, that it is positive and thus it is strictly greater than some number $c > 0$. Then, for every $x \in X$ there exists $\lambda_x \in \Lambda$ such that $L(x, \lambda_x) > c$. By continuity, there exists a neighborhood V_x of x in X (i.e., the intersection with X of an open set containing x) such that $L(x', \lambda_x) \geq c$ for all $x' \in V_x$. Since X is compact, we can find finitely many points $x_1, \ldots, x_p$ in X such that the union set given by $\bigcup_{i=1}^{p} V_{x_i}$ is exactly X. Then, we deduce $\max_{1 \leq i \leq p} L(x, \lambda_{x_i}) \geq c$ for every $x \in X$.

Now, define
$$\bar{\Lambda} := \mathrm{Conv}\left\{\lambda_{x_1}, \ldots, \lambda_{x_n}\right\}.$$

Then, $\bar{\Lambda}$ is compact by design and $\max_{\lambda \in \bar{\Lambda}} L(x, \lambda) \geq c$ for every $x \in X$, so that $c \leq \min_{x \in X} \max_{\lambda \in \bar{\Lambda}} L(x, \lambda)$. We can now apply Theorem 22.3 to L and the convex *compact* sets X and $\bar{\Lambda}$ to get the equality in the following chain:

$$c \leq \min_{x \in X} \max_{\lambda \in \bar{\Lambda}} L(x, \lambda) = \max_{\lambda \in \bar{\Lambda}} \min_{x \in X} L(x, \lambda) \leq \sup_{\lambda \in \Lambda} \min_{x \in X} L(x, \lambda) = 0.$$

This implies $c \leq 0$ and gives the desired contradiction (recall that $c > 0$). ∎

In the case when one of the sets, say Λ, is unbounded, a slightly stronger premise of Theorem 22.7 allows us to replace (22.4) with the existence of a saddle point.

Theorem 22.8 [Existence of saddle point in convex–concave saddle point problem (Sion–Kakutani, semi-bounded case)] *Let X and Λ be nonempty closed convex sets in $\mathbf{R}^n$ and $\mathbf{R}^m$, respectively, and suppose that X is compact. Let*

$$L(x, \lambda) : X \times \Lambda \to \mathbf{R}$$

be a continuous function which is convex in $x \in X$ for every fixed $\lambda \in \Lambda$ and is concave in $\lambda \in \Lambda$ for every fixed $x \in X$. Furthermore, assume that for every $a \in \mathbf{R}$, there exist a collection of points $x_1^a, \ldots, x_{n_a}^a \in X$ such that the set

$$\{\lambda \in \Lambda : L(x_i^a, \lambda) \geq a,\ 1 \leq i \leq n_a\}$$

is bounded.[1] *Then, L has saddle points on $X \times \Lambda$.*

Proof Since X is compact and L is continuous, the function

$$\underline{L}(\lambda) = \min_{x \in X} L(x, \lambda)$$

is real-valued and continuous on Λ. Furthermore, for every $a \in \mathbf{R}$, the set $\{\lambda \in \Lambda : \underline{L}(\lambda) \geq a\}$ is clearly contained in the set $\{\lambda : L(x_i^a, \lambda) \geq a,\ 1 \leq i \leq n_a\}$ and thus is bounded. Thus, $\underline{L}(\lambda)$ is a continuous function on a closed set Λ, and its superlevel sets $\{\lambda \in \Lambda : \underline{L}(\lambda) \geq a\}$ are bounded. Therefore, $\underline{L}$ attains its maximum on Λ. Then, by Theorem 22.7 we deduce that $\inf_{x \in X}[\overline{L}(x) := \sup_{\lambda \in \Lambda} L(x, \lambda)]$ is finite. Hence, the function $\overline{L}(\cdot)$ is not $+\infty$ identically in $x \in X$. Also, as L is continuous, $\overline{L}$ is lower semicontinuous. Thus, $\overline{L} : X \to \mathbf{R} \cup \{+\infty\}$ is a lower semicontinuous proper (i.e., not identically $+\infty$) function on X, and, since we additionally have that X is compact, we conclude that $\overline{L}$ attains its minimum on X. Thus, both the problems $\max_{\lambda \in \Lambda} \underline{L}(\lambda)$ and $\min_{x \in X} \overline{L}(x)$ are solvable, and their optimal objective

---

[1] This is definitely the case when $L(\bar{x}, \lambda)$ is *coercive* in λ for some $\bar{x} \in X$, meaning that the sets $\{\lambda \in \Lambda : L(\bar{x}, \lambda) \geq a\}$ are bounded for every $a \in \mathbf{R}$, or, equivalently, whenever $\lambda_i \in \Lambda$ and $\|\lambda_i\|_2 \to \infty$ as $i \to \infty$ we have $L(\bar{x}, \lambda_i) \to -\infty$ as $i \to \infty$.

values are equal by Theorem 22.7. Then, by Theorem 22.2, we conclude that L has a saddle point. ∎

22.5 Proof of Fact

Fact 22.4 *Let X and Λ be nonempty compact sets in $\mathbf{R}^n$ and $\mathbf{R}^m$, respectively, and let $L(x, \lambda) : X \times \Lambda \to \mathbf{R}$ be a continuous function. Then, the problems (P) and (D) are solvable.*

Proof Since X and Λ are compact sets and L is continuous on $X \times \Lambda$, owing to the well-known Analysis theorem (Theorem B.29 and Remark B.30) L is uniformly continuous on $X \times \Lambda$. That is, for every $\epsilon > 0$ there exists $\delta(\epsilon) > 0$ such that

$$\|x - x'\|_2 + \|\lambda - \lambda'\|_2 \leq \delta(\epsilon) \quad \Longrightarrow \quad |L(x, \lambda) - L(x', \lambda')| \leq \epsilon. \tag{22.5}$$

In particular, for every $\lambda = \lambda' \in \Lambda$ we have

$$\|x - x'\|_2 \leq \delta(\epsilon) \quad \Longrightarrow \quad |L(x, \lambda) - L(x', \lambda)| \leq \epsilon,$$

which immediately implies

$$\|x - x'\|_2 \leq \delta(\epsilon) \quad \Longrightarrow \quad |\overline{L}(x) - \overline{L}(x')| \leq \epsilon,$$

so that the function $\overline{L}$ is continuous on X. This, together with the fact that X is compact, implies that the problem (P) is solvable. The same reasoning applied to the variables λ leads to the conclusion that problem (D) is solvable. ∎

23

Exercises for Part IV

23.1 Around Conic Duality

Exercise IV.1 Given the linear dynamic system

$$x_0 = 0,$$
$$x_{t+1} = Ax_t + Bu_t, \; t = 0, 1, \ldots, N-1 \quad \text{(LDS)}$$

($A : n \times n, B : n \times m$), with controls u_t subject to the "energy constraints"

$$\|u_t\|_2 \leq 1, \quad 0 \leq t < N, \quad \text{(EN)}$$

pose the problem of maximizing $f^\top x_N$ (f is a given vector) as a conic problem on the product of Lorentz cones, write down the conic dual of this problem, and answer the following questions:

1. Is the problem essentially strictly feasible?
2. Is the problem bounded?
3. Is the problem solvable?
4. Is the dual problem feasible?
5. Is the dual problem solvable?
6. Are the optimal values equal to each other?
7. What do the optimality conditions say?

Exercise IV.2 [TrYs] Consider the conic constraint $Ax - b \in K$ where $K \subset \mathbf{R}^m$ is a regular cone and matrix A is of full column rank (i.e., has linearly independent columns, or, which is the same, has a trivial kernel). Suppose that the constraint is feasible. Show that the following properties are all equivalent to each other:

1. the feasible region $\{x \in \mathbf{R}^n : Ax - b \in K\}$ is bounded;
2. $\text{Im}(A) \cap K = \{0\}$, where $\text{Im}(A) := \{Ax : x \in \mathbf{R}^n\}$;
3. the following system of vector inequalities is solvable:

$$A^\top \lambda = 0, \quad \lambda \in \text{int } K_*.$$

Using these conclude that the property of whether a given conic optimization problem has a bounded feasible region or not is independent of the choice of b, provided that the problem is feasible.

Exercise IV.3 [EdYs] Given a cone K in a Euclidean space E with inner product $\langle \cdot, \cdot \rangle$, we call a pair of elements $x \in K$ and $y \in K_*$ *complementary* if they satisfy $\langle x, y \rangle = 0$.

In this exercise, we will examine complementarity relations for the second-order cones $\mathbf{L}^n$ and the positive semidefinite cone $\mathbf{S}^n_+$.

1. Consider $\mathbf{L}^n := \{x = [\tilde{x}; x_n] \in \mathbf{R}^{n-1} \times \mathbf{R} : x_n \geq \|\tilde{x}\|_2\}$; as we know, this cone is self-dual (Example 6.9). Prove that $x, s \in \mathbf{L}^n$ satisfy $\langle x, s \rangle = 0$ if and only if $x_n \tilde{s} + s_n \tilde{x} = 0$ holds.

2. Consider the space of $n \times n$ symmetric matrices, i.e., $E = \mathbf{S}^n$ equipped with the Frobenius inner product $\langle X, Y \rangle = \text{Tr}(XY) = \sum_{i=1}^n \sum_{j=1}^n X_{ij} Y_{ij}$. Let $K = \mathbf{S}^n_+ := \{X \in \mathbf{S}^n : x^\top X x \geq 0, \forall x \in \mathbf{R}^n\}$ be the positive semidefinite cone; recall that this cone is self-dual (Example 6.10). Prove that $X, Y \in \mathbf{S}^n_+$ are complementary, i.e., $\langle X, Y \rangle = 0$, if and only if their matrix product is zero, i.e., $XY = YX = 0$. In particular, matrices from a complementary pair commute and therefore share a common orthonormal eigenbasis.

Exercise IV.4 [EdYs] By General Theorem of the Alternative, a system of m scalar linear constraints $Ax \geq b$ in variables $x \in \mathbf{R}^n$ (or, which is the same, the conic inequality $Ax \geq_{\mathbf{R}^m_+} b$) has no solutions if and only if it can be led to contradiction by aggregation: there exist nonnegative weights $\lambda_1, \ldots, \lambda_m$ such that the associated weighted sum $\lambda^\top Ax \geq \lambda^\top b$ of inequalities from the system is a contradictory inequality, that is, $A^\top \lambda = 0$ and $b^\top \lambda > 0$. For a general conic constraint of the form

$$Ax \geq_{\mathbf{K}} b, \qquad (I)$$

where $\mathbf{K} \subset \mathbf{R}^m$ is a regular cone, a similar recipe for certifying infeasibility would read

$$\exists \lambda \in \mathbf{K}_* : A^\top \lambda = 0 \quad \text{and} \quad b^\top \lambda > 0. \qquad (II)$$

The goal of this exercise is to investigate the relation between the feasibility statuses of (I) and of (II).

The first task is easy:

1. Prove that if (II) is feasible, then (I) is infeasible.

The remaining effort is aimed at investigating to what extent this item can be inverted: if and when is it true that when (II) has no solutions, then (I) is feasible? The General Theorem of the Alternative says that this indeed is the case when $\mathbf{K}$ is the nonnegative orthant $\mathbf{R}^m_+$. In the general case, the situation is different.

2. Let (I) be the univariate conic inequality

$$Ax := [1; 0; 1]x \geq_{\mathbf{L}^3} b := [0; 1; 0] \qquad (i)$$

where $\mathbf{L}^3$ is the three-dimensional Lorentz cone. Write down the associated system (II) and check that both this system and (i) are infeasible. Conclude from this example that, in general, the solvability of (II) is only a sufficient, not a necessary, condition for the infeasibility of (I).

3. Prove that (II) is infeasible if and only if (I) is *nearly feasible*, meaning that for every $\epsilon > 0$ there exists b' such that $\|b' - b\|_2 \leq \epsilon$ and the conic constraint $Ax \geq_{\mathbf{K}} b'$ is feasible. Equivalently: (II) is infeasible if and only if b belongs to the closure $\overline{B}$ of the set $B = A\mathbf{R}^n - \mathbf{K}$ of those right-hand side vectors in (I) for which (I) is feasible.

Conclusion: *The solvability of* (II) *is necessary and sufficient for the infeasibility of* (I) *if and only if the set* $B = A\mathbf{R}^n - \mathbf{K}$ *of the right-hand sides in the conic constraint* (I) *resulting in the constraint's solvability is closed; in fact, the solvability of* (II) *is a necessary and sufficient condition for b to belong to the closure* $\overline{B}$ *of B*. Now, when $\mathbf{K} = \{y : Py \geq 0\}$ is a polyhedral cone, e.g., $\mathbf{R}_+^m$, B is polyhedral (since its definition in the case under consideration is its polyhedral representation as well) and therefore closed, which explains why, when the cone $\mathbf{K}$ is polyhedral, the infeasibility of (II) is equivalent to the solvability of (I). At the same time, when $\mathbf{K}$ is not polyhedral, B can be non-closed, as is the case in the example from item 2 above. Let us look at the geometry of this example. The inequality (i) requires that we find a point in the intersection of the cone $\mathbf{L}^3 = \{x \in \mathbf{R}^3 : x_3 \geq \sqrt{x_1^2 + x_2^2}\}$ with the line $\ell = \{[t; -1; t] \in \mathbf{R}^3 : t \in \mathbf{R}\}$. The line ℓ belongs to the two-dimensional plane $L = \{x \in \mathbf{R}^3 : x_2 = -1\}$, and the intersection of $\mathbf{L}^3$ with this plane is the set $\{[x_1; -1; x_3] : x_3^2 - x_1^2 \geq 1, x_3 \geq 0\}$, or, which is the same, the set $\{[x_1; -1; x_3] : (x_3 - x_1)(x_3 + x_1) \geq 1, x_3 - x_1 \geq 0\}$; introducing the coordinates $u = x_1 + x_3$, $v = x_1 - x_3$ on the two-dimensional plane L, the intersection of L and $\mathbf{L}^3$ in these coordinates becomes the inner part $H = \{[u; v] : u \geq 1/v, v > 0\}$ of the branch $\Gamma = \{[u; v] : uv = 1, v > 0\}$ of a hyperbola. In u, v-coordinates the line ℓ is just the line $v = 0$. Thus, geometrically the situation is as follows: to intersect ℓ and $\mathbf{L}^3$ is the same as to intersect H with the v-axis of the $[u; v]$-plane; the intersection is clearly empty, so that (i) is infeasible. At the same time, our line is an asymptote of Γ, so that the shift $v = \epsilon$ of the line $v = 0$ makes the intersection of the shifted line with H nonempty, however small $\epsilon > 0$ is. The outlined shift of ℓ in our original x-coordinates reduces to passing from $b = [0; 1; 0]$ to $b_\epsilon = [0; 1; -\epsilon]$. The bottom line is that $b \notin B$ and $b \in \overline{B}$, since $b = \lim_{\epsilon \to +0} b_\epsilon$ and $b_\epsilon \in B$.

The result of item 3 attracts our attention to the following question: *what are natural sufficient conditions which guarantee the closedness of the set* $A\mathbf{R}^n - \mathbf{K}$? Here is a simple answer:

4. Prove that when the only common point of the image space $L := \{y \in \mathbf{R}^m : \exists x : y = Ax\}$ of A and of $\mathbf{K}$ is the origin, the set $B := A\mathbf{R}^n - \mathbf{K} = L - \mathbf{K}$ is closed. Prove that the same holds true when the condition $L \cap \mathbf{K} = \{0\}$ is "heavily violated," meaning that $L \cap \text{int}\,\mathbf{K} \neq \emptyset$.

Exercise IV.5 [EdYs] [Follow-up to Exercise IV.4] Let $\mathbf{K} \subset \mathbf{R}^m$ be a regular cone, $P \in \mathbf{R}^{m \times n}$, $Q \in \mathbf{R}^{m \times k}$, and $p \in \mathbf{R}^m$. Consider the set

$$\overline{K} := \{x \in \mathbf{R}^n : \exists u \in \mathbf{R}^k : Px + Qu + p \in \mathbf{K}\}.$$

This set is clearly convex. When the cone $\mathbf{K}$ is polyhedral, the above description of $\overline{K}$ is its polyhedral representation, so that the set $\overline{K}$ is polyhedral and as such is closed.

The goal of this exercise is to understand what happens to the closedness of $\overline{K}$ when $\mathbf{K}$ is a general-type regular cone.

1. Is it true that $\overline{K}$ is closed whenever $\mathbf{K}$ is a regular cone?
   Hint: Look what happens when $\mathbf{K} = \mathbf{L}^3$, $P = I_3$, $Q = [0; 1; 1] \in \mathbf{R}^{3 \times 1}$, and $p = [0; 0; 0]$.
2. Prove that when $\mathbf{K}$ is a regular cone and $\text{Im}\,Q \cap \mathbf{K} = \{0\}$, $\overline{K}$ is closed.

Exercise IV.6 [EdYs] Let $\mathfrak{n}(x)$ be a norm on $\mathbf{R}^n$ such that it is continuously differentiable outside of the origin, and let

$$\mathfrak{n}_*(y) = \max_x \{y^\top x : \mathfrak{n}(x) \leq 1\}$$

be the norm conjugate to $\mathfrak{n}$ (see Fact 13.4), so that $\mathfrak{n}_*(\cdot)$ is a norm such that

$$x^\top y \leq \mathfrak{n}(x)\mathfrak{n}_*(y), \quad \forall x, y \in \mathbf{R}^n$$

and $(\mathfrak{n}_*)_* = \mathfrak{n}$, implying that for every $x \neq 0$ there exists $y \neq 0$ such that

$$x^\top y = \mathfrak{n}(x)\mathfrak{n}_*(y).$$

Here are the tasks:

1. Let M be a $d \times d$ matrix, $d \geq 2$, with diagonal entries equal to 1. Assume that $M\lambda \leq 0$ for some nonzero vector $\lambda \geq 0$. How large could $\min_{i,j} M_{ij}$ be?
2. For $d \geq 2$, suppose $p_1, \ldots, p_d$ are $\mathfrak{n}_*(\cdot)$-unit vectors, and $w_1, \ldots, w_d$ are $\mathfrak{n}(\cdot)$-unit vectors such that $p_i^\top w_i = 1$ holds for all $1 \leq i \leq d$. Assume that $0 \in \text{Conv}\{p_1, \ldots, p_d\}$. How small could $\max_{i \neq j} \mathfrak{n}(w_i - w_j)$ be?
3. Let $x \in \mathbf{R}^n$ be nonzero.
    - Let $g = \nabla \mathfrak{n}(x)$.
        (i) What is $\mathfrak{n}_*(g)$?
        (ii) What is $g^\top x$?
        (iii) Let e be such that $\mathfrak{n}_*(e) \leq \mathfrak{n}_*(g)$ and $e^\top x = g^\top x$. Is it true that $e = g$?
    - Given N points $y_i \in \mathbf{R}^n$, consider the problem of finding the smallest $\mathfrak{n}(\cdot)$-ball containing $y_1, \ldots, y_N$.
        (i) Write down this problem as a conic problem (P), and write down the conic dual of this problem. Are both problems solvable with equal optimal values?
        (ii) Assume that the data are such that the optimal value in (P) is equal to 1. How small can $\max_{i,j} \mathfrak{n}(y_i - y_j)$ be?
        Hint: Write down and analyze the optimality conditions.
        (iii) In the situation of the previous item, assume that $\mathfrak{n}(x) = \|x\|_2$ is the standard Euclidean norm. How small can $\max_{i,j} \mathfrak{n}(y_i - y_j)$ be now?

23.1.1 ★ Geometry of Primal–Dual Pair of Conic Problems

Exercise IV.7 [EdYs] [Geometry of primal–dual pair of conic problems] The goal of this exercise is to reveal the notable geometry of a primal–dual pair of conic problems.

It is convenient to work with the primal problem in the form

$$\text{Opt}(P) = \min_x \left\{ c^\top x : Ax - b \geq_\mathbf{K} 0, Px = p \right\} \tag{P}$$

23.1 Around Conic Duality

where $\mathbf{K}$ is a regular cone in a certain $\mathbf{R}^N$. As can immediately be seen, the conic dual of (P) reduces to the problem[1]

$$\mathrm{Opt}(D) = \max_{y,z} \left\{ b^\top y + p^\top z : y \in \mathbf{K}_*, \ A^\top y + P^\top z = c \right\}. \tag{D}$$

From now on we make the following, in fact rather weak, assumption.

Assumption: *The systems of linear equality constraints in (P) and (D) are solvable.*

Let us fix $\bar{x}$ and $(\bar{y}, \bar{z})$ such that

$$P\bar{x} = p \quad \text{and} \quad A^\top \bar{y} + P^\top \bar{z} = c. \tag{\#}$$

The first task is as follows:

1. Pass in (P) from variables x to the *primal slack* $\xi := Ax - b$. Specifically, prove that, in terms of the primal slack, (P) becomes the problem

$$\begin{array}{c} \mathrm{Opt}(\mathcal{P}) = \min_\xi \left\{ \bar{y}^\top \xi : \xi \in \mathbf{K} \cap [\mathcal{L} - \bar{\xi}] \right\} \\ [\mathcal{L} = \{\xi : \exists x : \xi = Ax, Px = 0\}, \ \bar{\xi} = b - A\bar{x}] \end{array} \tag{$\mathcal{P}$}$$

namely, prove that

(i) Every feasible solution x to (P) induces a feasible solution $\xi = Ax - b$ to $(\mathcal{P})$, and the value of the objective of (P) at x differs from the value of the objective of $(\mathcal{P})$ at $\xi = Ax - b$ by a constant that is independent of x:

$$\bar{y}^\top \xi = c^\top x - \left[\bar{y}^\top b + \bar{z}^\top p \right]. \tag{A}$$

(ii) Vice versa, every feasible solution ξ to $(\mathcal{P})$ is of the form $Ax - b$ for some feasible solution x to (P).

The bottom line is that (P) can be reformulated equivalently as $(\mathcal{P})$, and the optimal values of these two problems are linked by the relation

$$\mathrm{Opt}(\mathcal{P}) = \mathrm{Opt}(P) - \left[\bar{y}^\top b + \bar{z}^\top p \right].$$

The next task is as follows:

2. Pass from problem (D) in variables y, z to the problem

$$\begin{array}{c} \max_{y} \left\{ \bar{\xi}^\top y : y \in \mathbf{K}_* \cap [\mathcal{L}^\perp + \bar{y}] \right\} \\ [\mathcal{L}^\perp := \{ y : y^\top \xi = 0, \ \forall \xi \in \mathcal{L} \} = \{ y : \exists z : A^\top y + P^\top z = 0 \}] \end{array} \tag{$\mathcal{D}$}$$

in the variable y only. Specifically, prove that

---

[1] Building the conic dual to a conic problem is a purely mechanical process; however, this process as presented in section 18.4 operates with a conic problem in a form slightly different from that of (P), namely, with linear inequality constraints instead of linear equalities. To apply this process to (P), it suffices to represent the linear equalities $Px = p$ by a pair of opposite linear inequalities $Px - p \geq 0, -Px + p \geq 0$. Applying the recipe from section 18.4 to the resulting problem, the dual reads

$$\max_{y,z',z''} \left\{ b^\top y + p^\top [z' - z''] : A^\top y + P^\top [z' - z''] = c, \ y \in \mathbf{K}_*, z' \geq 0, z'' \geq 0 \right\}.$$

Passing from z', z'' to $z = z' - z''$, we reduce the latter problem to (D).

(i) The orthogonal complement $\mathcal{L}^\perp$ of $\mathcal{L}$ is indeed the linear subspace $\{y : \exists z : A^\top y + P^\top z = 0\}$.

(ii) The y-component of a feasible solution (y, z) to (D) is a feasible solution to $(\mathcal{D})$, and vice versa – every feasible solution y to $(\mathcal{D})$ can be augmented by z to yield a feasible solution (y, z) to (D). Besides this, whenever (y, z) is feasible for (D), we have

$$b^\top y + p^\top z = \bar{\xi}^\top y + c^\top \bar{x}. \tag{B}$$

The bottom line is that (D) can be reformulated equivalently as $(\mathcal{D})$, and the optimal values of these two problems are linked by the relation

$$\text{Opt}(\mathcal{D}) = \text{Opt}(D) - c^\top \bar{x}.$$

A summary of items 1 and 2 is as follows:

- The primal–dual pair (P), (D) of conic problems reduces to the pair of problems $(\mathcal{P})$, $(\mathcal{D})$, "reduces" meaning that feasible solutions x and (y, z) to (P), (D) induce feasible solutions $\xi = Ax - b$ and y to $(\mathcal{P})$, $(\mathcal{D})$, and every pair of feasible solutions to the latter problems can be obtained, in the fashion just described, from a pair of feasible solutions to (P), (D).

- Geometrically, $(\mathcal{P})$, $(\mathcal{D})$ are as follows:
  - The data for the problems are (a) primal–dual pair of regular cones $\mathbf{K}$, $\mathbf{K}_*$ in some $\mathbf{R}^N$, (b) a pair of linear subspaces $\mathcal{L}_\mathcal{P}$, $\mathcal{L}_\mathcal{D}$ in $\mathbf{R}^N$ which are orthogonal complements to each other, and (c) a pair of vectors $\bar{y}, \bar{\xi}$ in $\mathbf{R}^N$.
  - $(\mathcal{P})$ is the problem of minimizing the linear objective $\bar{y}^\top \xi$ over the intersection of the *primal feasible plane* $\mathcal{M}_\mathcal{P} := \mathcal{L}_\mathcal{P} - \bar{\xi}$ with the cone $\mathbf{K}$, while $(\mathcal{D})$ is the problem of maximizing the linear objective $\bar{\xi}^\top y$ over the intersection of the *dual feasible plane* $\mathcal{M}_\mathcal{D} := \mathcal{L}_\mathcal{D} + \bar{y}$ and the dual cone $\mathbf{K}_*$.

Note the "nearly perfect" primal–dual symmetry; the only asymmetry is that, in the primal feasible plane, the shift vector is $-\bar{\xi}$, that is, minus the vector of coefficients of the objective in $(\mathcal{D})$, while in the dual feasible plane the shift vector is $\bar{y}$, the vector of coefficients of the objective in $(\mathcal{P})$. This minor asymmetry stems from the fact that by tradition one of the problems (in our presentation, $(\mathcal{P})$) is written as a minimization program, and the other problem from the pair as a maximization program.

In fact, the symmetry can be made perfect, and the objectives are not mentioned at all.

3. Consider pairs of problems (P), (D) along with problems $(\mathcal{P})$, $(\mathcal{D})$, let x, (y, z) be feasible solutions to (P), (D) and let ξ, y be the feasible solutions to $(\mathcal{P})$, $(\mathcal{D})$ induced by x and (y, z), respectively. Prove that the *duality gap*

$$\text{DualityGap}(x; y, z) := c^\top x - [b^\top y + p^\top z],$$

which is the difference between the objective of primal problem (P) evaluated at primal feasible solution x and the objective of the dual problem (D) evaluated at the dual feasible solution (y, z), is just the inner product $\xi^\top y$ of ξ and y.

Since solving the primal–dual pair (P), (D) is the same as minimizing the duality gap over pairs $(x, (y, z))$ of their feasible solutions, we conclude that

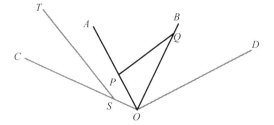

Figure 23.1 Geometry of primal–dual conic pair: $\angle AOB$, cone $\mathbf{K}$; $\angle COD$, cone $\mathbf{K}_*$; segment $[P, Q]$, feasible set of $(\mathcal{P})$; ray $[S, T)$, feasible set of $(\mathcal{D})$; Q is the primal, and S is the dual optimal solution. Note the orthogonality of $\overrightarrow{PQ}$ to $\overrightarrow{ST}$, $\overrightarrow{OA}$ to $\overrightarrow{OD}$, and of $\overrightarrow{OQ}$ to $\overrightarrow{OS}$.

Solving $(P), (D)$ is the same as finding, in the feasible set $\mathcal{M}_\mathcal{P} \cap \mathbf{K}$ of $(\mathcal{P})$ and the feasible set $\mathcal{M}_\mathcal{D} \cap \mathbf{K}_*$ of $(\mathcal{D})$, a pair of vectors ξ, y as close to orthogonality as possible.

Note that for every pair $\xi \in \mathcal{M}_\mathcal{P} \cap \mathbf{K}$, $y \in \mathcal{M}_\mathcal{D} \cap \mathbf{K}_*$ we have $\xi \in \mathbf{K}$, $y \in \mathbf{K}_*$, that is, $\xi^\top y \geq 0$; the zero value of the latter inner product means a zero duality gap for the feasible solutions x, (y, z) to (P), (D) associated with ξ, y. The latter, by Weak Duality, implies the optimality of x in (P) and of (y, z) in (D). By the Conic Duality Theorem, the desired $\xi \in \mathcal{M}_\mathcal{P} \cap \mathbf{K}$ and $y \in \mathcal{M}_\mathcal{D} \cap \mathbf{K}_*$, which are orthogonal to each other, definitely exist, provided that the primal and the dual feasible planes intersect the interiors of the respective cones.

We comment that the geometric formulation of $(\mathcal{P})$, $(\mathcal{D})$ – find vectors, which are orthogonal to each other, in the intersections of given affine planes with given cones[2] – sounds to the authors, who in their high-school years were taught traditional geometry, like a problem from their old geometry textbooks (modulo the fact that the dimension now is arbitrary, and not 2 or 3). It is very surprising that in spite of its quite old-fashioned appearance, this geometric problem is responsible for an extremely wide spectrum of applications, ranging from feeding poultry and cattle to decision making, signal and image processing, engineering design, etc.

23.2 Around the $\mathcal{S}$-Lemma

Exercise IV.8 Recall that the $\mathcal{S}$-Lemma guarantees that the validity of the implication

$$x^\top A x \geq 0 \implies x^\top B x \geq 0 \qquad [A, B \in \mathbf{S}^n]$$

is the same as the existence of $\lambda \geq 0$ such that $B \succeq \lambda A$, but only under the assumption that the inequality $x^\top A x \geq 0$ is strictly feasible. Does the lemma remain true when this assumption is lifted?

Exercise IV.9 [EdYs] Given $A \in \mathbf{S}^n$, consider the set $Q_A = \{x \in \mathbf{R}^n : x^\top A x \leq 0\}$.

1. Let $B \in \mathbf{S}^n$ be such that $B \neq A$ and $Q_B = Q_A$. Then, is it always true that there exists $\rho > 0$ such that $B = \rho A$?

---

[2] These are not *arbitrary* planes and *arbitrary* cones: the planes should be shifts of linear subspaces which are orthogonal complements of each other; the cones should be duals of each other.

2. Suppose that $A \in \mathbf{S}^n$ satisfies $A_{ij} \geq 0$ for all i, j. Under this condition, does your answer to item 1 change?

3. Suppose that $A \in \mathbf{S}^n$ satisfies $\lambda_{\min}(A) < 0 < \lambda_{\max}(A)$. Under this condition, does your answer to item 1 change?

Exercise IV.10 [EdYs] For two nonzero reals a, b, one has $2|ab| = \min_{\lambda > 0} [\lambda^{-1} a^2 + \lambda b^2]$, implying by the Schur Complement Lemma that $2|ab| \leq c$ if and only if there exists $\lambda > 0$ such that $\begin{bmatrix} c - \lambda b^2 & a \\ a & \lambda \end{bmatrix} \succeq 0$. Assuming $b \neq 0$, we have also that $2|ab| \leq c$ if and only if there exists $\lambda \geq 0$ such that $\begin{bmatrix} c - \lambda b^2 & a \\ a & \lambda \end{bmatrix} \succeq 0$. Note also that $c \geq 2|ab|$ is the same as $c \geq 2a\delta b$ for all $\delta \in [-1, 1]$.

Prove the following matrix analogy of the above observation:[3]

Let $A \in \mathbf{R}^{p \times r}$ and $B \in \mathbf{R}^{p \times s}$, let $B \neq 0$, and let $\mathcal{D} = \{\Delta \in \mathbf{R}^{r \times s} : \|\Delta\| \leq 1\}$, where $\|\cdot\|$ is the spectral norm. Then $C \succeq [A \Delta B^\top + B \Delta^\top A^\top]$ for all $\Delta \in \mathcal{D}$ if and only if there exists $\lambda \geq 0$ such that $\begin{bmatrix} C - \lambda B B^\top & A \\ A^\top & \lambda I_r \end{bmatrix} \succeq 0$. In particular, when $a, b \in \mathbf{R}^p$ and $b \neq 0$, one has $C \succeq \pm [ab^\top + ba^\top]$ if and only if there exists $\lambda \geq 0$ such that $\begin{bmatrix} C - \lambda bb^\top & a \\ a^\top & \lambda \end{bmatrix} \succeq 0$.

Exercise IV.11 [EdYs] [Robust TTD][4] Let us come back to the TTD problem (5.2) in section 5.3. Assume we have solved this problem and have at our disposal the resulting *nominal truss* withstanding best of all, when the total truss volume is a given $W > 0$, the load of interest f. Now, we cannot ignore the possibility that "in real life" the truss can be affected by, apart from the load of interest f, perhaps small, but still nonzero, occasional loads composed of forces acting at the free nodes of the nominal truss (think of a railroad bridge and wind). In order for our truss to be useful, it should withstand all small enough occasional loads of this type. Note that our design gave no guarantees of this type – when building the nominal truss, we took into account just one loading scenario f.

1. To get impression of potential dangers of "small occasional loads," run a numerical study as follows:
   - Compute the optimal console t^* (see "console design" Exercise I.16).
   - Looking one by one at the free nodes $p^1, \ldots, p^\mu$ actually used by the nominal console, associate with every one of them a single-force occasional load, the corresponding force acting at the node under consideration; then generate this force as a random two-dimensional vector of Euclidean length 0.01 (that is, 1% of the magnitude of the single nonzero force in the load of interest) and compute the compliance of the nominal truss w.r.t. to the resulting occasional load. Conclude that the nominal console can be crushed by small occasional loads and is therefore completely impractical.

---

[3] S. Boyd, L. El Ghaoui, E. Feron, and V. Balakrishnan, *Linear matrix inequalities in system and control theory*, SIAM, 1994.
[4] The preceding exercises in the TTD series are I.16, I.18, and III.9.

2. The proposed cure is, of course, to use Robust Optimization methodology – to immunize the truss against small occasional loads, that is, to control its compliance w.r.t. the load of interest *and* all small occasional loads. An immediate question is where the occasional loads should be applied. There is no sense in allowing them to act at all free nodes from the original set of tentative nodes – we have every reason to believe that some, if not most, of these nodes will not be used in the optimal truss, so that we should not bother about forces acting at these nodes. On the other hand, we should take into account occasional loads acting at the nodes actually used by the optimal robust truss, and we do *not* know in advance which these nodes are. A reasonable compromise here is as follows. After the nominal optimal truss is built, we can reduce the nodal set to the nodes actually used in this truss, allow for all pair connection of these nodes, and resolve the TTD problem on this reduced set of tentative nodes and tentative bars, now taking into account not only the load of interest but all small occasional loads distributed along the nodes of our new nodal set. This approach can be implemented as follows.

- We specify $\overline{\mathcal{V}}$ as the set of virtual displacements of nodes of our reduced nodal set, preserving the original status (fixed or free) of these nodes, and denote by $\overline{f}$ the natural projection of the load of interest onto $\overline{\mathcal{V}}$; note that all nonzero blocks in f – those representing nonzero physical forces from the collection specifying f – are inherited by $\overline{f}$, since the free nodes where these nonzero forces are applied should clearly be used by the nominal truss.
- We specify $\mathcal{F}$ as the "ellipsoidal envelope" of $\overline{f}$ and all loads from $\overline{\mathcal{V}}$ that are small in magnitude (measured in the $\|\cdot\|_2$-norm). Specifically, we use $\overline{f}$ as one of the half-axes of $\mathcal{F}$; the other $\overline{M} - 1$ half-axes of $\mathcal{F}$ ($\overline{M} := \dim \overline{\mathcal{V}}$) are orthogonal to each other and to the $\overline{f}$ vectors from $\overline{\mathcal{V}}$ of $\|\cdot\|_2$-norm $\rho\|\overline{f}\|_2$, where the "uncertainty level" $\rho \in [0, 1]$ is a parameter of our construction. Note that

$$\mathcal{F} = \{g = Ph : h^\top h \leq 1\}$$

for a properly selected $\overline{M} \times \overline{M}$ matrix P.
- We define the *robust compliance* $\overline{\mathcal{C}}(\overline{t})$ of a truss $\overline{t} \in \mathbf{R}_+^{\overline{N}}$ ($\overline{N}$ is the number of bars in our new – reduced – set of tentative bars) as the supremum, over $g \in \mathcal{F}$, of the usual compliances (computed for the new nodal set) of $\overline{t}$ w.r.t. load g, and we pose the robust counterpart of the TTD problem as the problem of minimizing this robust compliance over trusses $\overline{t} \geq 0$ of total volume W. Solving this problem, we arrive at the *robust truss*.

An immediate question is how to solve for the robust counterpart. Readers who have already solved Exercise I.16.3 know that, as stated right now, the robust counterpart is the *semi-infinite* – with infinitely many convex constraints – optimization program

$$\overline{\mathrm{Opt}} = \min_{\overline{t}, \tau} \left\{ \tau : \overline{t} \in \mathbf{R}_+^{\overline{N}}, \sum_{i=1}^{\overline{N}} \overline{t}_i = W, \begin{bmatrix} \overline{B}\,\mathrm{Diag}\{\overline{t}\}\overline{B}^\top & g \\ g^\top & 2\tau \end{bmatrix} \succeq 0, \forall g \in \mathcal{F} \right\} \quad (\#)$$

where $\overline{B}$ is the matrix built for the new TTD data in the same fashion as the matrix B was built for the original data.

Here are the tasks:

(i) Reformulate (#) as a "normal" convex optimization problem – one with an efficiently computable convex objective and finitely many explicitly verifiable convex constraints.

(ii) Solve the console design version of the latter problem and subject the resulting robust truss to the same tests as those proposed above for quantifying the "real-life" quality of the nominal truss.

23.3 Miscellaneous Exercises

Exercise IV.12 [TrYs] Find the minimizer of a linear function

$$f(x) = c^\top x$$

on the set

$$V_p = \left\{ x \in \mathbf{R}^n : \sum_{i=1}^{n} |x_i|^p \leq 1 \right\};$$

here p, $1 < p < \infty$, is a parameter. What happens to the solution when the parameter becomes 0.5?

Exercise IV.13 [TrYs] Each of three random variables ξ_1, ξ_2, ξ_3 takes values 0 and 1 with probabilities 0.5, and every two of these three variables are independent. Is it true that all three variables are mutually independent? If not, how large could the probability of the event $\xi_1 = \xi_2 = \xi_3 = 1$ be?

Exercise IV.14 [TrYs] [Computational study] Consider a situation as follows: at discrete time instants $t = 1, 2, \ldots, T$ we observe the states $y_t \in \mathbf{R}^\kappa$ of a dynamical system; our observations are

$$y_t + \sigma \xi_t, \quad t = 1, 2, \ldots, T,$$

where $\sigma > 0$ is a given noise intensity and ξ_t are zero-mean Gaussian noises that are independent across t and have unit covariance matrix. All we know about the trajectory of the system is that

$$\|y_{t+1} - 2y_t + y_{t-1}\|_2 \leq dt^2 \alpha \qquad (!)$$

where $dt > 0$ is the continuous time interval between consecutive discrete time instants; in other words, the Euclidean norm of the (finite-difference approximation of the) acceleration of the system is $\leq \alpha$. Given a time delay d, we want to estimate the linear form $f^\top y_{T+d}$ of the system's state at time $T + d \geq 1$, and we intend to use a linear estimate

$$\widehat{y} = \sum_{t=1}^{T} h_t^\top \omega_t.$$

1. Write down an optimization problem specifying the minimum-risk linear estimate, with the risk of an estimate defined as

$$\text{Risk}[\widehat{y}] = \sqrt{\sup_{y \in \mathcal{Y}} \mathbf{E}\{|\widehat{y} - f^\top y_{T+d}|^2\}},$$

where $\mathcal{Y}$ is the set of all trajectories $y = \{y_t, -\infty < t < \infty\}$ satisfying all the constraints (!) for all $t \in \{0, \pm 1, \pm 2, \ldots, \pm \overline{T}\}$, where $\overline{T} := \max\{T, T+d\}$.

2. Use conic duality to convert the problem from the previous item into a conic quadratic problem.
3. Carry out numerical experimentation with the minimum-risk linear estimate.

Exercise IV.15 [TrYs] [Computational study] Consider the following problem:

> A particle is moving through $\mathbf{R}^d$. Given the positions and velocities of the particle at times $t = 0$ and $t = 1$, find the trajectory of the particle on $[0, 1]$ with minimum possible (upper bound on) acceleration.

1. Formulate the (discretized-in-time version of the) problem as a conic quadratic problem and write down its conic dual. Are the problems solvable? Are the optimal values equal to each other? What is said by the optimality conditions?
2. Run numerical experiments in two and three dimensions and look at the results.

Exercise IV.16 [TrYs] [Computational study] This exercise is aimed at answering the following question:

> A steel rod is heated at time $t = 0$, the magnitude of the temperature being $\leq R$, and is left to cool, the temperature at the endpoints being kept at 0 the whole time. We measure the temperature of the rod at locations s_i and times $t_i > 0$, $1 \leq i \leq m$; the measurements are affected by Gaussian noise with zero mean and covariance matrix $\sigma^2 I_m$. Given the measurements, we want to recover the distribution of temperature of the rod at time $\bar{t} > 0$.

Building the model. With properly selected units of temperature and length (so that the rod becomes the segment $[0, 1]$), the evolution of the temperature $u(t, s)$ ($t \geq 0$ is time, $s \in [0, 1]$ is location) is governed by the *heat equation*

$$\frac{\partial}{\partial t} u(t, s) = \frac{\partial^2}{\partial s^2} u(t, s). \qquad [\text{where } u(t, 0) = u(t, 1) \equiv 0]$$

It is convenient to represent functions on $[0, 1]$ as

$$f(s) = \sum_{k=1}^{\infty} f_k \phi_k(s), \quad \phi_k(s) = \sqrt{2} \sin(\pi k s).$$

The functions ϕ_k form an orthonormal basis in the space $L_2 = L_2[0, 1]$ of square summable real-valued functions on $[0, 1]$ equipped with the inner product

$$\langle f, g \rangle = \int_0^1 f(s) g(s) ds,$$

the corresponding norm being $\|f\|_2 = \sqrt{\int_0^1 f^2(s) ds}$.

The claim that the functions ϕ_k form an orthonormal basis in L_2 means that a series

$$\sum_{k=1}^{\infty} f_k \phi_k(s),$$

converges in $\|\cdot\|_2$ to some $f \in L_2$ if and only if $\sum_k f_k^2 < \infty$, and in this case $f_k = \langle f, \phi_k \rangle$; moreover,

$$\left\langle \sum_k f_k \phi_k(\cdot), \sum_k g_k \phi_k(\cdot) \right\rangle = \sum_k f_k g_k$$

for all square-summable sequences $\{f_k\}$ and $\{g_k\}$. In particular,

$$u(t,s) = \sum_{k=1}^{\infty} u_k(t)\phi_k(s), \quad u_k(t) = \int_0^1 u(t,s)\phi_k(s)\,ds.$$

Assuming $|u(0,\cdot)| \le R$, we have

$$\sum_k u_k^2(0) \le R^2, \tag{23.1}$$

and in terms of the coefficients $u_k(t)$ of the rod's temperature, the heat equation becomes very simple:

$$\frac{d}{dt} u_k(t) = -\pi^2 k^2 u_k(t) \implies u_k(t) = \exp\{-\pi^2 k^2 t\}\, u_k(0).$$

As a result, when $t > 0$, the coefficients $u_k(t)$ go to 0 exponentially fast as $k \to \infty$, so that the series

$$\sum_k u_k(t)\phi_k(s)$$

converges to the solution (t,s) of the heat equation not only in $\|\cdot\|_2$, but uniformly on $[0,1]$ as well, implying, due to $\phi_k(0) = \phi_k(1) = 0$, that the series does satisfy the boundary conditions $u(t,0) = u(t,1) = 0$, $t > 0$.

Now our problem can be posed as follows:

The sequence of coefficients $\{u_k^t\}_{k=1}^{\infty}$ of $u(t,\cdot)$ in the orthonormal basis $\{\phi_k(\cdot)\}_{k\ge 1}$ of L_2 evolves according to

$$u_k^t = \exp\{-\pi^2 k^2 t\}\, u_k^0,$$

with

$$u^0 := \{u_k^0\}_{k\ge 1} \in \mathbf{B} := \left\{ \{c_k\}_{k\ge 1} : \sum_k c_k^2 \le R^2 \right\}.$$

Given m noisy observations

$$\omega_i = \Omega_i[u^0] + \sigma \xi_i, \quad \Omega_i[u^0] = \sum_{k=1}^{\infty} \exp\{-\pi^2 k^2 t_i\}\, u_k^0 \phi_k(s_i),$$

where $\xi_1, \ldots, \xi_m$ are $\mathcal{N}(0,1)$ observation noises that are independent of each other, and $t_i > 0$, $s_i \in [0,1]$ are given, we want to recover the sequence $\{u_k^{\bar{t}}\}_{k\ge 1}$. We quantify the performance of a candidate estimate $\omega := (\omega_1, \ldots, \omega_m) \mapsto \widehat{u} := \{\widehat{u}_k(\omega)\}_{k\ge 1}$ by the risk

$$\mathrm{Risk}[\widehat{u}] = \sqrt{\max_{u^0 \in \mathbf{B}} \mathbf{E}_\xi \left\{ \sum_{k\ge 1} \left[\widehat{u}_k(\Omega_1[u^0] + \xi_1, \ldots, \Omega_m[u^0] + \xi_m) - \exp\{-\pi^2 k^2 \bar{t}\}\, u_k^0 \right]^2 \right\}}$$

that is, the square of Risk *is the supremum, taken over all distributions* $u(0, \cdot)$ *of temperature at time* $t = 0$ *with* $\|u(0, \cdot)\|_2 \leq R$, *of the expected squared Euclidean norm of the recovery error associated with the distribution.*

Our last modeling step is to replace the infinite sequences $\{u_k^0\}_{k \geq 1}$ with their finite initial segments $\{u_k^0\}_{1 \leq k \leq K}$, that is, to approximate the situation by one where $u_0^k = 0$ when $k > K$. The simplest way to do it is as follows. Let $\underline{t} = \min[\min_i t_i, \bar{t}]$, so that $\underline{t} > 0$. For $u_0 \in \mathbf{B}$ and $K \geq 1$, the magnitude of the total contribution of the coefficients u_0^k, $k > K$, to $u(t, s)$ with $t \geq \underline{t}$ does not exceed

$$\sum_{k=K+1}^{\infty} \max_s |\phi_k(s)| \exp\{-\pi^2 k^2 \underline{t}\} |u_0^k| \leq \delta := \sqrt{2} R \sum_{k=K+1}^{\infty} \exp\{-\pi^2 k^2 \underline{t}\}.$$

Given a very small tolerance $\bar{\delta} > 0$, say, $\bar{\delta} = 10^{-10}$, we can easily find $K = K(\bar{\delta})$ such that $\delta \leq \bar{\delta}$. Thus, as far as the temperatures we measure and the temperatures we want to recover are concerned, zeroing-out coefficients u_0^k with $k > K(\bar{\delta})$ changes these temperatures by at most $\bar{\delta}$. Common sense (which can be easily justified by formal analysis) says, that with $\bar{\delta}$ as small as 10^{-10}, these changes have no effect on the quality of our recovery, at least when $\sigma \gg \bar{\delta}$.

Now, here is the task:

1. Assuming $u_0^k = 0$ for $k > K$, model the problem of interest as the following estimation problem:

   There exists a K-dimensional signal u known to belong to the Euclidean ball $B^R = \{u \in \mathbf{R}^K : u^\top u \leq R^2\}$ of a given radius R and centered at the origin. Given noisy observations

   $$\omega = Au + \sigma \xi, \qquad [\text{where } A \in \mathbf{R}^{m \times K} \text{ and } \xi \sim \mathcal{N}(0, I_m)]$$

   we want to recover Bu, quantifying the recovery error of a candidate estimate $\omega \mapsto \widehat{u}(\omega)$ by its risk

   $$\text{Risk2}[\widehat{u}] = \sqrt{\sup_{u \in B^R} \mathbf{E}_{\xi \sim \mathcal{N}(0, I_m)} \left\{ [\widehat{u}(Au + \sigma \xi) - Bu]^\top [\widehat{u}(Au + \sigma \xi) - Bu] \right\}}$$

   where B is a given $K \times K$ matrix.

   Write down the expressions for the matrices A and B.

2. Build a convex optimization problem for the minimum-risk *linear estimate* – an estimate of the form $\widehat{u}(\omega) = H^\top \omega$.

3. Compute the minimum-risk linear estimate and run simulations to test its performance. Recommended setup:
   - $\bar{t} \in \{0.01, 0.001, 0.0001, 0.00001\}$;
   - $m = 100$, the t_i are drawn at random from the uniform distribution on $[\bar{t}, 2\bar{t}]$, and the s_i are drawn at random from the uniform distribution on $[0, 1]$;
   - $R = 10^4$, $\sigma = 10$, $\bar{\delta} = 10^{-10}$;
   - to accelerate computation, truncate $K(\bar{\delta})$ at the level 100.

Exercise IV.17 [TrYs] Given positive definite $A \in \mathbf{S}^n$, define
$$P[A] := \{X \in \mathbf{S}^n : X \succeq 0, X^2 \preceq A\}, \qquad Q[A] := \{X \in \mathbf{S}^n : X \succeq 0, X \preceq A^{1/2}\}.$$
From the $\succeq$-monotonicity of the matrix square root on $\mathbf{S}^n_+$ (Example 20.5 in section 20.2) it follows that $P[A] \subseteq Q[A]$. The task is to answer the following question:

Are $P[A]$ and $Q[A]$ comparable, meaning that does there exist $c > 0$ independent of A (but perhaps depending on n) such that
$$Q[A] \subseteq c \cdot P[A]?$$

Exercise IV.18 Find the optimal value in the convex optimization problem
$$\mathrm{Opt}(a) = \min_x \left\{ \sum_{i=1}^n [-(1+a_i)x_i + x_i \ln x_i] : x \geq 0, \sum_i x_i \leq 1 \right\}$$
where $0 \ln 0 = 0$ by definition, so that the function $x \ln x$ is well defined and continuous on the nonnegative ray $x \geq 0$.

Exercise IV.19 [EdYs] Given an $m \times n$ matrix A with trivial kernel, consider the matrix-valued function $F(X) = [A^\top X^{-1} A]^{-1} : \mathrm{Dom}\, F := \{X \in \mathbf{S}^m, X \succ 0\} \to \mathbf{S}^n_+$. Prove that F is $\succeq$-concave on its domain.

23.4 Around Convex Cone-Constrained and Conic Problems

Exercise IV.20 [EdYs] [Cone-constrained semidefinite problems]

1. Let $X, Y \in \mathbf{S}^m_+$. Prove that $\mathrm{Tr}(XY) = 0$ if and only if $XY = YX = 0$.
2. Given an ordered collection $\nu = \{n_1, \ldots, n_k\}$ of positive integers, let $\mathbf{S}^\nu$ be the space of block-diagonal symmetric matrices with k diagonal blocks of sizes $n_1 \times n_1, \ldots, n_k \times n_k$, and let $\mathbf{S}^\nu_+$ be the cone of positive semidefinite matrices from $\mathbf{S}^\nu$. Equipping $\mathbf{S}^\nu$ with the Frobenius inner product, $\mathbf{S}^\nu_+$ is clearly a self-dual regular cone in the resulting Euclidean space.
   A convex cone-constrained problem on the cone $\mathbf{S}^\nu_+$ is of the generic form
$$\mathrm{Opt}(\mathrm{SDP}) = \min_{x \in X} \left\{ f(x) : \overline{g}(x) := Ax - b \leq 0, \widehat{g}(x) := -\mathrm{Diag}\{g_1(x), \ldots, g_k(x)\} \leq_{\mathbf{S}^\nu_+} 0 \right\} \quad (\mathrm{SDP})$$
where X is a nonempty convex set in some $\mathbf{R}^n$, the function $f : X \to \mathbf{R}$ is convex, and the mapping $\widehat{g} : X \to \mathbf{S}^\nu$ is $\mathbf{S}^\nu_+$-convex.

Prove that in the case of the convex cone-constrained semidefinite problem (SDP) Theorem 19.7 reads as follows:

Theorem 19.7 *Consider a convex cone-constrained semidefinite problem* (SDP), *let $x^* \in X$ be a feasible solution to the problem, and let f and $\widehat{g}$ be differentiable at x^*.*

(i) *If x^* is a KKT point of* (SDP), *the Lagrange multipliers being $\overline{\lambda}^* \geq 0$ and $\widehat{\lambda}^* \in \mathbf{S}^\nu_+$, meaning that*
$$\overline{\lambda}^*_i [\overline{g}(x^*)]_i = 0, \forall i \quad \text{and} \quad \widehat{\lambda}^* \widehat{g}(x^*) = 0 \qquad \text{[SDP complementary (slackness)]}$$
$$\nabla_x \left[f(x) + [\overline{\lambda}^*]^\top \overline{g}(x) + \mathrm{Tr}(\widehat{\lambda}^* \widehat{g}(x)) \right] \Big|_{x=x^*} \in -N_X(x^*) \qquad \text{[KKT equation]}$$

23.4 Around Convex Cone-Constrained and Conic Problems

(here, as always, $N_X(x^*)$ is the normal cone of X at x^*, see (11.5)), then x^* is an optimal solution to (SDP).

(ii) If x^* is an optimal solution to (SDP) and, if addition to the above premise, (SDP) satisfies the cone-constrained relaxed Slater condition, then x^* is an SDP KKT point, as defined in (i).

Exercise IV.21 [TrYs] [Follow-up to Exercise IV.20] In the sequel, we fix the dimension n of the embedding space and denote by $E_C = \{x \in \mathbf{R}^n : x^\top C x \leq 1\}$ the ellipsoid centered at the origin and associated with the positive definite $n \times n$ matrix C. Given a positive integer K, and K ellipsoids $E_{A_k}, k \leq K$, consider two optimization problems:

- $\mathcal{O}$: Find the smallest volume centered at the origin ellipsoid containing $\cup_{k \leq K} E_{A_k}$.
- $\mathcal{I}$: Find the largest volume centered at the origin ellipsoid contained in $\cap_{k \leq K} E_{A_k}$.

1. Pose $\mathcal{O}$ and $\mathcal{I}$ as solvable convex cone-constrained semidefinite programs.
2. Prove that problems $\mathcal{O}$ and $\mathcal{I}$ reduce to each other at the cost of appropriate modification of the data.
3. Prove that there exist matrices $\Lambda_k \succeq 0$ such that $\Lambda := \sum_k \Lambda_k \succ 0$ and
$$\Lambda_k = \Lambda_k A_k \Lambda, \quad \forall k \leq K.$$

Exercise IV.22 [TrYs] Recall the convex cone-constrained problem in Example 18.1, section 18.1:
$$\text{Opt}(P) = \min_{x=(t,y) \in \mathbf{R} \times \mathbf{S}^n} \{t : \underbrace{t \geq \text{Tr}(y)}_{\iff \langle y, I_n \rangle - t \leq 0}, y^2 \preceq B\} \tag{18.1}$$

where B is a positive definite matrix.

1. Verify (18.2).
2. Find Lagrange multipliers certifying that $t_* = -\text{Tr}(B^{1/2})$, $y_* = -B^{1/2}$ is a cone-constrained KKT point of problem (18.1) (and thus, by Theorem 19.7, it is an optimal solution to the problem).
3. Consider the parametric family
$$\text{Opt}(p := (v, w)) = \min_{t \in \mathbf{R}, y \in \mathbf{S}^n} \{t : t \geq \text{Tr}(y), yv^{-1}y \preceq w\} \tag{P[p]}$$
of convex cone-constrained problems, with $p \in P = \{p = (v, w) : v \in \text{int } \mathbf{S}_+^n, w \in \text{int } \mathbf{S}_+^n\}$, so that (18.1) is problem (P[$\bar{p}$]) corresponding to
$$\bar{p} = (I_n, B).$$
Prove that Opt(p) is a convex function of $p \in P$ and find a subgradient of this function at the point $\bar{p}$.

Exercise IV.23 [TrYs] [Follow-up to Exercise IV.4] Given positive integers m, n, consider two parametric families of convex sets:

- $S_1[P] = \{(X, Y) \in \mathcal{R}_1 := \mathbf{S}^m \times \mathbf{S}^n : \begin{bmatrix} X & P \\ P^\top & Y \end{bmatrix} \succeq 0\}$, where the "parameter" P runs through the space $\mathbf{R}^{m \times n}$ of $m \times n$ matrices, which we temporarily denote by $\mathcal{P}_1$;

- $S_2[P] = \{(X,Y) \in \mathcal{R}_2 := \mathbf{S}^m \times \mathbf{R}^{m \times n} : \begin{bmatrix} X & Y \\ \hline Y^\top & P \end{bmatrix} \succeq 0\}$, where the "parameter" P runs through the positive semidefinite cone $\mathbf{S}^n_+$, which we temporarily denote by $\mathcal{P}_2$.

Prove that for $\chi = 1, 2$ the set-valued mappings $P \to S_\chi[P]$ are super–additive on their domains:

$$P, Q \in \mathcal{P}_\chi \implies P + Q \in \mathcal{P}_\chi \text{ and } \underbrace{S_\chi[P] + S_\chi[Q] \subseteq S_\chi[P+Q]}_{(*)},$$

and that the concluding inclusion

- is not necessarily equality for $\chi = 1$, and
- is equality for $\chi = 2$.

Exercise IV.24 In the simplest Steiner problem, one is given m distinct points $a_1, \ldots, a_m$ in $\mathbf{R}^n$ and looks for a point x_* such that the sum of Euclidean distances between the points and x_* is as small as possible (think, e.g., about m oil wells on a two-dimensional plane and the problem of locating the collector to be linked to the wells by pipes in a way that minimizes the total length of the pipes).

1. Pose the problem as conic problem, the cone being the direct product of m Lorentz cones.
2. Build the dual problem. Are the primal and the dual problems solvable? Are the primal and the dual optimal values equal to each other?
3. Write down optimality conditions and see what they say.
   Hint: You are advised to consider separately the case where the optimal solution differs from all the points $a_1, \ldots, a_m$, and the case when it is one of the points.
4. Solve the problem in the case $n = 2$, $m = 3$ and a_1, a_2, a_3 are the vertices of a triangle on the two-dimensional plane.
   Note: The point on the plane minimizing the sum of distances to the vertices of a given triangle is called the *Fermat point* of the triangle. Quoting from "Fermat point" in Wikipedia, "This question [to minimize the sum of distances from a point to the vertices of triangle] was proposed by Fermat, as a challenge to Evangelista Torricelli. He solved the problem in a similar way to Fermat's [...] His pupil, Viviani, published the solution in 1659."

Exercise IV.25 [TrYs] Consider a primal–dual pair of conic problems

$$\text{Opt}(P) = \min_x \left\{ c^\top x : Ax \geq_\mathbf{K} b \right\}, \qquad (P)$$

$$\text{Opt}(D) = \max_y \left\{ b^\top y : y \geq_{\mathbf{K}_*} 0, \ A^\top y = c \right\} \qquad (D)$$

($\mathbf{K} \subset \mathbf{R}^n$ is a regular cone) and assume that both problems are feasible.

1. Find the recessive cones $\text{Rec}(P)$ and $\text{Rec}(D)$ of the primal and the dual feasible sets.
2. Prove that the feasible set of at least one of the problems is unbounded.

Exercise IV.26 (Semidefinite duality) A *semidefinite program* is a conic program involving the positive semidefinite cone. As a matter of fact, *Semidefinite programming* – the family of semidefinite programs – possesses extremely powerful "expressive abilities." It is prudent to say that for all practical purposes, whatever that means, Semidefinite Programming is "the same" as the whole of Convex Programming. In this exercise we would like to acquaint the reader with the specific form taken by conic duality when the cone involved is the positive semidefinite cone.

Formally, a semidefinite program is of the form

$$\text{Opt}(P) = \min_{x \in \mathbf{R}^n} \left\{ c^\top x : \begin{array}{l} Ax - b := \sum_j a_j x_j - b \geq 0, \\ \mathcal{A}x - B := x_1 A_1 + \cdots + x_n A_n - B \succeq 0 \end{array} \right\}, \qquad (P)$$

where a_j, b are vectors from some $\mathbf{R}^p$, and A_j, B are matrices from some $\mathbf{S}^q$. The real-life form of a semidefinite program is usually a bit different, namely,

$$\text{Opt}(\mathcal{P}) = \min_{x \in \mathbf{R}^n} \left\{ c^\top x : \begin{array}{l} Ax - b := \sum_j a_j x_j - b \geq 0, \\ \mathcal{A}_i x - B^i := x_1 A_1^i + \cdots + x_n A_n^i - B^i \succeq 0, \; \forall i \leq m \end{array} \right\}, \qquad (\mathcal{P})$$

where $A_j^i, B^i \in \mathbf{S}^{q_i}$. In formulation $(\mathcal{P})$ as opposed to formulation (P) we have a collection of positive semidefinite cone constraints, i.e., $\mathcal{A}_i x - B^i \succeq 0$, $i \leq m$, instead of a single constraint $\mathcal{A}x - B \succeq 0$. We can always rewrite $(\mathcal{P})$ in the form of (P) by assembling A_j^i, B^i into block-diagonal matrices $A_j = \text{Diag}\{A_j^1, \ldots, A_j^m\}$, $B = \text{Diag}\{B^1, \ldots, B^m\}$. Taking into account that a block-diagonal symmetric matrix is positive semidefinite if and only if all the diagonal blocks are positive semidefinite, we deduce that $(\mathcal{P})$ is equivalent to the problem

$$\min_{x \in \mathbf{R}^n} \left\{ c^\top x : Ax - b := \sum_j x_j a_j - b \geq 0, \; \mathcal{A}x - B := \sum_j x_j A_j - B \succeq 0 \right\}$$

of the form (P). When proving theorems, it is usually better to work with programs in the form of (P) – it saves notation; in contrast, when working with real-life semidefinite programs, it is usually better to operate with problems in the more detailed form $(\mathcal{P})$.

The task is as follows:

1. Verify that the conic dual of $(\mathcal{P})$ is the semidefinite program

$$\max_{\lambda, \{\Lambda_i, i \leq m\}} \left\{ b^\top \lambda + \sum_{i=1}^m \text{Tr}(\Lambda_i B^i) : \begin{array}{l} \lambda \in \mathbf{R}_+^p, \Lambda_i \in \mathbf{S}_+^{q_i}, i \leq m, \\ A^\top \lambda + \sum_{i=1}^m \mathcal{A}_i^* \Lambda_i = c, \end{array} \right\}, \qquad (\mathcal{D})$$

where for the linear mapping $x \mapsto \sum_j x_j A_j : \mathbf{R}^n \to \mathbf{S}^q$ its conjugate linear mapping $X \mapsto \mathcal{A}^* X : \mathbf{S}^q \to \mathbf{R}^n$ is given by the identity

$$\text{Tr}(X[\mathcal{A}x]) \equiv [\mathcal{A}^* X]^\top x, \quad \forall (x \in \mathbf{R}^n, X \in \mathbf{S}^q),$$

or, which is the same,

$$\mathcal{A}^* X = [\text{Tr}(A_1 X); \ldots; \text{Tr}(A_n X)].$$

Exercise IV.27 [EdYs] [Example of semidefinite relaxation] Let $T_k \succeq 0, k \leq K$, be positive semidefinite $m \times m$ matrices such that $\sum_k T_k \succ 0$, let $\mathcal{T} \subset \mathbf{R}_+^K$ be a convex compact set intersecting the interior of $\mathbf{R}_+^K$, and let A be a symmetric $m \times m$ matrix. Also let $\phi_\mathcal{T}(z) = \max_{t \in \mathcal{T}} z^\top t$ be the support function of $\mathcal{T}$. Prove that

$$\text{Opt} := \min_z \{\phi_{\mathcal{T}}(z) : z \geq 0, A \preceq \sum_k z_k T_k\} \quad (a)$$
$$= \max_{\Lambda, t} \{\text{Tr}(A\Lambda) : \Lambda \succeq 0, t \in \mathcal{T}, \text{Tr}(T_k \Lambda) \leq t_k, k \leq K\} \quad (b)$$

and that both these minimization and maximization problems are solvable.

Comment. Exercise IV.27 tells us an interesting story. Consider the following problem:

$$\text{Opt}_* := \max_{x \in \mathbf{R}^m} \{x^\top A x : x \in \mathcal{X}\},$$

$$\mathcal{X} := \{x \in \mathbf{R}^m : \exists t \in \mathcal{T} : x^\top T_k x \leq t_k, k \leq K\},$$

with the above restrictions on T_k and $\mathcal{T}$. Such a set is called an *ellitope*, and this notion covers many interesting sets, e.g.:

- The finite and bounded intersections of ellipsoids and/or elliptic cylinders centered at the origin, i.e., when $\mathcal{T} = \{[1; \ldots; 1]\}$ and $\mathcal{X}$ is bounded. In particular, the intersection of K stripes that are symmetric with respect to the origin (such stripes are sets of the form $\{x : |a_k^\top x| \leq 1\}$) is an ellitope provided that the intersection is bounded, e.g., these are the set $\mathcal{X}$ obtained from $\mathcal{T} = \{[1; \ldots; 1]\}$ and $T_k = a_k a_k^\top$ for all k. For example, the unit box $\{x \in \mathbf{R}^m : \|x\|_\infty \leq 1\}$ is an example of an ellitope obtained in this manner.
- $\|\cdot\|_p$-balls with $2 \leq p \leq \infty$: $\{x \in \mathbf{R}^m : \|x\|_p \leq 1\} = \{x \in \mathbf{R}^m : \exists t \in \mathcal{T} : x_k^2 \leq t_k, k \leq K = m\}$ with $\mathcal{T} = \{t \in \mathbf{R}_+^m : \|t\|_{p/2} \leq 1\}$.

It is known that computing the maximum $\text{Opt}_* = \max_{x \in \mathcal{X}} x^\top A x$ of a quadratic form over an ellitope, even as simple as the unit box, is a computationally intractable problem, even when A is restricted to be positive semidefinite. However, the difficult-to-compute quantity Opt_* admits a *semidefinite relaxation bound*, built as follows: whenever $z \in \mathbf{R}_+^K$ is such that $A \preceq \sum_k z_k T_k$ and $x \in \mathcal{X}$, there exists $t \in \mathcal{T}$ such that $x^\top T_k x \leq t_k, k \leq K$, and we have

$$x^\top A x \leq x^\top \left[\sum_k z_k T_k\right] x = \sum_k z_k x^\top T_k x \leq \sum_k z_k t_k \leq \phi_{\mathcal{T}}(z),$$

implying that the optimal value Opt of (a) is an upper bound on Opt_*. For reasonable $\mathcal{T}$, in particular those in the above examples, Opt, in contrast with Opt_*, is efficiently computable.

Now, the fact that Opt is the optimal value not only of (a), but of (b) as well also admits a useful interpretation. Namely, instead of maximizing $x^\top A x$ over $x \in \mathcal{X}$, let us maximize the *expectation* of $\xi^\top A \xi$ over *randomized* solutions ξ, that is, random vectors ξ which satisfy the constraints specifying $\mathcal{X}$ on average, which are precisely the random vectors ξ with distributions P satisfying the condition

$$\exists t = t[P] \in \mathcal{T} : \mathbf{E}_{\xi \sim P}\{\xi^\top T_k \xi\} \leq t_k, k \leq K. \quad (\#)$$

Setting $\Lambda = \mathbf{E}_{\xi \in P}\{\xi \xi^\top\}$, $(\#)$ implies that $\Lambda \succeq 0$ and $\text{Tr}(\Lambda T_k) \leq t_k, k \leq K$, for some $t \in \mathcal{T}$; that is, (Λ, t) is feasible for (b). Vice versa, if (Λ, t) is feasible for (b), then, representing Λ as the covariance matrix of random vector ξ (which can always be done, in many ways, since $\Lambda \succeq 0$), we get a randomized solution ξ such that (ξ, t) satisfies $(\#)$. The bottom line is that the equality in (a), (b) allows for an alternative interpretation of the semidefinite relaxation upper bound Opt on Opt$_*$, as the maximal expected value of $\xi^\top A\xi$ over all random vectors ξ belonging to $\mathcal{X}$ "on average." This interpretation underlies all known results on the *tightness* of semidefinite relaxation, such as the relation

$$\text{Opt}_* \leq \text{Opt} \leq 3\ln(\sqrt{3}K)\,\text{Opt}_*.$$

The derivation of this bound, which is in fact unimprovable up to absolute constants without additional restrictions on A and $\mathcal{X}$, on the quality of semidefinite relaxations, while not being too difficult, goes beyond the scope of this textbook. It should be added that in special cases better approximation bounds can be found. For example, when $A \succeq 0$ and $\mathcal{X} = \{x : \|x\|_\infty \leq 1\}$ is the unit box (or, more generally, matrices $T_1, \ldots, T_K$ that commute with each other), the "tightness factor" $3\ln(\sqrt{3}K)$ can be improved to $\frac{\pi}{2} \approx 1.571$ (Nesterov's $\frac{\pi}{2}$ Theorem), and when A is further restricted to have nonpositive off-diagonal entries and zero row sums it can be improved to 1.138 (the MAXCUT Theorem of Goemans and Williamson).

Exercise IV.28 [EdYs][5] What follows is the concluding exercise in the truss topology design (TDD) series. We have already used the TTD problem to present instructive real-life illustrations of the power of several results of Convex Analysis, specifically, Carathéodory's Theorem (Exercise I.18), the epigraph description of convexity and Helly's Theorem (Exercise III.9), and the $\mathcal{S}$-Lemma (Exercise IV.11), not to speak of the Schur Complement Lemma, which was instrumental in all these exercises. Now it is time to illustrate the power of conic duality.

In the sequel, we assume that the reader is reasonably well acquainted with the truss topology design story as told in Exercise I.16 and we will use without additional comments the notions, notation, and results presented in this exercise, including the default assumption $\mathfrak{R}$ which remains in force below. In addition, we assume from now on that the load of interest f is nonzero – this is the only nontrivial case in TTD.

Recall that the TTD problem as posed in Exercise I.16.2 reads

$$\text{Opt} = \min_{\tau,r}\left\{\tau : \begin{bmatrix} B\,\text{Diag}\{t\}B^\top & f \\ f^\top & 2\tau \end{bmatrix} \succeq 0,\, t \geq 0,\, \sum_i t_i = W\right\}. \quad (P)$$

In our present language, this is a semidefinite program, and we know from Exercise I.16 that this problem is solvable.

The first task is easy:

1. Build the semidefinite dual of (P) and prove that the dual problem is solvable with the same optimal value Opt as the primal problem (P).

    Since passing from a semidefinite problem to its dual is a purely mechanical process, and the subsequent tasks will be formulated in terms of the dual problem, here is the dual as given by conic duality:

$$\max_{V,g,\theta,\lambda,\mu}\left\{-2f^\top g - W\mu : 2\theta = 1,\, \mathfrak{b}_i^\top V \mathfrak{b}_i + \lambda_i - \mu = 0\,\forall i,\, \lambda \geq 0,\, \begin{bmatrix} V & g \\ g^\top & \theta \end{bmatrix} \succeq 0\right\}.$$

---

[5] The preceding exercises in the TTD series are I.16, I.18, III.9, IV.11.

Eliminating the variable θ (which is fixed by the corresponding constraint), we rewrite the dual as

$$\max_{V,g,\lambda,\mu} \left\{ -2f^\top g - W\mu : b_i^\top V b_i + \lambda_i - \mu = 0 \,\forall i, \lambda \geq 0, \begin{bmatrix} V & g \\ g^\top & \frac{1}{2} \end{bmatrix} \succeq 0 \right\}. \quad (D)$$

What remains to be done, is to verify the derivation and to prove that (D) is solvable with the same optimal value Opt as (P).

The next task is still easy:

2. Verify that eliminating, by partial optimization, of the variables V and λ, problem (D) reduces to the problem

$$\max_{g,\mu} \left\{ -2f^\top g - W\mu : \begin{bmatrix} \mu & b_i^\top g \\ b_i^\top g & \frac{1}{2} \end{bmatrix} \succeq 0 \,\forall i \right\} \quad (\overline{D})$$

and the latter problem is solvable with the same optimal value Opt as (P) and (D).

Note the first surprising fact: semidefinite constraints in $(\overline{D})$ involve the cone $\mathbf{S}_+^2$ of 2×2 positive semidefinite matrices, and this cone, as we know, is, up to one-to-one linear transformation, just the Lorentz cone $\mathbf{L}^3$. Thus, $(\overline{D})$ is a conic quadratic problem.

The next task is the following:

3. Pass from problem $(\overline{D})$ to its semidefinite dual $(\overline{P})$ and prove that the latter problem is solvable with optimal value Opt.

At first glance, the task seems crazy: the dual of the dual is the primal! Note, however, that $(\overline{D})$ is *not* exactly the plain conic dual to (P) given in problem (D) – it is obtained from (D) by eliminating some variables, and nobody told us that this elimination keeps the dual to $(\overline{D})$ equivalent to the dual of (D), that is, to (P).

By the same reasons as in item 1, let us write down $(\overline{P})$ as follows:

$$\min_{s,t,q} \left\{ \frac{1}{2} \sum_i s_i : \sum_i t_i = W, \sum_i q_i b_i = f, \begin{bmatrix} t_i & q_i \\ q_i & s_i \end{bmatrix} \succeq 0 \,\forall i \right\}. \quad (\overline{P})$$

What remains is to prove that $(\overline{P})$ is solvable with optimal value Opt.

Now – the main surprise:

4. Verify that $(\overline{P})$ allows the elimination, by partial minimization, of the variables t_i and s_i, which reduces $(\overline{P})$ to the solvable optimization problem

$$\min_q \left\{ \frac{1}{2W} \left(\sum_i |q_i| \right)^2 : \sum_i q_i b_i = f \right\} \quad (\#1)$$

with the same optimal value Opt as all the preceding problems, (P) included.

This is indeed a great surprise – $(\#1)$ is equivalent to the *linear programming* problem

$$G = \min_q \left\{ \|q\|_1 : \sum_i q_i b_i = f \right\}, \quad (\#2)$$

the optimal value in this problem being

$$G = \sqrt{2W\text{Opt}}.$$

The challenge is, of course, to extract from optimal solution to (#2) an optimal truss t^* – one with total bar volume W and compliance, w.r.t. load f, equal to Opt, and this is the final task:

5. Extract from optimal solution to (#2) an optimal truss.

23.5 ★ Cone-Convexity

Exercise IV.29 [EdYs] [Elementary properties of cone-convex functions] The goal of this exercise is to extend the elementary properties of convex functions to cone-convex mappings.

A. Let $\mathcal{X}, \mathcal{Y}$ be Euclidean spaces equipped with norms $\|\cdot\|_\mathcal{X}, \|\cdot\|_\mathcal{Y}$. Next let $\mathbf{X}$ be a closed pointed cone in $\mathcal{X}$, $\mathbf{Y}$ be a closed *and* pointed cone in $\mathcal{Y}$, and $f : X \to \mathcal{Y}$ be a mapping defined on a nonempty convex set $X \subseteq \mathcal{X}$. Recall that for a closed and pointed cone $\mathbf{K}$ in Euclidean space $\mathcal{K}$ and $x, x' \in \mathcal{K}$, the relation $x \leq_\mathbf{K} x'$, which is the same as $x' \geq_\mathbf{K} x$, means that $x' - x \in \mathbf{K}$.

Recall that f is called

- $(\mathbf{X}, \mathbf{Y})$-monotone on X, if
$$\{x, x' \in X \text{ and } x \leq_\mathbf{X} x'\} \implies f(x) \leq_\mathbf{Y} f(x');$$

- $\mathbf{Y}$-convex on X, if
$$f(\lambda x + (1 - \lambda)x') \leq_\mathbf{Y} \lambda f(x) + (1 - \lambda) f(x')$$

for every $x, x' \in X$ and $\lambda \in [0, 1]$. For example,
- an affine mapping $f(x) = Ax + a : \mathcal{X} \to \mathcal{Y}$ is $\mathbf{Y}$-convex, no matter what the closed pointed cone $\mathbf{Y}$ is;
- when $\mathcal{Y} = \mathbf{R}$ and $\mathbf{Y} = \mathbf{R}_+$, the functions that are $\mathbf{Y}$-convex on X are simply the usual convex (in the sense of the standard definition of convexity) real-valued functions on X.

A.1 In the situation of **A**, let $\mathbf{Y}^*$ be the cone dual to $\mathbf{Y}$. For $e \in \mathcal{Y}$, let $f_e(x) = \langle e, f(x) \rangle_\mathcal{Y} : X \to \mathbf{R}$. Prove that f is
- $\mathbf{Y}$-convex on X if and only if the function f_e is convex on X whenever $e \in \mathbf{Y}^*$;
- $(\mathbf{X}, \mathbf{Y})$-monotone on X if and only if the function f_e is $\mathbf{X}$-monotone on X (i.e., $x, x' \in X, x \leq_\mathbf{X} x' \implies f_e(x) \leq f_e(x'))$ for every $e \in \mathbf{Y}^*$.

A.2 In the situation of **A**, let f be $\mathbf{Y}$-convex. Prove that f is locally bounded and locally Lipschitz continuous on the interior of X, meaning that if $\bar{X} \subset \text{int } X$ is a closed and bounded set, then there exists $M < \infty$ such that $\|f(x)\|_\mathcal{Y} \leq M$ holds for all $x \in \bar{X}$ (this is local boundedness) and there exists $L < \infty$ such that $\|f(x) - f(z')\|_\mathcal{Y} \leq L \|x - x'\|_\mathcal{X}$ holds for all $x, x' \in \bar{X}$ (this is local Lipschitz continuity).

B. Now let us look at elementary operations preserving cone convexity. From now on, $\text{Lin}(\mathcal{X}, \mathcal{Y})$ denotes the linear space of linear mappings acting from Euclidean space $\mathcal{X}$ to Euclidean space $\mathcal{Y}$. Prove the following statements:

B.1 ["nonnegative" linear combinations] Let X be a nonempty convex subset of Euclidean space $\mathcal{X}$, let $\mathcal{Y}_j$, $j \leq J$, and $\mathcal{Y}$ be Euclidean spaces equipped with pointed closed cones

$\mathbf{Y}_j$ and $\mathbf{Y}$, and let $\alpha_j \in \mathrm{Lin}(\mathcal{Y}_j, \mathcal{Y})$ be nonnegative coefficients, meaning that $\alpha_j y_j \in \mathbf{Y}$ whenever $y_j \in \mathbf{Y}_j$. When mappings $f_j(x) : X \to \mathcal{Y}_j$. are $\mathbf{Y}_j$-convex, $j \leq J$, their linear combination with coefficients α_j, i.e., the mapping

$$f(x) = \sum_j \alpha_j f_j(x) : X \to \mathcal{Y},$$

is $\mathbf{Y}$-convex.

B.2 [affine substitution of variables] In the situation of **A**, let $z \mapsto Az + a : \mathcal{Z} \to \mathcal{X}$ be an affine mapping, and let f be $\mathbf{Y}$-convex on X. Then, the function $g(z) := f(Az + a)$ is $\mathbf{Y}$-convex on the set $Z = \{z : Az + a \in X\}$.

B.3 [monotone composition] Let the $\mathcal{U}_j$, $j \leq J$, be Euclidean spaces equipped with closed pointed cones $\mathbf{U}_j$, let $\mathcal{U} = \mathcal{U}_1 \times \cdots \times \mathcal{U}_J$, let $\mathbf{U} = \mathbf{U}_1 \times \cdots \times \mathbf{U}_J$, and let $\mathcal{Y}$ be an Euclidean space equipped with closed pointed cone $\mathbf{Y}$. Next, let X be a nonempty convex set in Euclidean space $\mathcal{X}$, let U be a nonempty convex set in $\mathcal{U}$, and let $f_j(x) : X \to \mathcal{U}_j$ be $\mathbf{U}_j$-convex functions, $j \leq J$, such that $f(x) = [f_1(x); \ldots; f_J(x)] \in U$ whenever $x \in X$. Finally, consider a mapping $F : U \to \mathcal{Y}$ that is $(\mathbf{U}, \mathbf{Y})$-monotone and $\mathbf{Y}$-convex on U. Then the composition

$$G(x) = F(f(x)) : X \to \mathcal{Y}$$

is $\mathbf{Y}$-convex on X.

Note: We do allow some $\mathbf{U}_j$ to be the trivial cones $\{0\}$. When $\mathbf{U}_j$ is trivial, the $(\mathbf{U}_j, \mathbf{Y})$-monotonicity of $F(u_1, \ldots, u_J)$ with respect to u_j automatically holds true, while the $\mathbf{U}_j$-convexity of f_j holds true when f_j is affine, compare the convex monotone superposition rule in section 10.1.

C. The gradient inequality and the existence of directional derivatives can be extended from the usual convex functions (i.e., $\mathbf{R}_+$-convex functions taking values in $\mathbf{R}$) to cone-convex mappings. Prove the following statements:

C.1 [gradient inequality] In the situation of **A**, let $\bar{x} \in X$ and let f be $\mathbf{Y}$-convex on X and differentiable at $\bar{x}$. Then

$$\forall y \in X : f(y) \geq_\mathbf{Y} f(\bar{x}) + f'(\bar{x})(y - \bar{x}),$$

where $f'(\bar{x})$ is the Jacobian of f at $\bar{x}$.

C.2 [existence of directional derivatives] In the situation of **A**, let f be $\mathbf{Y}$-convex on X, let $\bar{x} \in \mathrm{int}\, X$, and let $d \in \mathcal{X}$. Then

$$\exists Df(\bar{x})[d] := \lim_{t \to +0} \frac{f(\bar{x} + td) - f(\bar{x})}{t}$$

and

$$(t \geq 0 \text{ and } \bar{x} + td \in X) \implies f(\bar{x} + td) \geq_\mathbf{Y} f(\bar{x}) + tDf(\bar{x})[d]. \qquad (\#)$$

Besides this, as a function of $d \in \mathcal{X}$, $Df(\bar{x})[d]$ is positively homogeneous of degree 1 (i.e., $Df(\bar{x})[td] = tDf(\bar{x})[d]$ when $t \geq 0$) and $\mathbf{Y}$-convex.

D. Subdifferentials of the usual convex functions admit natural extensions to cone-convex mappings. Specifically, in the situation of **A**, let $\bar{x} \in X$. Let us say that $g \in \text{Lin}(\mathcal{X}, \mathcal{Y})$ is a *sub-Jacobian* of f at $\bar{x}$, if

$$\forall y \in X : f(y) \geq_{\mathbf{Y}} f(\bar{x}) + g[y - x].$$

For example, C.1 says that if f is **Y**-convex on X and differentiable at $\bar{x} \in X$, then the Jacobian $f'(\bar{x})$ of f, taken at x, is a sub-Jacobian of f at $\bar{x}$. Clearly, for a usual convex function its sub-Jacobians at a point are exactly the linear forms on $\mathcal{X}$ given by the subgradients $f'(x)$ of f at x according to

$$gh = \langle f'(x), h \rangle_{\mathcal{X}}, \qquad h \in \mathcal{X}.$$

Let $\mathcal{J}f(x)$ be the set of all sub-Jacobians of f at $x \in X$. Prove the following statements:

D.1 In the situation of **A**, for $x \in X$ one has $g \in \mathcal{J}f(x)$ if and only if for every $e \in \mathbf{Y}^*$ the vector $g^* e \in \mathcal{X}$ is a subgradient of f_e at x; here, for $g \in \text{Lin}(\mathcal{X}, \mathcal{Y})$, $g^* \in \text{Lin}(\mathcal{Y}, \mathcal{X})$ is the conjugate of g: $\langle gu, v \rangle_{\mathcal{Y}} = \langle u, g^* v \rangle_{\mathcal{X}}$ for all $u \in \mathcal{X}, v \in \mathcal{Y}$.

D.2 In the situation of **A**, let f be **Y**-convex on X. Then
- D.2.1 For every $x \in X$, the set $\mathcal{J}f(x)$ is a closed convex subset of $\text{Lin}(\mathcal{X}, \mathcal{Y})$.
- D.2.2 The mapping $x \mapsto \mathcal{J}f(x)$ is locally bounded on the interior of X; that is, for every closed and bounded set $\bar{X} \subset \text{int } X$, the induced norms $\|g\|_{\mathcal{X}, \mathcal{Y}} = \max_z \{\|gz\|_{\mathcal{Y}} : \|z\|_{\mathcal{X}} \leq 1\}$ of the linear mappings $g \in \mathcal{J}f(x)$, $x \in \bar{X}$ are bounded away from $+\infty$.
- D.2.3 The multivalued mapping $x \mapsto \mathcal{J}f(x)$ is closed on $\text{int } X$: if the $x_i \in \text{int } X$ converge as $i \to \infty$ to $\bar{x} \in \text{int } X$ and the linear mappings $g_i \in \mathcal{J}f(x_i)$ converge as $i \to \infty$ to some $\bar{g} \in \text{Lin}(\mathcal{X}, \mathcal{Y})$, then $\bar{g} \in \mathcal{J}f(\bar{x})$.

The most attractive property of subgradients of the usual convex functions is their existence, at least at interior points of the function's domain. This fact extends to cone-convex mappings. Prove the following statements:

D.3 [existence of sub-Jacobians] In the situation of **A**, let $\bar{x} \in \text{int } X$ and let f be **Y**-convex on X. Then $\mathcal{J}f(\bar{x})$ is nonempty.

For a real-valued convex function f and $x \in \text{int Dom } f$, $d \in \mathcal{X}$, one has

$$Df(x)[d] = \max_{y \in \partial f(x)} \langle y, d \rangle_{\mathcal{X}}.$$

A similar fact holds true for cone-convex functions.

D.4 In the situation of **A**, let f be **Y**-convex on X. Also let $\bar{x} \in \text{int } X$ and $d \in \mathcal{X}$. Then for properly selected $g \in \mathcal{J}f(\bar{x})$ one has

$$Df(\bar{x})[d] = gd,$$

while for every $g' \in \mathcal{J}f(\bar{x})$ one has

$$Df(\bar{x})[d] \geq_{\mathbf{Y}} g'd.$$

There is a natural relation between the sub-Jacobians of the **Y**-convex function f and the subgradients of functions $f_e = \langle e, f \rangle_{\mathcal{Y}}, e \in \mathbf{Y}^*$.

D.5 In the situation of **A**, let f be $\mathbf{Y}$-convex on X and $\bar{x} \in \text{int } X$. For $e \in \mathbf{Y}^*$, $h \in \partial f_e(\bar{x})$ (that is, $f_e(y) \geq f_e(\bar{x}) + \langle h, y - \bar{x}\rangle_{\mathcal{X}}$ for all $y \in X$) if and only if $h = g^* e$ for some $g \in \mathcal{J}f(\bar{x})$.

Finally, the chain rule:

D.6 [Chain rule] Let $\mathcal{U}_j$, $j \leq J$, be Euclidean spaces equipped with closed pointed cones $\mathbf{U}_j$, let $\mathcal{U} = \mathcal{U}_1 \times \cdots \times \mathcal{U}_J$, $\mathbf{U} = \mathbf{U}_1 \times \cdots \times \mathbf{U}_J$, and let $\mathcal{Y}$ be an Euclidean space equipped with closed pointed cone $\mathbf{Y}$. Next, let X be a nonempty convex set in Euclidean space $\mathcal{X}$, let U be a nonempty convex set in $\mathcal{U}$, and let $f_j(x) : X \to \mathcal{U}_j$ be $\mathbf{U}_j$-convex on X functions, $j \leq J$, such that $f(x) = [f_1(x); \ldots; f_J(x)] \in U$ whenever $x \in X$. Finally, let the mapping $F : U \to \mathcal{Y}$ be $(\mathbf{U}, \mathbf{Y})$-monotone and $\mathbf{Y}$-convex on U. As we know from B.3, the composition

$$G(x) = F(f(x)) : X \to \mathcal{Y}$$

is $\mathbf{Y}$-convex on X. Now let $\bar{x} \in \text{int } X$, $\bar{u}_j = f_j(\bar{x})$ be such that $\bar{u} = [\bar{u}_1; \ldots; \bar{u}_J] \in \text{int } U$. Finally, let $g_j \in \mathcal{J}f_j(\bar{x})$, $j \leq J$, and $g \in \mathcal{J}F(\bar{u})$. Then the linear mapping $[u_1; \ldots; u_J] \mapsto g[u_1; \ldots; u_J]$ is $(\mathbf{U}, \mathbf{Y})$-monotone, and the linear mapping

$$h \mapsto \widehat{g}h := g[g_1 h; \ldots; g_J h] : \mathcal{X} \to \mathcal{Y}$$

is a sub-Jacobian of G at $\bar{x}$.

Exercise IV.30 The univariate function $f(x) = x^{-1/2} : \{x > 0\} \to \mathbf{R}$ is nonincreasing and convex, and $\nabla f(x) = -x^{-3/2}/2$, $x > 0$. Let P be $m \times n$ matrix with trivial kernel.

1. Prove that the mapping $F(X) = [PXP^\top]^{-1/2} : \mathbf{S}^n_{++} \to \mathbf{S}^m$, where $\mathbf{S}^n_{++} = \text{int } \mathbf{S}^n_+ = \{X \in \mathbf{S}^n : X \succ 0\}$, is $(\mathbf{S}^n_+, \mathbf{S}^m_+)$-antimonotone and $\mathbf{S}^m_+$-convex.

2. Assuming $P = I_2$, compute numerically $F(X)$ and $dF(X)[dX]$ for $X = \begin{bmatrix} 2 & -1 \\ -1 & 1 \end{bmatrix}$ and $dX = \begin{bmatrix} 0 & 1 \\ 1 & 0 \end{bmatrix}$. For the above X, also compute the Jacobian J of F at X – the matrix of the linear mapping $dX \mapsto DF(X)[dX] : \mathbf{S}^2 \to \mathbf{S}^2$ – in the basis $[1, 0; 0, 0]$, $[0, 0; 0, 1]$, $[0, 1/\sqrt{2}; 1/\sqrt{2}, 0]$ of $\mathbf{S}^2$.

3. What does the "gradient inequality" (Exercise IV.29.C.1) for the $\mathbf{S}^n_+$-convex mapping F look like?

23.6 ★ Around Conic Representations of Sets and Functions

23.6.1 Conic Representations: Definitions

Let $\mathfrak{K}$ be a family of regular cones in Euclidean space which contains the nonnegative ray $\mathbf{R}_+$ and is closed with respect to taking finite direct products and passing from a cone to its dual. Instructive examples are the families $\mathfrak{R}$ of nonnegative orthants, $\mathfrak{L}$ of finite direct products of Lorentz cones, and $\mathfrak{S}$ of finite direct products of semidefinite cones.

- A $\mathfrak{K}$-*representation* ($\mathfrak{K}$-r.) of a set $X \subseteq \mathbf{R}^n$ is its representation of the form

$$X = \{x \in \mathbf{R}^n : \exists u : Px + Qu - r \in \mathbf{K}\} \tag{23.2}$$

where $\mathbf{K} \in \mathfrak{K}$ is a representation of X as the projection of the solution set of the conic inequality $Px + Qu \geq_\mathbf{K} r$ in variables x, u onto the plane of x-variables where X lives. A set X admitting a conic representation with a cone from $\mathfrak{K}$ is called $\mathfrak{K}$-*representable* ($\mathfrak{K}$-r for short).

- A $\mathfrak{K}$-*representation of a function* $f : \mathbf{R}^n \to \mathbf{R} \cup \{+\infty\}$ is, by definition, a $\mathfrak{K}$-representation of the epigraph of f:

$$[t; x] \in \mathrm{epi}\{f\} := \{[x; t] : t \geq f(x)\} \iff \exists u : Px + tp + Qu - r \in \mathbf{K} \text{ with } \mathbf{K} \in \mathfrak{K}.$$

Functions admitting a $\mathfrak{K}$-representation are called $\mathfrak{K}$-*representable* ($\mathfrak{K}$-r for short)

We are already acquainted with $\mathfrak{R}$-representability – it is what was called polyhedral representability. By Fourier–Motzkin elimination, polyhedral representable sets $X \subseteq \mathbf{R}^n$ admit polyhedral representations not involving additional variables u, and similarly for $\mathfrak{R}$-representable functions; this is not the case for more general families $\mathfrak{K}$, such as the families $\mathfrak{L}$ of Lorentz-representable sets and $\mathfrak{S}$ of semidefinite-representable sets.

The following exercise explains the rationale underlying the above restrictions on $\mathfrak{K}$ and why we are interested in $\mathfrak{K}$-representations.

Exercise IV.31 [EdYs] Check that

1. Every finite system $P_0 y \geq r_0$, $P_i y - r_i \in \mathbf{K}_i$, $i \leq I$, of scalar linear inequalities and of conic inequalities, involving cones from $\mathfrak{K}$, in variables y is equivalent to a single conic inequality, with a cone from $\mathfrak{K}$, in these variables:

$$\{P_0 y - r_0 \geq 0, P_i y - r_i \in \mathbf{K}_i, 1 \leq i \leq I\}$$
$$\iff \left\{ ([P_0; P_1; \ldots; P_I] y - [r_0; r_1; \ldots; r_I]) \in \mathbf{K} := \underbrace{\mathbf{R}_+ \times \cdots \times \mathbf{R}_+}_{\dim r_0 \text{ times}} \times \mathbf{K}_1 \times \mathbf{K}_2 \times \cdots \times \mathbf{K}_I \right\}$$

where $\mathbf{K} \in \mathfrak{K}$ (since $\mathbf{R}_+ \in \mathfrak{K}$ and $\mathfrak{K}$ is closed with respect to taking finite direct products). As a result, the representation of a set X given by:

$$X = \{x : \exists u : P_0 x + Q_0 u - r^0 \geq 0, P_i x + Q_i u - r_i \in \mathbf{K}_i, 1 \leq i \leq I\} \quad [\mathbf{K}_i \in \mathfrak{K}] \quad (!)$$

that is, as the projection of the solution set of a finite system of linear and $\mathfrak{K}$-conic inequalities in variables x, u onto the plane of x-variables where X lives, can be straightforwardly converted into a $\mathfrak{K}$-representation of X.

Important: Item 1 allows us from now on to refer to representations of the form (!) as $\mathfrak{K}$-representations of X, skipping the conversion (always straightforward and purely mechanical) of such a representation into the "canonical" representation (23.2).

2. A $\mathfrak{K}$-representation of a function straightforwardly induces $\mathfrak{K}$-representations of its sublevel sets:

$$\left\{ \{t \geq f(x)\} \iff \{\exists u : Px + tp + Qu - r \in \mathbf{K}\} \right\} \qquad [a \in \mathbf{R}, \mathbf{K} \in \mathfrak{K}]$$
$$\implies X_a := \{x : f(x) \leq a\} = \{x : \exists u : Px + Qu - [r - ap] \in \mathbf{K}\}.$$

3. Given $\mathfrak{K}$-representations of a set $X \subseteq \mathbf{R}^n$ and a function $f : \mathbf{R}^n \to \mathbf{R} \cup \{+\infty\}$:

$$X = \{x \in \mathbf{R}^n : \exists u : P_X x + Q_X u - r_X \in \mathbf{K}_X\},$$
$$\mathrm{epi}\{f\} = \{[x; t] : \exists v : P_f x + t p_f + Q_f v - r_f \in \mathbf{K}_f\} \qquad [\mathbf{K}_X \in \mathfrak{K}, \mathbf{K}_f \in \mathfrak{K}]$$

we can straightforwardly convert the optimization problem

$$\min_{x \in X} f(x) \qquad (*)$$

into a conic problem on a cone from $\mathfrak{K}$, namely, the problem

$$\min_{x,t,u,v} \left\{ t : A[x;t;u;v] - b := [P_X x + Q_X u; P_f x + t p_f + Q_f v] - [r_X; r_f] \in \underbrace{\mathbf{K} := \mathbf{K}_X \times \mathbf{K}_f}_{\in \mathfrak{K}} \right\}.$$

As a result, a solver $\mathcal{S}$ capable of solving conic problems on cones from $\mathfrak{K}$ can be straightforwardly utilized when solving problems $(*)$ with X and f given by $\mathfrak{K}$-representations.

4. Given a conic problem

$$\min_x \left\{ c^\top x : Ax - b \in \mathbf{K}, Rx \geq r \right\} \qquad (P)$$

on a cone from $\mathfrak{K}$, its conic dual – the conic problem

$$\max_{y,z} \left\{ \langle b, y \rangle + r^\top z : A^* y + R^\top z = c, y \in \mathbf{K}_*, z \geq 0 \right\}$$

$$\begin{bmatrix} \langle \cdot, \cdot \rangle \text{ is the inner product in the Euclidean space where } \mathbf{K} \text{ lives,} \\ \mathbf{K}_* \text{ is the cone dual to } \mathbf{K}, \\ A^* \text{ is the conjugate of } A: \langle Ax, y \rangle \equiv x^\top A^* y \; \forall x, y \end{bmatrix} \qquad (D)$$

is also a conic problem on a cone from $\mathfrak{K}$ (since $\mathfrak{K}$ is closed with respect to passing from a cone to its dual and contains nonnegative orthants).

Note that the option mentioned in the last item of Exercise IV.31 is implemented in "CVX: MATLAB software for disciplined convex programming" by M. Grant and S. Boyd, http://cvxr.com/cvx. This tool is second to none in its scope and user-friendliness for the numerical processing of well-structured convex problems, the underlying family $\mathfrak{K}$ being the semidefinite family $\mathfrak{S}$. We conclude that it makes sense to develop a kind of calculus allowing one to recognize $\mathfrak{K}$-representability of sets and functions and to build, when possible, their $\mathfrak{K}$-representations. The desired calculus exists and is fairly simple, general, and fully algorithmic. The goal of subsequent exercises is to acquaint the reader with the most frequently used elements of this calculus; for more on this subject, see [BTN].

23.6.2 Conic Representability: Elementary Calculus

The elementary calculus of conic representability is very similar to the calculus of polyhedral representations from section 3.3.

Exercise IV.32 [Elementary calculus of $\mathfrak{K}$-representable sets] Check that the following basic convexity-preserving[6] operations with sets preserve $\mathfrak{K}$-representability.

1. The finite intersection of $\mathfrak{K}$-r sets $X_i = \{x \in \mathbf{R}^n : \exists u^i : P_i x + Q_i u^i - r_i \in \mathbf{K}_i\}, i \leq I$ (here and in what follows all cones involved are from $\mathfrak{K}$), is $\mathfrak{K}$-r. It is given by

---

[6] "Convexity-preserving" is crucial – clearly, $\mathfrak{K}$-r sets and functions must be convex!

23.6 ★ Around Conic Representations of Sets and Functions

$$\bigcap_{i \leq I} X_i = \Big\{ x \in \mathbf{R}^n : \exists u = [u^1; \ldots; u^I] :$$
$$Px + Qu - r := [P_1 x + Q_1 u^1; \ldots; P_I x + Q_I u^I] - [r_1; \ldots; r_I] \in$$
$$\underbrace{\mathbf{K} := \mathbf{K}_1 \times \cdots \times \mathbf{K}_I}_{\in \mathfrak{K}} \Big\}.$$

2. The direct product of finitely many $\mathfrak{K}$-r sets $X_i = \{x \in \mathbf{R}^n : \exists u^i : P_i x + Q_i u^i - r_i \in \mathbf{K}_i\}$, $i \leq I$ is $\mathfrak{K}$-r:

$$X_1 \times \cdots \times X_I = \Big\{ x = [x^1; \ldots; x^I] : \exists u = [u^1; \ldots; u^I] :$$
$$Px + Qu - r := [P_1 x^1 + Q_1 u^1; \ldots; P_I x^I + Q_I u^I] - [r_1; \ldots; r_I] \in$$
$$\underbrace{\mathbf{K} := \mathbf{K}_1 \times \cdots \times \mathbf{K}_I}_{\in \mathfrak{K}} \Big\}.$$

3. The affine image $Y = \{y = Ax + b : x \in X\}$ of the $\mathfrak{K}$-r set $X = \{x \in \mathbf{R}^n : \exists u : Px + Qu - r \in \mathbf{K}\}$ is $\mathfrak{K}$-r:

$$Y = \{y : \exists [x; u] : Ax + b = y, Px + Qu - r \in \mathbf{K}\}.$$

This set is the projection onto the y-plane of a set given by an explicit finite system of linear and $\mathfrak{K}$-conic inequalities and as such admits an explicit $\mathfrak{K}$-representation by item 1 of Exercise IV.31.

4. The inverse affine image $Y = \{y : Ay + b \in X\}$ of the $\mathfrak{K}$-r set $X = \{x \in \mathbf{R}^n : \exists u : Px + Qu - r \in \mathbf{K}\}$ is $\mathfrak{K}$-r:

$$Y = \{y : \exists u : PAy + Qy - [r - Pb] \in \mathbf{K}\}.$$

5. The arithmetic sum $X = X_1 + \cdots + X_I$ of $\mathfrak{K}$-r sets $X_i = \{x \in \mathbf{R}^n : \exists u^i : P_i x + Q_i u^i - r_i \in \mathbf{K}_i\}$, $i \leq I$, is $\mathfrak{K}$-r. Specifically,

$$X = \Big\{ x : \exists [x^1; \ldots; x^I; u^1; \ldots; u^I] : x - \sum_i x^i = 0, P_i x^i + Q_i u^i - r_i \in \mathbf{K}_i, i \leq I \Big\},$$

and we can apply item 1 of Exercise IV.31.

Exercise IV.33 [EdYs] [Elementary calculus of $\mathfrak{K}$-representable functions] Check that the following convexity-preserving operations with functions preserve $\mathfrak{K}$-representability:

1. Restricting onto a $\mathfrak{K}$-r set: A $\mathfrak{K}$-representation of a function $f : \mathbf{R}^n \to \mathbf{R} \cup \{+\infty\}$ given by

$$t \geq f(x) \iff \exists u : P_f x + t_f p + Q_f u - r_f \in \mathbf{K}_f$$

together with a $\mathfrak{K}$-representation of a set $X \subseteq \mathbf{R}^n$ given by

$$X = \{x \in \mathbf{R}^n : \exists v : P_X x + Q_X v - r_X \in \mathbf{K}_X\}$$

induce a $\mathfrak{K}$-representation of the restriction of f onto X, i.e., the function

$$f|_X(x) = \begin{cases} f(x), & \text{if } x \in X \\ +\infty, & \text{if } x \notin X \end{cases},$$

as follows:
$$t \geq f|_X(x) \iff \exists u, v : P_f x + t_f p + Q_f u - r_f \in \mathbf{K}_f, P_X x + Q_X v - r_X \in \mathbf{K}_X.$$

2. Taking a linear combination $\sum_{i=1}^{I} \lambda_i f_i$ with positive coefficients of $\mathfrak{K}$-r functions f_i:
$$t_i \geq f_i(x) \iff \exists u^i : P_i x + t_i p_i + Q_i u^i - r_i \in \mathbf{K}_i, i \leq I$$
$$\Downarrow$$
$$t \geq f(x) := \sum_{i=1}^{I} \lambda_i f_i(x) \iff \begin{cases} \exists [t_1; \ldots; t_I; u^1; \ldots; u^i] : \\ t \geq \sum_i \lambda_i t_i, \\ P_i x + t_i p_i + Q_i u^i - r_i \in \mathbf{K}_i, i \leq I. \end{cases}$$

3. Direct summation:
$$t_i \geq f_i(x^i) \iff \exists u^i : P_i x^i + t_i p_i + Q_i u^i - r_i \in \mathbf{K}_i, i \leq I$$
$$\Downarrow$$
$$t \geq f(x^1, \ldots, x^I) := \sum_{i=1}^{I} f_i(x^i) \iff \begin{cases} \exists [t_1; \ldots; t_I; u^1; \ldots; u^i] : \\ t \geq \sum_i t_i, \\ P_i x^i + t_i p_i + Q_i u^i - r_i \in \mathbf{K}_i, i \leq I. \end{cases}$$

4. Taking finite maxima:
$$t \geq f_i(x) \iff \exists u^i : P_i x + t p_i + Q_i u^i - r_i \in \mathbf{K}_i, i \leq I$$
$$\Downarrow$$
$$t \geq f(x) := \max_{i \leq I} f_i(x) \iff \exists [u^1; \ldots; u^i] : P_i x + t p_i + Q_i u^i - r_i \in \mathbf{K}_i, i \leq I.$$

5. The affine substitution of variables:
$$t \geq f(x) \iff \exists u : Px + tp + Qu - r \in \mathbf{K},$$
$$\Downarrow$$
$$\tau \geq g(y) := f(Ax + b) \iff \exists u : PAu + \tau p + Qu - [r - Pb] \in \mathbf{K}.$$

In fact, the claims in items 2–5 are special cases of the following observation:

6. Monotone superposition: Let functions $f_i(x)$, $i \leq I$, be $\mathfrak{K}$-r, where the first K functions are affine, and let $F(y) : \mathbf{R}^I \to \mathbf{R} \cup \{+\infty\}$ be $\mathfrak{K}$-r and monotonically nondecreasing in $y_{K+1}, \ldots, y_I$, i.e.,
$$y, y' \in \mathbf{R}^I, y \geq y', y_i = y'_i, i \leq K \implies F(y) \geq F(y').$$

Then the function
$$g(x) = \begin{cases} F(f_1(x), \ldots, f_I(x)), & \text{if } f_i(x) < \infty, \forall i, \\ +\infty, & \text{otherwise,} \end{cases}$$

23.6 ★ Around Conic Representations of Sets and Functions

is $\mathfrak{K}$-r; specifically,

$$\left\{\begin{array}{l} f_i \text{ are affine}, i \leq K, \& t_i \geq f_i(x) \iff \exists u^i : P_i x + t_i p_i + Q_i u^i - r_i \in \mathbf{K}_i, K < i \leq I, \\ t \geq F(y) \iff \exists u : Py + tp + Qu - r \in \mathbf{K}, \end{array}\right\}$$

$$\Downarrow$$

$$t \geq g(x) \iff \exists t_i, 1 \leq i \leq I, u^i, K < i \leq I, u : \left\{\begin{array}{ll} \underbrace{t_i - f_i(x) = 0,}_{\text{linear equations}} & \forall i \leq K, \\ P_i x + t_i p_i + Q_i u^i - r_i \in \mathbf{K}_i, & \forall i, K < i \leq I, \\ P[t_1; \ldots; t_I] + tp + Qu - r \in \mathbf{K}. & \end{array}\right.$$

23.6.3 $\mathfrak{R}/\mathfrak{L}/\mathfrak{S}$ Hierarchy

Exercise IV.34 [EdYs]

1. Let $\mathfrak{K}$ and $\mathfrak{M}$ be two families of regular cones, each containing nonnegative rays and closed with respect to taking finite direct products and to passing from a cone to its dual cone. Assume that every cone $\mathbf{M} \in \mathfrak{M}$ admits a $\mathfrak{K}$-representation:

$$\mathbf{M} = \{y : \exists v : P_{\mathbf{M}} y + Q_{\mathbf{M}} v - r_{\mathbf{M}} \in \underbrace{\mathbf{K}_{\mathbf{M}}}_{\in \mathfrak{K}}\}.$$

Show that an $\mathfrak{M}$-representation of a set X given by $X = \{x \in \mathbf{R}^n : \exists u : Px + Qu - r \in \underbrace{\mathbf{M}}_{\in \mathfrak{M}}\}$ can be straightforwardly converted into a $\mathfrak{K}$-representation of X.

2. [Cf. Exercise IV.35] Note that $\mathbf{R}^n_+$ belongs to $\mathfrak{L}$ (like every other family of cones we are considering here – all these families contain nonnegative rays and are closed with respect to taking finite direct products), thus, every polyhedral-representable set or function is Lorentz-representable as well, by item 1. Check that the Lorentz cone $\mathbf{L}^m$ is semidefinite-representable as well; specifically,

$$\mathbf{L}^m := \left\{ x \in \mathbf{R}^m : x_m \geq \sqrt{\sum_{i=1}^{m-1} x_i^2} \right\}$$

$$= \left\{ x \in \mathbf{R}^m : \text{Arrow}(x) := \begin{bmatrix} x_m & x_1 & \cdots & x_{m-1} \\ x_1 & x_m & & \\ \vdots & & \ddots & \\ x_{m-1} & & & x_m \end{bmatrix} \succeq 0 \right\}$$

implying by item 1 that cones from $\mathfrak{L}$ admit explicit $\mathfrak{S}$-representations and thus that Lorentz-representable sets and functions are semidefinite representable as well, with $\mathfrak{S}$-representations readily given by $\mathfrak{L}$-representations.

Exercise IV.35 [EdYs] It is easy to see that the nonnegative orthant $\mathbf{R}^n_+$ in the positive semidefinite cone $\mathbf{S}^n_+$ is simply given by the intersection of $\mathbf{S}^n_+$ with the linear subspace L of diagonal matrices from $\mathbf{S}^n$. Formally: Let A be the embedding of $\mathbf{R}^n$ into $\mathbf{S}^n$ which maps any vector ξ into the diagonal matrix $\text{Diag}\{\xi\}$; then $z \in \mathbf{R}^n_+$ if and only if $Az \in \mathbf{S}^n_+$. Alternatively, one can get $\mathbf{R}^n_+$ as the linear image of the positive semidefinite cone. In particular, $\mathbf{R}^n_+$ is the image of $\mathbf{S}^n_+$ under the linear mapping which maps a symmetric $n \times n$ matrix Z into the vector $\text{Dg}(Z)$ composed of diagonal entries of Z. As a result, a linear programming problem $\min_{x \in \mathbf{R}^n} \{c^\top x : Ax \leq b\}$ can be converted into the equivalent semidefinite problem $\min_{X \in \mathbf{S}^n} \{\sum_i c_i X_{ii} : X \succeq 0, A \text{Dg}(X) \leq b\}$. Indeed, similar possibilities exist for the

Lorentz cone $\mathbf{L}^n$, including the possibility of reformulating a conic problem involving direct products of Lorentz cones as a semidefinite program. Specifically,

1. Prove that $x \in \mathbf{L}^n$ if and only if the "arrow" matrix

$$\text{Arrow}(x) := \begin{bmatrix} x_n & x_1 & x_2 & \cdots & x_{n-1} \\ x_2 & x_n & & & \\ \vdots & & \ddots & & \\ x_{n-1} & & & & x_n \end{bmatrix}$$

is positive semidefinite.

2. Represent $\mathbf{L}^n$ as the image of $\mathbf{S}^n_+$ under a linear mapping.

23.6.4 More Calculus

The following calculus rules are less trivial.

Exercise IV.36 [EdYs] [Passing from a set to its support function and polar] Let $X \subseteq \mathbf{R}^n$ be a nonempty closed convex set given by an essentially strictly feasible $\mathfrak{K}$-representation:

$$X = \{x \in \mathbf{R}^n : \exists u : Ax + Bu - c \geq 0, \ Px + Qu - r \in \mathbf{K}\},$$
$$\text{and } \exists \bar{x}, \bar{u} : A\bar{x} + B\bar{u} - c \geq 0, \ P\bar{x} + Q\bar{u} - r \in \text{int } \mathbf{K}. \tag{$*$}$$

This representation induces a $\mathfrak{K}$-representation of the support function $\phi_X(y) = \sup_{x \in X} y^\top x$; specifically,

$$t \geq \phi_X(y) \iff \exists (\lambda, \xi) : \begin{array}{l} A^\top \lambda + P^* \xi + y = 0, \ B^\top \lambda + Q^* \xi = 0, \\ c^\top \lambda + \langle r, \xi \rangle + t \geq 0, \ \lambda \geq 0, \ \xi \in \mathbf{K}_*, \end{array}$$

where $\langle \cdot, \cdot \rangle$ is the inner product in the Euclidean space where $\mathbf{K}$ lives and, as always, $\mathbf{K}_*$ is the cone dual to $\mathbf{K}$. In addition, $(*)$ induces a $\mathfrak{K}$-representation of Polar (X):

$$\text{Polar}(X) := \{y : y^\top x \leq 1, \forall x \in X\}$$
$$= \left\{ y : \exists (\lambda, \xi) : \begin{array}{l} A^\top \lambda + P^* \xi + y = 0, B^\top \lambda + Q^* \xi = 0, \\ c^\top \lambda + \langle r, \xi \rangle + 1 \geq 0, \lambda \geq 0, \xi \in \mathbf{K}_* \end{array} \right\}.$$

Exercise IV.37 [EdYs] Let $f : \mathbf{R}^n \to \mathbf{R} \cup \{+\infty\}$ be a proper convex lower semicontinuous function with an essentially strictly feasible $\mathfrak{K}$-representation:

$$t \geq f(x) \iff \exists u : Ax + tq + Bu \geq c, \ Px + tp + Qu - r \in \mathbf{K},$$
$$\text{and } \exists \bar{x}, \bar{t}, \bar{u} : A\bar{x} + \bar{t}q + B\bar{u} \geq c, \ P\bar{x} + \bar{t}p + Q\bar{u} - r \in \text{int } \mathbf{K}.$$

Build a $\mathfrak{K}$-representation of the Legendre transform of f, i.e., of the function

$$f^*(y) = \sup_x \left[y^\top x - f(x) \right].$$

23.6.5 Raw Materials

The rules of grammar become useful only after we have at our disposal words in "dictionary form" which we can combine using these rules. Similarly, the calculus of conic

23.6 ★ Around Conic Representations of Sets and Functions

representations becomes useful only after we have a rich enough dictionary built of "raw materials," that is, "atoms" – specific $\mathfrak{K}$-representable sets and functions. In contrast with the rules of calculus, which are basically independent of the family $\mathfrak{K}$ of cones in question, the raw materials do depend on $\mathfrak{K}$. Here we restrict ourselves to a few instructive examples of Lorentz- and semidefinite-representable sets and functions; for an in-depth acquaintance with this topic, we refer the reader to [BTN].

We already understand well "atomic" $\mathfrak{R}$-representable functions and sets – these are half-spaces and affine functions. Other polyhedrally representable sets are intersections of finite families of half-spaces, and other polyhedrally representable functions are the maxima of finitely many affine functions restricted to a polyhedral domain. In other words, all $\mathfrak{R}$-representable functions and sets are obtained from the above atoms via the calculus we have just outlined.

In the next two exercises we present instructive examples of $\mathfrak{L}$-r functions and sets.

Exercise IV.38 [TrYs] [$\mathfrak{L}$-representability of $\|\cdot\|_2$ and $\|\cdot\|_2^2$] Check that the functions $\|x\|_2$ and $x^\top x$ on $\mathbf{R}^n$ admit $\mathfrak{L}$-representations as follows:

$$\{[x;t] \in \mathbf{R}^n \times \mathbf{R} : t \geq \|x\|_2\} = \left\{[x;t] \in \mathbf{R}^n \times \mathbf{R} : [x;t] \in \mathbf{L}^{n+1}\right\},$$

$$\{[x;t] \in \mathbf{R}^n \times \mathbf{R} : t \geq x^\top x\} = \left\{[x;t] \in \mathbf{R}^n \times \mathbf{R} : [2x;t-1;t+1] \in \mathbf{L}^{n+2}\right\}.$$

Exercise IV.39 [EdYs] [$\mathfrak{L}$-representability of power functions] Justify the following claims.

1. Let k be a positive integer. Then the set

$$\mathfrak{G}_k = \left\{[t;x_1;x_2;\ldots;x_{2^k}] \in \mathbf{R}_+^{1+2^k} : t \leq \left[\prod_{i=1}^{2^k} x_i\right]^{1/2^k}\right\},$$

which is the intersection of the *hypograph* (see section B.3.4 in the appendices) of the geometric mean of the 2^k nonnegative variables $x_1,\ldots,x_{2^k}$ with the half-space $\{[t;x] \in \mathbf{R}_x^{2^k} \times \mathbf{R}_t : t \geq 0\}$, admits an $\mathfrak{L}$-representation; specifically,

$$\mathfrak{G}_k = \left\{[t;x_1;x_2;\ldots;x_{2^k}] \in \mathbf{R}_+^{1+2^k} : \begin{array}{l} \exists\{u_{i,\ell} \geq 0, 1 \leq \ell \leq k, 1 \leq i \leq 2^\ell\} : \\ u_{ik} = x_i, \; 1 \leq i \leq 2^k, \\ {[2u_{i\ell}; u_{2i-1,\ell+1} - u_{2i,\ell+1}; u_{2i-1,\ell+1} + u_{2i,\ell+1}] \in \mathbf{L}^3,} \\ \quad 1 \leq i \leq 2^\ell, 1 \leq \ell < k, \\ {[2t; u_{1,1} - u_{2,1}; u_{1,1} + u_{2,1}] \in \mathbf{L}^3} \end{array}\right\}.$$

(∗)

Surprisingly, item 1 paves the road to $\mathfrak{L}$-representations of power functions.

2. Build explicit $\mathfrak{L}$-representations of the following univariate functions.
   2.1 $f(x) = (\max\{0,x\})^\theta$ with rational $\theta = p/q \geq 1$.
   2.2 $f(x) = \begin{cases} x^{p_+/q_+}, & x \geq 0, \\ |x|^{p_-/q_-}, & x \leq 0, \end{cases}$ where $p_\pm, q_\pm$ are positive integers with $p_+/q_+ \geq 1$, $p_-/q_- \geq 1$.
   2.3 $f(x) = \begin{cases} -x^{p/q}, & x \geq 0, \\ +\infty, & x < 0, \end{cases}$ with positive integers p,q such that $p/q \leq 1$.

2.4 $f(x) = \begin{cases} x^{-p/q}, & x > 0, \\ +\infty, & x \leq 0. \end{cases}$ with positive integers p, q.

3. Build $\mathfrak{L}$-representations of the following sets:

    3.1 The hypograph
    $$\{[x; t] \in \mathbf{R}^n_+ \times \mathbf{R} : t \leq f(x) := x_1^{\pi_1} x_2^{\pi_2} \cdots x_n^{\pi_n}\}$$
    of an algebraic monomial of n nonnegative variables, where π_i are positive rationals such that $\sum_i \pi_i \leq 1$ (the latter inequality for nonnegative π_i is a necessary and sufficient condition for f to be concave on $\mathbf{R}^n_+$).

    3.2 The epigraph of an algebraic monomial $f(x) = x_1^{-\pi_1} x_2^{-\pi_2} \cdots x_n^{-\pi_n}$ of n positive variables, where π_i are positive rationals.

    3.3 The epigraph of $\|\cdot\|_\pi$ on $\mathbf{R}^n$ with rational $\pi \geq 1$.

By Exercise IV.34, the expressive abilities of semidefinite representations are at least as strong as those of Lorentz representability. In fact, $\mathfrak{S}$-representability is strong enough to bring, for all practical purposes, the entirety of Convex Optimization within the grasp of Semidefinite Optimization. In our next exercise we are just touching the tip of the "semidefinite iceberg."

Exercise IV.40 [EdYs]

1. As a start, build $\mathfrak{S}$-representations of the maximum eigenvalue of a symmetric matrix and of the spectral norm $\|\cdot\|_{2,2}$ (the maximum singular value) of a rectangular matrix. *Hint:* Note that, for a $p \times q$ matrix A, the eigenvalues of the symmetric $(p+q) \times (p+q)$ matrix $\begin{bmatrix} & A \\ \hline A^\top & \end{bmatrix}$ are the singular values of A, the negatives of these singular values, and perhaps a number of zeros.

As a matter of fact, the single most valuable $\mathfrak{S}$-representation is that for the sums $S_k(X)$ of the k largest eigenvalues of a symmetric matrix X; the convexity of these sums in X was established in chapter 14.

2. Build an $\mathfrak{S}$-representation of the sum $S_k(X)$ of the $k \leq m$ largest eigenvalues of an $m \times m$ symmetric matrix X.
    Hint: Recall the polyhedral representation, built in Exercise I.29, of the vector analogy of $S_k(X)$, that is, the sum $s_k(x)$ of the k largest entries in an m-dimensional vector x:
    $$t \geq s_k(x) \iff \exists z \geq 0, s : x \leq z + s\mathbf{1}, \sum_i z_i + ks \leq t,$$
    where $\mathbf{1}$ is the "all-ones" vector.

The importance of the $\mathfrak{S}$-representability of $S_k(\cdot)$ becomes clear from the following fact.

3. Let $f(x) : \mathbf{R}^m \to \mathbf{R} \cup \{+\infty\}$ be a convex function symmetric with respect to permutations of entries in the argument, and let
    $$F(X) = f(\lambda(X)) : \mathbf{S}^m \to \mathbf{R} \cup \{+\infty\}.$$

Recall that F is convex by Proposition 14.3. Show that $F(X)$ admits the following representation:

$$t \geq F(x) \iff \exists u \in \mathbf{R}^m : \begin{array}{ll} f(u) \leq t & (a) \\ u_1 \geq u_2 \geq \cdots \geq u_m & (b) \\ S_k(X) \leq u_1 + \cdots + u_k, \ 1 \leq k < m & (c_k) \\ \mathrm{Tr}(X) = u_1 + \cdots + u_m & (c_m) \end{array} \quad (23.3)$$

Combine this fact with the $\mathfrak{S}$-representability of $S_k(\cdot)$ to arrive at the following corollary.

Corollary 23.1 *In the situation of item 3, assume that f is not just symmetric, but is $\mathfrak{S}$-representable as well. An $\mathfrak{S}$-representation of f gives rise to an explicit $\mathfrak{S}$-representation of $F(X)$.*

This corollary underlies the $\mathfrak{S}$-representations of numerous highly important functions and sets, e.g., the *Shatten norms* of rectangular matrices, that is, the p-norms of the vector of a matrix's singular values, or the hypograph $t \leq \mathrm{Det}^{1/m}(X)$ of the (appropriate power of the) determinant of $X \in \mathbf{S}_+^m$, or the epigraph of the function $\mathrm{Det}^{-1}(X)$ of $X \succ 0$.

Exercise IV.41 [EdYs] An interesting example of $\mathfrak{S}$-representable sets deals with matrix squares and matrix square roots:

1. [$\succeq$-epigraph of the matrix square] Prove that the function $F(X) = X^\top X : \mathbf{R}^{m \times n} \to \mathbf{S}^n$ is $\succeq$-convex and find an $\mathfrak{S}$-representation of its $\succeq$-epigraph $\{(X, Y) \in \mathbf{R}^{m \times n} \times \mathbf{S}^n : Y \succeq X^\top X\}$.
2. [$\succeq$-hypograph of the matrix square root] Prove that the set $\{(X, Y) \in \mathbf{S}^n \times \mathbf{S}^n : X \succeq 0, Y \preceq X^{1/2}\}$ is convex and find its $\mathfrak{S}$-representation.

Note: The solutions to items 1 and 2 provide us with $\mathfrak{S}$-representations of the sets $\{(X, Y) \in \mathbf{S}^n \times \mathbf{S}^n : X \succeq 0, 0 \preceq X \preceq Y^{1/2}\}$ and $\{(X, Y) \in \mathbf{S}^n \times \mathbf{S}^n : X \succeq 0, X^2 \preceq Y\}$. These sets are different, and the second is "essentially smaller" than the first; see Exercise IV.17.

Exercise IV.42 [EdYs] [An important example of $\mathfrak{S}$-representation] Consider a situation as follows. Given a *basic set* $\mathcal{B} \subseteq \mathbf{R}^n$ which is the solution set of a strictly feasible quadratic inequality,

$$\mathcal{B} = \left\{ u \in \mathbf{R}^n : u^\top Q u + 2 q^\top u + \kappa \leq 0 \right\}, \qquad [Q \in \mathbf{S}^n, q \in \mathbf{R}^n, \kappa \in \mathbf{R}]$$

we consider a *target set*

$$\mathcal{Q} = \{ x \in \mathbf{R}^m : x^\top S x + 2 s^\top x + \sigma \leq 0 \} \qquad [S \in \mathbf{S}^m, s \in \mathbf{R}^m, \sigma \in \mathbf{R}]$$

and an affine mapping

$$u \mapsto P(x) := Pu + p : \mathbf{R}^n \to \mathbf{R}^m.$$

We are interested in the case when the image of the basic set under the mapping $P(\cdot)$ is contained in the target set, and we want to describe this in terms of the parameters S, s, σ, P, p.

The task is as follows. Set

$$\mathcal{M}(S,s,\sigma;P,p;\lambda) = [P,p]^\top S[P,p] + \begin{bmatrix} -\lambda Q & P^\top s - \lambda q \\ \hline s^\top P - \lambda q^\top & 2s^\top p + \sigma - \lambda \kappa \end{bmatrix}.$$

Prove that the inclusion $P(\mathcal{B}) \subseteq \mathcal{Q}$ is equivalent to the existence of $\lambda \geq 0$ such that

$$\mathcal{M}(S,s,\sigma;P,p;\lambda) \preceq 0. \qquad (!)$$

Why this is important: Note that

1. When (S,s,σ) is fixed and $S \succeq 0$, the matrix $\mathcal{M}(S,s,\sigma;P,p;\lambda)$ is $\succeq$-convex in (P,p,λ); moreover, introducing a $\succeq$-upper bound V on $[P,p]^\top S[P,p]$, we have

$$\mathcal{M}(S,s,\sigma;P,p;\lambda) \preceq 0 \iff \exists V \in \mathbf{S}^{n+1} : \begin{bmatrix} V & [P,p]^\top S^{1/2} \\ S^{1/2}[P,p] & I_m \end{bmatrix} \succeq 0,$$

$$\text{and } V + \begin{bmatrix} -\lambda Q & P^\top s - \lambda q \\ \hline s^\top P - \lambda q^\top & 2s^\top p + \sigma - \lambda \kappa \end{bmatrix} \preceq 0,$$

and we arrive at an $\mathfrak{S}$-representation of the set of parameters P,p of affine mappings mapping $\mathcal{B}$ into $\mathcal{Q}$.

2. When P,p are fixed, (!) is a linear matrix inequality in variables S,s,σ; as a result, we get an $\mathfrak{S}$-representation of the set of parameters S,s,σ of quadratic forms resulting in $P(\mathcal{B}) \subseteq \mathcal{Q}$ for fixed $P(\cdot)$.

Item 1 here allows one to pose as an explicit semidefinite program the problem of finding the largest-volume ellipsoid contained in the intersection $\mathcal{S}$ of finitely many ellipsoids (or, more generally, sublevel sets of convex quadratic functions, e.g., linear functions); this problem arises in numerous applications. To this end it suffices to specify the basic set as the unit Euclidean ball in $\mathbf{R}^m$ and to note that an ellipsoid ("flat" or full-dimensional) in $\mathbf{R}^m$ is the image of this ball under an affine transformation $u \mapsto Pu + p$ with symmetric positive semidefinite P. Item 1 says that the set of parameters $(P \succeq 0, p)$ of ellipsoids contained in $\mathcal{S}$ admits an explicit $\mathfrak{S}$-representation. Taking into account that the volume of the ellipsoid $P\mathcal{B} + p$ with $P \succeq 0$ is proportional to $\text{Det}(P)$ and that the function $-\text{Det}^{1/m}(P)$ of $P \in \mathbf{S}^m_+$ admits an explicit $\mathfrak{S}$-representation, see Exercise IV.40.3, we arrive at the semidefinite reformulation of the problem of interest.

Similarly, item 2 allows one to handle another problem with a wide spectrum of applications – the problem of finding the smallest-volume ellipsoid containing the union of finitely many given ellipsoids. Indeed, representing these ellipsoids as the images of the unit ball in $\mathbf{R}^m$ under given affine mappings, representing the ellipsoid of interest as

$$E = \{x \in \mathbf{R}^n : (x-c)^\top S(x-c) \leq 1\}$$

with $S \succ 0$, and passing from variables S,c to variables $S, s := -Sc$, we get

$$E = \{x \in \mathbf{R}^n : x^\top Sx + 2s^\top x + \sigma(S,s) \leq 0\},$$
$$S \succ 0, \ \sigma(S,s) := s^\top S^{-1}s - 1, \qquad (!!)$$

23.6 ★ Around Conic Representations of Sets and Functions

so that the fact that E contains a given ellipsoid $P\mathcal{B}+p$ is equivalent to the existence of $\lambda \geq 0$ such that

$$\mathcal{M}(S,s,\sigma(S,s);P,p;\lambda) = [P,p]^\top S[P;p] + \left[\begin{array}{c|c} -\lambda Q & P^\top s - \lambda q \\ \hline s^\top P - \lambda q^\top & 2s^\top p + s^\top S^{-1}s - 1 - \lambda \kappa \end{array}\right] \preceq 0,$$

or, which is the same, to the existence of $\lambda \geq 0$ and μ such that

$$[P;p]^\top S[P;p] + \left[\begin{array}{c|c} -\lambda Q & P^\top s - \lambda q \\ \hline s^\top P - \lambda q^\top & 2s^\top p + \mu - 1 - \lambda \kappa \end{array}\right] \preceq 0, \quad \left[\begin{array}{c|c} S & s \\ \hline s^\top & \mu \end{array}\right] \succeq 0.$$

Thus, we can point out an explicit $\mathfrak{S}$-representation of the set of parameters $(S \succ 0, s)$ of ellipsoids containing all ellipsoids from a given finite collection. The volume of the ellipsoid (!!) is proportional to $\mathrm{Det}^{-1/2}(S)$ so that the problem of interest is to maximize $\mathrm{Det}(S)$ over a subset of $\mathbf{S}_+^m$ given by an $\mathfrak{S}$-representation. Recalling that the function $-\mathrm{Det}^{1/m}(S)$ of $S \succeq 0$ admits an explicit $\mathfrak{S}$-representation, we again reduce the problem of interest to an explicit semidefinite program.

As a simple illustration, consider the *inscribed ellipsoid* algorithm.[7] This algorithm is designed to solve the optimization problem

$$\mathrm{Opt} = \min_{x \in \mathbf{R}^n} \{f(x) : \|x\|_\infty \leq 1\},$$

where f is a convex continuous function on the unit box $B_n = \{x \in \mathbf{R}^n : \|x\|_\infty \leq 1\}$. The algorithm is applicable when one can compute the value and a subgradient of f at any desired point $x \in \mathrm{int}\, B_n$ and works as follows. At the beginning of iteration $i = 1, 2, \ldots$ we have at our disposal a *localizer* G_i – a polytope known to belong to B_n and to contain the optimal set X_* of the problem, with $G_1 = B_n$. At iteration i we proceed as follows:

- compute the ith *search point* x_i – the center of the largest-volume ellipsoid E_i contained in G_i; we already know that this requires solving an auxiliary semidefinite problem and can be done efficiently;
- compute $f(x_i)$ and a subgradient g_i of f at x_i. We define the *approximate solution* x^i generated at the first i iterations as the best – that is, with the smallest value of the objective – among the search points $x_1, \ldots, x_i$, and set $f^i = f(x^i)$;
- update the localizer by setting

$$G_{i+1} := \{x \in G_i : f(x_i) + g_i^T(x - x_i) \leq f^i\}.$$

Note that the polytope G_{i+1} is indeed a localizer. Indeed, $X_* \subseteq G_i$ since G_i is a localizer; the only way in which the inclusion $X_* \subseteq G_{i+1}$ can be violated is for some $x_* \in X_*$ to satisfy $f(x_i) + g_i^T(x_* - x_i) > f^i$, which is impossible – the left-hand side in this inequality is $\leq f(x_*)$ due to $g_i \in \partial f(x_i)$, and $f(x_*) \leq f^i = f(x^i)$.

The above recurrence is terminated when the average linear size $\mathrm{Vol}^{1/n}(E_i)$ of E_i becomes less than ϵ, where $\epsilon \in (0, 1/2)$ is a prescribed tolerance. The theory says that

- the number I of iterations before termination is bounded by $O(1)n \ln(1/\epsilon)$, and

---

[7] S.P. Tarasov, L.G. Khachiyan, I.I. Erlikh, "The method of inscribed ellipsoids," *Dokl. Akad. Nauk SSSR*, 298:5 (1988), 1081–1085; English translation: *Dokl. Math.*, 37:1 (1988), 226–230.

- the resulting approximate solution $x^I \in B_n$ satisfies the error bound

$$f(x^I) - \text{Opt} \leq \epsilon \left[\max_{B_n} f - \min_{B_n} f \right].$$

Moreover, the outlined algorithm possesses provably optimal complexity (in a certain precise sense that we do not discuss here).

The interested reader is highly recommended to implement and run this algorithm, preferably on a low-dimensional problem (say, with $n = 5$ or $n = 10$) in order to ensure that solving the auxiliary problems is not too time-consuming.

Recommended setup:

- $n = 5$ or $n = 10$
- $f(x) = \max_{i \leq 1000n} |a_i^T x + b_i|^3$ with i.i.d. $\mathcal{N}(0, 1)$ entries in a_i and b
- $\epsilon = 10^{-6}$.

Appendix A

Prerequisites from Linear Algebra

Note: Appendices A–D reproduce, courtesy of World Scientific Publishing Co., Appendices A–C in the textbook A. Nemirovski, *Introduction to Linear Optimization*, World Scientific, 2024.

Regarded as mathematical entities, the objective and the constraints in a mathematical programming problem are functions of several real variables; therefore before entering into optimization theory, we need to recall several basic notions and facts about the spaces $\mathbf{R}^n$ where these functions live and about the functions themselves.

The following material is considered basic; we expect the reader to be already familiar with it, and thus we use a "cook book" style here.

A.1 Space $\mathbf{R}^n$: Algebraic Structure

Basically all events and constructions to be considered take place in the *space $\mathbf{R}^n$* of *n*-dimensional real vectors. This space can be described as follows.

A.1.1 A Point in $\mathbf{R}^n$

A *point* in $\mathbf{R}^n$ (called also an *n-dimensional vector*) is an ordered collection $x = [x_1; \ldots; x_n]$ of n reals, called the *coordinates*, or *components*, or *entries* of vector x; the space $\mathbf{R}^n$ itself is the set of all collections of this type. Note that we follow MATLAB notation for representing vectors and matrices. That is, the notation $x = [x_1; \ldots; x_n]$ represents a column vector, while $x = [x_1, \ldots, x_n]$ represents a row vector.

A.1.2 Linear Operations

The space $\mathbf{R}^n$ is equipped with two *basic operations*:

- *Addition of vectors.* This operation takes as input two vectors $x = [x_1; \ldots; x_n]$ and $y = [y_1; \ldots; y_n]$ and produces from them a new vector with entries which are sums of the corresponding entries in x and in y, i.e.,

$$x + y = [x_1 + y_1; \ldots; x_n + y_n].$$

- *Multiplication of vectors by reals.* This operation takes as input a real λ and an n-dimensional vector $x = [x_1; \ldots; x_n]$ and produces from them a new vector with entries which are λ times the entries of x, i.e.,

$$\lambda x = [\lambda x_1; \ldots; \lambda x_n].$$

As far as addition and multiplication by reals are concerned, the arithmetic of $\mathbf{R}^n$ inherits most of the common rules of Real Arithmetic, such as $x+y = y+x$, $(x+y)+z = x+(y+z)$, $(\lambda + \mu)(x + y) = \lambda x + \mu x + \lambda y + \mu y$, $\lambda(\mu x) = (\lambda \mu)x$, etc.

Above, it was tacitly assumed that n is a positive integer. To avoid unnecessary trivial comments in the sequel, it makes sense to define $\mathbf{R}^0$ as the linear space with exactly one element, denoted by 0, and the only possible linear operations in this case are $0 + 0 = 0$, $\lambda \cdot 0 = 0$, for any $\lambda \in \mathbf{R}$.

A.2 Linear Subspaces

Definition A.1 *[Linear subspace] A linear subspace in $\mathbf{R}^n$ is, by definition, a nonempty subset of $\mathbf{R}^n$ which is closed with respect to addition of vectors and multiplication of vectors by reals. That is,*

$$L \subseteq \mathbf{R}^n \text{ is a linear subspace} \iff \begin{cases} L \neq \emptyset; \\ x, y \in L \implies x + y \in L; \\ x \in L, \lambda \in \mathbf{R} \implies \lambda x \in L. \end{cases}$$

A.2.1 Examples of Linear Subspaces

We have some immediate examples of linear subspaces.

Example A.2 Each of the following is a linear subspace:

1. The entire space $\mathbf{R}^n$.
2. The *trivial* subspace containing the single zero vector $0 = [0; \ldots; 0]$ (this vector or point is called also the *origin*).
   Here, consider carefully the notation: we use the same symbol 0 to denote the real number zero and the n-dimensional vector with all coordinates equal to zero; these two zeros are not the same, and one should understand from the context (this is always very easy) which zero is meant.
3. The set $\{x \in \mathbf{R}^n : x_1 = 0\}$ of all vectors x with the first coordinate equal to zero. ∎

In fact, the last example in the above list admits a natural extension:

Example A.3 The set of all solutions to a *homogeneous* system of linear equations (i.e., with zero right-hand side),

$$\left\{ x \in \mathbf{R}^n : \begin{array}{c} a_{11}x_1 + \cdots + a_{1n}x_n = 0 \\ a_{21}x_1 + \cdots + a_{2n}x_n = 0 \\ \cdots \\ a_{m1}x_1 + \cdots + a_{mn}x_n = 0 \end{array} \right\}, \tag{A.1}$$

is always a linear subspace in $\mathbf{R}^n$.

In fact, we will see in Proposition A.46 that this example is "generic," that is, *every* linear subspace in $\mathbf{R}^n$ is the solution set of a (finite) system of homogeneous linear equations.

Definition A.4 *[Linear combination] Given a set of vectors $x^1, \ldots, x^N \in \mathbf{R}^n$ and a set of real numbers $\lambda_1, \ldots, \lambda_N$, the vector $\sum_{i=1}^{N} \lambda_i x^i$ is called a* linear combination *of the vectors $x^1, \ldots, x^N$, and the reals $\lambda_1, \ldots, \lambda_N$ are called the* coefficients *of the combination.*

A.2 Linear Subspaces

Definition A.5 *[Linear span] Given a nonempty set $X \subseteq \mathbf{R}^n$ of vectors, the* linear span *of X [notation: Lin(X)] is the linear subspace that consists of all vectors x which can be represented as a linear combination of vectors from X. That is,*

$$\mathrm{Lin}(X) := \left\{ x \in \mathbf{R}^n : \exists N, x^i \in X, i = 1, \ldots, N, \lambda \in \mathbf{R}^N : x = \sum_{i=1}^{N} \lambda_i x^i \right\}.$$

By definition, the linear span of the empty set is the trivial linear subspace, the origin, i.e., $\mathrm{Lin}(\varnothing) = \{0\}$.

Example A.6 Given a nonempty set $X \subseteq \mathbf{R}^n$ of vectors, their linear span $\mathrm{Lin}(X)$ is a linear subspace.

The definition of the linear span also immediately points to an "outer" characterization as follows (for an explanation of the terms "inner" and "outer", see the start of section 7.3).

Fact A.7 *For $X \subset \mathbf{R}^n$, $\mathrm{Lin}(X)$ is the smallest linear subspace which contains X: if L is a linear subspace such that $L \supseteq X$, then $L \supseteq \mathrm{Lin}(X)$.*

We will see in Theorem A.16 that the "linear span" example is generic as well. That is, *every linear subspace in $\mathbf{R}^n$ is the linear span of an appropriately chosen finite set of vectors from $\mathbf{R}^n$*. Note that this latter characterization of the linear span is of an "inner" description type.

A.2.2 Sums and Intersections of Linear Subspaces

Let $\{L_\alpha\}_{\alpha \in I}$ be a family (finite or infinite) of linear subspaces of $\mathbf{R}^n$. From this family, we can build two sets:

1. *the sum* $\sum_\alpha L_\alpha$ of the subspaces L_α, which consists of all vectors which can be represented as finite sums of vectors each taken from its own subspace of the family;
2. *the intersection* $\bigcap_\alpha L_\alpha$ of the subspaces from the family.

Theorem A.8 *Let $\{L_\alpha\}_{\alpha \in I}$ be a family of linear subspaces of $\mathbf{R}^n$. Then:*

(i) *The sum $\sum_\alpha L_\alpha$ of the subspaces from the family is itself a linear subspace of $\mathbf{R}^n$; it is the smallest of those subspaces of $\mathbf{R}^n$ which contain every subspace L_α from the family.*
(ii) *The intersection $\bigcap_\alpha L_\alpha$ of the subspaces from the family is itself a linear subspace of $\mathbf{R}^n$; it is the largest of those subspaces of $\mathbf{R}^n$ which are contained in every subspace L_α from the family.*

A.2.3 Linear Independence, Bases, Dimensions

Definition A.9 *[Linear independence] A collection $X := \{x^1, \ldots, x^N\}$ of vectors from $\mathbf{R}^n$ is called* linearly independent *if no nontrivial linear combination of vectors from X (i.e., with at least one nonzero coefficient) is zero. That is,*

$$\sum_{i=1}^{N} \lambda_i x^i = 0 \implies \lambda_1 = \cdots = \lambda_N = 0.$$

Example A.10 [Example of a linearly independent set] The collection of n *standard basis vectors*, (a.k.a. *standard basic orths*) or simply *basic orths* in $\mathbf{R}^n$, i.e., the vectors $e_1 := [1; 0; \ldots; 0]$, $e_2 := [0; 1; 0; \ldots; 0]$, ..., $e_n := [0; \ldots; 0; 1]$, is linearly independent.

Example A.11 [Examples of linearly dependent sets] The following sets are all linearly dependent:

1. $X = \{0\}$;
2. $X = \{e_1, e_1\}$;
3. $X = \{e_1, e_2, e_1 + e_2\}$.

Definition A.12 [Basis] *A collection of vectors $f^1, \ldots, f^m$ is called a* basis *in $\mathbf{R}^n$, if*
- *the collection is linearly independent, and*
- *every vector from $\mathbf{R}^n$ is a linear combination of vectors from the collection (i.e., $\mathrm{Lin}\{f^1, \ldots, f^m\} = \mathbf{R}^n$).*

Example A.13 (Basis) The collection of standard basis vectors $e_1, \ldots, e_n$ is a basis in $\mathbf{R}^n$.

Example A.14 (Non-basis) None of the following examples forms a basis of $\mathbf{R}^n$.

1. The collection $\{e_2, \ldots, e_n\}$: this collection is linearly independent, but not every vector in $\mathbf{R}^n$ is a linear combination of the vectors from the collection.
2. The collection $\{e_1, e_1, e_2, \ldots, e_n\}$: every vector in $\mathbf{R}^n$ is a linear combination of vectors from this collection, but the collection is not linearly independent.

We can indeed talk about the bases of any linear subspace, not just the entire space $\mathbf{R}^n$.

Definition A.15 [Basis of a linear subspace] *A collection $\{f^1, \ldots, f^m\}$ of vectors is called a* basis of a linear subspace L, *if*
- *the collection is linearly independent, and*
- *$L = \mathrm{Lin}\{f^1, \ldots, f^m\}$, i.e., all vectors f^i belong to L, and every vector from L is a linear combination of the vectors $f^1, \ldots, f^m$.*

In order to avoid trivial remarks, we follow the standard convention:

An empty set of vectors is linearly independent, and an empty linear combination of vectors $\sum_{i \in \varnothing} \lambda_i x^i$ has a value, specifically, it is equal to zero.

With this convention, the trivial linear subspace $L = \{0\}$ also has a basis, specifically, an empty set of vectors. This convention also is fully compatible with our convention $\mathrm{Lin}(\varnothing) = \{0\}$.

Theorem A.16 *Let L be a linear subspace of $\mathbf{R}^n$. Then, L admits a (finite) basis, and all bases of L are composed of the same number of vectors.*

Definition A.17 [Dimension of a linear subspace] *Given a linear subspace L, the number of elements in its basis is called the* dimension of L *[notation: $\dim(L)$].*

We have seen that $\mathbf{R}^n$ admits a basis composed of n elements, i.e., the standard basis vectors e_i. From Theorem A.16, it follows that *every* basis of $\mathbf{R}^n$ contains exactly n vectors, and the dimension of $\mathbf{R}^n$ is n.

A.2 Linear Subspaces

Theorem A.18 *The larger is a linear subspace of $\mathbf{R}^n$, the larger is its dimension. Suppose $L \subseteq L'$ are linear subspaces of $\mathbf{R}^n$. Then, $\dim(L) \leq \dim(L')$. Moreover, $\dim(L) = \dim(L')$ holds if and only if $L = L'$.*

We have seen that $\dim(\mathbf{R}^n) = n$; according to the above convention, the trivial linear subspace $\{0\}$ of $\mathbf{R}^n$ admits an empty basis, so that its dimension is 0. Since every linear subspace L of $\mathbf{R}^n$ satisfies $\{0\} \subseteq L \subseteq \mathbf{R}^n$, it follows from Theorem A.18 that the dimension of a linear subspace in $\mathbf{R}^n$ is an integer between 0 and n.

Theorem A.19 *Let L be a linear subspace in $\mathbf{R}^n$. Then:*

(i) Every linearly independent subset of vectors from L can be extended to a basis for L.
(ii) From every spanning subset X for L, i.e., a set X such that $\mathrm{Lin}(X) = L$, we can extract a basis for L.

It follows from Theorem A.19 that

- every linearly independent subset of L contains at most $\dim(L)$ vectors, and if it contains exactly $\dim(L)$ vectors, it is a basis of L;
- every spanning set for L contains at least $\dim(L)$ vectors, and if it contains exactly $\dim(L)$ vectors, it is a basis of L.

Theorem A.20 *Let L be a linear subspace in $\mathbf{R}^n$. Suppose L is nontrivial (i.e., $L \neq \{0\}$) and $\{f^1, \ldots, f^m\}$ is a basis of L. Then, every vector $x \in L$ admits a unique representation*

$$x = \sum_{i=1}^{m} \lambda_i(x) f^i$$

as a linear combination of vectors from the basis. Moreover, the mapping

$$x \mapsto (\lambda_1(x), \ldots, \lambda_m(x)) : L \to \mathbf{R}^m$$

is a one-to-one mapping of L onto $\mathbf{R}^m$ which is linear, i.e., for every $i = 1, \ldots, m$ we have

$$\begin{aligned}\lambda_i(x + y) &= \lambda_i(x) + \lambda_i(y), \quad \forall (x, y \in L); \\ \lambda_i(\nu x) &= \nu \lambda_i(x), \quad \forall (x \in L, \nu \in \mathbf{R}).\end{aligned} \tag{A.2}$$

Definition A.21 *[Coordinates] Given a linear subspace L, a basis $f^1, \ldots, f^m$ of L, and a point $x \in L$, the reals $\lambda_i(x)$, $i = 1, \ldots, m$, in the unique representation of x given by $x = \sum_{i=1}^{m} \lambda_i(x) f^i$ are called the* coordinates *of $x \in L$ in the basis $f^1, \ldots, f^m$.*

For example, the coordinates of a vector $x \in \mathbf{R}^n$ in the *standard basis* $e_1, \ldots, e_n$ of $\mathbf{R}^n$ – the basis composed of the standard basis vectors – are exactly the entries of x.

Theorem A.22 *[Dimension formula] Let L_1, L_2 be linear subspaces of $\mathbf{R}^n$. Then,*

$$\dim(L_1 \cap L_2) + \dim(L_1 + L_2) = \dim(L_1) + \dim(L_2).$$

A.2.4 Linear Mappings and Matrices

A function $\mathcal{A}(x)$, also called a *mapping*, defined on $\mathbf{R}^n$ and taking values in $\mathbf{R}^m$, is called *linear* if it preserves linear operations:

$$\mathcal{A}(x+y) = \mathcal{A}(x) + \mathcal{A}(y), \quad \forall (x, y \in \mathbf{R}^n), \text{ and}$$
$$\mathcal{A}(\lambda x) = \lambda \mathcal{A}(x), \quad \forall (x \in \mathbf{R}^n, \lambda \in \mathbf{R}).$$

It can immediately be seen that a linear mapping from $\mathbf{R}^n$ to $\mathbf{R}^m$ can be represented as multiplication by an $m \times n$ matrix. That is, given a linear mapping $\mathcal{A} : \mathbf{R}^n \to \mathbf{R}^m$, there exists a matrix $A \in \mathbf{R}^{m \times n}$ such that

$$\mathcal{A}(x) = Ax.$$

Moreover, this matrix A is uniquely defined by the mapping: the columns A_j of A are just the images of the standard basis vectors e_j under the mapping $\mathcal{A}$, i.e.,

$$A_j = \mathcal{A}(e_j).$$

Linear mappings from $\mathbf{R}^n$ into $\mathbf{R}^m$ can be added to each other, i.e.,

$$(\mathcal{A} + \mathcal{B})(x) = \mathcal{A}(x) + \mathcal{B}(x),$$

and multiplied by reals, i.e.,

$$(\lambda \mathcal{A})(x) = \lambda \mathcal{A}(x), \quad \forall \lambda \in \mathbf{R}.$$

Moreover, the results of these operations are again linear mappings from $\mathbf{R}^n$ to $\mathbf{R}^m$. The addition of linear mappings and the multiplication of these mappings by reals correspond to the same operations with matrices representing the mappings: when adding or multiplying mappings by reals, we add, respectively multiply, the corresponding matrices by reals.

Given two linear mappings $\mathcal{A}(x) : \mathbf{R}^n \to \mathbf{R}^m$ and $\mathcal{B}(y) : \mathbf{R}^m \to \mathbf{R}^k$, we can build their superposition

$$\mathcal{C}(x) \equiv \mathcal{B}(\mathcal{A}(x)) : \mathbf{R}^n \to \mathbf{R}^k,$$

which is again a linear mapping, now from $\mathbf{R}^n$ to $\mathbf{R}^k$. In the language of the matrices representing the mappings, superposition corresponds to matrix multiplication; the $k \times n$ matrix C representing the mapping $\mathcal{C}$ is the product of the matrices representing $\mathcal{A}$ and $\mathcal{B}$:

$$\mathcal{A}(x) = Ax, \quad \mathcal{B}(y) = By \quad \Longrightarrow \quad \mathcal{C}(x) \equiv \mathcal{B}(\mathcal{A}(x)) = B \cdot (Ax) = \underbrace{(BA)}_{=C}x.$$

A.2.5 Determinant and Rank

Let us recall some basic facts about ranks and determinants of matrices.

A.2.6 Determinant

Let $A = [a_{ij}]_{\substack{1 \le i \le n \\ 1 \le j \le n}}$ be a square matrix. A *diagonal* of the matrix A is the collection of n cells with indices $(1, j_1), (2, j_2), \ldots, (n, j_n)$, where $j_1, j_2, \ldots, j_n$ are distinct from each other, so that the mapping $\sigma : \{1, \ldots, n\} \to \{1, \ldots, n\}$ given by $\sigma(i) = j_i$, $1 \le i \le n$, is a *permutation*, i.e., a one-to-one mapping of the set $\{1, \ldots, n\}$ onto itself. There are $n!$ different permutations of $\{1, \ldots, n\}$, and these permutations form the group Σ_n with group operation the product $(\sigma^1 \sigma^2)(i) = \sigma^1(\sigma^2(i))$, $1 \le i \le n$. Any permutation $\sigma \in \Sigma_n$ can be assigned a *sign*, where $\text{sign}(\sigma) \in \{\pm 1\}$, in such a way that the sign of the identity permutation $\sigma(i) \equiv i$ is +1, the sign of the product of two permutations is the product of their signs, that is, $\text{sign}(\sigma^1 \sigma^2) = \text{sign}(\sigma^1)\text{sign}(\sigma^2)$ for all σ^1, σ^2, and the sign of a *transposition* – a permutation which swaps two distinct indices and keeps all other indices intact – is -1; these properties specify the sign of a permutation in a unique fashion.

Then, for any $n \times n$ real or complex matrix $A = [a_{ij}]_{1 \le i,j \le n}$, the quantity

$$\text{Det}(A) := \sum_{\sigma \in \Sigma_n} \text{sign}(\sigma) \prod_{i=1}^{n} a_{i\sigma(i)}$$

is called the *determinant* of A. We have the following main properties of the determinant:

1. $\text{Det}(A) = \text{Det}(A^\top)$.
2. [polylinearity] $\text{Det}(A)$ is linear in the rows of A: when all rows but one are fixed, $\text{Det}(A)$ is a linear function of the remaining row. An analogous property holds for the columns.
3. [antisymmetry] When swapping rows with two distinct indices, the determinant is multiplied by -1. An analogous property holds for the columns.
4. $\text{Det}(I_n) = 1$.

Note: The last three properties uniquely define $\text{Det}(\cdot)$.

5. [multiplicativity] For two $n \times n$ matrices A, B, $\text{Det}(AB) = \text{Det}(A)\text{Det}(B)$.
6. An $n \times n$ matrix A is *nonsingular*, that is, $AB = I_n$ for properly selected B, if and only if $\text{Det}(A) \neq 0$.
7. [Cramer's rule] For a nonsingular $n \times n$ matrix A, the linear system $Ax = b$ in variables x has a unique solution, and the entries of this solution are given by

$$x_i := \frac{\text{Det}(A, i, b)}{\text{Det}(A)}, \quad 1 \le i \le n,$$

where $\text{Det}(A, i, b)$ is the determinant of the matrix obtained from A by replacing its ith column with the right-hand side vector b.

8. [decomposition] Let A be an $n \times n$ matrix with $n > 1$. Then, for every $i \in \{1, \ldots, n\}$, we have

$$\text{Det}(A) = \sum_{j=1}^{n} a_{ij} C^{ij},$$

where C^{ij} is the *algebraic complement* of the (i, j)th entry in A, i.e., $C^{ij} := (-1)^{i+j}\text{Det}(A^{ij})$, and here A^{ij} is the $(n-1) \times (n-1)$ matrix obtained from A by eliminating the ith row and jth column.

9. An affine mapping $x \mapsto Ax + b : \mathbf{R}^n \to \mathbf{R}^n$ multiplies n-dimensional volumes by $|\text{Det}(A)|$. In particular, for a real $n \times n$ matrix A the quantity $|\text{Det}(A)|$ is the n-dimensional volume of the parallelotope

$$X = \left\{ x = \sum_j s_j A_j : 0 \le s_j \le 1, \, \forall j = 1, \ldots, n \right\}$$

spanned by the columns $A_1, \ldots, A_n$ of A.

A.2.7 Rank

Let A be an $m \times n$ matrix. Every matrix is associated with two linear subspaces, namely its image space and kernel. The *image space* of A [notation: Im(A) or Im A] is defined as the linear span of the columns of A, i.e., the subspace of the destination space $\mathbf{R}^m$ formed by the vectors that admit a representation of the form Ax for some $x \in \mathbf{R}^n$. The *kernel* (a.k.a. *nullspace*) of A [notation: Ker(A)] is given by Ker(A) := $\{x : Ax = 0\}$. Note that Ker(A) is the orthogonal complement of Im($A^\top$).

Let $R_p := \{i_1 < i_2 < \cdots < i_p\}$ be a collection of $p \geq 1$ indices, distinct from each other, of rows of A, and let $C_q := \{j_1 < j_2 < \cdots < j_q\}$ be a collection of $q \geq 1$ indices, distinct from each other, of columns of A. The matrix $B \in \mathbf{R}^{p \times q}$ with entries $B_{k\ell} = A_{i_k, j_\ell}$, $1 \leq k \leq p, 1 \leq \ell \leq q$ is called the submatrix of A with row indices R_p and column indices C_q; this is precisely what we get from the matrix A as the intersection of rows with indices from R_p and columns with indices from C_q.

The *rank* of A (which is denoted by rank(A)) is, by definition, the largest row sizes, or, which is the same, the column size of the *nonsingular* square submatrices of A. When no such submatrix exists, that is, when A is the zero matrix, the rank of A is by definition 0.

The main properties of the rank of a matrix are as follows:

1. rank(A) is the dimension of the image space Im A, i.e., rank(A) = dim(Im A). Equivalently, rank(A) is the maximum cardinality of linearly independent collections of columns of A. Moreover, a collection of columns of A with rank(A) distinct indices is linearly independent if and only if its intersection with a collection of rank(A) properly selected rows is a nonsingular submatrix of A.

2. rank(A) = rank($A^\top$), so that rank(A) is the maximum cardinaly of linearly independent collections of rows of A. Thus, rank(A) is the codimension of the kernel of A. That is, for a given $A \in \mathbf{R}^{m \times n}$,

$$\dim(\mathrm{Ker}(A)) = n - \mathrm{rank}(A).$$

   A collection of rows of A with rank(A) distinct indices is linearly independent if and only if its intersection with a collection of rank(A) properly selected columns is a nonsingular submatrix of A.

3. Whenever the product AB of two matrices makes sense, we have

$$\mathrm{rank}(AB) \leq \min\{\mathrm{rank}(A), \mathrm{rank}(B)\}.$$

   For two matrices A, B of the same size, we have

$$|\mathrm{rank}(A) - \mathrm{rank}(B)| \leq \mathrm{rank}(A + B) \leq \mathrm{rank}(A) + \mathrm{rank}(B).$$

4. An $n \times n$ matrix B is nonsingular if and only if rank(B) = n.

A.3 Space $\mathbf{R}^n$: Euclidean Structure

So far, we have been interested solely in the *algebraic structure* of $\mathbf{R}^n$, or, equivalently, in the properties of the *linear* operations (the addition of vectors and multiplication of vectors by scalars) with which the space is endowed. Now let us consider another structure on $\mathbf{R}^n$ – the *standard Euclidean structure* – which allows us to speak about distances, angles, convergence, etc., and thus makes the space $\mathbf{R}^n$ a much richer mathematical entity.

A.3.1 Euclidean Structure

The standard Euclidean structure on $\mathbf{R}^n$ is given by the *standard inner product* – an operation which takes as input two vectors x, y and produces from them a real number, specifically, the real number given by

$$\langle x, y \rangle := x^\top y = \sum_{i=1}^n x_i y_i.$$

The basic properties of the inner product are as follows:

1. [bi-linearity]: The real-valued function $\langle x, y \rangle$ of two vector arguments $x, y \in \mathbf{R}^n$ is linear with respect to each argument when the other argument is fixed:

$$\langle \lambda u + \mu v, y \rangle = \lambda \langle u, y \rangle + \mu \langle v, y \rangle, \quad \forall (u, v, y \in \mathbf{R}^n, \lambda, \mu \in \mathbf{R}),$$
$$\langle x, \lambda u + \mu v \rangle = \lambda \langle x, u \rangle + \mu \langle x, v \rangle, \quad \forall (x, u, v \in \mathbf{R}^n, \lambda, \mu \in \mathbf{R}).$$

2. [symmetry]: The function $\langle x, y \rangle$ is symmetric:

$$\langle x, y \rangle = \langle y, x \rangle, \quad \forall (x, y \in \mathbf{R}^n).$$

3. [positive definiteness]: The function $\langle x, x \rangle$ is always nonnegative, and it is zero if and only if x is zero.

Remark A.23 The three properties outlined above (i.e., bi-linearity, symmetry, and positive definiteness) form a definition of an *Euclidean inner product*. In fact, there are infinitely many different ways to satisfy these properties. In other words, there are infinitely many different inner products on $\mathbf{R}^n$. The standard inner product $\langle x, y \rangle = x^\top y$ is just a particular case of this general notion. Although in this book we normally work with the standard inner product, the reader should remember that the facts we are about to recall are valid for all Euclidean inner products, and not only for the standard one.

The notion of an inner product underlies a number of purely algebraic constructions, in particular, those of the *inner product representation of linear forms* and of the *orthogonal complement*.

A.3.2 Inner Product Representation of Linear Forms on $\mathbf{R}^n$

A *linear form* on $\mathbf{R}^n$ is a real-valued function $f(x)$ on $\mathbf{R}^n$ which is *additive*, i.e., $f(x+y) = f(x) + f(y)$ for all $x, y \in \mathbf{R}^n$, and *homogeneous*, i.e., $f(\lambda x) = \lambda f(x)$ for all $x \in \mathbf{R}^n$ and $\lambda \in \mathbf{R}$.

Example A.24 (Example of a linear function) The function $f(x) := \sum_{j=1}^n j x_j$ is linear.

Example A.25 (Examples of nonlinear functions) Each of the following functions is nonlinear:

1. $f(x) := x_1 + 1$;
2. $f(x) := x_1^2 - x_2^2$;
3. $f(x) := \sin(x_1)$.

When we add two linear forms or when we multiply a linear form by a real number, we again get a linear form (mathematically speaking: "linear forms on $\mathbf{R}^n$ form a linear space"). A *Euclidean structure allows us to identify linear forms on $\mathbf{R}^n$ with vectors from $\mathbf{R}^n$*, as follows.

Theorem A.26 *Let $\langle \cdot, \cdot \rangle$ be a Euclidean inner product on $\mathbf{R}^n$.*

(i) *Let $f(x)$ be a linear form on $\mathbf{R}^n$. Then, there exists a uniquely defined vector $f \in \mathbf{R}^n$ such that the linear form $f(x)$ is just the inner product with f and x:*

$$f(x) = \langle f, x \rangle, \quad \forall x \in \mathbf{R}^n.$$

(ii) *Vice versa: every vector $f \in \mathbf{R}^n$ defines, via the formula*

$$f(x) := \langle f, x \rangle,$$

a linear form on $\mathbf{R}^n$.

(iii) *The above one-to-one correspondence between the linear forms and vectors on $\mathbf{R}^n$ is linear: adding linear forms (or multiplying a linear form by a real number), we add (respectively, multiply by the real number) the vector(s) representing the form(s).*

A.3.3 Orthogonal Complement

The Euclidean structure allows us to associate with a linear subspace $L \subseteq \mathbf{R}^n$ another linear subspace, namely the *orthogonal complement* (or *annulator*) of L [notation: $L^\perp$]. By definition, $L^\perp$ consists of all vectors which are orthogonal to every vector from L. That is,

$$L^\perp := \{ f \in \mathbf{R}^n : \langle f, x \rangle = 0, \quad \forall x \in L \}.$$

Theorem A.27 (i) *Whenever L is a linear subspace of $\mathbf{R}^n$, we have*

$$(L^\perp)^\perp = L.$$

(ii) *The larger a linear subspace L is, the smaller is its orthogonal complement: if $L_1 \subset L_2$ are linear subspaces of $\mathbf{R}^n$, then $L_1^\perp \supset L_2^\perp$.*

(iii) *The intersection of a subspace and its orthogonal complement is trivial, and the sum of these subspaces is the entire $\mathbf{R}^n$:*

$$L \cap L^\perp = \{0\}, \qquad L + L^\perp = \mathbf{R}^n.$$

Remark A.28 From Theorem A.27(iii) and the dimension formula (Theorem A.22) it follows that for every subspace L in $\mathbf{R}^n$ we have

$$\dim(L) + \dim(L^\perp) = n.$$

Moreover, every vector $x \in \mathbf{R}^n$ admits a unique decomposition as a sum of two vectors:

$$x = x_L + x_{L^\perp},$$

where x_L belongs to L and $x_{L^\perp}$ belongs to $L^\perp$. This decomposition is called the *orthogonal decomposition* of x taken with respect to $L, L^\perp$. In this decomposition, x_L is the *orthogonal projection* of x onto L, and $x_{L^\perp}$ is the orthogonal projection of x onto the orthogonal complement of L. Both projections depend on x linearly. That is, we have

$$(x+y)_L = x_L + y_L, \ \forall x, y \in \mathbf{R}^n, \quad \text{and} \quad (\lambda x)_L = \lambda x_L, \ \forall x \in \mathbf{R}^n, \forall \lambda \in \mathbf{R}.$$

The mapping $x \mapsto x_L$ is called the *orthogonal projector* onto L.

A.3.4 Orthonormal Bases

A collection of vectors $f^1, \ldots, f^m$ is *orthonormal* w.r.t. a Euclidean inner product $\langle \cdot, \cdot \rangle$ if each vector from the collection is orthogonal to every other vector from it, i.e.,

$$i \neq j \implies \langle f^i, f^j \rangle = 0,$$

and the inner product of every vector f^i with itself is equal to 1, i.e.,

$$\langle f^i, f^i \rangle = 1, \ i = 1, \ldots, m.$$

Theorem A.29 *An orthonormal collection of vectors $f^1, \ldots, f^m$ is always linearly independent and is thus a basis of its linear span $L = \mathrm{Lin}(f^1, \ldots, f^m)$ (such a basis in a linear subspace is called orthonormal). The coordinates of a vector $x \in L$ w.r.t. an orthonormal basis $f^1, \ldots, f^m$ of L are given by explicit formulas:*

$$x = \sum_{i=1}^m \lambda_i(x) f^i \iff \lambda_i(x) = \langle x, f^i \rangle, \ \forall i = 1, \ldots, m.$$

Proof Consider any $x \in \mathbf{R}^n$ and any $i = 1, \ldots, m$. Starting from the representation $x = \sum_{j=1}^m \lambda_j(x) f^j$, by taking the inner product of both sides of this equality with f^i, we get

$$\langle x, f_i \rangle = \left\langle \sum_{j=1}^m \lambda_j(x) f^j, f^i \right\rangle$$

$$= \sum_{j=1}^m \lambda_j(x) \langle f^j, f^i \rangle \qquad \text{[bilinearity of inner product]}$$

$$= \lambda_i(x) \qquad \text{[orthonormality of } \{f^i\}\text{]}.$$

Plugging $x = 0$ into this representation results in $\lambda_i(0) = 0$ for all i, i.e., all the coefficients are zero. Hence, an orthonormal system is linearly independent. ■

Example A.30 (An orthonormal basis in $\mathbf{R}^n$) The standard basis $\{e_1, \ldots, e_n\}$ is orthonormal *with respect to the standard inner product $\langle x, y \rangle = x^\top y$ on $\mathbf{R}^n$* (but is not orthonormal w.r.t. other Euclidean inner products on $\mathbf{R}^n$).

Theorem A.31 *(i) If $f^1, \ldots, f^m$ is an orthonormal basis in a linear subspace L, then the inner product of two vectors $x, y \in L$ in the coordinates $\lambda_i(\cdot)$ w.r.t. this basis is given by the standard formula*

$$\langle x, y \rangle = \sum_{i=1}^m \lambda_i(x) \lambda_i(y).$$

(ii) Every linear subspace L of $\mathbf{R}^n$ admits an orthonormal basis. Moreover, every orthonormal system $f^1, \ldots, f^m$ of vectors from L can be extended to an orthonormal basis in L.

Proof To prove (i), consider any two vectors $x, y \in L$ along with their basis representations, i.e., $x = \sum_{i=1}^{m} \lambda_i(x) f^i$ and $y = \sum_{i=1}^{m} \lambda_i(y) f^i$. Then,

$$\langle x, y \rangle = \left\langle \sum_{i=1}^{m} \lambda_i(x) f^i, \sum_{i=1}^{m} \lambda_i(y) f^i \right\rangle$$

$$= \sum_{i=1}^{m} \sum_{j=1}^{m} \lambda_i(x) \lambda_j(y) \langle f^i, f^j \rangle \qquad \text{[bilinearity of the inner product]}$$

$$= \sum_{i=1}^{m} \lambda_i(x) \lambda_i(y). \qquad \text{[orthonormality of the vectors } \{f^i\}\text{]}$$

The proof of (ii) is given by the *Gram–Schmidt orthogonalization process* (which is important by its own right) as follows. We start with an arbitrary basis $h^1, \ldots, h^m$ in L and step by step convert it into an orthonormal basis $f^1, \ldots, f^m$. At the beginning of step t of the construction, we already have an orthonormal collection $f^1, \ldots, f^{t-1}$ such that $\text{Lin}\{f^1, \ldots, f^{t-1}\} = \text{Lin}\{h^1, \ldots, h^{t-1}\}$. For any $t = 1, \ldots, m$, at step t we proceed as follows:

1. Build the vector

$$g^t := h^t - \sum_{j=1}^{t-1} \langle h^t, f^j \rangle f^j.$$

It is easily seen (check it!) that
- we have

$$\text{Lin}\{f^1, \ldots, f^{t-1}, g^t\} = \text{Lin}\{h^1, \ldots, h^t\}; \qquad (A.3)$$

- moreover, $g^t \neq 0$ (derive this fact from (A.3) and the linear independence of the collection $h^1, \ldots, h^m$);
- also, g^t is orthogonal to $f^1, \ldots, f^{t-1}$.

2. Since $g^t \neq 0$, the quantity $\langle g^t, g^t \rangle$ is positive (by the positive definiteness of the inner product), so the vector

$$f^t := \frac{1}{\sqrt{\langle g^t, g^t \rangle}} g^t$$

is well defined. It can be easily seen (check it!) that the collection $f^1, \ldots, f^t$ is orthonormal and

$$\text{Lin}\{f^1, \ldots, f^t\} = \text{Lin}\{f^1, \ldots, f^{t-1}, g^t\} = \text{Lin}\{h^1, \ldots, h^t\},$$

which completes the step t of the orthogonalization process.

After m steps of the orthogonalization process, we end up with an orthonormal system $f^1, \ldots, f^m$ of vectors from L such that

$$\text{Lin}\{f^1, \ldots, f^m\} = \text{Lin}\{h^1, \ldots, h^m\} = L,$$

so that $f^1, \ldots, f^m$ is an orthonormal basis in L.

The construction can be easily modified (do it!) to extend a given orthonormal system of vectors from L to an orthonormal basis of L. ∎

Theorem A.31 admits an important corollary which states that

All Euclidean spaces of the same dimension are "the same."

Corollary A.32 *Suppose L is an m-dimensional subspace in a space $\mathbf{R}^n$ equipped with an Euclidean inner product $\langle \cdot, \cdot \rangle$. Then, there exists a one-to-one mapping $x \mapsto A(x)$ of L onto $\mathbf{R}^m$ such that*

(i) the mapping preserves linear operations:
$$A(x+y) = A(x) + A(y), \quad \forall (x, y \in L) \quad \text{and}$$
$$A(\lambda x) = \lambda A(x), \quad \forall (x \in L, \lambda \in \mathbf{R});$$

(ii) the mapping converts the $\langle \cdot, \cdot \rangle$ inner product on L into the standard inner product on $\mathbf{R}^m$:
$$\langle x, y \rangle = (A(x))^\top A(y), \quad \forall x, y \in L.$$

Proof Indeed, by Theorem A.31(ii) L admits an orthonormal basis $f^1, \ldots, f^m$; using Theorem A.31(i), we can immediately check that the mapping
$$x \mapsto A(x) = [\lambda_1(x); \ldots; \lambda_m(x)]$$
which maps $x \in L$ into the m-dimensional vector composed of the coordinates of x in the basis $f^1, \ldots, f^m$, meets all the requirements. ∎

Fact A.33 *(i) Let L be a linear subspace of $\mathbf{R}^n$, and $f^1, \ldots, f^m$ be an orthonormal basis in L. Then, for every $x \in \mathbf{R}^n$, the orthoprojection x_L of x onto L is given by the formula*
$$x_L := \sum_{i=1}^m (x^\top f^i) f^i.$$

(ii) Let L_1, L_2 be linear subspaces in $\mathbf{R}^n$. Then
$$(L_1 + L_2)^\perp = L_1^\perp \cap L_2^\perp \quad \text{and} \quad (L_1 \cap L_2)^\perp = L_1^\perp + L_2^\perp.$$

A.4 Affine Subspaces in $\mathbf{R}^n$

In this book, many events that we consider take place not in the entire $\mathbf{R}^n$, but in its *affine subspaces*. Geometrically, affine subspaces are planes of different dimensions in $\mathbf{R}^n$. Let us become acquainted with these subspaces.

A.4.1 Affine Subspaces and Affine Hulls

In geometry, a linear subspace L of $\mathbf{R}^n$ is a special plane – it passes through the origin of the space (i.e., it contains the zero vector). To obtain an arbitrary plane M, it suffices to *translate* an appropriate linear subspace L, i.e., add a fixed *shifting vector* $\bar{a}$ to all points from L. This geometric intuition leads to the following definition.

Definition A.34 *[Affine subspace]* An affine subspace (a.k.a. affine plane) *in $\mathbf{R}^n$ is a set of the form*

$$M = \bar{a} + L = \{\bar{a} + x : x \in L\}, \tag{A.4}$$

where L is a linear subspace in $\mathbf{R}^n$ and $\bar{a}$ is a vector from $\mathbf{R}^n$.

According to our convention on the arithmetic of sets, in (A.4) we should have written $\{\bar{a}\} + L$ instead of $\bar{a} + L$ as we did not define the arithmetic sum of a vector and a set. It is usual to ignore this difference and omit the braces when writing down singleton sets in similar expressions: thus, we shall write $\bar{a} + L$ instead of $\{\bar{a}\} + L$, $\mathbf{R}d$ instead of $\mathbf{R}\{d\}$, etc.

Example A.35 Let L be the linear subspace composed of vectors with first entries equal to zero. Suppose that we shift L by a vector $\bar{a} = [\bar{a}_1; \ldots; \bar{a}_n]$. Then, we obtain the set $M := \bar{a} + L$ of all vectors x with $x_1 = \bar{a}_1$. According to our terminology, this set M is an affine subspace.

An immediate question about the notion of an affine subspace is what "degrees of freedom" we may have in the decomposition (A.4). For example, would M uniquely determine the linear subspace L and the shifting vector $\bar{a}$? The next proposition provides an answer to this question.

Proposition A.36 *Given an affine subspace M in $\mathbf{R}^n$, the linear subspace L in its decomposition (A.4) is uniquely determined by M. Specifically, L is the set of all differences of the vectors from M, i.e.,*

$$L = M - M = \{x - y : x, y \in M\}. \tag{A.5}$$

In contrast, the shifting vector $\bar{a}$ is not uniquely defined by M and can be chosen as an arbitrary vector from M.

In this book, given an affine subspace M, we refer to the linear subspace $L := M - M$ as the subspace *parallel* to the affine subspace M.

A.4.2 Intersections of Affine Subspaces, Affine Combinations, and Affine Hulls

An immediate conclusion of Proposition A.36 is as follows:

Corollary A.37 *Let $\{M_\alpha\}$ be an arbitrary family of affine subspaces in $\mathbf{R}^n$. Whenever the set $M := \bigcap_\alpha M_\alpha$ is nonempty, M is an affine subspace.*

From Corollary A.37 it immediately follows that for every nonempty subset Y of $\mathbf{R}^n$ there exists the smallest affine subspace containing Y; this is the intersection of all affine subspaces containing Y. This smallest affine subspace containing Y is called the *affine hull* (a.k.a. *affine span*) of Y [notation: $\mathrm{Aff}(Y)$].

All this resembles the story about linear spans. In fact, we can further extend this analogy and get an "inner" description (see Remark A.48 below) of the affine hull Aff(Y) in terms of elements of Y similar to that of the linear span (recall that the linear span of X is also characterized as the set of all linear combinations of vectors from X).

Given a nonempty set Y, let us choose an arbitrary point $y^0 \in Y$, and consider the set

$$X := Y - y^0.$$

All affine subspaces containing Y will also contain y^0. Therefore, by Proposition A.36, Aff(Y) can be represented as $M = y^0 + L$, where L is a linear subspace. It is absolutely evident that an affine subspace $M = y^0 + L$ contains Y if and only if the subspace L contains X, and that the larger is L, the larger is M:

$$L \subset L' \implies M = y^0 + L \subset M' = y^0 + L'.$$

Thus, in order to find the smallest among the *affine subspaces containing Y*, it suffices to find the smallest among the *linear subspaces containing X* and then translate the latter space by y^0:

$$\text{Aff}(Y) = y^0 + \text{Lin}(X) = y^0 + \text{Lin}(Y - y^0). \tag{A.6}$$

Now, recall that by definition Lin($Y - y^0$) is the set of all linear combinations of vectors from $Y - y^0$, so that a generic element of Lin($Y - y^0$) is

$$x = \sum_{i=1}^{k} \mu_i (y^i - y^0) \qquad \text{[here k may depend on x]}$$

with $y^i \in Y$ and coefficients $\mu_i \in \mathbf{R}$ for $i = 1, \ldots, k$. Then, a generic element of Aff(Y) is given by

$$y = y^0 + \sum_{i=1}^{k} \mu_i (y^i - y^0) = \sum_{i=0}^{k} \lambda_i y^i,$$

where

$$\lambda_0 := 1 - \sum_{i} \mu_i, \quad \text{and} \quad \lambda_i := \mu_i, \ i = 1, \ldots, k.$$

We deduce that a generic element of Aff(Y) is a linear combination of vectors from Y. Note, however, that in this combination the coefficients λ_i are not completely arbitrary: their sum is equal to 1. Linear combinations of this type – with a unit sum of coefficients – have a special name; they are called *affine combinations*.

We have seen that every vector from Aff(Y) is an affine combination of vectors from Y. In fact, the reverse is also true, i.e., Aff(Y) contains all affine combinations of vectors from Y. Indeed, if

$$y = \sum_{i=1}^{k} \lambda_i y^i$$

is an affine combination of vectors from Y, then, using the equality $\sum_i \lambda_i = 1$, we can write it also as

$$y = y^0 + \sum_{i=1}^{k} \lambda_i (y^i - y^0),$$

where y^0 is the "arbitrarily selected" vector we used in our previous reasoning. And any vector y of this form, as we already know, belongs to $\mathrm{Aff}(Y)$. Thus, we arrive at the following result.

Proposition A.38 *[Structure of affine hull] For any nonempty set Y, we have*

$$\mathrm{Aff}(Y) = \{\textit{the set of all affine combinations of vectors from } Y\}.$$

When Y itself is an affine subspace, it coincides, of course, with its affine hull, and the previous proposition leads to the following consequence.

Corollary A.39 *An affine subspace M is closed with respect to taking affine combinations of its members, i.e., every combination of this type is a vector from M. Vice versa, a nonempty set which is closed with respect to taking affine combinations of its members is an affine subspace.*

A.4.3 Affinely Spanning Sets, Affinely Independent Sets, Affine Dimension

Affine subspaces are closely related to linear subspaces, and the basic notions associated with linear subspaces have natural and useful affine analogies. Here, we introduce these notions and discuss their basic properties.

Affinely spanning sets. Consider an affine subspace $M = \bar{a} + L$. We say that a subset Y of M is *affinely spanning* for M (we also say that Y spans M affinely, or that M is affinely spanned by Y) if $M = \mathrm{Aff}(Y)$, or equivalently, due to Proposition A.38, if every point of M is an affine combination of points from Y. Hence, we arrive at the following immediate consequence of section A.4.2.

Proposition A.40 *Let $M = \bar{a} + L$ be an affine subspace, let Y be a subset of M, and consider any $y^0 \in Y$. Then, the set Y affinely spans M, i.e., $M = \mathrm{Aff}(Y)$, if and only if the set*

$$X := Y - y^0$$

spans the linear subspace L, i.e., $L = \mathrm{Lin}(X)$.

Affinely independent sets. A linearly independent set of vectors $x^1, \ldots, x^k$ is such that no nontrivial linear combination of $x^1, \ldots, x^k$ is equal to zero (see Definition A.9). An equivalent definition is given by Theorem A.16(iv). That is, the vectors $x^1, \ldots, x^k$ are linearly independent if in any linear combination

$$x = \sum_{i=1}^{k} \lambda_i x^i,$$

the coefficients λ_i are *uniquely* determined by the vector x. This equivalent form reflects the essence of the matter – what we indeed need is the uniqueness of the coefficients in expansions. Accordingly, this equivalent form is the prototype for the notion of an affinely independent set: we want to introduce this notion in such a way that the coefficients λ_i in an *affine* combination

$$y = \sum_{i=0}^{k} \lambda_i y^i$$

of an "affinely independent" set of vectors $y^0, \ldots, y^k$ would be uniquely defined by y. *Non*-uniqueness would mean that we can write y as an affine combination of the vectors $y^0, \ldots, y^k$ using two different sets of coefficients λ_i and λ_i' such that $\sum_{i=0}^{k} \lambda_i = \sum_{i=0}^{k} \lambda_i' = 1$. That is,

$$y = \sum_{i=0}^{k} \lambda_i y^i = \sum_{i=0}^{k} \lambda_i' y^i.$$

In such a case, we arrive at

$$\sum_{i=0}^{k} (\lambda_i - \lambda_i') y^i = 0,$$

which implies that the vectors $y^0, \ldots, y^k$ are linearly dependent. Moreover, there exists a nontrivial combination of these vectors which represents the zero vector and the sum of the coefficients in this representation satisfies $\sum_i (\lambda_i - \lambda_i') = \sum_i \lambda_i - \sum_i \lambda_i' = 1 - 1 = 0$. Our reasoning can be reversed: if there exists a nontrivial linear combination of the vectors y^i with a zero sum of coefficients which results in the zero vector, then the coefficients in the representation of any vector as an affine combination of the vectors y^i are not uniquely defined. Thus, in order to get uniqueness we need to forbid relations

$$\sum_{i=0}^{k} \mu_i y^i = 0$$

with nontrivial coefficients μ_i satisfying $\sum_{i=0}^{k} \mu_i = 0$. Thus, this discussion motivates the following definition.

Definition A.41 [Affine independence] *A collection $y^0, \ldots, y^k$ of vectors in $\mathbf{R}^n$ is called affinely independent if no nontrivial linear combination of the vectors with zero sum of coefficients is the zero vector, i.e.,*

$$\sum_{i=0}^{k} \lambda_i y^i = 0 \text{ and } \sum_{i=0}^{k} \lambda_i = 0 \implies \lambda_0 = \lambda_1 = \cdots = \lambda_k = 0.$$

(To compare with the definition of linear independence, see Definition A.9.)

With this definition of an affinely independent set of vectors, we arrive at the following result, which is analogous to Theorem A.20.

Corollary A.42 Let $y^0, \ldots, y^k$ be affinely independent. Then, the coefficients λ_i in any affine combination

$$y = \sum_{i=0}^{k} \lambda_i y^i \qquad \left[where \sum_{i=0}^{k} \lambda_i = 1 \right]$$

of the vectors $y^0, \ldots, y^k$ are uniquely defined by the vector y that is the value of the combination.

Verification of the affine independence of a collection can be immediately reduced to verification of the linear independence of a closely related collection.

Proposition A.43 *It holds that $k + 1$ vectors $y^0, \ldots, y^k$ are affinely independent if and only if the k vectors given by $(y^1 - y^0), (y^2 - y^0), \ldots, (y^k - y^0)$ are linearly independent.*

On the basis of this last proposition we deduce for example that the set of vectors $0, e_1, \ldots, e_n$ composed of the origin and the standard basis vectors is affinely independent. Note that this collection is linearly dependent (as every set of vectors containing zero is linearly dependent). The difference between the two notions of independence with which we are dealing is important to keep in mind: linear independence means that no nontrivial linear combination of the vectors can be zero, while affine independence means that no nontrivial linear combination *from a certain restricted class of them* (with zero sum of coefficients) can be zero. Therefore, there are more affinely independent sets than linearly independent sets: a linearly independent set is affinely independent, but not vice versa.

Affine bases and affine dimension. Propositions A.38 and A.40 reduce the notions of affine spanning and affinely independent sets to the notions of spanning and linearly independent sets. Combined with Theorem A.16, they result in the following analogies of the latter two statements:

Proposition A.44 *[Affine dimension] Let $M = \bar{a} + L$ be an affine subspace in $\mathbf{R}^n$. Then, the following two quantities are finite integers which are equal to each other:*

(i) *the minimal number of elements in the subsets of M which affinely span M;*
(ii) *the maximal number of elements in affinely independent subsets of M.*
    The common value of these two integers is exactly equal to $\dim(L) + 1$.

By definition, the *affine dimension* of an affine subspace $M = \bar{a} + L$ is the dimension $\dim(L)$ of L, i.e., $\dim(M) := \dim(L)$. Thus, if M is of affine dimension k, then the minimal cardinality of sets affinely spanning M, same as the maximal cardinality of affinely independent subsets of M, is $k + 1$.

Theorem A.45 *[Affine bases] Let $M = \bar{a} + L$ be an affine subspace in $\mathbf{R}^n$.*

(i) *For any $Y \subseteq M$, the following three properties of Y are equivalent:*
    (a) *Y is an affinely independent set which affinely spans M;*
    (b) *Y is affinely independent and contains $\dim(L) + 1$ elements;*
    (c) *Y affinely spans M and contains $\dim(L) + 1$ elements.*
        A subset Y of M possessing these preceding equivalent properties is called an affine basis *of M. Affine bases of M are exactly the collections $y^0, \ldots, y^{\dim(L)}$ such that $y^0 \in M$ and $(y^1 - y^0), \ldots, (y^{\dim(L)} - y^0)$ is a basis of L.*

(ii) Every affinely independent collection of vectors of M either is itself an affine basis of M or can be extended to such a basis by adding new vectors. In particular, every affine subspace M admits an affine basis.
(iii) If Y affinely spans M, then we can always extract from Y an affine basis of M.

We already know that the standard basis vectors $e_1, \ldots, e_n$ form a basis of the entire space $\mathbf{R}^n$. And what about affine bases in $\mathbf{R}^n$? According to Theorem A.45(i), we can choose such an affine basis for an affine subspace M as any collection of vectors $e^0, e^0 + e_1, \ldots, e^0 + e_n$, where e^0 is an arbitrary vector from M.

Barycentric coordinates. Let M be an affine subspace, and let $y^0, \ldots, y^k$ be an affine basis of M. Since the basis, by definition, affinely spans M, every vector y from M is an affine combination of the vectors of the basis, i.e.,

$$y = \sum_{i=0}^{k} \lambda_i y^i \quad \left[\text{where } \sum_{i=0}^{k} \lambda_i = 1 \right].$$

Moreover, since the vectors of the affine basis are affinely independent, the coefficients of this combination are uniquely defined by y (Corollary A.42). These coefficients are called *barycentric coordinates* of y with respect to the affine basis in question. In contrast with the usual coordinates with respect to a (linear) basis, barycentric coordinates cannot be quite arbitrary: their sum must be equal to 1.

A.4.4 Dual Description of Linear Subspaces and Affine Subspaces

So far, we have introduced the notions of linear subspaces and affine subspaces and have presented a scheme for generating these entities. For example, to get a linear subspace, we start from an arbitrary nonempty set $X \subseteq \mathbf{R}^n$ and add to it all linear combinations of the vectors from X. By replacing linear combinations with affine ones, we obtain a way to generate affine subspaces.

The way just indicated of generating linear or affine subspaces resembles the approach of a worker building a house: he starts with the base and then adds new elements to it until the house is ready. There is yet another way to generate such subspaces, which resembles the approach of an artist creating a sculpture: the artist takes something large and then deletes extra parts of it. The "artist's way" to represent linear subspaces and affine subspaces is instructive as well and we will examine it next.

A.4.5 Affine Subspaces and Systems of Linear Equations

Let L be a linear subspace in $\mathbf{R}^n$. According to Theorem A.27(i) it is an orthogonal complement itself, namely, L is the orthogonal complement to the linear subspace $L^\perp$. Now let the vectors $a_1, \ldots, a_m$ be a finite spanning set in $L^\perp$. A vector x which is orthogonal to $a_1, \ldots, a_m$ is orthogonal to the entire subspace $L^\perp$ (since every vector from $L^\perp$ is a linear combination of $a_1, \ldots, a_m$ and the inner product is bilinear). Of course, the reverse is also true: a vector orthogonal to the entire subspace $L^\perp$ is orthogonal to every vector in $a_1, \ldots, a_m$. Therefore, we arrive at

$$L = (L^\perp)^\perp = \left\{ x \in \mathbf{R}^n : a_i^\top x = 0,\ i = 1, \ldots, m \right\}. \tag{A.7}$$

Thus, we get the following very important, although simple, result.

Proposition A.46 ["Outer" description of a linear subspace] *Every linear subspace L in $\mathbf{R}^n$ is the set of all solutions to a homogeneous system of linear equations*

$$a_i^\top x = 0, \quad i = 1, \ldots, m, \tag{A.8}$$

given by a properly chosen m and vectors $a_1, \ldots, a_m \in \mathbf{R}^n$.

Recall from Example A.3 that the solution set to a homogeneous system of linear equations with n variables is always a linear subspace in $\mathbf{R}^n$. Thus, Proposition A.46 is indeed an "if and only if" statement.

From Proposition A.46 and the facts about the dimensions of linear subspaces we can easily derive several important consequences:

- The systems of equations (A.8) which define a given linear subspace L are exactly the systems given by the vectors $a_1, \ldots, a_m$ which span $L^\perp$.[1]
- If m is the smallest possible number of equations in (A.8), then m is also the dimension of $L^\perp$. That is, by Remark A.28, we have $\operatorname{codim}(L) := n - \dim(L)$.[2]

Now, an affine subspace M is, by definition, a translation of a linear subspace, i.e., $M = \bar{a} + L$. Recall that the vectors x from L are exactly the solutions of a certain *homogeneous* system of linear equations,

$$a_i^\top x = 0, \quad i = 1, \ldots, m.$$

It is absolutely clear that adding to these vectors a fixed vector $\bar{a}$, we get exactly the set of solutions to the *inhomogeneous* feasible system of linear equations

$$a_i^\top x = b_i, \quad i = 1, \ldots, m,$$

where $b_i := a_i^\top \bar{a}$ for all i. The reverse is also true, i.e., the set of solutions to a *feasible* system of linear equations

$$a_i^\top x = b_i, \quad i = 1, \ldots, m,$$

with n variables is the sum of a particular solution to the system and the solution set to the underlying homogeneous system of linear equations (the latter set, as we already know, is a linear subspace in $\mathbf{R}^n$). That is, the set of solutions to a *feasible* system of linear equations is an affine subspace. Thus, we arrive at the following result.

Proposition A.47 ["Outer" description (see Remark A.48 below) of an affine subspace] *Every affine subspace $M = a + L$ in $\mathbf{R}^n$ is the set of all solutions to a feasible system of linear equations*

$$a_i^\top x = b_i, \quad i = 1, \ldots, m, \tag{A.9}$$

given by a properly chosen m and vectors $a_1, \ldots, a_m$.

---

[1] The reasoning which led us to Proposition A.46 states that if $a_1, \ldots, a_m$ span $L^\perp$ then this implies that (A.8) defines L. Here, we claim that the reverse is also true.

[2] Note that this statement holds true also in the extreme case when $L = \mathbf{R}^n$ (i.e., when $\operatorname{codim}(L) = 0$) due to the fact that the solution set of an *empty* set of equations or inequalities in variables $x \in \mathbf{R}^n$ is the entire space; indeed, were it not the case, there would exist a vector in $\mathbf{R}^n$ violating one or more equalities or inequalities composing the system, which clearly is not the case when the system is empty.

Vice versa, the set of all solutions to a feasible system of linear equations with n variables is an affine subspace in $\mathbf{R}^n$.

The linear subspace L associated with M is exactly the set of solutions of the homogeneous (with the right-hand side set to 0) version of system (A.9).

We see, in particular, that an affine subspace is always closed.

Remark A.48 The "outer" description of a linear or affine subspace (the artist's description) is in many cases much more useful than the "inner" description via linear or affine combinations (the worker's description). For example, using the outer description, it is very easy to check whether a given vector belongs to a given linear (or affine) subspace. In contrast, this task is not particularly easy with the inner description.[3] In fact these descriptions are complementary to each other and work perfectly well in parallel: what is difficult to see with using one description is clear using the other. The idea of using "inner" and "outer" descriptions of the entities we meet (e.g., linear subspaces, affine subspaces, convex sets, optimization problems) – the general idea of *duality* – is, in our humble opinion, the main driving force of Convex Analysis and Optimization, and so, not surprisingly, in this book we meet with different implementations of this fundamental idea all the time.

A.4.6 Structure of the Simplest Affine Subspaces

Here, we mainly introduce some terminology. According to their dimension, affine subspaces in $\mathbf{R}^n$ are named as follows:

- Subspaces of dimension 0 are translations of the only 0-dimensional linear subspace (i.e., $\{0\}$), that is, they are singleton sets – vectors from $\mathbf{R}^n$. These subspaces are called *points*; a point is a solution to a system of n linear equations in n unknowns with a nonsingular matrix.
- Subspaces of dimension 1 are *lines*. These subspaces are translations of one-dimensional linear subspaces of $\mathbf{R}^n$. A one-dimensional linear subspace has a single-element basis given by a nonzero vector d and is simply the set of all possible multiples of this vector. Consequently, a line l is a set of the form

$$l := \{\bar{a} + td : t \in \mathbf{R}\}$$

given by a pair of vectors $\bar{a}$ (the origin of the line) and d (the direction of the line), $d \neq 0$. Naturally, this is the inner description of the line l. Note that in this description the origin of the line and its direction are not uniquely defined by the line; you can choose as origin any point on the line and you can also multiply a particular direction by any nonzero real number.

In barycentric coordinates a line l is described as follows:

$$l = \{\lambda_0 y^0 + \lambda_1 y^1 : \lambda_0 + \lambda_1 = 1\} = \{\lambda y^0 + (1-\lambda) y^1 : \lambda \in \mathbf{R}\},$$

where y^0, y^1 is an affine basis of l (such a basis can be chosen as any pair of distinct points on the line).

---

[3] In principle it is not difficult to certify that a given point belongs to, say, a linear subspace given as the linear span of some set – it suffices to point out a representation of the point as a linear combination of vectors from the set. But how can you certify that the point does *not* belong to the subspace?

The outer description of a line is as follows: it is the set of solutions to a system of $n-1$ linearly independent linear equations in n variables (unknowns).
- Subspaces of dimension k where $2 \le k < n-1$ have no special names; sometimes they are called affine planes of dimension k.
- Affine subspaces of dimension $n-1$, owing to the important role they play in Convex Analysis, have a special name; they are called *hyperplanes*. The outer description of a hyperplane is that it is the solution set of a *single* linear equation

$$a^\top x = b$$

with nontrivial left-hand side ($a \ne 0$). In other words, a hyperplane is the level set $a(x) = b$ where $b \in \mathbf{R}$ is fixed of a nonconstant linear form $a(x) = a^\top x$.
- The "largest possible" affine subspace in $\mathbf{R}^n$ – that of dimension n – is unique and it is the entire space $\mathbf{R}^n$. This subspace is given by an empty system of linear equations.

A.5 Proof of Fact

Fact A.33

(i) *Let L be a linear subspace of $\mathbf{R}^n$, and $f^1, \ldots, f^m$ be an orthonormal basis in L. Then, for every $x \in \mathbf{R}^n$, the orthoprojection x_L of x onto L is given by the formula*

$$x_L := \sum_{i=1}^m (x^\top f^i) f^i.$$

(ii) *Let L_1, L_2 be linear subspaces in $\mathbf{R}^n$. Then*

$$(L_1 + L_2)^\perp = L_1^\perp \cap L_2^\perp \quad \text{and} \quad (L_1 \cap L_2)^\perp = L_1^\perp + L_2^\perp.$$

Proof (i): Note that x_L belongs to L and therefore $x_L = \sum_i \lambda_i f^i$ for some λ_i, and $x - x_L$ is orthogonal to L. Hence, we have

$$x = (x - x_L) + \sum_{i=1}^m \lambda_i f^i.$$

Then, for any j, by taking inner products of both sides of the above equality with f^j, we arrive at

$$(f^j)^\top x = (f^j)^\top (x - x_L) + \sum_{i=1}^m \lambda_i (f^j)^\top f^i = \sum_{i=1}^m \lambda_i (f^j)^\top f^i = \lambda_j.$$

The result follows by plugging the expressions for λ_j into the equation $x_L = \sum_i \lambda_i f^i$.

(ii): Since $L_1 \subseteq L_1 + L_2$ and $L_2 \subseteq L_1 + L_2$, a vector orthogonal to $L_1 + L_2$ is orthogonal to both L_1 and L_2. Vice versa, a vector with the latter property is orthogonal to sums of vectors from L_1 and from L_2, that is, it is orthogonal to $L_1 + L_2$. Thus, the vectors orthogonal to $L_1 + L_2$ are exactly the vectors orthogonal to both L_1 and L_2. This proves the first equality. Applying this equality to the orthogonal complements of L_1 and L_2 in the roles of L_1, L_2, we get the second equality. ∎

A.6 Exercises

Exercise A.1

1. Mark in the list below those subsets of $\mathbf{R}^n$ which are linear subspaces. For the ones that are linear subspaces, find out their dimensions and point out bases. For the ones that are not linear subspaces provide counterexamples.
   - $\mathbf{R}^n$
   - $\{0\}$
   - $\emptyset$
   - $\left\{ x \in \mathbf{R}^n : \sum_{i=1}^{n} i x_i = 0 \right\}$
   - $\left\{ x \in \mathbf{R}^n : \sum_{i=1}^{n} i x_i^2 = 0 \right\}$
   - $\left\{ x \in \mathbf{R}^n : \sum_{i=1}^{n} i x_i = 1 \right\}$
   - $\left\{ x \in \mathbf{R}^n : \sum_{i=1}^{n} i x_i^2 = 1 \right\}$

2. Suppose that we know L is a subspace of $\mathbf{R}^n$ with exactly one basis. What is L?

Exercise A.2 Consider the sets given in Exercise A.1 and identify those that are affine subspaces. For those that are affine subspaces, find their affine dimensions and point out their linear subspaces that are parallel to them. For those that are not affine subspaces, provide counterexamples.

Exercise A.3
1. What is the orthogonal complement (w.r.t. the standard inner product) of the subspace $\left\{ x \in \mathbf{R}^n : \sum_{i=1}^{n} x_i = 0 \right\}$ in $\mathbf{R}^n$?
2. Find an orthonormal basis (w.r.t. the standard inner product) in the linear subspace $\{x \in \mathbf{R}^n : x_1 = 0\}$ of $\mathbf{R}^n$.

Exercise A.4 Suppose $a \in \mathbf{R}^n$ where $a_i > 0$ for all $i = 1, \ldots, n$, and consider the affine subspace

$$M = \left\{ x \in \mathbf{R}^n : \sum_{i=1}^{n} a_i x_i = 1 \right\}.$$

Point out the linear subspace parallel to M and find an affine basis in M.

Exercise A.5 Let $\emptyset \neq C \subseteq \mathbf{R}^n$ and $x \in \mathbf{R}^n$ be given.

1. Is it always true that $\mathrm{Aff}(C - \{x\}) = \mathrm{Aff}(C) - \{x\}$?
2. Is it always true that $\mathrm{Lin}(C - \{x\}) = \mathrm{Aff}(C) - \{x\}$?
3. Do your answers to the previous questions change if you further assume $x \in \mathrm{Aff}(C)$?

Exercise A.6 Suppose that we are given n sets $E_1, E_2, \ldots, E_n$ in $\mathbf{R}^{100}$ that are distinct from each other and that they satisfy

$$E_1 \subset E_2 \subset \cdots \subset E_n.$$

How large can n be, if

1. Every E_i is a linear subspace?
2. Every E_i is an affine subspace?
3. Every E_i is a convex set?

Exercise A.7 Prove that the *triangle inequality in the Euclidean norm*, i.e., $\|x + y\|_2 \leq \|x\|_2 + \|y\|_2$, holds true as an *equality* if and only if x and y are nonnegative multiples of some vector (which always can be taken to be $x + y$).

Appendix B

Prerequisites from Real Analysis

B.1 Space $\mathbf{R}^n$: Metric Structure and Topology

A Euclidean structure on the space $\mathbf{R}^n$ gives rise to a number of extremely important metric notions – distances, convergence, etc. For the sake of definiteness, we associate these notions with the standard inner product $\langle x, y \rangle = x^\top y$.

B.1.1 Euclidean Norm and Distances

By positive definiteness, the quantity $x^\top x$ is always nonnegative, so that the quantity

$$\|x\|_2 = \sqrt{x^\top x} = \sqrt{x_1^2 + x_2^2 + \cdots + x_n^2}$$

is well defined. This quantity is called the (standard) *Euclidean norm* of vector x (or simply the norm of x) and is treated as the distance from the origin to x. The distance between two arbitrary points $x, y \in \mathbf{R}^n$ is, by definition, the norm $d_2(x, y) = \|x - y\|_2$ of the difference $x - y$. These notions we have just introduced indeed satisfy all the basic requirements on the general notions of a norm $\|\cdot\| : \mathbf{R}^n \to \mathbf{R}$ and a distance $d(x, y) : \mathbf{R}^m \times \mathbf{R}^n \to \mathbf{R}$. Specifically:

1. *Positivity of norms:* The norm of a vector is always nonnegative. Furthermore, it is zero if and only if the vector is zero:

   $$\|x\| \geq 0 \quad \forall x; \qquad \|x\| = 0 \iff x = 0.$$

2. *Homogeneity of norms:* When a vector is multiplied by a real, its norm is multiplied by the absolute value of the real:

   $$\|\lambda x\| = |\lambda| \cdot \|x\| \quad \forall (x \in \mathbf{R}^n, \lambda \in \mathbf{R}).$$

3. *Triangle inequality:* The norm of the sum of two vectors is less than or equal to the sum of their norms:

   $$\|x + y\| \leq \|x\| + \|y\| \quad \forall (x, y \in \mathbf{R}^n).$$

In contrast with the properties of positivity and homogeneity, which are absolutely evident, the triangle inequality is nontrivial and definitely requires a proof. Its proof involves a fact which is extremely important in its own right – the *Cauchy inequality*, which perhaps is the most frequently used inequality in mathematics.

Theorem B.1 *[Cauchy inequality] For any $x, y \in \mathbf{R}^n$, we have*

$$|x^\top y| \leq \|x\|_2 \|y\|_2.$$

Moreover, the above relation holds as an equality if and only if one of the vectors is proportional to the other:

$$|x^\top y| = \|x\|_2 \|y\|_2$$
$$\iff \exists \alpha \in \mathbf{R} \text{ such that } x = \alpha y \text{ or } \exists \beta \in \mathbf{R} \text{ such that } y = \beta x.$$

Proof Without loss of generality we may assume that both x and y are nonzero (otherwise the Cauchy inequality is clearly an equality, and one vector is a constant times (specifically, zero times) the other, as desired). Assuming $x, y \neq 0$, consider the function

$$f(\lambda) = (x - \lambda y)^\top (x - \lambda y) = x^\top x - 2\lambda x^\top y + \lambda^2 y^\top y.$$

By positive definiteness of the inner product, this function – which is a second-order polynomial – is nonnegative on the entire axis; hence the discriminant $(x^\top y)^2 - (x^\top x)(y^\top y)$ of f is nonpositive:

$$(x^\top y)^2 \leq (x^\top x)(y^\top y).$$

By taking the square roots of both sides, we arrive at the Cauchy inequality. We also see that the inequality holds as an equality if and only if the discriminant of the second-order polynomial $f(\lambda)$ is zero, i.e., if and only if the polynomial has a (multiple) real root. But, owing to the positive definiteness of the inner product, $f(\cdot)$ has a root λ if and only if $x = \lambda y$, which proves the second part of the theorem. ∎

From Cauchy's inequality to the triangle inequality. Let $x, y \in \mathbf{R}^n$. Then,

$$\begin{aligned}
\|x+y\|_2^2 &= (x+y)^\top (x+y) && \text{[by the definition of the Euclidean norm]} \\
&= x^\top x + y^\top y + 2x^\top y && \text{[by opening the parentheses]} \\
&\leq \underbrace{x^\top x}_{=\|x\|_2^2} + \underbrace{y^\top y}_{=\|y\|_2^2} + 2\|x\|_2 \|y\|_2 && \text{[by Cauchy's inequality]} \\
&= (\|x\|_2 + \|y\|_2)^2 \\
\implies \|x+y\|_2 &\leq \|x\|_2 + \|y\|_2.
\end{aligned}$$
∎

We also have the following simple and useful fact.

Fact B.2 *For every norm $\|\cdot\|$ on $\mathbf{R}^n$ and any $x, y \in \mathbf{R}^n$, we have*

$$|\|x\| - \|y\|| \leq \|x - y\|.$$

Proof Indeed, $x = (x - y) + y$, and so, by the triangle inequality, $\|x\| \leq \|x - y\| + \|y\|$. That is, $\|x\| - \|y\| \leq \|x - y\|$. Swapping x and y, we get $\|y\| - \|x\| \leq \|y - x\| = \|x - y\|$, as well. Thus, the result follows. ∎

B.1 Space $\mathbf{R}^n$: Metric Structure and Topology

Euclidean distance between a pair of points in $\mathbf{R}^n$. The properties of a norm (i.e., of the distance to the origin) that we have established induce properties of the distances between pairs of arbitrary points in $\mathbf{R}^n$. Specifically, $d_2(x, y) = \|x - y\|_2$, as for any other norm-induced distance $d_{\|\cdot\|}(x, y) = \|x - y\|$, possesses the following standard properties of a general distance $d(x, y) : \mathbf{R}^n \times \mathbf{R}^n \to \mathbf{R}$:

1. *Positivity:* The distance $d(x, y)$ between two points $x, y \in \mathbf{R}^n$ is positive, except for the case when the points coincide ($x = y$) in which case the distance between x and y is zero:
$$d(x, y) \geq 0, \ \forall (x, y \in \mathbf{R}^n); \qquad d(x, y) = 0 \iff x = y.$$

2. *Symmetry:* For any $x, y \in \mathbf{R}^n$, the distance from x to y is the same as the distance from y to x:
$$d(x, y) = d(y, x), \quad \forall (x, y \in \mathbf{R}^n).$$

3. *Triangle inequality:* For every $x, y, z \in \mathbf{R}^n$, the distance from x to z does not exceed the sum of the distances between x and y and between y and z:
$$d(x, z) \leq d(x, y) + d(y, z), \quad \forall (x, y, z \in \mathbf{R}^n).$$

A norm-induced distance $d(x, y) = \|x - y\|$ possesses the following additional properties:

4. *Shift invariance:* $d(x + h, y + h) = d(x, y)$ for all $x, y, h \in \mathbf{R}^n$.
5. *Homogeneity:* $d(\lambda x, \lambda y) = |\lambda| d(x, y)$ for all $x, y \in \mathbf{R}^n$ and $\lambda \in \mathbf{R}$.

Equivalence of norms on $\mathbf{R}^n$. As can immediately be seen, any distance $d(x, y) = \|x - y\|$ induced by an arbitrary norm $\|\cdot\|$ on $\mathbf{R}^n$ possesses the above properties 1–5. In fact (check it!), every distance $d(x, y)$ on $\mathbf{R}^n$ satisfying properties 1–5. is indeed induced by a norm, specifically, the norm $d(0, x)$ (in the case of properties 1–5, $d(0, x)$ is indeed a norm).

We next discuss a very important fact (characteristic of *finite-dimensional* linear spaces) which states that all these distances are within constant factors of each other.

Proposition B.3 *Any two norms $\|\cdot\|, \|\cdot\|'$ on $\mathbf{R}^n$ are equivalent, i.e., there exists a positive constant c (which depends on the given pair of norms) such that*
$$c \leq \frac{\|x\|'}{\|x\|} \leq \frac{1}{c}, \quad \forall (x \in \mathbf{R}^n, x \neq 0). \tag{B.1}$$

Proof Clearly it suffices to justify the equivalence of any norm to a fixed one, say, the Euclidean norm $\|\cdot\|_2$. Thus, let us take $\|\cdot\|' \equiv \|\cdot\|_2$. It suffices to prove that there exist positive constants c and C such that both the following conditions hold:
$$\begin{array}{ll}(a) \ \|x\| \leq C\|x\|_2, & \forall x \neq 0; \\ (b) \ \|x\|_2 \leq c\|x\|, & \forall x \neq 0.\end{array} \tag{B.2}$$

To prove (a), all we need to do is to set $C := \sum_{i=1}^n \|e_i\|$, where $e_1, \ldots, e_n$ are the standard basis vectors. This is so because
$$\|x\| = \left\|\sum_{i=1}^n x_i e_i\right\| \leq \sum_{i=1}^n |x_i| \|e_i\| \leq \left(\max_{i=1,\ldots,n} |x_i|\right) \sum_{i=1}^n \|e_i\| \leq C\|x\|_2.$$

In order to prove (b), we will show that $0 < \inf_x \{\|x\|/\|x\|_2 : x \neq 0\}$. Assume, on the contrary, that the infimum in question is 0. Then, there exists a sequence of nonzero vectors x^t, $t = 1, 2, \ldots$ such that $\|x^t\|/\|x^t\|_2 \to 0$ as $t \to \infty$. Since both $\|\cdot\|$ and $\|\cdot\|_2$ are homogeneous, we lose nothing by scaling x^t so that $\|x^t\|_2 = 1$ for all t, whence we have $\|x^t\| \to 0$ as $t \to \infty$. As $\|x^t\|_2 = 1$ for all t, for any $1 \leq i \leq n$ the sequences x_i^t of the ith entries in x^t are bounded. Passing to a subsequence, we can assume w.l.o.g. that all these sequences have limits as $t \to \infty$, i.e., $\lim_{t \to \infty} x_i^t = z_i$ for some $z \in \mathbf{R}^n$. Passing to the limit in the equality $\sum_{i=1}^n (x_i^t)^2 = 1$, we get $\|z\|_2 = 1$. Clearly, $\|x^t - z\|_2 \to 0$ as $t \to \infty$. Moreover, from (a) we have $\|x^t - z\| \leq C\|x^t - z\|_2$, which implies that $\|x^t - z\| \to 0$ as $t \to \infty$. Then, by the triangle inequality, we deduce that $\|z\| \leq \|z - x^t\| + \|x^t\| \to 0$ as $t \to \infty$, and thus $\|z\| = 0$. This is the desired contradiction since $z \neq 0$ and therefore $\|z\| > 0$. ∎

In the preceding proof, we took for granted the following fundamental property of the real line (stemming from the centuries-long development of the rigorous theory of real numbers):

> (✠) *Every bounded sequence of real numbers $\{x_t\}_{t \geq 1}$ has a converging subsequence, that is, for properly selected $t_1 < t_2 < \cdots$ and $\bar{x} \in \mathbf{R}$ the sequence x_{t_i} converges to $\bar{x}$ as $i \to \infty$: for every $\epsilon > 0$ we have $|\bar{x} - x_{t_i}| < \epsilon$ for all but finitely many indices i.*

As an immediate consequence of this statement, from every bounded sequence $\{x^t\}_{t \geq 1}$ of vectors from $\mathbf{R}^n$ (boundedness meaning that sequences of the ith entries in x^t are bounded for every $i \leq n$) we can extract a subsequence $\{x^{t_s} : s = 1, 2, \ldots, t_1 < t_2 < \cdots\}$ such that for every $i \leq n$ the sequences of the ith coordinates $x_i^{t_s}$ of vectors x^{t_s} converge as $s \to \infty$ to some reals – the fact we used in the proof of Proposition B.3. Indeed, the sequences $\{x_i^t\}_{t \geq 1}$ are bounded for every i. By (✠) we can extract from $\{x^t\}_{t \geq 1}$ a subsequence $\{x^{t_s}, s = 1, 2, \ldots, t_1 < t_2 < \cdots\}$ with converging first entries, from this subsequence we can extract a subsequence with converging second entries, and so on. In n steps of this process we get a subsequence of the original sequence of vectors such that the ith entries in the vectors from our subsequence converge to some reals $\bar{x}_i$, $i \leq n$. This is formalized as Theorem B.15 in Section B.1.4.

B.1.2 Convergence

Equipped with distances, we can define the fundamental notion of the *convergence of a sequence of vectors*. Specifically, we say that a *sequence $x^1, x^2, \ldots$ of vectors from $\mathbf{R}^n$ converges to a vector $\bar{x}$*, or, equivalently, that *$\bar{x}$ is the limit of the sequence $\{x^i\}$* [notation: $\bar{x} = \lim_{i \to \infty} x^i$], if the distances from $\bar{x}$ to x^i go to 0 as $i \to \infty$:

$$\bar{x} = \lim_{i \to \infty} x^i \iff \|\bar{x} - x^i\|_2 \to 0 \text{ as } i \to \infty,$$

or, equivalently, for every $\epsilon > 0$ there exists $i = i(\epsilon)$ such that the distance between every point x^i, $i \geq i(\epsilon)$, and $\bar{x}$ does not exceed ϵ:

$$\left\{\|\bar{x} - x^i\|_2 \to 0 \text{ as } i \to \infty\right\} \iff \left\{\forall \epsilon > 0, \exists i(\epsilon) : i \geq i(\epsilon) \implies \|\bar{x} - x^i\|_2 \leq \epsilon\right\}.$$

B.1 Space $\mathbf{R}^n$: Metric Structure and Topology

Note that by Proposition B.3 every two norm-induced distances on $\mathbf{R}^n$ are within multiplicative factors from each other, meaning that *replacing in the definition of convergence the Euclidean distance with any other norm-induced distance, we do not affect convergence per se – the fact that a sequence converges, and the limit of such a sequence.*

Fact B.4 *All the following statements are correct:*

(i) *For any $\bar{x} \in \mathbf{R}^n$, we have $\bar{x} = \lim_{t \to \infty} x^t$ if and only if for every index $i = 1, \ldots, n$ the ith coordinates of the vectors x^t converge to the ith coordinate of the vector $\bar{x}$ as $t \to \infty$.*
(ii) *If a sequence converges, its limit is uniquely defined.*
(iii) *Convergence is compatible with linear operations:*
- *if $x^t \to x$ and $y^t \to y$ as $t \to \infty$, then $x^t + y^t \to x + y$ as $t \to \infty$;*
- *if $x^t \to x$ and $\lambda_t \to \lambda$ as $t \to \infty$, then $\lambda_t x^t \to \lambda x$ as $t \to \infty$.*

We also introduce two other notions related to sequences in $\mathbf{R}$, in order to analyze their "extremes." Given a sequence $\{x_t\} \subset \mathbf{R}$, its *limit inferior* (lower limit) [notation: $\liminf_{t \to \infty} x_t$] is defined by

$$\liminf_{t \to \infty} x_t := \lim_{t \to \infty} \left(\inf_{m \geq t} x_m \right).$$

That is,

$$\liminf_{t \to \infty} x_t = \sup_{t \geq 0} \inf_{m \geq t} x_m = \sup \{\inf\{x_m : m \geq t\} : t \geq 0\}.$$

In other words, if $\{x_t\}$ is a sequence of reals, $a = \liminf_{t \to \infty} x_t$ is

- $-\infty$, if there is a subsequence diverging to $-\infty$,
- or $+\infty$, if $x_t \to \infty, t \to \infty$,
- or the smallest real a which can be represented as the limit of a subsequence of $\{x_t\}$.

In contrast with the usual limit, lim inf is well defined for every sequence of reals, e.g., $\liminf_{t \to \infty} (-1)^t = -1$.

Similarly, the *limit superior* (upper limit) [notation: $\limsup_{t \to \infty} a_t$] is defined by

$$\limsup_{t \to \infty} x_t := \lim_{t \to \infty} \left(\sup_{m \geq t} x_m \right),$$

which is simply

$$\limsup_{t \to \infty} x_t = \inf_{t \geq 0} \sup_{m \geq t} x_m = \inf \{\sup\{x_m : m \geq t\} : t \geq 0\}.$$

B.1.3 Closed and Open Sets

Now that we have at our disposal the notions of distance and convergence, we can speak about *closed* and *open* sets.

Definition B.5 *[Closed set] A set $X \subseteq \mathbf{R}^n$ is called* closed, *if it contains the limits of all converging sequences of elements from X:*

$$\left\{x^i \in X,\ x = \lim_{i \to \infty} x^i\right\} \implies x \in X.$$

Example B.6 (Examples of closed sets) Each of the following sets is closed:

1. $\mathbf{R}^n$.
2. $\varnothing$.
3. The set composed of the points $x^i = (i, 0, \ldots, 0)$, $i = 1, 2, 3, \ldots$
4. Any finite subset of $\mathbf{R}^n$.
5. Any linear subspace in $\mathbf{R}^n$, i.e., any set of the form
$$\left\{x \in \mathbf{R}^n : \sum_{i=1}^n a_{ij} x_j = 0,\ i = 1, \ldots, m\right\} \text{ (see Proposition A.46).}$$
6. Any affine subspace of $\mathbf{R}^n$, i.e., any nonempty set of the form
$$\left\{x \in \mathbf{R}^n : \sum_{i=1}^n a_{ij} x_j = b_i,\ i = 1, \ldots, m\right\} \text{ (see Proposition A.47).}$$

Example B.7 (Examples of non-closed sets) Each of the following sets is non-closed, provided $n > 0$:

1. $\mathbf{R}^n \setminus \{0\}$.
2. The set composed of points $x^i = (1/i, 0, \ldots, 0)$, for $i = 1, 2, 3, \ldots$
3. The set $\left\{x \in \mathbf{R}^n : x_j > 0,\ \forall j = 1, \ldots, n\right\}$.
4. The set $\left\{x \in \mathbf{R}^n : \sum_{i=1}^n x_j > 5\right\}$.

Definition B.8 *[Open set] A set $X \subseteq \mathbf{R}^n$ is called* open, *if whenever x belongs to X, all points close enough to x also belong to X:*

$$\forall (x \in X)\ \exists (\delta > 0) : \ \|x' - x\|_2 < \delta \implies x' \in X.$$

Definition B.9 *[Neighborhood, interior point] An open set containing a point $x \in \mathbf{R}^n$ is called a* neighborhood *of x. A point $x \in \mathbf{R}^n$ from a set X is called an* interior point *of X if X contains a neighborhood of x.*

Note that on the basis of these definitions, open sets are exactly the sets such that every point of the set is its interior point.

Definition B.10 *[Interior of a set] Given a set $X \subseteq \mathbf{R}^n$, the* interior *of X [notation: $\text{int}(X)$ or $\text{int } X$] is defined as the set of all interior points of X:*

$$\text{int}(X) := \left\{x \in X : \exists \delta > 0 \text{ such that } x' \in X\ \forall (x' : \|x - x'\|_2 < \delta)\right\}.$$

Note that a set $X \subseteq \mathbf{R}^n$ is open if and only if $X = \text{int}(X)$.

Example B.11 (Examples of open sets) Each of the following sets is open:

1. $\mathbf{R}^n$.
2. $\varnothing$.

3. The set $\left\{x \in \mathbf{R}^n : \sum_{j=1}^n a_{ij} x_j > b_i, \forall i = 1, \ldots, m \right\}$.
4. The complement of any finite set.

Example B.12 (Examples of non-open sets) Each of the following sets in $\mathbf{R}^n$, $n > 0$, is non-open:

1. A nonempty finite set.
2. The sequence $x^i = (1/i, 0, \ldots, 0)$, $i = 1, 2, 3, \ldots$
3. The set composed of points from a sequence $x^i \in \mathbf{R}^n$, $i = 1, 2, 3, \ldots$
4. The set $\left\{ x \in \mathbf{R}^n : \sum_{i=1}^n x_j \geq 5 \right\}$.

From these examples, we see that $\mathbf{R}^n$ and $\emptyset$ are both open and closed. Also, note that there are sets that are neither closed nor open, e.g.,

$$\left\{ x \in \mathbf{R}^2 : x_1 > 0, \, x_2 \geq 0 \right\}.$$

Fact B.13

(i) A set $X \subseteq \mathbf{R}^n$ is closed if and only if its complement $\overline{X} := \mathbf{R}^n \setminus X$ is open.
(ii) The intersection of every (finite or infinite) family of closed sets is closed. The union of every (finite or infinite) family of open sets is open.
(iii) The union of finitely many closed sets is closed. The intersection of finitely many open sets is open.

We close this section with another important definition.

Definition B.14 *[Closure of a set] The closure of a set* $X \subseteq \mathbf{R}^n$ *[notation:* cl X *or* cl(X)*] is the smallest closed set which contains* X, *i.e., it is the intersection of all closed sets containing* X *(this intersection is indeed closed by Fact B.13(ii)).*

Equivalently (the equivalence is established in Fact 1.23), cl X *is the set composed of the limits of all converging sequences of points from* X:

$$\text{cl } X = \left\{ x \in \mathbf{R}^n : \exists \left\{ x^i \right\}_i \in X \text{ such that } x = \lim_{i \to \infty} x^i \right\}.$$

Note that a set $X \subseteq \mathbf{R}^n$ is closed if and only if $X = \text{cl}(X)$.

B.1.4 Local Compactness of $\mathbf{R}^n$

The following is a fundamental fact about convergence in $\mathbf{R}^n$, which in certain sense is characteristic of this series of spaces.

Theorem B.15 *From every bounded sequence* $\{x^i\}_{i=1}^\infty$ *of points from* $\mathbf{R}^n$ *we can extract a converging subsequence* $\{x^{i_j}\}_{j=1}^\infty$.

Proof Let $\{x^i \in \mathbf{R}^n\}_{i \geq 1}$ be a bounded sequence. Since every bounded sequence of reals has a converging subsequence, we can extract from $\{x^i\}$ a subsequence with the first entries of its members converging to a limit. Note that this subsequence will itself be bounded. Then, from this subsequence, we can extract a subsequence with the second entries of the members converging to a limit. Extracting in this fashion subsequences from those already generated, after n steps we get a subsequence of the original sequence with every entry of the selected members converging to a limit. That is, the selected final subsequence converges entrywise (and in the Euclidean norm as well) to the vector composed of the above limits. ∎

We next introduce an important concept related to closed sets.

Definition B.16 *[Compact set] A set X is called* compact *if every sequence in X has a subsequence that converges to an element that is also contained in X.*

Theorem B.17 *Any set $X \subseteq \mathbf{R}^n$ that is closed and bounded is compact. Moreover, any compact set $X \subseteq \mathbf{R}^n$ is closed and bounded.*

Proof Let X be a closed and bounded subset of $\mathbf{R}^n$. Consider any sequence $\{x^i\}$ of points from X. Note that this sequence is bounded, as X is bounded. Then, by Theorem B.15 there exists a subsequence of our sequence which has a limit. Since X is closed, this limit belongs to X, as required in the definition of compact sets.

To prove the second statement in this theorem, let X be a compact set in $\mathbf{R}^n$. Note that if the set X were unbounded, it would be possible to select a sequence of points from X with norms diverging to $+\infty$, and such a sequence has no converging subsequences, which contradicts X being compact. Furthermore, if X were not closed, we could find a sequence of points from X converging to a point *not* from X. For such a sequence, no subsequence converges to a point from X (since the limits of all these subsequences are the same as the limit of the entire sequence, and the latter does not belong to X), which once again contradicts X being compact. ∎

Given a set $X \subseteq \mathbf{R}^n$, we refer to a family $\mathcal{C}$ of open sets such that $X \subseteq \bigcup_{U \in \mathcal{C}} U$ as an *open covering of X*. We first establish a general statement regarding open coverings, and then show that a much stronger version of this property gives an alternative characterization of compact sets.

Proposition B.18 *Given a set $X \subseteq \mathbf{R}^n$ and its open covering $\mathcal{C}$, we can extract a countable subcovering: there exists a sequence $U_1, U_2, \ldots$ of members of $\mathcal{C}$ such that $X \subseteq \bigcup_{i \geq 1} U_i$.*

Proof Observe, first, that the family $\mathcal{V}$ of all open balls $\{x \in \mathbf{R}^n : \|x - \bar{x}\|_2 < r\}$ where the radius r and all coordinates of the centers $\bar{x}$ are rational is countable, i.e., all balls from $\mathcal{V}$ can be arranged into a sequence $V_1, V_2, \ldots$ This is so because the rational data – the radius and the coordinates of the center – of a ball from $\mathcal{V}$ form a collection of $n+1$ rational numbers, or, which is the same, a collection of $2(n+1)$ integers, and all collections of the latter type can be arranged into a sequence: we first list all collections of $2(n+1)$ integers for which the total of their magnitudes does not exceed 1 (there are finitely many collections of this type), then list yet unlisted collections for which the total of the magnitudes of their members does not exceed 2, and so on. This construction clearly arranges into a sequence all collections of $2(n+1)$ integers, that is, all balls with rational centers and positive rational radii.

Now let us build a countable subcovering $U_1, U_2, \ldots$ of a given open covering $\mathcal{C}$ of $X \subseteq \mathbf{R}^n$. We will process the balls $V_1, V_2, \ldots$ one by one. When processing V_i, we check whether this set is covered by a member of $\mathcal{C}$. If V_i is covered by a member of $\mathcal{C}$, we add this member to the sequence $U_1, U_2, \ldots$ that we are building. Otherwise, we add nothing to the U-sequence and pass to processing V_{i+1}.

Now, every point x of X belongs to a certain member W_x of the covering $\mathcal{C}$, as $X \subseteq \bigcup_{C \in \mathcal{C}} C$. Since W_x is open, among the balls V_i we can find a ball, let us call its index $i(x)$, such that the center of $V_{i(x)}$ is close to x and the radius of $V_{i(x)}$ is small enough to ensure that $x \in V_{i(x)} \subseteq W_x$. Then, at the step $i(x)$ of processing $V_1, V_2, \ldots$, the family $\mathcal{C}$ contains a member, namely, W_x, which covers $V_{i(x)}$. Therefore, at this step we will add to the U-sequence a set which contains $V_{i(x)}$ and hence contains x. We conclude that every $x \in X$ is contained in some set of the sequence $U_1, U_2, \ldots$ of members of $\mathcal{C}$, and thus this sequence is a countable subcovering of X as desired. ∎

We close this section by providing a characterization of compact sets through their open coverings.

Theorem B.19 *A set $X \subseteq \mathbf{R}^n$ is compact if and only if it possesses the following property. From every open covering $\mathcal{C}$ of X we can extract a* finite *subcovering, i.e., a finite subfamily $\mathcal{C}' \subseteq \mathcal{C}$ such that $\mathcal{C}'$ still is a covering of X.*

Proof First, suppose that X satisfies the stated property. We will prove that X is closed and bounded, which by Theorem B.17 will imply that X is compact. Suppose X is unbounded, and consider the covering of X given by all open balls centered at points from X. Then, we would be unable to extract a finite subcovering of X from this open covering. Now, assume for contradiction that X is not closed, and consider the covering of X given by the open sets $\{x : \|x - \bar{x}\|_2 > 1/i\}$, $i = 1, 2, \ldots$ associated with a point $\bar{x} \in \mathrm{cl}(X) \setminus X$. Then, we would be unable to extract a finite subcovering of X from this open covering of X, which is a contradiction.

Now, let X be closed and bounded (and thus compact by Theorem B.17). Consider an open covering $\mathcal{C}$ of X. By Proposition B.18, we can extract from $\mathcal{C}$ a countable subcovering: $X \subset \bigcup_{i \geq 1} U_i, U_i \in \mathcal{C}$. We claim that, in fact, *finitely* many of the sets U_i suffice to cover X. This, of course, is enough to show that X obeys the desired property stated in the theorem. Now, assume for contradiction that for every $i = 1, 2, \ldots$ there is a point $x^i \in X \setminus \bigcup_{j \leq i} U_j$. Since X is compact, by definition we can extract from $\{x^i\}$ a subsequence converging to some $\bar{x} \in X$. Since the sets U_i are open and cover X, there exists i_* such that $\bar{x} \in U_{i_*}$. Also, as U_{i_*} is open and the subsequence in question converges to $\bar{x}$, the subsequence (and therefore the entire sequence $\{x^i\}$) visits U_{i_*} infinitely many times. This gives us the desired contradiction, since by construction none of the points x^i with $i > i_*$ belong to U_{i_*}. ∎

B.2 Continuous Functions on $\mathbf{R}^n$

B.2.1 Continuity of a Function

In this section, we consider $X \subseteq \mathbf{R}^n$ and examine a function (also called a *mapping*) $f(x) : X \to \mathbf{R}^m$ defined on X and taking values in $\mathbf{R}^m$. We start with basic definitions.

Definition B.20 *[Continuity at a point] A function $f(x) : X \to \mathbf{R}^m$ is called* continuous at a point $\bar{x} \in X$, *if, for every sequence x^i of points of X converging to $\bar{x}$, the sequence $f(x^i)$ converges to $f(\bar{x})$. Equivalently, $f : X \to \mathbf{R}^m$ is continuous at $\bar{x} \in X$ if for every $\epsilon > 0$ there exists $\delta > 0$ such that*

$$x \in X, \|x - \bar{x}\|_2 < \delta \implies \|f(x) - f(\bar{x})\|_2 < \epsilon.$$

Definition B.21 *[Continuous function] A function $f(x) : X \to \mathbf{R}^m$ is called continuous on X if f is continuous at every point from X. Equivalently, f is continuous if f preserves convergence: whenever a sequence of points $x^i \in X$ converges to a point $x \in X$, the sequence $f(x^i)$ converges to $f(x)$.*

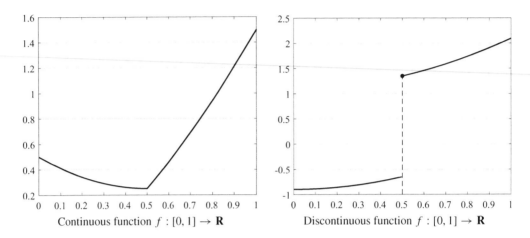

Figure B.2.1 Continuity versus discontinuity.

Example B.22 (Continuous mappings) An *affine* mapping
$$f(x) := \left[\sum_{j=1}^{m} A_{1j}x_j + b_1; \ldots; \sum_{j=1}^{m} A_{mj}x_j + b_m\right] \equiv Ax + b : \mathbf{R}^n \to \mathbf{R}^m$$
is continuous on the entire space $\mathbf{R}^n$ (and thus on every subset of $\mathbf{R}^n$) (check this!).

Fact B.23 *The norm $\|x\|_2$ is a real-valued function and it is continuous on $\mathbf{R}^n$ (and thus on every subset of $\mathbf{R}^n$) (check this!). In fact, every norm $\|\cdot\|$ on $\mathbf{R}^n$ is continuous.*

B.2.2 Elementary Continuity-Preserving Operations

All "elementary" operations with mappings preserve continuity.

Theorem B.24 *Let $X \subseteq \mathbf{R}^n$.*

(i) [stability of continuity w.r.t. linear operations] If $f_1(x), f_2(x)$ are continuous functions on X taking values in $\mathbf{R}^m$ and $\lambda_1(x), \lambda_2(x)$ are continuous real-valued functions on X, then the function
$$f(x) = \lambda_1(x)f_1(x) + \lambda_2(x)f_2(x) : X \to \mathbf{R}^m$$
is continuous on X.

(ii) [stability of continuity w.r.t. superposition] Let
- *$X \subseteq \mathbf{R}^n, Y \subseteq \mathbf{R}^m$;*
- *$f : X \to \mathbf{R}^m$ be a continuous mapping such that $f(x) \in Y$ for every $x \in X$;*
- *$g : Y \to \mathbf{R}^k$ be a continuous mapping.*

Then, the composite mapping

$$h(x) := g(f(x)) : X \to \mathbf{R}^k$$

is continuous on X.

All these claims are self-evident.

B.2.3 Basic Properties of Continuous Functions on $\mathbf{R}^n$

The basic properties of continuous functions on $\mathbf{R}^n$ can be summarized in the next three theorems.

Theorem B.25 *Let X be a nonempty compact subset of $\mathbf{R}^n$. If a mapping $f : X \to \mathbf{R}^m$ is continuous on X, it is bounded on X, i.e., there exists $C \in \mathbf{R}$ such that $\|f(x)\|_2 \leq C$ for all $x \in X$. Moreover, its image $f(X) := \{f(x) : x \in X\}$ is closed. Hence, the image $f(X)$ of a compact set $X \subset \mathbf{R}^n$ under a continuous mapping f is compact.*

Proof Assume for contradiction that f is unbounded. Then, for every $i \in \mathbf{N}$ there exists a point $x^i \in X$ such that $\|f(x^i)\|_2 > i$. Since X is bounded, by Theorem B.15 we can extract from the sequence $\{x^i\}$ a subsequence $\{x^{i_j}\}_{j=1}^{\infty}$ which converges to a point $\bar{x} \in X$. The real-valued function $g(x) = \|f(x)\|_2$ is continuous (as the superposition of two continuous mappings, see Theorem B.24(ii)) and therefore its values at the points x^{i_j} should converge, as $j \to \infty$, to its value at $\bar{x}$. On the other hand, $g(x^{i_j}) \geq i_j \to \infty$ as $j \to \infty$, and we get the desired contradiction.

In order to prove that $f(X)$ is closed, we need to prove that if a sequence $y^i := f(x^i)$, $x^i \in X$, of points from $f(X)$ converges to $\bar{y}$, as $i \to \infty$, then $\bar{y} \in f(X)$. Indeed, since X is compact, we can extract from $\{x^i\}$ a subsequence converging to some $\bar{x} \in X$. By continuity of f, the corresponding subsequence of the sequence $\{y^i = f(x^i)\}$ converges to $f(\bar{x})$. Thus, $\bar{y} = f(\bar{x}) \in f(X)$, as claimed. ■

A stronger notion of continuity is defined as follows:

Definition B.26 *[Uniformly continuous function] A function $f(x) : X \to \mathbf{R}^m$ is called uniformly continuous on X, if for every $\epsilon > 0$ there exists $\delta > 0$ such that*

$$x, y \in X, \ \|x - y\|_2 < \delta \implies \|f(x) - f(y)\|_2 < \epsilon.$$

Remark B.27 It is important to note the difference between the usual type of continuity and uniform continuity. The usual continuity of a function f means that given $\epsilon > 0$ *and a point $x \in X$*, it is possible to choose $\delta(\epsilon, x) > 0$ such that for all $y \in X$, $\|x - y\|_2 < \delta(\epsilon, x) \implies \|f(x) - f(y)\|_2 < \epsilon$; here the appropriate value of δ can depend on ϵ and on x. Uniform continuity means that this positive δ may be chosen as a function of *only* ϵ.

Example B.28 (Uniformly continuous functions) All the following mappings are uniformly continuous on $X := [1, 100]$:

- $f(x) = ax + b$
- $f(x) = |x|$
- $f(x) = x^2$
- $f(x) = 1/x$

Theorem B.29 *Let $X \subset \mathbf{R}^n$ be compact. If a mapping $f : X \to \mathbf{R}^m$ is continuous on X then it is uniformly continuous.*

Proof Assume for contradiction that there exists $\epsilon > 0$ such that for every $\delta > 0$ we can find a pair of points $x, y \in X$ such that $\|x - y\|_2 < \delta$ and $\|f(x) - f(y)\|_2 \geq \epsilon$. In particular, for every $i = 1, 2, \ldots$, we can find two points x^i, y^i in X such that $\|x^i - y^i\|_2 \leq 1/i$ and $\|f(x^i) - f(y^i)\|_2 \geq \epsilon$. As X is compact, by Theorem B.15 we can extract from the sequence $\{x^i\}$ a subsequence $\{x^{i_j}\}_{j=1}^\infty$ which converges to a certain point $\bar{x} \in X$. Since $\|y^{i_j} - x^{i_j}\|_2 \leq 1/i_j \to 0$ as $j \to \infty$, the sequence $\{y^{i_j}\}_{j=1}^\infty$ converges to the same point $\bar{x}$ as the sequence $\{x^{i_j}\}_{j=1}^\infty$ (why?). Since f is continuous, we have

$$\lim_{j \to \infty} f(y^{i_j}) = f(\bar{x}) = \lim_{j \to \infty} f(x^{i_j}),$$

whence $\lim_{j \to \infty} (f(x^{i_j}) - f(y^{i_j})) = 0$. But this contradicts the assumption that $\|f(x^{i_j}) - f(y^{i_j})\|_2 \geq \epsilon > 0$ for all j. ∎

Remark B.30 The fact that a function that is continuous on a compact set is automatically uniformly continuous on that set is one of the most useful features of compact sets. Note also that the compactness – closedness and boundedness – of the domain is crucial here: each of the functions $f(x) = 1/x : (0, 1] \to \mathbf{R}$, $f(x) = x^2 : \mathbf{R} \to \mathbf{R}$ is continuous on the respective domain, yet neither is uniformly continuous on it.

Theorem B.31 *Let $X \subset \mathbf{R}^n$ be nonempty and compact and let f be a real-valued continuous function on X. Then, f attains both its minimum and its maximum on X, i.e.,*

$$\operatorname*{Argmin}_X f := \left\{ x \in X : f(x) = \inf_{y \in X} f(y) \right\} \neq \varnothing,$$

$$\operatorname*{Argmax}_X f := \left\{ x \in X : f(x) = \sup_{y \in X} f(y) \right\} \neq \varnothing.$$

Proof We will prove that f attains its maximum on X; the proof for the minimum is completely analogous. Since f is bounded on X by Theorem B.25, the quantity

$$f^* := \sup_{x \in X} f(x)$$

is finite. Thus, we can find a sequence $\{x^i\}$ of points from X such that $f^* = \lim_{i \to \infty} f(x^i)$. As X is compact, we can extract from the sequence $\{x^i\}$ a subsequence $\{x^{i_j}\}_{j=1}^\infty$ which converges to a certain point $\bar{x} \in X$. Since f is continuous on X, we have

$$f(\bar{x}) = \lim_{j \to \infty} f(x^{i_j}) = \lim_{i \to \infty} f(x^i) = f^*,$$

so that the maximum of f on X is indeed achieved (e.g., at the point $\bar{x}$). ∎

Theorem B.31 admits the following useful generalization for unbounded sets:

Theorem B.32 *Let $X \subseteq \mathbf{R}^n$ be a nonempty and closed set, and let f be a real-valued continuous function on X which "goes to $+\infty$ at ∞," i.e., $f(x^t) \to \infty$, $t \to \infty$, along every sequence $\{x^t \in X, t \geq 1\}$ with $\|x^t\|_2 \to \infty$, $t \to \infty$, or, which is the same, such that* the sublevel sets $X_a := \{x \in X : f(x) \leq a\}$ of f on X are bounded for all $a \in \mathbf{R}$. Then, f attains its minimum on X:

$$\operatorname*{Argmin}_{X} f \neq \varnothing.$$

Needless to say, Theorem B.32 admits a maximization version; the reader is strongly advised to formulate this version on their own.

Proof Since $X \neq \varnothing$, there exists a real number $a \in \mathbf{R}$ such that the set X_a is nonempty. As X is closed and f is continuous on X, X_a is closed. Moreover, under the premise of the theorem, this set is also bounded, so that X_a is compact by Theorem B.17. By Theorem B.31, f attains its minimum on X_a, and due to the definition of X_a the minimizers of f on X_a are just the minimizers of f on X. ∎

B.3 Semicontinuity

Theorem B.32 admits a useful extension in which we can relax the requirement for f to be continuous. The related framework is as follows.

B.3.1 Functions with Values in the Extended Real Axis

So far, when speaking about scalar functions, we have dealt with real-valued functions defined on subsets of $\mathbf{R}^n$. However, in various minimization-related situations it is more convenient to speak about functions on $\mathbf{R}^n$ taking values in the *extended real axis* $\mathbf{R} \cup \{+\infty\}$. To this end, we will

- augment $\mathbf{R}$ with a "fictitious" point, denoted $+\infty$, and extend the usual orders $<$, $\leq$ onto the resulting *extended real line* $\mathbf{R} \cup \{+\infty\}$ by the natural convention that a real number is $< +\infty$ (and therefore is $\leq +\infty$), and, of course, $+\infty \leq +\infty$;
- consider functions $f(x) : \mathbf{R}^n \to \mathbf{R} \cup \{+\infty\}$; for such a function, its *domain* Dom f is defined as the set where the values of f are reals:

$$\operatorname{Dom} f := \{x \in \mathbf{R}^n : f(x) \in \mathbf{R}\}.$$

Note that our "old" partially defined real-valued functions on $\mathbf{R}^n$, i.e., the functions $f : X \to \mathbf{R}$ with $X \subseteq \mathbf{R}^n$, can be extended by the value $+\infty$ outside X to become functions on the *entire* space $\mathbf{R}^n$, taking values in $\mathbf{R} \cup \{+\infty\}$. Such an extended function clearly "remembers its origin" – its domain is X, and its restriction onto X coincides with the original f. Thus, two entities – the set X on which the original f was defined, and the original function f itself – are now encoded in a single entity, i.e., the function on $\mathbf{R}^n$ taking values in $\mathbf{R} \cup \{+\infty\}$. While there is nothing deep in this encoding (it is just a convention), it saves a lot of words.

In the sequel, if not explicitly stated otherwise, "function" means a function on $\mathbf{R}^n$ taking values in $\mathbf{R} \cup \{+\infty\}$.

Note that the function with *empty* domain – one which is identically equal to $+\infty$ – is a fully legitimate inhabitant of our new world. In this world, functions with nonempty domains have a special name; they are called *proper*.

B.3.2 Epigraph of a Function

Given a function $f : \mathbf{R}^n \to \mathbf{R} \cup \{+\infty\}$, we associate with it its *epigraph* – the set

$$\mathrm{epi}\{f\} := \{[x;t] \in \mathbf{R}^n \times \mathbf{R} : t \geq f(x)\}.$$

Clearly, a set E in $\mathbf{R}^n \times \mathbf{R}$ is an epigraph if and only if for every $x \in \mathbf{R}^n$ the intersection of E with the "vertical" line $\ell_x := \{[x;t] : t \in \mathbf{R}\}$ is either empty, or is a ray of the form $\{t \in \mathbf{R} : t \geq t_x\}$ with some real number t_x. Moreover, a set with these properties remembers the underlying function of which it is epigraph, i.e., the value of this function at a point $x \in \mathbf{R}^n$ is $+\infty$ when ℓ_x does not intersect E, and is t_x otherwise. The domain of a function is just the projection of its epigraph onto the x-plane:

$$\mathrm{Dom}\, f := \{x \in \mathbf{R}^n : \exists t \text{ such that } [x;t] \in \mathrm{epi}\{f\}\}.$$

B.3.3 Lower Semicontinuity

Functions with *closed* epigraphs have a name – they are called *lower semicontinuous* (*lsc* for short). Here is their characterization:

Theorem B.33 *Let $f(x) : \mathbf{R}^n \to \mathbf{R} \cup \{+\infty\}$. The following properties of f are equivalent to each other:*

(i) *[lower semicontinuity]* $\mathrm{epi}\{f\}$ *is closed;*
(ii) *[closedness of sublevel sets] For every $\alpha \in \mathbf{R}$, the sublevel set of f defined as*

$$\mathrm{lev}_\alpha(f) := \{x \in \mathbf{R}^n : f(x) \leq \alpha\}$$

is closed;
(iii) *Whenever a sequence $\{x^i \in \mathbf{R}^n, i \geq 1\}$ has a limit, say x, we have*

$$f(x) \leq \liminf_{i \to \infty} f(x^i). \tag{B.3}$$

Proof (i) $\Longrightarrow$ (ii): We want to prove that if $\mathrm{epi}\{f\}$ is closed and $\alpha \in \mathbf{R}$, then $\mathrm{lev}_\alpha(f)$ is closed. This is immediate: if $x^i \in \mathrm{lev}_\alpha(f)$ and $x^i \to x$, $i \to \infty$, then $f(x^i) \leq \alpha$ for all i, so that the sequence $\bar{x}^i := [x^i; \alpha] \in \mathbf{R}^n \times \mathbf{R}$ belongs to $\mathrm{epi}\{f\}$; as $i \to \infty$, this sequence converges to $[x; \alpha]$, and since $\mathrm{epi}\{f\}$ is closed, $[x; \alpha] \in \mathrm{epi}\{f\}$, implying that $f(x) \leq \alpha$, that is, $x \in \mathrm{lev}_\alpha(f)$.

(ii) $\Longrightarrow$ (iii): Assume that the sublevel sets of f are closed, and let us prove that whenever $x^i \to x$ as $i \to \infty$, (B.3) holds true. Passing to a subsequence of $\{x^i\}$, it suffices to consider the situation when $\liminf_{i \to \infty} f(x^i) = \lim_{i \to \infty} f(x^i) =: \beta$. Note that β is $+\infty$, or a real number, or $-\infty$. In the first case we clearly have $f(x) \leq \beta$. Suppose the second or the third case holds, and assume for the contradiction that $\beta < f(x)$. Let γ be a real number such that $\beta < \gamma < f(x)$. Since $\gamma > \beta$, for all large enough i we have $f(x^i) \leq \gamma$, that is, $x^i \in \mathrm{lev}_\gamma(f)$, implying that $x = \lim_{i \to \infty} x^i$ is the limit of a converging sequence of points from $\mathrm{lev}_\gamma(f)$. As the sublevel sets of f are closed, we conclude that $x \in \mathrm{lev}_\gamma(f)$, that is, $f(x) \leq \gamma$, which contradicts the definition of γ.

(iii) $\Longrightarrow$ (i): Assume that f obeys (iii), and let us prove that $\mathrm{epi}\{f\}$ is closed. Suppose the points $[x^i; t_i] \in \mathrm{epi}\{f\}$ converge as $i \to \infty$ to $[x; t]$; we need to prove that

$[x; t] \in \text{epi}\{f\}$. As $[x^i; t_i] \in \text{epi}\{f\}$, we have $f(x^i) \leq t_i$, and as $t_i \to t, i \to \infty$, we deduce $\liminf_{i \to \infty} f(x^i) \leq t$. Since f obeys (iii), we conclude that $f(x) \leq t$, that is, $[x; t] \in \text{epi}\{f\}$. ∎

Clearly, all finite-valued continuous functions are also lower semicontinuous. Moreover, there are many finite-valued discontinuous functions that are lower semicontinuous, i.e., precisely those with closed epigraphs. From the definition of lower semicontinuity, we also see that there are finite-valued discontinuous function that are not lsc. For example, the discontinuous function in Figure B.2.1 is not lsc.

The following corollary shows that a large class of continuous functions with extended values are also lsc.

Corollary B.34 *Let $f : \mathbf{R}^n \to \mathbf{R} \cup \{+\infty\}$ be a function that is continuous on its domain Dom f and let its domain be closed. Then, f is lower semicontinuous.*

Proof For any $\alpha \in \mathbf{R}$, consider the sublevel set of f given by $\text{lev}_\alpha(f) = \{x \in \text{Dom } f : f(x) \leq \alpha\}$. Under the premise of the corollary we will show that these sublevel sets are all closed, and then the result will follow from Theorem B.33(ii). Consider a converging sequence $\{x^i\}$ contained in $\text{lev}_\alpha(f)$ and let $x := \lim_{i \to \infty} x^i$. As Dom f is closed and $x^i \in$ Dom f, we have $x \in $ Dom f. Since f is continuous on Dom f, we deduce that $f(x) = \lim_{i \to \infty} f(x^i)$. Finally, as $f(x^i) \leq \alpha$ for all i, we get $f(x) \leq \alpha$, that is, $x \in \text{lev}_\alpha(f)$. This proves that $\text{lev}_\alpha(f)$ is closed as desired. ∎

The closedness of the domain in Corollary B.33 is indeed crucial – look at what happens with the univariate function which is zero on the positive ray and $+\infty$ on the nonpositive ray.

Another useful example of lsc function is the pointwise supremum

$$f(x) = \sup_{\alpha \in \mathcal{A}} f_\alpha(x)$$

of an arbitrary family of lsc functions. Indeed, we clearly have $\text{epi}\{f\} = \bigcap_{\alpha \in \mathcal{A}} \text{epi}\{f_\alpha\}$, and the intersection of a family of closed sets is closed.

We are now ready to state and prove the promised extension of Theorem B.32.

Theorem B.35 *Let $f : \mathbf{R}^n \to \mathbf{R} \cup \{+\infty\}$ be a proper lower semicontinuous function with bounded sublevel sets. Then, f is below bounded and attains its minimum.*

Proof Since f is proper, there exists $\alpha \in \mathbf{R}$ such that the set $\text{lev}_\alpha(f)$ is nonempty. Since f is lsc, $\text{lev}_\alpha(f)$ is closed. Moreover, under the premise of the theorem $\text{lev}_\alpha(f)$ is also bounded, and thus it is compact. Now, let $x^i \in \text{lev}_\alpha(f)$ be a sequence such that $f(x^i) \to \inf_{x \in \text{lev}_\alpha(f)} f(x)$ (note also that $\inf_{x \in \text{lev}_\alpha(f)} f(x) = \inf_{x \in \mathbf{R}^n} f(x)$). Since $\text{lev}_\alpha(f)$ is compact, passing to a subsequence we can assume that $x^i \to \bar{x}$ as $i \to \infty$, whence, by Theorem B.33(iii), $f(\bar{x}) \leq \liminf_{i \to \infty} f(x^i) = \inf_{x \in \mathbf{R}^n} f(x)$, implying that $\inf_{x \in \mathbf{R}^n} f(x)$ is $f(\bar{x}) \in \mathbf{R}$ and therefore $\inf_{x \in \mathbf{R}^n} f(x)$ is achieved and is finite. ∎

B.3.4 Hypograph and Upper Semicontinuity

While in the preceding sections we focused on minimizers of the functions and stated conditions under which such minimizers exist, there are of course natural "maximization" analogies of these developments. In the maximization analogies, the functions are allowed to take values in $\mathbf{R} \cup \{-\infty\}$, and the role of the epigraph is played by the *hypograph* of a function, i.e., the set

$$hypo\{f\} = \{[x;t] \in \mathbf{R}^n \times \mathbf{R} : t \leq f(x)\},$$

and lower semicontinuity is replaced with *upper semicontinuity*, i.e., the closedness of the hypograph. For example, the discontinuous function in Figure B.2.1 is upper semicontinuous.

In fact, a function $f : \mathbf{R}^n \to (\mathbf{R} \cup \{\pm\infty\})$ is upper semicontinuous if and only if the function $g(x) := -f(x)$ f is lower semicontinuous; and such a function f is continuous if and only if it is both upper and lower semicontinuous.

B.4 Proofs of Facts

Fact B.4 *All the following statements are correct:*

(i) *For any $\bar{x} \in \mathbf{R}^n$, we have $\bar{x} = \lim_{t\to\infty} x^t$ if and only if for every index $i = 1, \ldots, n$ the ith coordinates of the vectors x^t converge to the ith coordinate of the vector $\bar{x}$ as $t \to \infty$.*

(ii) *If a sequence converges, its limit is uniquely defined.*

(iii) *Convergence is compatible with linear operations:*
- *if $x^t \to x$ and $y^t \to y$ as $t \to \infty$, then $x^t + y^t \to x + y$ as $t \to \infty$;*
- *if $x^t \to x$ and $\lambda_t \to \lambda$ as $t \to \infty$, then $\lambda_t x^t \to \lambda x$ as $t \to \infty$.*

Proof (i): Indeed, $\|\bar{x} - x^i\|_2 = \sqrt{\sum_{j=1}^n (\bar{x}_j - (x^i)_j)^2}$ converges to 0 as $i \to \infty$ if and only if $\bar{x}_j - (x^i)_j$ converges to 0 as $i \to \infty$ for every $j \leq n$.

(ii): Suppose x' and x'' both are limits of a converging sequence $\{x^i\}$. Then, for every $\epsilon > 0$ and all large enough i it holds that $\|x' - x^i\|_2 \leq \epsilon$, $\|x'' - x^i\|_2 \leq \epsilon$, and so by the triangle inequality $\|x' - x''\|_2 \leq 2\epsilon$. Since $\epsilon > 0$ is arbitrary, we conclude that $\|x' - x''\|_2 \leq 0$, and thus $x' = x''$.

(iii): By elementary calculus, both claims are valid for sequences of real numbers; this combines with item (i) to imply the validity of the claims for vectors. ∎

Fact B.13

(i) *A set $X \subseteq \mathbf{R}^n$ is closed if and only if its complement $\overline{X} := \mathbf{R}^n \setminus X$ is open.*
(ii) *The intersection of every (finite or infinite) family of closed sets is closed. The union of every (finite or infinite) family of open sets is open.*
(iii) *The union of finitely many closed sets is closed. The intersection of finitely many open sets is open.*

Proof

(i): Clearly, a point $\bar{x} \in \mathbf{R}^n$ is a limit of a sequence of points from a set $X \subseteq \mathbf{R}^n$ if and only if every ball centered at $\bar{x}$ and of positive radius contains points from X. Consequently,
- when X is closed and $\bar{x} \notin X$, there is a ball of positive radius centered at $\bar{x}$ and not intersecting X, implying that the complement $\overline{X}$ of a closed set X is open;

- when X is open, none of the points from X is the limit of a sequence of points from the complement $\overline{X}$ of X, implying that the limit of every converging sequence of points from $\overline{X}$ belongs to $\overline{X}$, that is, $\overline{X}$ is closed.

(ii): This is evident.

(iii): When $X_1,\ldots,X_N$ are closed and $\{x^i\}$ is a converging sequence of points from $X = \cup_{i\leq N} X_i$, the sequence visits a certain X_j infinitely many times, that is, the sequence has a subsequence where all terms are from that X_j. The limit of this subsequence (which is the limit of $\{x^i\}$ as well) belongs to X_j since this set is closed, and therefore the limit of $\{x^i\}$ belongs to X. Thus, X is closed. Passing from the sets to their complements and invoking (i), we extract from what just have been proved that the intersection of finitely many open sets is open. ∎

Fact B.23 *The norm $\|x\|_2$ is a real-valued function and it is continuous on $\mathbf{R}^n$ (and thus – on every subset of $\mathbf{R}^n$) (check this!). In fact, every norm $\|\cdot\|$ on $\mathbf{R}^n$ is continuous.*

Proof By Fact B.2, we have $|\|x\|_2 - \|y\|_2| \leq \|x-y\|_2$, so that $\lim_{t\to\infty} x^t = x$ implies that $0 \leq |\|x\|_2 - \|x^t\|_2| \leq \|x - x^t\|_2 \to 0, t \to \infty$; hence $\lim_{t\to\infty} \|x^t\|_2 = \|x\|_2$, implying the continuity of $\|\cdot\|_2$. This reasoning can be repeated word for word for an arbitrary norm $\|\cdot\|$ and the convergence induced by this norm in the role of $\|\cdot\|_2$ and the convergence coming from the latter norm. This combines with the the fact that the convergence of a sequence from $\mathbf{R}^n$ and its limit are independent of the norm-induced distance on $\mathbf{R}^n$ that we use, see section B.1.2, to imply the continuity of every norm on $\mathbf{R}^n$. ∎

B.5 Exercises

Exercise B.1 Mark in the list below those sets which are closed and those which are open (the sets are in $\mathbf{R}^n$, $\|\cdot\|$ is a norm on $\mathbf{R}^n$, where $n > 0$):

1. All vectors with integer coordinates.
2. All vectors with rational coordinates.
3. All vectors with positive coordinates.
4. All vectors with nonnegative coordinates.
5. $\{x \in \mathbf{R}^n : \|x\| < 1\}$
6. $\{x \in \mathbf{R}^n : \|x\| = 1\}$
7. $\{x \in \mathbf{R}^n : \|x\| \leq 1\}$
8. $\{x \in \mathbf{R}^n : \|x\| \geq 1\}$
9. $\{x \in \mathbf{R}^n : \|x\| > 1\}$
10. $\{x \in \mathbf{R}^n : 1 < \|x\| \leq 2\}$

Exercise B.2 Consider the function $f(x_1,x_2) : \mathbf{R}^2 \to \mathbf{R}$ defined as

$$f(x_1,x_2) = \begin{cases} \frac{x_1^2 - x_2^2}{x_1^2 + x_2^2}, & \text{if } (x_1,x_2) \neq 0 \\ 0, & \text{if } x_1 = x_2 = 0. \end{cases}$$

Check whether this function is continuous on the following sets:

1. $\mathbf{R}^2$
2. $\mathbf{R}^2 \setminus \{0\}$
3. $\{x \in \mathbf{R}^2 : x_1 = 0\}$
4. $\{x \in \mathbf{R}^2 : x_2 = 0\}$
5. $\{x \in \mathbf{R}^2 : x_1 + x_2 = 0\}$
6. $\{x \in \mathbf{R}^2 : x_1 - x_2 = 0\}$
7. $\{x \in \mathbf{R}^2 : |x_1 - x_2| \leq x_1^4 + x_2^4\}$

Exercise B.3 Let $f : \mathbf{R}^n \to \mathbf{R}^m$ be a continuous mapping. Among the following statements, mark those which are always true:

1. If U is an open set in $\mathbf{R}^m$, then so is the set $f^{-1}(U) := \{x \in \mathbf{R}^n : f(x) \in U\}$.
2. If U is an open set in $\mathbf{R}^n$, then so is the set $f(U) = \{f(x) : x \in U\}$.
3. If F is a closed set in $\mathbf{R}^m$, then so is the set $f^{-1}(F) = \{x \in \mathbf{R}^n : f(x) \in F\}$.
4. If F is a closed set in $\mathbf{R}^n$, then so is the set $f(F) = \{f(x) : x \in F\}$.

Exercise B.4 Prove that in general *none* of Theorems B.25, B.29, and B.31 remains valid when X is closed, but not bounded; the same holds when X is bounded, but not closed.

Appendix C

Prerequisites from Calculus

C.1 Differentiable Functions on $\mathbf{R}^n$

We next turn to prerequisites from calculus, specifically differentiable functions.

C.1.1 The Derivative

Recall the notion of the *derivative* of a real-valued function $f : \mathbf{R} \to \mathbf{R}$ of a real variable x[1]:

$$f'(x) := \lim_{\Delta x \to 0} \frac{f(x + \Delta x) - f(x)}{\Delta x}.$$

This definition does not work when we pass from functions of a single real variable to functions of several real variables, or, which is the same, to functions with vector arguments. Indeed, in this case the shift in the argument Δx is a vector, and we do not know what it means to *divide* by a vector...

A proper way to extend the notion of the derivative to real- and vector-valued functions of vector arguments is to realize what in fact the meaning of the derivative is in the univariate case: $f'(x)$ gives us the precise description of *how to approximate f in a neighborhood of x by a linear function*. Specifically, if $f'(x)$ exists, then the linear function $f'(x)\Delta x$ of Δx approximates the change $f(x + \Delta x) - f(x)$ in f up to a remainder which is of higher order as compared to Δx as $\Delta x \to 0$:

$$\left| f(x + \Delta x) - f(x) - f'(x)\Delta x \right| \leq \bar{o}(|\Delta x|) \text{ as } \Delta x \to 0.$$

In the above formula, we meet with the notation $\bar{o}(|\Delta x|)$, and here is the explanation of this notation:

> The terminology $\bar{o}(|\Delta x|)$ is a common name for functions $\phi(\Delta x)$ of Δx which are well defined in a neighborhood of the point $\Delta x = 0$ on the axis, vanish at the point $\Delta x = 0$, and are such that
>
> $$\frac{\phi(\Delta x)}{|\Delta x|} \to 0 \text{ as } \Delta x \to 0.$$

---

[1] Here and in what follows, when speaking about limits like that on the right-hand side of the definition of f', we exclude the value $\Delta x = 0$ (at which the quantity we are looking at is undefined). In other words, we are speaking about a limit taken w.r.t. a "percolated neighborhood" of the origin: by definition, the relation $a = \lim_{\Delta x \to 0} g(\Delta x)/\Delta x$ means that for every $\epsilon > 0$ there exists $\delta > 0$ such that $|g(\Delta x)/\Delta x - a| \leq \epsilon$ for all nonzero Δx satisfying $|\Delta x| \leq \delta$.

For example:

1. $(\Delta x)^2 = \bar{o}(|\Delta x|)$, $\Delta x \to 0$.
2. $|\Delta x|^{1.01} = \bar{o}(|\Delta x|)$, $\Delta x \to 0$.
3. $\sin^2(\Delta x) = \bar{o}(|\Delta x|)$, $\Delta x \to 0$.
4. $\Delta x \neq \bar{o}(|\Delta x|)$, $\Delta x \to 0$.

Later we shall meet the notation "$\bar{o}(|\Delta x|^k)$ as $\Delta x \to 0$", where k is a positive integer. The definition of $\bar{o}(|\Delta x|^k)$ is completely similar to that for the case of $k=1$:

The terminology $\bar{o}(|\Delta x|^k)$ is a common name for functions $\phi(\Delta x)$ of Δx which are well-defined in a neighborhood of the point $\Delta x = 0$ on the axis, vanish at the point $\Delta x = 0$, and are such that

$$\frac{\phi(\Delta x)}{|\Delta x|^k} \to 0 \text{ as } \Delta x \to 0.$$

Note that if $f(\cdot)$ is a function defined in a neighborhood of a point x on the axis, then there perhaps are many linear functions $a\Delta x$ of Δx which well approximate $f(x+\Delta x) - f(x)$, in the sense that the remainder in the approximation

$$f(x+\Delta x) - f(x) - a\Delta x$$

tends to 0 as $\Delta x \to 0$. Among these approximations, however, there exists *at most one* which approximates $f(x+\Delta x) - f(x)$ "very well" – so that the remainder is $\bar{o}(|\Delta x|)$, and not merely tending to 0 as $\Delta x \to 0$. Indeed, if

$$f(x+\Delta x) - f(x) - a\Delta x = \bar{o}(|\Delta x|),$$

then, dividing both sides by Δx, we get

$$\frac{f(x+\Delta x) - f(x)}{\Delta x} - a = \frac{\bar{o}(|\Delta x|)}{\Delta x}.$$

By definition of $\bar{o}(\cdot)$, the right-hand side in this equality tends to 0 as $\Delta x \to 0$, whence

$$a = \lim_{\Delta x \to 0} \frac{f(x+\Delta x) - f(x)}{\Delta x} = f'(x).$$

Thus, *if* a linear function $a\Delta x$ of Δx approximates the change $f(x+\Delta x) - f(x)$ in f up to a remainder which is $\bar{o}(|\Delta x|)$ as $\Delta x \to 0$, *then* a is the derivative of f at x. We can easily verify that the inverse statement is also true: *if* the derivative of f at x exists, *then* the linear function $f'(x)\Delta x$ of Δx approximates the change $f(x+\Delta x) - f(x)$ in f up to a remainder which is $\bar{o}(|\Delta x|)$ as $\Delta x \to 0$.

The advantage of the "$\bar{o}(|\Delta x|)$"-definition of the derivative is that it can be naturally extended to vector-valued functions of vector arguments (by just replacing the "axis" with $\mathbf{R}^n$ in the definition of $\bar{o}$) and sheds light on the *essence* of the notion of the derivative: when it exists, this is exactly *the linear function of Δx which approximates the change $f(x+\Delta x) - f(x)$ in f up to a remainder which is $\bar{o}(|\Delta x|)$.* The precise definition is as follows:

C.1 Differentiable Functions on $\mathbf{R}^n$

Definition C.1 *[Frechet differentiability] Let f be a function which is well defined in a neighborhood of a point $x \in \mathbf{R}^n$ and takes values in $\mathbf{R}^m$. We say that f is differentiable at x, if there exists a linear function $Df(x)[\Delta x]$ of $\Delta x \in \mathbf{R}^n$ taking values in $\mathbf{R}^m$ which approximates the change $f(x + \Delta x) - f(x)$ in f up to a remainder which is $\bar{o}(\|\Delta x\|_2)$, i.e.,*

$$\|f(x + \Delta x) - f(x) - Df(x)[\Delta x]\|_2 \leq \bar{o}(\|\Delta x\|_2). \tag{C.1}$$

Equivalently, a function f which is well defined in a neighborhood of a point $x \in \mathbf{R}^n$ and takes values in $\mathbf{R}^m$ is called differentiable *at x, if there exists a linear function $Df(x)[\Delta x]$ of $\Delta x \in \mathbf{R}^n$ taking values in $\mathbf{R}^m$ such that for every $\epsilon > 0$ there exists $\delta > 0$ satisfying the relation*

$$\|\Delta x\|_2 \leq \delta \implies \|f(x + \Delta x) - f(x) - Df(x)[\Delta x]\|_2 \leq \epsilon \|\Delta x\|_2.$$

Note that due to the equivalence of norms on finite-dimensional spaces, in this definition the standard Euclidean norms in which we measure changes in f and in the magnitude of Δx can be replaced with any other pair of norms on $\mathbf{R}^m$ and $\mathbf{R}^n$ without affecting differentiability and $Df(x)[\cdot]$.

C.1.2 Derivative and Directional Derivatives

We have defined what it means for a function $f : \mathbf{R}^n \to \mathbf{R}^m$ to be differentiable at a point x but have not stated yet what the *derivative* is. The reader may guess that the derivative is exactly "the linear function $Df(x)[\Delta x]$ of $\Delta x \in \mathbf{R}^n$ taking values in $\mathbf{R}^m$ which approximates the change $f(x + \Delta x) - f(x)$ in f up to a remainder which is less than or equal to $\bar{o}(\|\Delta x\|_2)$" participating in the definition of differentiability. While this guess is correct, we cannot merely call the entity participating in the definition the derivative – how do we know that this entity is unique? Perhaps there are many different linear functions of Δx approximating the change in f up to a remainder which is $\bar{o}(\|\Delta x\|_2)$? In fact there is no more than a single linear function with this property, owing to the following observation.

Proposition C.2 *Let $f : \mathbf{R}^n \to \mathbf{R}^m$ be differentiable at x, and $Df(x)[\Delta x]$ be a linear function participating in the definition of differentiability. Then,*

$$Df(x)[\Delta x] = \lim_{t \to +0} \frac{f(x + t\Delta x) - f(x)}{t}, \qquad \forall \Delta x \in \mathbf{R}^n. \tag{C.2}$$

In particular, the derivative $Df(x)[\cdot]$ is uniquely defined by f and x.

Proof For all $\Delta x \in \mathbf{R}^n$ and for any $t > 0$, we have

$$\|f(x + t\Delta x) - f(x) - Df(x)[t\Delta x]\|_2 \leq \bar{o}(\|t\Delta x\|_2)$$

$$\implies \left\| \frac{f(x + t\Delta x) - f(x)}{t} - \frac{Df(x)[t\Delta x]}{t} \right\|_2 \leq \frac{\bar{o}(\|t\Delta x\|_2)}{t}$$

$$\iff \left\| \frac{f(x + t\Delta x) - f(x)}{t} - Df(x)[\Delta x] \right\|_2 \leq \frac{\bar{o}(\|t\Delta x\|_2)}{t}$$

$$\implies Df(x)[\Delta x] = \lim_{t \to +0} \frac{f(x + t\Delta x) - f(x)}{t},$$

where the second line is obtained by dividing by $t > 0$, the third line follows since $Df(x)[\cdot]$ is linear, and the last line is obtained by passing to the limit as $t \to +0$ and noting that $\dfrac{\bar{o}(\|t\Delta x\|_2)}{t} \to 0$ as $t \to +0$. ∎

We conclude this section with three important remarks as follows:

1. The right-hand side limit in (C.2) is an important entity called the *directional derivative of f taken at x along (a direction) Δx*; note that this quantity is defined in the "purely univariate" fashion – by dividing the change in f by the magnitude of the shift in the direction Δx and passing to the limit as the magnitude of the shift approaches 0. The relation (C.2) states that the derivative, if it exists, is, at every Δx, just the directional derivative of f taken at x along Δx. Note, however, that differentiability is much more than the existence of directional derivatives along all directions Δx. In particular, differentiability requires also *the directional derivatives to be "well organized"* – to depend linearly on the direction Δx. It is easily seen that the mere existence of directional derivatives does not imply that they are "well organized." For example, the Euclidean norm

$$f(x) = \|x\|_2$$

at $x = 0$ possesses directional derivatives along all directions:

$$\lim_{t \to +0} \frac{f(0 + t\Delta x) - f(0)}{t} = \|\Delta x\|_2.$$

These derivatives, however, depend *nonlinearly* on Δx, so that the Euclidean norm is *not* differentiable at the origin (although it is differentiable everywhere outside the origin, but this is another story).

2. It should be stressed that the derivative, if it exists, is *a linear function of $\Delta x \in \mathbf{R}^n$ taking values in $\mathbf{R}^m$*. As we shall see in a while, we can *represent* this function by something "tractable," like a vector or a matrix, and we can understand how to compute such a representation. However, a careful reader should bear in mind that *a* representation is not exactly the same as *the* represented entity. Sometimes the difference between derivatives and the entities which represent them is reflected in the terminology: what we call the *derivative* may be called by some the *differential* while the word "derivative" is reserved by them for the vector or matrix representing the differential.

3. Sometimes we need to speak about differentiability of a mapping $f : X \to \mathbf{R}^m$, $X \subset \mathbf{R}^n$, at a point $x \in X$ which is not in the interior of X, so that there is possibly no neighborhood of x where f is well-defined. In these cases, we shall say that f is differentiable at x, if there exists a linear function of $\Delta x \in \mathbf{R}^n$, denoted by $Df(x)[\Delta x]$, taking values in $\mathbf{R}^m$, such that (C.1) holds for all Δx such that $x + \Delta x \in X$. Note that this extension affects neither differentiability nor the derivative of f at the interior points of X; as for $x \in X \setminus \text{int } X$, here the derivative may lose uniqueness (think about the case when $X \subset \mathbf{R}^2$ is a singleton, or the union of two singletons, or a line segment, and $f : X \to \mathbf{R}$ is identically zero). However, when X is convex, the restriction of $Df(x)[\cdot]$ onto the linear span $\text{Lin}(X - \{x\})$ of the set $X - \{x\}$ *is* uniquely defined by f. Indeed, the reasoning in the proof of Proposition C.2 shows that $Df(x)[\Delta x]$ is uniquely defined by f when $x + \Delta x \in X$, hence by linearity of the derivative in Δx, $Df(x)[\Delta x]$ is uniquely defined

by f when $\Delta x \in \mathrm{Lin}(X - \{x\})$. In particular, when $X \subset \mathbf{R}^n$ is convex and $\mathrm{Aff}(X) = \mathbf{R}^n$, the derivative of f at $x \in X$, if any, is uniquely defined by f.

C.1.3 Representations of the Derivative

By definition, the derivative of a mapping $f : \mathbf{R}^n \to \mathbf{R}^m$ at a point x is a linear function $Df(x)[\Delta x]$ taking values in $\mathbf{R}^m$. How could we represent such a function?

Case $m = 1$: the gradient. Let us start with real-valued functions (i.e., with the case $m = 1$); in this case the derivative is a *linear* real-valued function on $\mathbf{R}^n$. As we remember, the standard Euclidean structure on $\mathbf{R}^n$ allows us to represent every linear function on $\mathbf{R}^n$ as the inner product of the argument with a certain fixed vector. In particular, the derivative $Df(x)[\Delta x]$ of a scalar function can be represented as

$$Df(x)[\Delta x] = [\text{vector}]^\top \Delta x;$$

what is denoted by "vector" in this relation is called the *gradient* of f at x and is denoted by $\nabla f(x)$:

$$Df(x)[\Delta x] = (\nabla f(x))^\top \Delta x. \tag{C.3}$$

How to compute the gradient? In the case when $f : X \to \mathbf{R}$ and $x \in \mathrm{int}\, X$, the answer is given by (C.2). Indeed, let us look what (C.3) and (C.2) say when Δx is the ith standard basis vector. According to (C.3), $Df(x)[e_i]$ is the ith coordinate of the vector $\nabla f(x)$. Then, using (C.2), we arrive at

$$\left. \begin{aligned} Df(x)[e_i] &= \lim_{t \to +0} \frac{f(x + te_i) - f(x)}{t}, \\ Df(x)[e_i] &= -Df(x)[-e_i] = -\lim_{t \to +0} \frac{f(x - te_i) - f(x)}{t} \\ &= \lim_{t \to -0} \frac{f(x + te_i) - f(x)}{t} \\ \Longrightarrow Df(x)[e_i] &= \frac{\partial f(x)}{\partial x_i}. \end{aligned} \right\}$$

Thus, we conclude:

If a real-valued function $f : X \to \mathbf{R}$ is differentiable at $x \in \mathrm{int}\, X$, then the first-order partial derivatives of f at x exist, and the gradient of f at x is just the vector with coordinates which are the first-order partial derivatives of f taken at x:

$$\nabla f(x) := \begin{bmatrix} \partial f(x)/\partial x_1 \\ \vdots \\ \partial f(x)/\partial x_n \end{bmatrix}.$$

The derivative of f, taken at x, is the linear function of Δx given by

$$Df(x)[\Delta x] = (\nabla f(x))^\top \Delta x = \sum_{i=1}^n \frac{\partial f(x)}{\partial x_i} (\Delta x)_i.$$

Important: *from now and till the end of Appendix D, the functions under consideration are defined on an* open *domain in* $\mathbf{R}^n$, *and their differentiability and derivatives are understood according to Definition C.1, not in the "extended" sense discussed in section C.1.2.* With some abuse of notation, we write $f : \mathbf{R}^n \to \mathbf{R}^m$ to express the fact that f is well-defined on an open domain in $\mathbf{R}^n$ and takes values in $\mathbf{R}^m$. Needless to say, the derivatives of a function are taken at the points from its domain.

General case: the Jacobian. Now consider $f : \mathbf{R}^n \to \mathbf{R}^m$ with $m \geq 1$. In this case, $Df(x)[\Delta x]$, regarded as a function of Δx, is a linear mapping from $\mathbf{R}^n$ to $\mathbf{R}^m$. Recall that the standard way to represent a linear mapping from $\mathbf{R}^n$ to $\mathbf{R}^m$ is to represent it as multiplication by an $m \times n$ matrix:

$$Df(x)[\Delta x] = [m \times n \text{ matrix}] \cdot \Delta x. \tag{C.4}$$

What is denoted by "matrix" in (C.4) is called the *Jacobian* of f at x and is denoted by $f'(x)$. How to compute the entries of the Jacobian? Once again the answer is readily given by (C.2). Indeed, on the one hand, we have

$$Df(x)[\Delta x] = f'(x)\Delta x, \tag{C.5}$$

where

$$[Df(x)[e_j]]_i = [f'(x)]_{ij}, \quad i = 1, \ldots, m, \quad j = 1, \ldots, n.$$

On the other hand, by denoting

$$f(x) = \begin{bmatrix} f_1(x) \\ \vdots \\ f_m(x) \end{bmatrix},$$

the same computation as in the case of the gradient demonstrates that

$$[Df(x)[e_j]]_i = \frac{\partial f_i(x)}{\partial x_j}.$$

Thus, we arrive at the following conclusion:

If a vector-valued function $f(x) = [f_1(x); \ldots; f_m(x)]$ is differentiable at x, then the first-order partial derivatives of all f_i at x exist, and the Jacobian of f at x is just the $m \times n$ matrix with the entries $\left[\partial f_i(x)/\partial x_j\right]_{i,j}$ (so that the rows in the Jacobian are $[\nabla f_1(x)]^\top, \ldots, [\nabla f_m(x)]^\top$). The derivative of f, taken at x, is the linear vector-valued function of Δx given by

$$Df(x)[\Delta x] = f'(x)\Delta x = \begin{bmatrix} [\nabla f_1(x)]^\top \Delta x \\ \vdots \\ [\nabla f_m(x)]^\top \Delta x \end{bmatrix}.$$

Remark C.3 Note that for a real-valued function $f : \mathbf{R}^n \to \mathbf{R}$ we have defined both the gradient $\nabla f(x)$ and the Jacobian $f'(x)$. These two entities are nearly, but not exactly, the same. The Jacobian is a row vector and the gradient is a column vector; these two are linked by the relation

$$f'(x) = (\nabla f(x))^\top.$$

Of course, both these representations of the derivative of f yield the same linear approximation of the change in f:

$$Df(x)[\Delta x] = (\nabla f(x))^\top \Delta x = f'(x)\Delta x.$$

C.1.4 Existence of the Derivative

We have seen that the existence of the derivative of f at a point implies the existence of the first-order partial derivatives of the components $(f_1, \ldots, f_m)$ of f. The reverse statement is not exactly true. In particular, the existence of all first-order partial derivatives $\partial f_i(x)/\partial x_j$ for all i, j does not necessarily imply the existence of the derivative. In fact, we need a bit more, which is described next.

Theorem C.4 *[Sufficient condition for differentiability] Given a mapping $f = (f_1, \ldots, f_m) : \mathbf{R}^n \to \mathbf{R}^m$, f is differentiable at the point $\bar{x} \in \mathbf{R}^n$ if all the following conditions hold:*

(i) *the mapping f is well defined in a neighborhood U of the point $\bar{x} \in \mathbf{R}^n$;*
(ii) *the first-order partial derivatives of the components f_i of f exist everywhere in U; and*
(iii) *the first-order partial derivatives of the components f_i of f are continuous at the point $\bar{x}$.*

C.1.5 Calculus of Derivatives

The following elementary rules for the calculus of derivatives are useful to know.

Theorem C.5 (i) *[Differentiability and linear operations] Let $f_1(x)$, $f_2(x)$ be mappings defined in a neighborhood of a point $\bar{x} \in \mathbf{R}^n$ and taking values in $\mathbf{R}^m$, and let $\lambda_1(x), \lambda_2(x)$ be real-valued functions defined in a neighborhood of $\bar{x}$. Whenever $f_1, f_2, \lambda_1, \lambda_2$ are differentiable at $\bar{x}$, so is the function $f(x) := \lambda_1(x)f_1(x) + \lambda_2(x)f_2(x)$, and its derivative at $\bar{x}$ is given by*

$$Df(\bar{x})[\Delta x] = [D\lambda_1(\bar{x})[\Delta x]]f_1(\bar{x}) + \lambda_1(\bar{x})Df_1(\bar{x})[\Delta x]$$
$$+ [D\lambda_2(\bar{x})[\Delta x]]f_2(\bar{x}) + \lambda_2(\bar{x})Df_2(\bar{x})[\Delta x],$$
$$\implies f'(\bar{x}) = f_1(\bar{x})[\nabla \lambda_1(\bar{x})]^\top + \lambda_1(\bar{x})f_1'(\bar{x})$$
$$+ f_2(\bar{x})[\nabla \lambda_2(\bar{x})]^\top + \lambda_2(\bar{x})f_2'(\bar{x}).$$

(ii) *[Chain rule] Let a mapping $f : \mathbf{R}^n \to \mathbf{R}^m$ be differentiable at $\bar{x}$, and a mapping $g : \mathbf{R}^m \to \mathbf{R}^k$ be differentiable at $\bar{y} := f(\bar{x})$. Then, the superposition function given by $h(x) = g(f(x))$ is differentiable at $\bar{x}$, and its derivative at $\bar{x}$ is given by*

$$Dh(\bar{x})[\Delta x] = Dg(\bar{y})[Df(\bar{x})[\Delta x]],$$
$$\implies h'(\bar{x}) = g'(\bar{y})f'(\bar{x}).$$

If the outer function g is real-valued, then the latter formula implies that

$$\nabla h(\bar{x}) = [f'(\bar{x})]^\top \nabla g(\bar{y})$$

(recall that for a real-valued function ϕ, $\phi' = (\nabla \phi)^\top$).

C.1.6 Computing the Derivative

Representations of the derivative via first-order partial derivatives normally allow us to compute it by the standard calculus rules, in a completely mechanical fashion, not thinking at all of *what* we are computing. The examples to follow (especially Example C.8) demonstrate that it often makes sense to bear in mind exactly what the derivative is; this sometimes yields the result much faster than blindly implementing calculus rules.

Example C.6 (Gradient of an affine function) An *affine* function

$$f(x) = a + \sum_{i=1}^{n} g_i x_i \equiv a + g^\top x : \mathbf{R}^n \to \mathbf{R}$$

is differentiable at every point (Theorem C.4) and its gradient, of course, equals g:

$$(\nabla f(x))^\top \Delta x = \lim_{t \to +0} t^{-1}(f(x + t\Delta x) - f(x)) \qquad \text{[by (C.2)]}$$
$$= \lim_{t \to +0} t^{-1}(tg^\top \Delta x). \qquad \text{[by plugging in the definition of } f\text{]}$$

Hence, we arrive at

$$\boxed{\nabla(a + g^\top x) = g.}$$

Example C.7 (Gradient of a quadratic form) For now, let us define a homogeneous quadratic form on $\mathbf{R}^n$ as a function

$$f(x) = \sum_{i=1}^{n} \sum_{j=1}^{n} A_{ij} x_i x_j = x^\top A x,$$

where A is an $n \times n$ matrix. Note that the matrices A and $A^\top$ define the same quadratic form, and therefore the *symmetric* matrix $B := \frac{1}{2}(A + A^\top)$ also produces the same quadratic form as A and $A^\top$. Thus, we can always assume (and do assume from now on) that the matrix A producing the quadratic form in question is symmetric.

A quadratic form is a simple polynomial and as such is differentiable at every point (Theorem C.4). What is the gradient of f at a point x? Here is the computation:

$$(\nabla f(x))^\top \Delta x$$
$$= Df(x)[\Delta x]$$
$$= \lim_{t \to +0} t^{-1} \left((x + t\Delta x)^\top A(x + t\Delta x) - x^\top A x \right)$$

$$= \lim_{t \to +0} t^{-1}\left(x^\top Ax + t(\Delta x)^\top Ax + tx^\top A\Delta x + t^2(\Delta x)^\top A\Delta x - x^\top Ax\right)$$

$$= \lim_{t \to +0} t^{-1}\left(2t(Ax)^\top \Delta x + t^2(\Delta x)^\top A\Delta x\right)$$

$$= 2(Ax)^\top \Delta x,$$

where the second equality follows from (C.2), the third equality is obtained by opening the parentheses, and in the last line we used that A is symmetric.

Hence, we conclude that for a symmetric matrix A, we have

$$\boxed{\nabla(x^\top Ax) = 2Ax.}$$

Example C.8 (Derivative of X^{-1} on the domain of nonsingular $n \times n$ matrices) Define the mapping $F(X) := X^{-1}$ on the open set of nonsingular $n \times n$ matrices. Suppose $X \in \mathbf{R}^{n \times n}$ is nonsingular. Then, for any $\Delta X \in \mathbf{R}^{n \times n}$ we have

$$DF(X)[\Delta X] = \lim_{t \to +0} t^{-1}\left((X + t\Delta X)^{-1} - X^{-1}\right)$$

$$= \lim_{t \to +0} t^{-1}\left((X(I + tX^{-1}\Delta X))^{-1} - X^{-1}\right)$$

$$= \lim_{t \to +0} t^{-1}\left((I + tX^{-1}\Delta X)^{-1} X^{-1} - X^{-1}\right).$$

Thus, by defining $Y := X^{-1}\Delta X$, for all $\Delta X \in \mathbf{R}^{n \times n}$ we arrive at

$$DF(X)[\Delta X] = \left(\lim_{t \to +0} t^{-1}\left((I + tY)^{-1} - I\right)\right) X^{-1}$$

$$= \left(\lim_{t \to +0} t^{-1}\left(I - (I + tY)\right)(I + tY)^{-1}\right) X^{-1}$$

$$= \left(\lim_{t \to +0} (-Y(I + tY)^{-1})\right) X^{-1}$$

$$= -YX^{-1}$$

$$= -X^{-1}\Delta X X^{-1}.$$

Therefore, we arrive at the important relation

$$\boxed{D(X^{-1})[\Delta X] = -X^{-1}\Delta X X^{-1}.}$$

(Recall that the derivative of the univariate function x^{-1} at $x \neq 0$ is $-x^{-2}$.)

Example C.9 (Derivative of the log-det barrier) The *log-det barrier* is given by

$$F(X) = \ln \mathrm{Det}(X),$$

where X is an $n \times n$ matrix (or, if you prefer, an n^2-dimensional vector). The log-det barrier plays an extremely important role in modern optimization. In this example, we will compute its derivative.

Note that $F(X)$ is well defined and differentiable in a neighborhood of every point $\bar{X}$ with positive determinant. (Indeed, $\mathrm{Det}(X)$ is a polynomial of the entries of X and thus it is everywhere continuous and differentiable with continuous partial derivatives, while the function $\ln(t)$ is continuous and differentiable on the positive ray. Then, by Theorems B.24(ii) and C.5(ii), F is differentiable at every X such that $\mathrm{Det}(X) > 0$). While the computation of the derivative of F by the standard techniques would not be very pleasant, we next illustrate that this computation can be done easily by resorting to the fundamental definition of the derivative.

Let us consider a point $\bar{X}$ such that $\mathrm{Det}(\bar{X}) > 0$, and define $G(X) := \mathrm{Det}(X)$. Then, we have

$$
\begin{aligned}
DF(\bar{X})[\Delta X] &= D\ln(G(\bar{X}))\left(DG(\bar{X})[\Delta X]\right) \\
&= (G(\bar{X}))^{-1} DG(\bar{X})[\Delta X] \\
&= (\mathrm{Det}(\bar{X}))^{-1} \lim_{t \to +0} t^{-1} \left(\mathrm{Det}(\bar{X} + t\Delta X) - \mathrm{Det}(\bar{X})\right) \\
&= (\mathrm{Det}(\bar{X}))^{-1} \lim_{t \to +0} t^{-1} \left(\mathrm{Det}\left(\bar{X}(I + t\bar{X}^{-1}\Delta X)\right) - \mathrm{Det}(\bar{X})\right) \\
&= (\mathrm{Det}(\bar{X}))^{-1} \lim_{t \to +0} t^{-1} \left(\mathrm{Det}(\bar{X})\left(\mathrm{Det}(I + t\bar{X}^{-1}\Delta X) - 1\right)\right) \\
&= \lim_{t \to +0} t^{-1} \left(\mathrm{Det}(I + t\bar{X}^{-1}\Delta X) - 1\right) \\
&= \mathrm{Tr}(\bar{X}^{-1}\Delta X) = \sum_{i=1}^{n}\sum_{j=1}^{n} (\bar{X}^{-1})_{ji} (\Delta X)_{ij},
\end{aligned}
$$

where the first equality follows from the chain rule, the second from the fact that $\ln'(x) = x^{-1}$ for any $x > 0$, the third from the definition of G and (C.2), the fifth one follows from the relation $\mathrm{Det}(AB) = \mathrm{Det}(A)\mathrm{Det}(B)$ for every A, B of appropriate size, and the second to last equality, i.e.,

$$
\lim_{t \to +0} t^{-1}(\mathrm{Det}(I + tA) - 1) = \mathrm{Tr}(A) \equiv \sum_{i=1}^{n} A_{ii}, \tag{C.6}
$$

is immediately given by recalling the meaning of $\mathrm{Det}(I + tA)$: $\mathrm{Det}(I + tA)$ is a polynomial of t which is the sum of products, taken along all diagonals of an $n \times n$ matrix and assigned certain signs, of the entries of $I + tA$. At every one of these diagonals, except for the main diagonal, there are at least two cells with entries proportional to t, so that the corresponding products do not contribute to the constant term and the terms linear in t in $\mathrm{Det}(I + tA)$ and thus do not affect the limit in (C.6). The only product which does contribute to the linear and the constant terms in $\mathrm{Det}(I + tA)$ is the product $(1 + tA_{11})(1 + tA_{22}) \cdots (1 + tA_{nn})$ coming from the main diagonal. Moreover, it is clear that in this product the constant term is 1, and the term linear in t is $t(A_{11} + \cdots + A_{nn})$, and thus (C.6) follows.

C.2 Higher-Order Derivatives

Let $f : \mathbf{R}^n \to \mathbf{R}^m$ be a mapping which is well defined and differentiable at every point x from an open set U. The Jacobian of this mapping $J(x)$ is a mapping from U to the space $\mathbf{R}^{m \times n}$ of $m \times n$ matrices, i.e., it is a mapping taking values in certain $\mathbf{R}^M$ ($M = mn$).

The derivative of this mapping $J(x)$, if it exists, is called the *second derivative* of f, which again is a mapping from $\mathbf{R}^n$ to a certain $\mathbf{R}^M$ and as such can be differentiable, and so on, so that we can speak about the second, the third, ... derivatives of a vector-valued function with vector argument. A *sufficient* condition for the existence of k derivatives of f in U is that f is C^k in U, i.e., that all partial derivatives of f of orders $\leq k$ exist and are continuous everywhere in U (cf. Theorem C.4).

The preceding description explains what it means that f has k derivatives in U. Note, however, that according to this description, the highest-order derivatives at a point x are just long vectors; say, the second-order derivative of a scalar function f of two variables is the Jacobian of the mapping $x \mapsto f'(x) : \mathbf{R}^2 \to \mathbf{R}^2$, i.e., a mapping from $\mathbf{R}^2$ to $\mathbf{R}^{2\times 2} = \mathbf{R}^4$; the third-order derivative of f is therefore the Jacobian of a mapping from $\mathbf{R}^2$ to $\mathbf{R}^4$, i.e., a mapping from $\mathbf{R}^2$ to $\mathbf{R}^{4\times 2} = \mathbf{R}^8$, and so on. The question which should be addressed now is: *What is a natural and transparent way to represent the highest-order derivatives?*

The answer is as follows:

(∗) Let $f : \mathbf{R}^n \to \mathbf{R}^m$ be C^k on an open set $U \subseteq \mathbf{R}^n$. The derivative of order $\ell \leq k$ of f, taken at a point $x \in U$, can be naturally identified with a function

$$D^\ell f(x)[\Delta x^1, \Delta x^2, \ldots, \Delta x^\ell]$$

of ℓ vector arguments $\Delta x^i \in \mathbf{R}^n$, $i = 1, \ldots, \ell$, and taking values in $\mathbf{R}^m$. This function is linear in every one of the arguments Δx^i, the other arguments being fixed, and is symmetric with respect to permutation of arguments $\Delta x^1, \ldots, \Delta x^\ell$. In terms of f, the quantity $D^\ell f(x)[\Delta x^1, \Delta x^2, \ldots, \Delta x^\ell]$ (full name: "the ℓth derivative (or differential) of f taken at a point x along the directions $\Delta x^1, \ldots, \Delta x^\ell$") is given by

$$D^\ell f(x)[\Delta x^1, \Delta x^2, \ldots, \Delta x^\ell]$$
$$= \frac{\partial^\ell}{\partial t_\ell \partial t_{\ell-1} \ldots \partial t_1}\bigg|_{t_1=\cdots=t_\ell=0} f(x + t_1 \Delta x^1 + t_2 \Delta x^2 + \cdots + t_\ell \Delta x^\ell). \quad \text{(C.7)}$$

The explanation for our claims is as follows. Let $f : \mathbf{R}^n \to \mathbf{R}^m$ be C^k on an open set $U \subseteq \mathbf{R}^n$.

1. When $\ell = 1$, (∗) states that the first-order derivative of f, taken at x, is a linear function $Df(x)[\Delta x^1]$ of $\Delta x^1 \in \mathbf{R}^n$, taking values in $\mathbf{R}^m$, and that the value of this function at every Δx^1 is given by the relation

$$Df(x)[\Delta x^1] = \frac{\partial}{\partial t_1}\bigg|_{t_1=0} f(x + t_1 \Delta x^1) \quad \text{(C.8)}$$

(cf. (C.2)), which is in complete accordance with what we already know about the derivative.

2. To understand what the second derivative is, let us take the first derivative $Df(x)[\Delta x^1]$, *temporarily fix the argument* Δx^1 and treat the derivative as a function of x. As a function of x, Δx^1 being fixed, the quantity $Df(x)[\Delta x^1]$ is again a mapping which maps U into $\mathbf{R}^m$ and is differentiable by Theorem C.4 (provided, of course, that $k \geq 2$). The derivative of this mapping will be a certain linear function of $\Delta x \equiv \Delta x^2 \in \mathbf{R}^n$, depending on x as

a parameter; and of course it depends on Δx^1 as a parameter as well. Thus, the derivative of $Df(x)[\Delta x^1]$ in x is a certain function

$$D^2 f(x)[\Delta x^1, \Delta x^2]$$

of $x \in U$ and $\Delta x^1, \Delta x^2 \in \mathbf{R}^n$ and taking values in $\mathbf{R}^m$. What we know about this function is that it is linear in Δx^2. In fact, it is also linear in Δx^1, since it is the derivative in x of a certain function (namely, $Df(x)[\Delta x^1]$) *linearly depending on the parameter* Δx^1, so that the derivative of the function *in x* is linear in the parameter Δx^1 as well (differentiation is a linear operation with respect to the function we are differentiating: summing up functions or multiplying them by real constants, we are summing up, respectively multiplying by the same constants, the derivatives). Thus, $D^2 f(x)[\Delta x^1, \Delta x^2]$ is linear in Δx^1 when x and Δx^2 are fixed, and is linear in Δx^2 when x and Δx^1 are fixed. Moreover, we have

$$D^2 f(x)[\Delta x^1, \Delta x^2] = \left.\frac{\partial}{\partial t_2}\right|_{t_2=0} Df(x + t_2 \Delta x^2)[\Delta x^1] \qquad \text{[cf. (C.8)]}$$

$$= \left.\frac{\partial}{\partial t_2}\right|_{t_2=0} \left.\frac{\partial}{\partial t_1}\right|_{t_1=0} f(x + t_2 \Delta x^2 + t_1 \Delta x^1) \qquad \text{[by (C.8)]}$$

$$= \left.\frac{\partial^2}{\partial t_2 \partial t_1}\right|_{t_1=t_2=0} f(x + t_1 \Delta x^1 + t_2 \Delta x^2) \qquad (C.9)$$

as claimed in (C.7) for $\ell = 2$. The only piece of information about the second derivative which is contained in (*) and is not justified yet is that $D^2 f(x)[\Delta x^1, \Delta x^2]$ is symmetric in $\Delta x^1, \Delta x^2$. This fact is readily given by the representation (C.7), since, as it is proven in Calculus, if a function ϕ possesses *continuous* partial derivatives of orders $\leq \ell$ in a neighborhood of a point, then these derivatives in this neighborhood are independent of the order in which they are taken. Then, it follows that

$$D^2 f(x)[\Delta x^1, \Delta x^2] = \left.\frac{\partial^2}{\partial t_2 \partial t_1}\right|_{t_1=t_2=0} \underbrace{f(x + t_1 \Delta x^1 + t_2 \Delta x^2)}_{:=\phi(t_1, t_2)} \qquad \text{[by (C.9)]}$$

$$= \left.\frac{\partial^2}{\partial t_1 \partial t_2}\right|_{t_1=t_2=0} \phi(t_1, t_2)$$

$$= \left.\frac{\partial^2}{\partial t_1 \partial t_2}\right|_{t_1=t_2=0} f(x + t_2 \Delta x^2 + t_1 \Delta x^1)$$

$$= D^2 f(x)[\Delta x^2, \Delta x^1]. \qquad \text{[once again by (C.9)]}$$

3. Now it is clear how to proceed: to define $D^3 f(x)[\Delta x^1, \Delta x^2, \Delta x^3]$, in the second-order derivative $D^2 f(x)[\Delta x^1, \Delta x^2]$ we fix the arguments $\Delta x^1, \Delta x^2$ and treat it as a function of x only, thus arriving at a mapping which maps U into $\mathbf{R}^m$ and depends on $\Delta x^1, \Delta x^2$ as on parameters (linearly in each one). Differentiating the resulting mapping in x, we arrive at a function $D^3 f(x)[\Delta x^1, \Delta x^2, \Delta x^3]$ which by construction is linear in each of the arguments $\Delta x^1, \Delta x^2, \Delta x^3$ and satisfies (C.7); the latter relation, owing to the calculus result on the symmetry of partial derivatives, implies that $D^3 f(x)[\Delta x^1, \Delta x^2, \Delta x^3]$ is symmetric in $\Delta x^1, \Delta x^2, \Delta x^3$. After we have at our disposal the third derivative $D^3 f$, we

can build from it in the fashion already explained the fourth derivative, and so on, until the kth derivative is defined.

Remark C.10 Since $D^\ell f(x)[\Delta x^1, \ldots, \Delta x^\ell]$ is linear in each Δx^i, we can expand the derivative in a multiple sum:

$$\Delta x^i = \sum_{j=1}^n \Delta x^i_j \, e_j,$$

$$\implies D^\ell f(x)[\Delta x^1, \ldots, \Delta x^\ell] = D^\ell f(x) \left[\sum_{j_1=1}^n \Delta x^1_{j_1} e_{j_1}, \ldots, \sum_{j_\ell=1}^n \Delta x^\ell_{j_\ell} e_{j_\ell} \right] \quad \text{(C.10)}$$

$$= \sum_{1 \le j_1, \ldots, j_\ell \le n} D^\ell f(x) \left[e_{j_1}, \ldots, e_{j_\ell} \right] \Delta x^1_{j_1} \cdots \Delta x^\ell_{j_\ell}.$$

What is the origin of the coefficients $D^\ell f(x)[e_{j_1}, \ldots, e_{j_\ell}]$? According to (C.7), one has

$$D^\ell f(x)[e_{j_1}, \cdots, e_{j_\ell}] = \frac{\partial^\ell}{\partial t_\ell \partial t_{\ell-1} \cdots \partial t_1} \bigg|_{t_1 = \cdots = t_\ell = 0} f(x + t_1 e_{j_1} + t_2 e_{j_2} + \cdots + t_\ell e_{j_\ell})$$

$$= \frac{\partial^\ell}{\partial x_{j_\ell} \partial x_{j_{\ell-1}} \cdots \partial x_{j_1}} f(x),$$

so that the coefficients in (C.10) are just the partial derivatives, of order ℓ, of f.

Remark C.11 An important particular case of the relation (C.7) is the one when the vectors $\Delta x^1, \Delta x^2, \ldots, \Delta x^\ell$ are all the same. Let $d := \Delta x^1 = \cdots = \Delta x^\ell$. According to (C.7), we have

$$D^\ell f(x)[d, d, \ldots, d] = \frac{\partial^\ell}{\partial t_\ell \partial t_{\ell-1} \cdots \partial t_1} \bigg|_{t_1 = \cdots = t_\ell = 0} f(x + t_1 d + t_2 d + \cdots + t_\ell d).$$

This relation can be interpreted as follows: consider the function

$$\phi(t) := f(x + td)$$

of a real variable t. Then, (check this!)

$$\phi^{(\ell)}(0) = \frac{\partial^\ell}{\partial t_\ell \partial t_{\ell-1} \cdots \partial t_1} \bigg|_{t_1 = \cdots = t_\ell = 0} f(x + t_1 d + t_2 d + \cdots + t_\ell d) = D^\ell f(x)[d, \ldots, d].$$

In fact, $D^\ell f(x)[d, \ldots, d]$ has a special name. It is the *ℓth directional derivative of f taken at x along the direction d*. To define this quantity, we pass from the function f of several variables to the univariate function $\phi(t) := f(x + td)$, which restricts f to the line passing through x and directed by d, and then take the "usual" derivative of order ℓ of the resulting function of a single real variable t at the point $t = 0$ (which corresponds to the point x on our line).

Representation of higher-order derivatives. The kth-order derivative $D^k f(x)[\cdot, \ldots, \cdot]$ of a C^k function $f : \mathbf{R}^n \to \mathbf{R}^m$ is a symmetric k-linear mapping on $\mathbf{R}^n$ taking values in $\mathbf{R}^m$ and depending on x as a parameter. Choosing coordinates in $\mathbf{R}^n$ in some way, we can represent such a mapping in the form

$$D^k f(x)[\Delta x^1, \ldots, \Delta x^k] = \sum_{1 \leq i_1, \ldots, i_k \leq n} \frac{\partial^k f(x)}{\partial x_{i_k} \partial x_{i_{k-1}} \cdots \partial x_{i_1}} (\Delta x^1)_{i_1} \cdots (\Delta x^k)_{i_k}.$$

We say that the derivative can be represented by a k-index collection of m-dimensional vectors $\frac{\partial^k f(x)}{\partial x_{i_k} \partial x_{i_{k-1}} \cdots \partial x_{i_1}}$. This collection, however, is a difficult-to-handle entity, so that such a representation does not help. There is, however, a case when the collection becomes an entity we know to handle; this is the case of the second-order derivative of a scalar function ($k=2, m=1$). In this case, the collection in question is just a symmetric matrix

$$H(x) := \left[\frac{\partial^2 f(x)}{\partial x_i \partial x_j} \right]_{1 \leq i,j \leq n}.$$

This matrix is called the *Hessian* of f at x. Note that

$$D^2 f(x)[\Delta x^1, \Delta x^2] = [\Delta x^1]^\top H(x) \Delta x^2.$$

C.2.1 Calculus of C^k Mappings

The calculus of C^k mappings can be summarized as follows:

Theorem C.12 (i) *Let U be an open set in $\mathbf{R}^n$, $f_1(\cdot), f_2(\cdot) : \mathbf{R}^n \to \mathbf{R}^m$ be C^k in U, and let real-valued functions $\lambda_1(\cdot), \lambda_2(\cdot)$ be C^k in U. Then, the function*

$$f(x) = \lambda_1(x) f_1(x) + \lambda_2(x) f_2(x)$$

is C^k in U.

(ii) *Let U be an open set in $\mathbf{R}^n$, let V be an open set in $\mathbf{R}^m$, let a mapping $f : \mathbf{R}^n \to \mathbf{R}^m$ be such that it is C^k in U and that $f(x) \in V$ for $x \in U$, and, finally, let a mapping $g : \mathbf{R}^m \to \mathbf{R}^p$ be C^k in V. Then, the superposition*

$$h(x) = g(f(x))$$

is C^k in U.

Remark C.13 For higher-order derivatives, in contrast with first-order derivatives, there is no simple "chain rule" for computing the derivative of superposition. For example, the second-order derivative of the superposition $h(x) = g(f(x))$ of two C^2-mappings is given by the formula

$$Dh(x)[\Delta x^1, \Delta x^2]$$
$$= Dg(f(x))[D^2 f(x)[\Delta x^1, \Delta x^2]] + D^2 g(x)[Df(x)[\Delta x^1], Df(x)[\Delta x^2]]$$

(check this!). We see that both the first- and the second-order derivatives of f and g contribute to the second-order derivative of the superposition h.

The only case when there does exist a simple formula for the higher-order derivatives of a superposition is the case *when the inner function is affine*: if $f(x) = Ax + b$ and $h(x) = g(f(x)) = g(Ax + b)$ with a C^ℓ mapping g, then

$$D^\ell h(x)[\Delta x^1, \ldots, \Delta x^\ell] = D^\ell g(Ax+b)[A\Delta x^1, \ldots, A\Delta x^\ell]. \tag{C.11}$$

C.2.2 Examples of Higher-Order Derivatives

We next present some examples of computing higher-order derivatives.

Example C.14 (Second-order derivative of an affine function) Consider $f(x) = a + b^\top x$. Then, of course, its second-order derivative is identically zero. Indeed, as we have seen in Example C.6,

$$Df(x)[\Delta x^1] = b^\top \Delta x^1$$

is independent of x, and therefore the derivative of $Df(x)[\Delta x^1]$ in x, which should give us the second derivative $D^2 f(x)[\Delta x^1, \Delta x^2]$, is zero. Clearly, the third, the fourth, etc., derivatives of an affine function are zero as well.

Example C.15 (Second-order derivative of a homogeneous quadratic form) Consider $f(x) = x^\top A x$, where A is a symmetric $n \times n$ matrix. As we saw in Example C.7,

$$Df(x)[\Delta x^1] = 2x^\top A \, \Delta x^1.$$

Differentiating this mapping in x, we get

$$\begin{aligned} D^2 f(x)[\Delta x^1, \Delta x^2] &= \lim_{t \to +0} t^{-1} \left(2(x + t\Delta x^2)^\top A \, \Delta x^1 - 2x^\top A \, \Delta x^1 \right) \\ &= 2(\Delta x^2)^\top A \, \Delta x^1. \end{aligned}$$

Hence,

$$\boxed{D^2 f(x)[\Delta x^1, \Delta x^2] = 2(\Delta x^2)^\top A \, \Delta x^1.}$$

Note that the second derivative of a quadratic form is independent of x. Consequently, the third, fourth, etc., derivatives of a quadratic form are identically zero.

Example C.16 (Second-order derivative of the log-det barrier) Consider $F(X) = \ln \mathrm{Det}(X)$. As we have seen, this function of an $n \times n$ matrix is well defined and differentiable on the set U of matrices with positive determinant (which is an open set in the space $\mathbf{R}^{n \times n}$ of $n \times n$ matrices). In fact, this function is C^∞ in U.

Let us compute its second-order derivative. Recall from Example C.9 that

$$DF(X)[\Delta X^1] = \mathrm{Tr}(X^{-1} \Delta X^1). \tag{C.12}$$

To differentiate the right-hand side in X, we will need the derivative of the mapping $G(X) = X^{-1}$ which is defined on the open set of nonsingular $n \times n$ matrices. Recall from Example C.8 that we have the relation

$$D(X^{-1})[\Delta X] = -X^{-1} \Delta X X^{-1} \quad (X \in \mathbf{R}^{n \times n}, \mathrm{Det}(X) \neq 0),$$

which is the "matrix extension" of the standard relation $(x^{-1})' = -x^{-2}, x \in \mathbf{R}$.

Now we are ready to compute the second derivative of the log-det barrier $F(X) = \ln \mathrm{Det}(X)$. Starting from its first derivative $DF(X)[\Delta X^1] = \mathrm{Tr}(X^{-1} \Delta X^1)$, and differentiating we obtain

$$D^2 F(X)[\Delta X^1, \Delta X^2] = \lim_{t \to +0} t^{-1} \left(\mathrm{Tr}\left((X + t\Delta X^2)^{-1} \Delta X^1 \right) - \mathrm{Tr}(X^{-1} \Delta X^1) \right)$$

$$= \lim_{t \to +0} \mathrm{Tr}\left(t^{-1} \left((X + t\Delta X^2)^{-1} \Delta X^1 - X^{-1} \Delta X^1 \right) \right)$$

$$= \lim_{t \to +0} \mathrm{Tr}\left(t^{-1} \left((X + t\Delta X^2)^{-1} - X^{-1} \right) \Delta X^1 \right)$$

$$= \mathrm{Tr}\left(\left(\lim_{t \to +0} t^{-1} \left((X + t\Delta X^2)^{-1} - X^{-1} \right) \right) \Delta X^1 \right)$$

$$= \mathrm{Tr}\left(\left(-X^{-1} \Delta X^2 X^{-1} \right) \Delta X^1 \right),$$

where the last equation follows from $D(X^{-1})[\Delta X^2] = -X^{-1} \Delta X^2 X^{-1}$ for all nonsingular $X \in \mathbf{R}^{n \times n}$. Hence, we arrive at the formula

$$\boxed{D^2 F(X)[\Delta X^1, \Delta X^2] = -\mathrm{Tr}(X^{-1} \Delta X^2 X^{-1} \Delta X^1) \quad [X \in \mathbf{R}^{n \times n}, \mathrm{Det}(X) > 0].}$$

Since $\mathrm{Tr}(AB) = \mathrm{Tr}(BA)$ (check this!) for all matrices A, B such that the product AB makes sense and is square, the right-hand side in the above formula is symmetric in $\Delta X^1, \Delta X^2$, as it should be for the second derivative of a C^2 function.

C.2.3 Taylor Expansion

Assume that $f : \mathbf{R}^n \to \mathbf{R}^m$ is C^k in an open neighborhood U of a point $\bar{x}$. The *Taylor expansion of order* k of f, built at the point $\bar{x}$, is the function defined as

$$F_k(x) := f(\bar{x}) + \frac{1}{1!} Df(\bar{x})[x - \bar{x}] + \frac{1}{2!} D^2 f(\bar{x})[x - \bar{x}, x - \bar{x}] \qquad (C.13)$$

$$+ \frac{1}{3!} D^3 f(\bar{x})[x - \bar{x}, x - \bar{x}, x - \bar{x}] + \cdots + \frac{1}{k!} D^k f(\bar{x}) \underbrace{[x - \bar{x}, \ldots, x - \bar{x}]}_{k \text{ times}}.$$

We are already acquainted with the Taylor expansion of order 1,

$$F_1(x) = f(\bar{x}) + Df(\bar{x})[x - \bar{x}],$$

this is the affine function of x which approximates "very well" $f(x)$ in a neighborhood of $\bar{x}$, namely, within an approximation error of $\bar{o}(|x - \bar{x}|)$. We next state that a similar fact is true for Taylor expansions of higher order.

Theorem C.17 *Let $f : \mathbf{R}^n \to \mathbf{R}^m$ be C^k in a neighborhood of $\bar{x}$, and let $F_k(x)$ be the Taylor expansion of f at $\bar{x}$ of degree k. Then,*

(i) *$F_k(x)$ is a vector-valued polynomial of full degree less than or equal to k, i.e., each coordinate of the vector $F_k(x)$ is a polynomial of $x_1, \ldots, x_n$, and the sum of the powers of the x_i in every term of this polynomial does not exceed k.*
(ii) *$F_k(x)$ approximates $f(x)$ in a neighborhood of $\bar{x}$ up to a remainder which is $\bar{o}(\|x - \bar{x}\|_2^k)$ as $x \to \bar{x}$. That is, for every $\epsilon > 0$, there exists $\delta > 0$ such that*

$$\|x - \bar{x}\|_2 \leq \delta \implies \|F_k(x) - f(x)\|_2 \leq \epsilon \|x - \bar{x}\|_2^k.$$

$F_k(\cdot)$ is the unique polynomial with components of full degree less than or equal to k which approximates f up to a remainder which is $\bar{o}(\|x - \bar{x}\|^k)$.

(iii) *The value and the derivatives of F_k of orders $1, 2, \ldots, k$, taken at $\bar{x}$, are the same as the value and the corresponding derivatives of f taken at the same point.*

As stated in Theorem C.17, $F_k(x)$ approximates $f(x)$ for x close to $\bar{x}$ up to a remainder which is $\bar{o}(|x - \bar{x}|^k)$. In many cases, it is not enough to know that the reminder is $\bar{o}(\|x - \bar{x}\|_2^k)$ – we need an explicit bound on this remainder. The standard bound of this type is as follows:

Theorem C.18 *Let k be a positive integer, and let $f : \mathbf{R}^n \to \mathbf{R}^m$ be C^{k+1} in a ball $B_r := B_r(\bar{x}) = \{x \in \mathbf{R}^n : \|x - \bar{x}\|_2 < r\}$ of radius $r > 0$ centered at a point $\bar{x}$. Assume that the directional derivatives of order $k + 1$, taken at every point of B_r along every $\|\cdot\|_2$-unit direction, do not exceed certain $L < \infty$:*

$$\|D^{k+1} f(x)[d, \ldots, d]\|_2 \leq L, \quad \forall (x \in B_r), \ \forall (d \in \mathbf{R}^n : \|d\|_2 = 1).$$

Then, the following holds for the Taylor expansion F_k of order k of f taken at $\bar{x}$:

$$\|f(x) - F_k(x)\|_2 \leq \frac{L \|x - \bar{x}\|_2^{k+1}}{(k+1)!}, \quad \forall x \in B_r.$$

Thus, in a neighborhood of $\bar{x}$ the remainder of the *k*th-*order* Taylor expansion, taken at $\bar{x}$, is of order $L\|x - \bar{x}\|_2^{k+1}$, where L is the maximal (over all unit directions and all points from the neighborhood) $\|\cdot\|_2$-magnitude of the directional derivatives *of order $k + 1$ of f*.

Appendix D

Prerequisites: Symmetric Matrices and Positive Semidefinite Cone

D.1 Symmetric Matrices

Let $\mathbf{S}^m$ be the space of symmetric $m \times m$ matrices, and $\mathbf{R}^{m \times n}$ be the space of rectangular $m \times n$ matrices with real entries. From the viewpoint of their linear structure (i.e., the operations of addition and multiplication by real numbers) $\mathbf{S}^m$ is just the arithmetic linear space $\mathbf{R}^{m(m+1)/2}$ of dimension $\frac{m(m+1)}{2}$: by arranging the elements of a symmetric $m \times m$ matrix X in a single column, say, in row-by-row order, you get a usual m^2-dimensional column vector; the multiplication of a matrix by a real number and the addition of matrices correspond to the same operations with these "representing vector(s)." When X runs through $\mathbf{S}^m$, the vector representing X runs through the $m(m+1)/2$-dimensional subspace of $\mathbf{R}^{m^2}$ consisting of vectors satisfying the symmetry condition i.e., the coordinates coming from pairs of entries in X that are symmetric to each other are equal to each other. Similarly, $\mathbf{R}^{m \times n}$ as a linear space is just $\mathbf{R}^{mn}$, and it is natural to equip $\mathbf{R}^{m \times n}$ with the inner product defined as the usual inner product of the vectors representing the matrices:

$$\langle X, Y \rangle := \sum_{i=1}^{m}\sum_{j=1}^{n} X_{ij} Y_{ij} = \mathrm{Tr}(XY^\top) \quad \left[= \sum_{i=1}^{m}\sum_{j=1}^{n} Y_{ij} X_{ij} = \mathrm{Tr}(YX^\top) \right].$$

Here, Tr stands for the *trace* – the sum of the diagonal elements of a (square) matrix. With this inner product (called the *Frobenius inner product*), $\mathbf{R}^{m \times n}$ becomes a legitimate Euclidean space, and in connection with this space we may use all notions based upon the Euclidean structure, e.g., the *(Frobenius) norm* of a matrix

$$\|X\|_2 := \sqrt{\langle X, X \rangle} = \sqrt{\sum_{i=1}^{m}\sum_{j=1}^{n} X_{ij}^2} = \sqrt{\mathrm{Tr}(XX^\top)},$$

and likewise the notions of orthogonality, orthogonal complement of a linear subspace, etc. The same applies to the space $\mathbf{S}^m$ equipped with the Frobenius inner product; of course, the Frobenius inner product of symmetric matrices can be written without the transposition sign:

$$\langle X, Y \rangle = \mathrm{Tr}(XY), \quad X, Y \in \mathbf{S}^m.$$

The following simple fact is very useful:

Fact D.1 *Let X, Y be rectangular matrices such that XY makes sense and is a square matrix. Then, $\mathrm{Tr}(YX)$ also makes sense and*

$$\mathrm{Tr}(XY) = \mathrm{Tr}(YX).$$

D.1 Symmetric Matrices

Here is another equally simple and useful fact.

Fact D.2 *If $X, Y \in \mathbf{R}^{m \times n}$, the Frobenius inner product of X and Y is equal to the Frobenius inner product of $X^\top$ and $Y^\top$:*

$$\operatorname{Tr}(XY^\top) = \operatorname{Tr}\left((X^\top)(Y^\top)^\top\right).$$

Moreover, when U is an orthogonal $m \times m$ matrix (i.e., $UU^\top = U^\top U = I_m$, which is the same $U^{-1} = U^\top$), and V is an orthogonal $n \times n$ matrix (i.e., $VV^\top = V^\top V = I_n$), the Frobenius inner product of UXV and UYV is the same as the Frobenius inner product of X and Y:

$$\operatorname{Tr}(XY^\top) = \operatorname{Tr}\left((UXV)(UYV)^\top\right).$$

D.1.1 Main Facts on Symmetric Matrices

Here, we will focus on the space $\mathbf{S}^m$ of symmetric matrices. We start with the following most important property of these matrices.

Theorem D.3 *[Eigenvalue decomposition] An $n \times n$ matrix A is symmetric if and only if it admits an orthonormal system of eigenvectors, i.e., there exists an orthonormal basis $\{e_1, \ldots, e_n\}$ such that*

$$Ae_i = \lambda_i e_i, \quad i = 1, \ldots, n, \tag{D.1}$$

holds for some reals λ_i.

In connection with Theorem D.3, let us recall the following notions and facts.

D.1.1.A Eigenvectors and Eigenvalues

Definition D.4 *[Eigenvector and eigenvalue] An* eigenvector *of an $n \times n$ matrix A is a nonzero vector e (real or complex) such that $Ae = \lambda e$ holds for some (real or complex) scalar λ. This scalar is called the* eigenvalue *of A corresponding to the eigenvector e.*

The eigenvalues of A are exactly the roots of the *characteristic polynomial* of A given by

$$\pi(z) = \operatorname{Det}(zI - A) = z^n + b_1 z^{n-1} + b_2 z^{n-2} + \cdots + b_n.$$

In the sequel, we denote by $\sigma(A)$ the *spectrum* of $A \in \mathbf{S}^n$, i.e., the set of all real numbers which are eigenvalues of the matrix. To avoid misunderstanding, let us stress that $\sigma(A)$ has the same number of points as the distinct eigenvalues of A, irrespectively of the multiplicities of these eigenvalues; for example, $\sigma(I_n) = \{1\}$.

Theorem D.3 states, in particular, that, for a symmetric matrix A, all eigenvalues are real, and the corresponding eigenvectors can be chosen to be real and to form an orthonormal basis in $\mathbf{R}^n$.

Remark D.5 When Q is a square and nonsingular $n \times n$ matrix, the *similarity transformation* $A \mapsto QAQ^{-1}$ preserves the characteristic polynomial (and therefore the eigenvalues). Indeed,

$$\operatorname{Det}(zI - QAQ^{-1}) = \operatorname{Det}(Q[zI - A]Q^{-1}) = \operatorname{Det}(Q)\operatorname{Det}(zI - A)\operatorname{Det}(Q^{-1})$$
$$= \operatorname{Det}(zI - A).$$

D.1.1.B Eigenvalue Decomposition of a Symmetric Matrix

We will give an alternative characterization of symmetric matrices. To this end, we start with a definition and a fact that are important in their own right.

Definition D.6 *[Orthogonal matrix] An $n \times n$ matrix U is called* orthogonal *if $U^{-1} = U^\top$.*

There are several equivalent characterizations of orthogonal matrices.

Fact D.7 *An $n \times n$ matrix U is orthogonal if and only if any (or all) of the following holds:*

- $U^{-1} = U^\top$;
- $U^\top U = I$;
- $UU^\top = I$;
- *the columns of U form an orthonormal basis in $\mathbf{R}^n$;*
- *the rows of U form an orthonormal basis in $\mathbf{R}^n$.*

Theorem D.3 admits an equivalent reformulation as follows (check the equivalence!).

Theorem D.8 *An $n \times n$ matrix A is symmetric if and only if it can be represented in the form*

$$A = U \Lambda U^\top, \tag{D.2}$$

where

- *U is an orthogonal matrix,*
- *Λ is the diagonal matrix with diagonal entries $\lambda_1, \ldots, \lambda_n$.*

The representation (D.2) with orthogonal U and diagonal Λ is called the *eigenvalue decomposition* of A. In such a representation,

- the columns of U form an orthonormal system of eigenvectors of A, and
- the diagonal entries in Λ are the eigenvalues of A corresponding to these eigenvectors.

D.1.1.C Vector of Eigenvalues

When speaking about the eigenvalues $\lambda_i(A)$ of a symmetric $n \times n$ matrix A, we always arrange them in non-increasing order:

$$\lambda_1(A) \geq \lambda_2(A) \geq \cdots \geq \lambda_n(A).$$

Moreover, we use $\lambda(A) \in \mathbf{R}^n$ to denote the vector of eigenvalues of A taken in the above order. We let $\lambda_{\max}(A)$ (and respectively $\lambda_{\min}(A)$) correspond to the largest (smallest) eigenvalue of A.

D.1.1.D Freedom in Eigenvalue Decomposition

It is important to note that in the eigenvalue decomposition (D.2), part of the data Λ, U is uniquely defined by A, while the other data admit a certain freedom. Specifically, the sequence $\lambda_1, \ldots, \lambda_n$ of eigenvalues of A (i.e., the diagonal entries of Λ) is exactly the sequence of roots of the characteristic polynomial of A (every root is repeated according to its multiplicity) and thus is uniquely defined by A (provided that we arrange the entries of the sequence in non-increasing order). On the other hand, it is possible that the columns of U are not uniquely defined by A. What is uniquely defined, are the *linear spans* $E(\lambda)$ of the columns of U corresponding to all eigenvalues equal to a certain λ, and such a linear span is just the *spectral subspace* $\{x : Ax = \lambda x\}$ of A corresponding to the eigenvalue λ. Given $A \in \mathbf{S}^n$, there are as many spectral subspaces of A as the number of different eigenvalues of A. Moreover, the spectral subspaces corresponding to the different eigenvalues of a given symmetric matrix are orthogonal to each other, and their sum is the entire space. When building an orthogonal matrix U for the eigenvalue decomposition (D.2), one chooses an orthonormal eigenbasis in the spectral subspace corresponding to the largest element in $\sigma(A)$ and makes the vectors of this basis the first column in U, then chooses an orthonormal basis in the spectral subspace corresponding to the second largest element of $\sigma(A)$ and makes the vectors from this basis the next column of U, and so on.

D.1.1.E "Simultaneous" Decomposition of Commuting Symmetric Matrices

Given a number of symmetric matrices $A_1, \ldots, A_k \in \mathbf{S}^n$, it is useful to know when they commute with each other, i.e., $A_i A_j = A_j A_i$ for all i, j. It turns out that there is a complete characterization of this property, given as follows.

Fact D.9 *The matrices $A_1, \ldots, A_k \in \mathbf{S}^n$ commute with each other ($A_i A_j = A_j A_i$ for all i, j) if and only if they can be simultaneously diagonalized, i.e., if there exist a single orthogonal matrix U and diagonal matrices $\Lambda_1, \ldots, \Lambda_k$ such that*

$$A_i = U \Lambda_i U^\top, \qquad i = 1, \ldots, k.$$

The proof of Fact D.9 relies on two simple facts that are important in their own right. We state these facts next.

Fact D.10 *Let $A, B \in \mathbf{R}^{n \times n}$ be commuting and let λ be a real eigenvalue of A. Then, the spectral subspace $E = \{x \in \mathbf{R}^n : Ax = \lambda x\}$ of A corresponding to λ is an invariant subspace for B (i.e., $Be \in E$ for every $e \in E$).*

Fact D.11 *If A is an $n \times n$ matrix and L is an invariant subspace for A (i.e., L is a linear subspace such that $Ae \in L$ whenever $e \in L$), then the orthogonal complement $L^\perp$ of L is an invariant subspace for the matrix $A^\top$. In particular, if A is symmetric and L is an invariant subspace for A, then $L^\perp$ is an invariant subspace for A as well.*

D.1.2 Variational Characterization of Eigenvalues

To put what follows into a proper perspective, let us ask ourselves the following simple question: *for a vector x with coordinates $x_1, \ldots, x_n$, let $x^{(k)}$ be the kth largest of the coordinates – the one which will be in the kth place when the entries in x are rearranged in non-ascending order. For example, for $x = [2; 1; 2]$ we have $x^{(1)} = 2$, $x^{(2)} = 2$, $x^{(3)} = 1$. Now, the question is how to characterize $x^{(k)}$ without referring to rearranging the entries in x.* The answer is as follows. Let us think about n stones with weights $x_1, x_2, \ldots, x_n$. Then $x^{(k)}$ can be described as follows: let us throw away from the collection of our n stones $k - 1$ of them and look at the largest among the weights of the remaining stones; let this largest weight be w. The value of w depends on which $k - 1$ of the stones were thrown away. Now let us take the *minimum* of these weights w we can obtain in this fashion over all possible ways to select $k - 1$ stones to throw away; this minimum is exactly $x^{(k)}$. By the way, it is immediately self-evident that when $x, y \in \mathbf{R}^n$ and $y \geq x$, then $y^{(k)} \geq x^{(k)}$ for every

$k \leq n$. However, even self-evident facts need proof, and here is the proof: when increasing the entries x_i in x to y_i (which precisely corresponds to increasing the weights of every one of our stones) we clearly cannot decrease the above w's (that is, the largest of the weights in groups obtained by throwing away a particular collection of $k - 1$ stones). And since none of the w's decreases, their minimum does not decrease either.

The variational characterization of the eigenvalues of a symmetric matrix is an extremely useful matrix version of the elementary considerations above.

Theorem D.12 *[Variational characterization of eigenvalues (VCE)] For any $A \in \mathbf{S}^n$, we have*

$$\lambda_\ell(A) = \min_{E \in \mathcal{E}_\ell} \max_{x \in E, x^\top x = 1} x^\top A x, \quad \ell = 1, \ldots, n, \qquad (D.3)$$

where $\mathcal{E}_\ell$ is the family of all linear subspaces in $\mathbf{R}^n$ of the dimension $n - \ell + 1$.

Using VCE, in order to get the largest eigenvalue $\lambda_1(A)$ we maximize the quadratic form $x^\top A x$ over the unit sphere $S = \{x \in \mathbf{R}^n : x^\top x = 1\}$, and the maximum is exactly $\lambda_1(A)$. To get the second largest eigenvalue $\lambda_2(A)$, we act as follows: we choose a linear subspace E of dimension $n - 1$ and maximize the quadratic form $x^\top A x$ over the cross-section of S with this subspace; the maximum value of the quadratic form depends on E, and we minimize this maximum over linear subspaces E of the dimension $n - 1$; the resulting value is exactly $\lambda_2(A)$. To get $\lambda_3(A)$, we replace in the latter construction subspaces of dimension $n - 1$ by those of dimension $n - 2$, and so on. In particular, the smallest eigenvalue $\lambda_n(A)$ is just the minimum, over all linear subspaces E of dimension $n - n + 1 = 1$, i.e., over all lines passing through the origin, of the quantities $x^\top A x$, where $x \in E$ has unit norm ($x^\top x = 1$). In other words, $\lambda_n(A)$ is just the minimum of the quadratic form $x^\top A x$ over the unit sphere S.

Proof of the VCE Let $e_1, \ldots, e_n$ be an orthonormal eigenbasis of A, i.e., $A e_\ell = \lambda_\ell(A) e_\ell$ and e_ℓ are unit vectors for all $1 \leq \ell \leq n$. For $1 \leq \ell \leq n$, let

$$F_\ell := \mathrm{Lin}\{e_1, \ldots, e_\ell\} \quad \text{and} \quad G_\ell := \mathrm{Lin}\{e_\ell, e_{\ell+1}, \ldots, e_n\};$$

where "Lin" refers to a linear span (see Definition A.5). Finally, for $x \in \mathbf{R}^n$, let $\xi(x)$ be the vector of coordinates of x in the orthonormal basis $e_1, \ldots, e_n$. Note that

$$x^\top x = \xi(x)^\top \xi(x),$$

since $\{e_1, \ldots, e_n\}$ is an orthonormal basis, and that

$$x^\top A x = x^\top A \sum_{i=1}^n \xi_i(x) e_i = x^\top \sum_{i=1}^n \lambda_i(A) \xi_i(x) e_i = \sum_{i=1}^n \lambda_i(A) \xi_i(x) \underbrace{(x^\top e_i)}_{=\xi_i(x)} \qquad (D.4)$$

$$= \sum_{i=1}^n \lambda_i(A) \xi_i^2(x).$$

Now, consider a fixed ℓ such that $1 \leq \ell \leq n$, and set $E = G_\ell$. Note that E is a linear subspace of dimension $n - \ell + 1$. Moreover, from (D.4) we deduce that

$$x^\top A x = \sum_{i=\ell}^n \lambda_i(A) \xi_i(x)^2$$

holds for any $x \in E$. Therefore,

$$\max_x \{x^\top A x : x \in E, x^\top x = 1\} = \max_\xi \left\{ \sum_{i=\ell}^n \lambda_i(A) \xi_i^2 : \sum_{i=\ell}^n \xi_i^2 = 1 \right\}$$
$$= \max_{\ell \leq i \leq n} \lambda_i(A) = \lambda_\ell(A).$$

Thus, for appropriately chosen $E \in \mathcal{E}_\ell$, the inner maximum on the right-hand side of (D.3) equals $\lambda_\ell(A)$. Hence, we have

$$\min_{E \in \mathcal{E}_\ell} \max_{x \in E, x^\top x = 1} x^\top A x \leq \lambda_\ell(A).$$

It remains to prove the reverse inequality. To this end, consider a linear subspace E of dimension $n-\ell+1$ and observe that its intersection with the linear subspace F_ℓ of dimension ℓ is nontrivial (indeed, $\dim E + \dim F_\ell = (n - \ell + 1) + \ell > n$, so that $\dim(E \cap F) > 0$ by the dimension formula). Consider any unit vector $y \in E \cap F_\ell$. Since y is a unit vector from F_ℓ, it admits a representation of the form

$$y = \sum_{i=1}^\ell \eta_i e_i \quad \text{with} \quad \sum_{i=1}^\ell \eta_i^2 = 1,$$

whence, by (D.4),

$$y^\top A y = \sum_{i=1}^\ell \lambda_i(A) \eta_i^2 \geq \min_{1 \leq i \leq \ell} \lambda_i(A) = \lambda_\ell(A).$$

Since $y \in E$, we conclude that

$$\max_{x \in E, x^\top x = 1} x^\top A x \geq y^\top A y \geq \lambda_\ell(A).$$

Since E is an arbitrary subspace from $\mathcal{E}_\ell$, we conclude that the right-hand side in (D.3) is $\geq \lambda_\ell(A)$. ∎

Our reasoning in the proof of the VCE uses the relation (D.4), which is a simple and useful byproduct. Therefore, we state it formally as a corollary.

Corollary D.13 *For any $A \in \mathbf{S}^n$, the quadratic form $x^\top A x$ is exactly equal to the weighted sum of squares of the coordinates $\xi_i(x)$ of x taken with respect to an orthonormal eigenbasis of A. Moreover, the weights in this sum are exactly the eigenvalues of A. That is,*

$$x^\top A x = \sum_i \lambda_i(A) \xi_i^2(x).$$

D.1.3 Corollaries of the VCE

The VCE admits a number of extremely important corollaries as follows.

D.1.2.A Eigenvalue Characterization of Positive (Semi)definite Matrices

Definition D.14 *[Positive definite matrix]* A matrix A is called positive definite [notation: $A \succ 0$], *if it is symmetric and $x^\top A x > 0$ holds for all $x \neq 0$.*

Definition D.15 *[Positive semidefinite matrix]* A matrix A is called positive semidefinite [notation: $A \succeq 0$], *if it is symmetric and $x^\top A x \geq 0$ holds for all x.*

From the VCE we arrive at the following eigenvalue characterization of positive (semi) definite matrices.

Proposition D.16 *A symmetric matrix A is positive semidefinite if and only if its eigenvalues are nonnegative. Moreover, A is positive definite if and only if all eigenvalues of A are positive.*

Proof Indeed, A is positive definite (positive semidefinite) if and only if the minimum value of $x^\top A x$ over the unit sphere is positive (nonnegative). Recall also that by the VCE, the minimum value of $x^\top A x$ over the unit sphere is exactly the minimum eigenvalue of A. ∎

D.1.2.B $\succeq$-Monotonicity of the Vector of Eigenvalues

We write $A \succeq B$ ($A \succ B$) to express that A, B are symmetric matrices of the same size such that $A - B$ is positive semidefinite (respectively, positive definite).

Proposition D.17 *Consider $A, B \in \mathbf{S}^n$. If $A \succeq B$, then $\lambda(A) \geq \lambda(B)$. Also, if $A \succ B$, then $\lambda(A) > \lambda(B)$.*

Proof Indeed, when $A \succeq B$, we have, of course,
$$\max_{x \in E : x^\top x = 1} x^\top A x \geq \max_{x \in E : x^\top x = 1} x^\top B x$$
for every linear subspace E, whence
$$\lambda_\ell(A) = \min_{E \in \mathcal{E}_\ell} \max_{x \in E : x^\top x = 1} x^\top A x \geq \min_{E \in \mathcal{E}_\ell} \max_{x \in E : x^\top x = 1} x^\top B x = \lambda_\ell(B), \quad \ell = 1, \ldots, n,$$
i.e., $\lambda(A) \geq \lambda(B)$. The case of $A \succ B$ can be considered similarly. ∎

D.1.2.C Eigenvalue Interlacement Theorem

This is an extremely important result, and we formulate it as follows.

Theorem D.18 *[Eigenvalue Interlacement Theorem] For $A \in \mathbf{S}^n$, let $\bar{A}$ be the angular $(n-k) \times (n-k)$ submatrix of A where $k \leq n$. Then, for every $\ell \leq n - k$, the ℓth eigenvalue of $\bar{A}$ separates the ℓth and the $(\ell + k)$th eigenvalues of A:*
$$\lambda_\ell(A) \geq \lambda_\ell(\bar{A}) \geq \lambda_{\ell+k}(A). \tag{D.5}$$

Proof Recall that by the VCE,
$$\lambda_\ell(\bar{A}) = \min_{E \in \bar{\mathcal{E}}_\ell} \max_{x \in E : x^\top x = 1} x^\top A x,$$
where $\bar{\mathcal{E}}_\ell$ is the family of all linear subspaces of the dimension $n - k - \ell + 1$ contained in the linear subspace $\{x \in \mathbf{R}^n : x_{n-k+1} = x_{n-k+2} = \cdots = x_n = 0\}$. Since $\bar{\mathcal{E}}_\ell \subset \mathcal{E}_{\ell+k}$, we have
$$\lambda_\ell(\bar{A}) = \min_{E \in \bar{\mathcal{E}}_\ell} \max_{x \in E : x^\top x = 1} x^\top A x \geq \min_{E \in \mathcal{E}_{\ell+k}} \max_{x \in E : x^\top x = 1} x^\top A x = \lambda_{\ell+k}(A).$$
We have proved the left-hand inequality in (D.5). Applying this inequality to the matrix $-A$, we get
$$-\lambda_\ell(\bar{A}) = \lambda_{n-k-\ell}(-\bar{A}) \geq \lambda_{n-\ell}(-A) = -\lambda_\ell(A),$$
or, which is the same, $\lambda_\ell(\bar{A}) \leq \lambda_\ell(A)$. This then proves the first inequality in (D.5). ∎

D.1.4 Spectral Norm and Lipschitz Continuity of Vector of Eigenvalues

Spectral and induced norms of matrices. A useful example of a norm (to refresh your memory concerning norms, see section B.1) is the *spectral norm* of an $m \times n$ matrix A:

$$\|A\| = \max_{x: \|x\|_2 \leq 1} \|Ax\|_2.$$

The spectral norm is indeed a special case of the *induced norm of a linear mapping*. $x \mapsto y = Ax : E \to F$, where E and F are (finite-dimensional) linear spaces equipped with norms $\|\cdot\|_E$ and $\|\cdot\|_F$ respectively. The norm $\|A\|_{F,E}$ of A induced by these two norms is defined as

$$\|A\|_{F,E} := \max_{x} \{\|Ax\|_F : \|x\|_E \leq 1\}.$$

The definition of an induced norm immediately implies (check this!) that, first, for all $x \in E$, $y \in F$ we have

$$\|Ax\|_F \leq \|A\|_{F,E} \|x\|_E,$$

and, second, the latter inequality becomes an equality for properly selected nonzero $x \in E$. In other words, the induced norm is the largest "amplification factor" – the largest factor by which the norm $\|\cdot\|_F$ of Ax can be larger than the norm $\|\cdot\|_E$ of a nonzero vector x.

Clearly, when $E = \mathbf{R}^n$, $F = \mathbf{R}^m$, and $\|\cdot\|_E$, $\|\cdot\|_F$ are the standard Euclidean norms on the respective spaces, the induced norm of A is exactly the spectral norm of A.

Fact D.19 *(i) Let E and F be finite-dimensional linear spaces. The induced norm $\|\cdot\|_{F,E}$ is indeed a norm on the space $\mathrm{Lin}(E, F)$ of linear mappings from E to F.*
(ii) Let E, F, G be finite-dimensional linear spaces equipped with the norms $\|\cdot\|_E$, $\|\cdot\|_F$, $\|\cdot\|_G$ respectively. Let $y = Ax : E \to F$ and $z = By : F \to G$ be linear mappings. Then,

$$\|BA\|_{G,E} \leq \|B\|_{G,F} \|A\|_{F,E}.$$

We also have the following useful properties of the spectral norm $\|\cdot\|$ on $\mathbf{R}^{m \times n}$.

Fact D.20 *Let $\|\cdot\|$ be the spectral norm on $\mathbf{R}^{m \times n}$.*

(i) For any $A \in \mathbf{R}^{m \times n}$, we have

$$\|A\| = \max_{x,y} \left\{ y^\top A x : \|x\|_2 \leq 1, \|y\|_2 \leq 1 \right\},$$

hence also $\|A\| = \|A^\top\|$.
(ii) For any $A \in \mathbf{S}^n$, we have $\|A\| = \max\{|\lambda_{\max}(A)|, |\lambda_{\min}(A)|\}$. Moreover, for any $A \in \mathbf{R}^{m \times n}$ we have

$$\|A\|^2 = \|A^\top A\| = \lambda_{\max}(A^\top A) = \lambda_{\max}(AA^\top) = \|AA^\top\| = \|A^\top\|^2.$$

Lipschitz continuity of the vector of eigenvalues. Another useful consequence of VCE is the Lipschitz continuity of the vector of eigenvalues of a matrix as a function of the matrix.

Fact D.21 *The vector-valued function $A \mapsto \lambda(A) : \mathbf{S}^n \to \mathbf{R}^n$ is Lipschitz continuous; specifically, denoting by $\|\cdot\|$ the spectral norm, for all $k \leq n$ we have*

$$|\lambda_k(A) - \lambda_k(A')| \leq \|A - A'\|, \quad \forall (A, A' \in \mathbf{S}^n).$$

We also have the following immediate consequence of Fact D.21.

Corollary D.22 *Let S be a subset of $\mathbf{R}^n$, and $\mathcal{S}$ be the set of all matrices $A \in \mathbf{S}^n$ with $\lambda(A) \in S$. Whenever S is open, or closed, or bounded, so is $\mathcal{S}$.*

D.1.5 Functions of Symmetric Matrices

Let A be a symmetric $m \times m$ matrix, and f be a real-valued function defined on a subset of real line containing the spectrum $\sigma(A)$ of A. We define the $m \times m$ symmetric *matrix* $f(A)$ as follows:

Let $\sigma(A) = \{\mu_1, \ldots, \mu_s\}$, and let $\mathbf{R}^m = E_1 + \cdots + E_s$ be the decomposition of $\mathbf{R}^m$ into the sum of mutually orthogonal spectral subspaces of A, i.e., $E_i := \{x \in \mathbf{R}^m : Ax = \mu_i x\}$. For the matrix $f(A)$, every E_i is an invariant subspace, and $x \in E_i$ implies $Ax = f(\mu_i)x$ for all $i \leq s$.

Equivalently: given an eigenvalue decomposition of A, i.e.,
$$A = U \operatorname{Diag}\{\lambda_1, \ldots, \lambda_m\} U^\top,$$
we have
$$f(A) = U \operatorname{Diag}\{f(\lambda_1), \ldots, f(\lambda_m)\} U^\top.$$

From this definition it follows immediately that $f(A) = g(A)$ whenever f and g coincide with each other on the spectrum of A.

Moreover, it can immediately be seen (check this!) that the resulting "calculus of functions of a (fixed) symmetric matrix A" obeys very natural rules:

1. $f(\cdot) \mapsto f(A)$ is a mapping from the algebra $\mathbf{R}(\sigma(A))$ of real-valued functions on the subset $\sigma(A)$ of the real axis into the space $\mathbf{S}^m$ of $m \times m$ symmetric matrices which preserves linear operations and multiplication. That is, for any $f_1, \ldots, f_k \in \mathbf{R}(\sigma(A))$ and any $a_1, \ldots, a_k \in \mathbf{R}$, we have
$$[a_1 f_1(\cdot) + \cdots + a_k f_k(\cdot)](A) = a_1 f_1(A) + \cdots + a_k f_k(A)$$
and
$$[f_1(\cdot) \cdot f_2(\cdot)](A) = f_1(A) f_2(A).$$

   In particular, all matrices of the form $f(A)$, $f \in \mathbf{R}(\sigma(A))$, commute with each other.

   Besides this, if f is a real-valued function on $\sigma(A)$, then $\sigma(f(A)) = f(\sigma(A)) := \{f(s) : s \in \sigma(A)\}$, and if g is a real-valued function on $f(\sigma(A))$ and $g \circ f(s) = g(f(s))$, where $s \in \sigma(A)$, is the composition of g and f, then
$$(g \circ f)(A) = g(f(A)).$$

2. When $f(\cdot) \equiv 1$, we have $f(A) = I_m$, and when $f(x) \equiv x$, we have $f(A) = A$. More generally, whenever f is a real algebraic polynomial
$$f(x) = a_0 + \sum_{i=1}^{k} a_i x^i,$$

D.1 Symmetric Matrices

we have

$$f(A) = a_0 I_n + \sum_{i=1}^{k} a_i A^i.$$

3. The mapping $f(\cdot) \to f(A)$ "preserves nonnegativity," i.e., the matrix $f(A)$ is *positive semidefinite* whenever $f(\cdot)$ is nonnegative on $\sigma(A)$.
4. When a sequence of functions $f_i(\cdot) \in \mathbf{R}(\sigma(A))$ converges to $f(\cdot)$ pointwise on $\sigma(A)$ as $i \to \infty$, the matrices $f_i(A)$ converge to $f(A)$ as $i \to \infty$. Moreover, for functions f, g which are real-valued on $\sigma(A)$ we have

$$\|f(A) - g(A)\| \leq \max_{s \in \sigma(A)} |f(s) - g(s)|,$$

where $\|\cdot\|$ is the spectral norm.

Illustration: Square root of a positive semidefinite matrix. When A is a positive semidefinite matrix, its *matrix square root* $A^{1/2}$ is, by definition, $f(A)$, where $f(\mu) = \mu^{1/2}$ is the usual square root on the nonnegative ray. From the preceding rules, it follows that $A^{1/2}$ is symmetric and positive semidefinite, and it squares to A: $(A^{1/2})^2 = A$.

Similarly to the arithmetic square root of a nonnegative real number a, i.e., the unique *nonnegative* real number which squares to a, the square root $A^{1/2}$ of a positive semidefinite matrix A is the unique *positive semidefinite* matrix which squares to A.

Fact D.23 *Let A be a positive semidefinite $m \times m$ matrix, and B be a positive semidefinite matrix such that $B^2 = A$. Then, $B = A^{1/2}$.*

Fact D.24 *Let $f(x)$ be a continuously differentiable real-valued function on an interval (a, b) (where $-\infty \leq a < b \leq +\infty$) of the real axis. The function $\phi(X) = \mathrm{Tr}(f(X))$ is continuously differentiable on the open set $\mathcal{D}$ of $\mathbf{S}^m$ composed of all matrices with spectrum from (a, b), and*

$$\left.\frac{d}{dt}\right|_{t=0} \phi(X + tH) = \mathrm{Tr}(f'(X)H) \quad \forall X \in \mathcal{D}, H \in \mathbf{S}^m.$$

In other words, $\phi(\cdot)$ is continuously differentiable on the open set $\mathcal{D}$ and its gradient, taken w.r.t. the Frobenius Euclidean structure on $\mathbf{S}^m$, is $\nabla \phi(X) = f'(X)$.

Hint: In order to prove Fact D.24, it makes sense to verify its validity for algebraic polynomials and then use the fact that any continuously differentiable real-valued function on an interval (a, b) of the real axis is the limit, in the sense of uniform convergence along with the first-order derivative on compact subsets of (a, b), of a sequence of polynomials.

Remark D.25 Note that when a complex-valued function $f(z)$ of a complex-valued variable z can be represented as the sum of everywhere-converging power series (these functions are called *entire*),

$$f(z) = \sum_{\nu=0}^{\infty} c_\nu z^\nu \quad [\text{where } c_\nu \in \mathbf{C}]$$

we can "extend" f to the function

$$f(Z) = \sum_{\nu=0}^{\infty} c_\nu Z^\nu$$

defined on the space $\mathbf{C}^{n\times n}$ of $n \times n$ matrices with complex entries and taking values in the same $\mathbf{C}^{n\times n}$, ensuring that for entire functions f, g we have

$$f(z) \equiv z^\nu \implies f(Z) = Z^\nu, \quad \nu = 0, 1, \ldots,$$
$$(f+g)(Z) \equiv f(Z) + g(Z),$$
$$(f \cdot g)(Z) \equiv f(Z)g(Z),$$
$$(\lambda f)(Z) \equiv \lambda f(Z), \ \lambda \in \mathbf{C}.$$

In particular, for fixed Z, the matrices $f(Z)$ stemming from various entire functions f commute with each other.

Whenever an entire function f is real-valued on the real axis (this is the case if and only if the coefficients in the power series representing f are real), the matrix $f(Z)$, for a real $n \times n$ matrix Z, is also real, and in the case of a real symmetric Z $f(Z)$ is real symmetric as well. It is immediately seen that when f is an entire function real-valued on the real axis and Z is a real symmetric matrix, both definitions of $f(Z)$ – our original definition "keep the eigenbasis intact and replace eigenvalues λ with eigenvalues $f(\lambda)$" (this recipe works whenever f is a real-valued function well defined on the spectrum of Z) and the new definition via power series expansion – result in the same matrix $f(Z)$.

D.1.5.A Continuity of $f(Z)$ in Z

So far we have been interested in what happens to the matrix $f(Z)$ of a fixed symmetric Z when f changes. Now let us ask ourselves what happens to this matrix when Z changes. We are about to consider the simplest question – the continuity of $f(Z)$ in Z. Here is all we need in this respect:

Proposition D.26 *Let m be a positive integer, and let Σ be a nonempty closed subset of the real axis. Suppose $f(\cdot) : \Sigma \to \mathbf{R}$ is a continuous function. Then, the set $\mathcal{Z} := \mathcal{Z}_\Sigma$ composed of all symmetric $m \times m$ matrices with spectrum from Σ is closed, and the function $f(Z)$ of $Z \in \mathcal{Z}$ is continuous on $\mathcal{Z}$.*

Proof By Fact D.21, the vector $\lambda(Z)$ of eigenvalues of $Z \in \mathbf{S}^m$ is a Lipschitz continuous function of Z, and as Σ is closed we deduce that $\mathcal{Z}$ is closed. To prove continuity of $f(Z)$ in $Z \in \mathcal{Z}$, it is clearly sufficient to establish this fact when Σ is nonempty, closed and *bounded*. Thus, let Σ be nonempty closed and bounded, f be continuous on Σ, and let $Z_i \in \mathcal{Z}$ converge, as $i \to \infty$, to $\overline{Z}$; we should verify that $f(Z_i) \to f(\overline{Z})$ as $i \to \infty$. It is well known that a continuous function on a nonempty compact subset of the real axis can be approximated, within any desired accuracy in the uniform norm, by an algebraic polynomial. Therefore, given $\epsilon > 0$, we can find a univariate algebraic polynomial $p(s)$ such that $|p(s) - f(s)| \leq \epsilon/3$ for all $s \in \Sigma$, implying, by item 4 of Section D.1.5, that $\|p(Z) - f(Z)\| \leq \epsilon/3$ for all $Z \in \mathcal{Z}$, where $\|\cdot\|$ is the spectral norm. Since p is an algebraic polynomial, we clearly have $p(Z_i) \to p(\overline{Z})$ as $i \to \infty$, implying that there exists

D.2 Positive Semidefinite Matrices and Positive Semidefinite Cone

D.2.1 Positive Semidefinite Matrices

Recall that an $n \times n$ matrix A is called *positive semidefinite* [notation: $A \succeq 0$], if A is symmetric and produces a nonnegative quadratic form:

$$A \succeq 0 \iff \left\{ A = A^\top \quad \text{and} \quad x^\top A x \geq 0 \quad \forall x \right\}.$$

A is called *positive definite* [notation: $A \succ 0$], if it is positive semidefinite and the corresponding quadratic form is positive outside the origin:

$$A \succ 0 \iff \left\{ A = A^\top \quad \text{and} \quad x^\top A x > 0 \quad \forall x \neq 0 \right\}.$$

The class of positive semidefinite matrices plays an important role in Mathematics. Thus, we next list a number of equivalent definitions of a positive semidefinite matrix.

Theorem D.27 *For any $A \in \mathbf{S}^n$, the following properties of A are equivalent to each other:*

(i) $A \succeq 0$,
(ii) $\lambda(A) \geq 0$,
(iii) $A = D^\top D$ for a certain rectangular matrix D,
(iv) $A = \Delta^\top \Delta$ for a certain upper triangular $n \times n$ matrix Δ,
(v) $A = B^2$ for a certain symmetric matrix B,
(vi) $A = B^2$ for a certain $B \succeq 0$.

Moreover, for any $A \in \mathbf{S}^n$, we also have the following properties equivalent to each other:

(i′) $A \succ 0$,
(ii′) $\lambda(A) > 0$,
(iii′) $A = D^\top D$ for a certain rectangular matrix D of rank n,
(iv′) $A = \Delta^\top \Delta$ for a certain nonsingular upper triangular $n \times n$ matrix Δ,
(v′) $A = B^2$ for a certain nonsingular symmetric matrix B,
(vi′) $A = B^2$ for a certain $B \succ 0$.

Proof (i) $\iff$ (ii): This equivalence is given by Proposition D.16.

(ii) $\implies$ (vi): Suppose that we are in case (ii). Let $A = U \Lambda U^\top$ be the eigenvalue decomposition of A, so that U is orthogonal and Λ is diagonal with nonnegative diagonal entries $\lambda_i(A)$ (as we are in case (ii)). Let $\Lambda^{1/2}$ be the diagonal matrix with diagonal entries $\lambda_i^{1/2}(A)$. Note that $(\Lambda^{1/2})^2 = \Lambda$. Then, the matrix $B = U \Lambda^{1/2} U^\top$ is symmetric with nonnegative eigenvalues $\lambda_i^{1/2}(A)$, and thus $B \succeq 0$ by Proposition D.16. Moreover,

$$B^2 = U \Lambda^{1/2} \underbrace{U^\top U}_{=I} \Lambda^{1/2} U^\top = U (\Lambda^{1/2})^2 U^\top = U \Lambda U^\top = A,$$

as required in (vi).

(vi) $\Longrightarrow$ (v): Evident.

(v) $\Longrightarrow$ (iv): Suppose $A = B^2$ for a certain symmetric B. Let b_i be the ith column of B for all $i \leq n$. Applying the Gram–Schmidt orthogonalization process (see the proof of Theorem A.31(ii)), we can find an orthonormal system of vectors $u_1, \ldots, u_n$ and lower triangular matrix L such that

$$b_i = \sum_{j=1}^{i} L_{ij} u_j,$$

or, which is the same, $B^\top = LU$, where U is the orthogonal matrix with rows $u_1^\top, \ldots, u_n^\top$. Then, we have $A = B^2 = B^\top (B^\top)^\top = LUU^\top L^\top = LL^\top$. Recalling that L is a lower triangular matrix, we see that $A = \Delta^\top \Delta$, where the matrix $\Delta = L^\top$ is upper triangular.

(iv) $\Longrightarrow$ (iii): Evident.

(iii) $\Longrightarrow$ (i): If $A = D^\top D$, then $x^\top A x = (Dx)^\top (Dx) \geq 0$ for all x.

We have proved the equivalence of properties (i)–(vi). Slightly modifying the reasoning (try this!), one can prove the equivalence of the properties (i')–(vi'). ∎

Remark D.28 (i) [Checking positive semidefiniteness] Given a matrix $A \in \mathbf{S}^n$, we can check whether it is positive semidefinite by a purely algebraic finite algorithm (the so-called *Lagrange diagonalization of a quadratic form*), which requires at most $O(n^3)$ arithmetic operations. The positive definiteness of a matrix can be checked also by the *Cholesky factorization algorithm* which finds the decomposition in (iv'), if it exists, in approximately $\frac{1}{6}n^3$ arithmetic operations.

There exists another useful algebraic criterion (*Sylvester's criterion*) for the positive semidefiniteness of a matrix. According to this criterion, a symmetric matrix A is positive definite if and only if all its *angular (leading principal) minors* are positive, and A is positive semidefinite if and only if all its *principal minors* are nonnegative. For example, a symmetric 2×2 matrix

$$A = \begin{bmatrix} a & b \\ b & c \end{bmatrix}$$

is positive semidefinite if and only if $a \geq 0$, $c \geq 0$, and $\mathrm{Det}(A) := ac - b^2 \geq 0$. Another consequence is that when $A \succeq 0$, we have $A_{ii} A_{jj} \geq A_{ij}^2$ since the principal 2×2 minors of A should be nonnegative.

(ii) [Square root of a positive semidefinite matrix] By the first chain of equivalences in Theorem D.27, a symmetric matrix A is $\succeq 0$ if and only if A is the square of a positive semidefinite matrix B. The latter matrix is uniquely defined by $A \succeq 0$ and is called the *square root* of A [notation: $A^{1/2}$]; see section D.1.5.

A symmetric matrix A is called *negative semidefinite* [notation: $A \preceq 0$] if its negative, i.e., the matrix $-A$, is positive semidefinite. Similarly, A is called *negative definite* [notation: $A \prec 0$] if $-A$ is positive definite.

D.2.2 The Positive Semidefinite Cone

When adding symmetric matrices or multiplying them by real numbers, we add, respectively multiply by real numbers, the corresponding quadratic forms. Hence, we arrive at the following basic fact.

D.2 Positive Semidefinite Matrices and Positive Semidefinite Cone

Fact D.29 *Let $A, B \in \mathbf{S}^n$ be positive semidefinite. Then, $A + B$ is also positive semidefinite. For any $\alpha \in \mathbf{R}_+$, the matrix αA is also positive semidefinite.*

Thus, Fact D.29 is the same as the following fact, stated in section 1.2.4.

Fact D.30 *The set of $n \times n$ positive semidefinite matrices form a cone $\mathbf{S}^n_+$ in the Euclidean space $\mathbf{S}^n$ of symmetric $n \times n$ matrices equipped with the Frobenius inner product $\langle A, B \rangle = \mathrm{Tr}(AB) = \sum_{i,j} A_{ij} B_{ij}$.*

The cone $\mathbf{S}^n_+$ is called the *positive semidefinite cone* of size n. It is immediately seen that the cone $\mathbf{S}^n_+$ is quite nice. Specifically, it satisfies the following properties:

- $\mathbf{S}^n_+$ is closed: the limit of a converging sequence of positive semidefinite matrices is positive semidefinite.
- $\mathbf{S}^n_+$ is pointed: the only $n \times n$ matrix A such that both A and $-A$ are positive semidefinite is the $n \times n$ matrix of all zeros.
- $\mathbf{S}^n_+$ possesses a nonempty interior (the subset composed of all interior points of $\mathbf{S}^n_+$) which is exactly the set of positive definite matrices.

In fact, $\mathbf{S}^n_+$ is a *regular* cone; see section 16.3. Note that the relation $A \succeq B$ means exactly that $A - B \in \mathbf{S}^n_+$, while $A \succ B$ is equivalent to $A - B \in \mathrm{int}\, \mathbf{S}^n_+$. The "matrix inequalities" $A \succeq B$ ($A \succ B$) match the standard properties of the usual scalar inequalities, e.g.,

$$A \succeq A \qquad \text{[reflexivity]}$$
$$A \succeq B,\ B \succeq A \implies A = B \qquad \text{[antisymmetry]}$$
$$A \succeq B,\ B \succeq C \implies A \succeq C \qquad \text{[transitivity]}$$
$$A \succeq B,\ C \succeq D \implies A + C \succeq B + D \qquad \text{[compatibility with linear operations, I]}$$
$$A \succeq B,\ \lambda \geq 0 \implies \lambda A \succeq \lambda B \qquad \text{[compatibility with linear operations, II]}$$
$$A_i \succeq B_i,\ A_i \to A,\ B_i \to B \text{ as } i \to \infty \implies A \succeq B \qquad \text{[closedness]}$$

with evident modifications when $\succeq$ is replaced with $\succ$, for example,

$$A \succeq B,\ C \succ D \implies A + C \succ B + D,$$

etc. Along with these standard properties of inequalities, the inequality $\succeq$ possesses the following nice additional property.

Fact D.31 *For any $A, B \in \mathbf{S}^n$ such that $A \succeq B$, and for any $V \in \mathbf{R}^{n \times m}$, we always have*

$$V^\top A V \succeq V^\top B V.$$

Another important property of the positive semidefinite cone is that its dual cone is equal to itself, i.e., the positive semidefinite cone is *self-dual*.

Theorem D.32 *A matrix X has nonnegative Frobenius inner products with all positive semidefinite matrices if and only if X itself is positive semidefinite.*

Proof We first prove the "if" part. Assume that $X \succeq 0$, and let us prove that then $\mathrm{Tr}(XY) \geq 0$ for every $Y \succeq 0$. Using the eigenvalue decomposition of X, we can write it as

$$X = \sum_{i=1}^n \lambda_i(X) e_i e_i^\top,$$

where e_i are the orthonormal eigenvectors of X. Then, we arrive at

$$\text{Tr}(XY) = \text{Tr}\left(\left(\sum_{i=1}^{n} \lambda_i(X) e_i e_i^\top\right) Y\right) = \sum_{i=1}^{n} \lambda_i(X) \text{Tr}(e_i e_i^\top Y)$$

$$= \sum_{i=1}^{n} \lambda_i(X) \text{Tr}(e_i^\top Y e_i), \tag{D.6}$$

where the last equality is given by Fact D.1. Recall that the e_i are just vectors, and thus $\text{Tr}(e_i^\top Y e_i) = e_i^\top Y e_i$. Also, as $Y \succeq 0$, we have $e_i^\top Y e_i \geq 0$ for all i. Moreover, since $X \succeq 0$, by Proposition D.16 we have $\lambda_i(X) \geq 0$, and thus we conclude that

$$\text{Tr}(XY) = \sum_{i=1}^{n} \lambda_i(X) \text{Tr}(e_i^\top Y e_i) \geq 0.$$

In order to prove the "only if" part, suppose that X is such that $\text{Tr}(XY) \geq 0$ for all matrices $Y \succeq 0$. Note that for every vector y, the matrix $Y = yy^\top$ is positive semidefinite (Theorem D.27(iii)). Then, for any $y \in \mathbf{R}^n$ we have $0 \leq \text{Tr}(Xyy^\top) = \text{Tr}(y^\top X y) = y^\top X y$, where the first equality follows from Fact D.1. Thus, $X \succeq 0$ as desired. ■

Graphical illustration. The single-dimensional positive semidefinite cone $\mathbf{S}_+^1$ is just the nonnegative ray, i.e.,

$$\mathbf{S}_+^1 = \{x \in \mathbf{R} : x \geq 0\} = \mathbf{R}_+.$$

The cone

$$\mathbf{S}_+^2 = \left\{ \begin{bmatrix} x & y \\ y & z \end{bmatrix} \in \mathbf{S}^2 : \begin{bmatrix} x & y \\ y & z \end{bmatrix} \succeq 0 \right\}$$

is just the set

$$\left\{(x,y,z) \in \mathbf{R}^3 : x \geq 0,\ z \geq 0,\ xz \geq y^2\right\}$$

$$= \left\{(x,y,z) \in \mathbf{R}^3 : x + z \geq \sqrt{(x-z)^2 + 4y^2}\right\}.$$

After a linear invertible substitution of coordinates $w = x+z$, $u = x-z$, $v = 2y$, this last set becomes the *second-order*, or *Lorentz*, cone in $\mathbf{R}^3$, i.e.,

$$\left\{(u,v,w) \in \mathbf{R}^3 : w \geq \sqrt{u^2 + v^2}\right\}.$$

The smallest "genuine" – not belonging to a simpler family of cones – semidefinite cone is the cone $\mathbf{S}_+^3$ of positive semidefinite 3×3 matrices. This cone, however, lives in six-dimensional linear space $\mathbf{S}^3$, so that we cannot draw it in three dimensions. What we can draw are three-dimensional cross-sections of $\mathbf{S}_+^3$ by various three-dimensional planes. Several cross-sections of this type are shown in Figure D.1 illustrating that the range of convex sets that can be obtained by intersecting even the simplest semidefinite cone with affine planes is quite wide. In this respect it should be noted that the problems of minimizing linear objectives over the intersection of positive semidefinite cones and affine planes have

D.2 Positive Semidefinite Matrices and Positive Semidefinite Cone   417

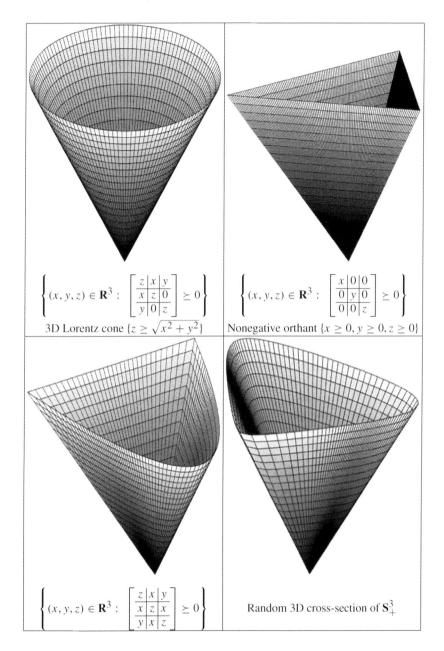

Figure D.1 Several three-dimensional cross-sections of $\mathbf{S}^3_+$.

a name – these are *semidefinite programs*, the subject of *Semidefinite Programming*, sometimes referred to as the "Linear Programming of the twenty-first century." We speak about Semidefinite Programming in more detail in sections 18.4 and 23.6 and exercises IV.20, IV.21, IV.26, IV.28, and IV.35, among others. It is important to note that the "expressive abilities" of Semidefinite Programming are extremely strong, and for all practical purposes, it covers basically all the convex problems arising in applications. The ultimate reason for this universality is the second-to-none richness of the family of convex sets which we obtain by intersecting positive semidefinite cones with affine planes.

Schur Complement Lemma. The Schur Complement Lemma is the following simple and extremely useful fact:

Proposition D.33 *[Schur Complement Lemma] Consider a symmetric 2×2 block matrix*

$$A = \begin{bmatrix} P & Q \\ \hline Q^\top & R \end{bmatrix}$$

with $R \succ 0$. Then, A is positive definite (semidefinite) if and only if the matrix

$$P - QR^{-1}Q^\top$$

is positive definite (respectively, semidefinite).

Proof Suppose $P \in \mathbf{S}^p$, $R \in \mathbf{S}^q$. Then, splitting the vector x from $\mathbf{R}^{p+q}$ into blocks $u \in \mathbf{R}^p$, $v \in \mathbf{R}^q$, we arrive at

$$A \succeq 0$$
$$\iff u^\top P u + 2 u^\top Q v + v^\top R v \geq 0, \quad \forall (u, v)$$
$$\iff \min_v \left\{ v^\top R v + 2 v^\top Q u + u^\top P u \right\} \geq 0, \quad \forall u$$
$$\iff \min_v \left\{ (v + R^{-1} Q u)^\top R (v + R^{-1} Q u) + u^\top P u - u^\top Q R^{-1} Q^\top u \right\} \geq 0, \quad \forall u$$
$$\iff u^\top P u - u^\top Q R^{-1} Q^\top u + \min_v \underbrace{\left\{ (v + R^{-1} Q u)^\top R (v + R^{-1} Q u) \right\}}_{\geq 0 \text{ as } R \succ 0} \geq 0, \quad \forall u$$
$$\iff P - Q R^{-1} Q^\top \succeq 0.$$

Here, in the last step we observe that for any $u \in \mathbf{R}^p$, by selecting $v = -R^{-1} Q u$ we see that the optimum value of the minimization problem is equal to zero. As we concluded $P - Q R^{-1} Q^\top \succeq 0$, this proves the "positive semidefinite" version of Schur Complement Lemma. The same reasoning, with evident modifications, justifies the "positive definite" version. ∎

D.3 Proofs of Facts

Fact D.1 *Let X, Y be rectangular matrices such that XY makes sense and is a square matrix. Then, $\mathrm{Tr}(YX)$ also makes sense and*

$$\mathrm{Tr}(XY) = \mathrm{Tr}(YX).$$

Proof In order for XY to make sense and be a square matrix, the sizes of the matrices should be $m \times n$ (for X) and $n \times m$ (for Y) with some m and n, We now have

$$\mathrm{Tr}(XY) = \sum_{i=1}^{m} \sum_{j=1}^{n} X_{ij} Y_{ji}, \quad \mathrm{Tr}(YX) = \sum_{j=1}^{n} \sum_{i=1}^{m} Y_{ji} X_{ij},$$

that is, both quantities are the same. ∎

Fact D.2 *If $X, Y \in \mathbf{R}^{m \times n}$, the Frobenius inner product of X and Y is equal to the Frobenius inner product of $X^\top$ and $Y^\top$:*

$$\mathrm{Tr}(XY^\top) = \mathrm{Tr}\left((X^\top)(Y^\top)^\top\right).$$

Moreover, when U is an orthogonal $m \times m$ matrix (i.e., $UU^\top = U^\top U = I_m$, or, which is the same, $U^{-1} = U^\top$), and V is an orthogonal $n \times n$ matrix (i.e., $VV^\top = V^\top V = I_n$), the Frobenius inner product of UXV and UYV is the same as the Frobenius inner product of X and Y:

$$\mathrm{Tr}(XY^\top) = \mathrm{Tr}\left((UXV)(UYV)^\top\right).$$

Proof We have $\mathrm{Tr}((X^\top)(Y^\top)^\top) = \mathrm{Tr}(X^\top Y) = \mathrm{Tr}((X^\top Y)^\top) = \mathrm{Tr}(Y^\top X) = \mathrm{Tr}(XY^\top)$, where the concluding inequality is given by Fact D.1. Similarly, given orthogonal matrices $U \in \mathbf{R}^{m \times m}$ and $V \in \mathbf{R}^{n \times n}$, we have

$$\mathrm{Tr}((UXV)(UYV)^\top) = \mathrm{Tr}(UX(VV^\top)Y^\top U^\top) = \mathrm{Tr}(XY^\top(U^\top U)) = \mathrm{Tr}(XY^\top),$$

where the second and the last equalities follow from the fact that U, V are orthogonal matrices and thus $V^\top V = VV^\top = I_n$ and $U^\top U = UU^\top = I_m$, and the second equality also uses Fact D.1. ∎

Fact D.9 *The matrices $A_1, \ldots, A_k \in \mathbf{S}^n$ commute with each other ($A_i A_j = A_j A_i$ for all i, j) if and only if they can be simultaneously diagonalized, i.e., if there exist a single orthogonal matrix U and diagonal matrices $\Lambda_1, \ldots, \Lambda_k$ such that*

$$A_i = U \Lambda_i U^\top, \qquad i = 1, \ldots, k.$$

Proof The fact that when the matrices are simultaneously diagonalizable, then they commute, is evident: for diagonal matrices D, E and an orthogonal matrix U we have

$$(U^\top D U)(U^\top E U) = U^\top D (U^\top U) E U = U^\top D E U,$$

and similarly $(U^\top E U)(U^\top D U) = U^\top E D U$; it remains to note that $ED = DE$ as D, E are diagonal matrices.

In the opposite direction, let us carry out an induction on the number k of commuting symmetric matrices. There is nothing to prove when $k = 1$, or, more precisely, the case of $k = 1$ is covered by the results on the eigenvalue decomposition of a symmetric matrix. Inductive step: Assuming the statement is valid for some k and given $k + 1$ commuting matrices $A_1, \ldots, A_{k+1}$, let $\lambda_1, \ldots, \lambda_s$ be distinct eigenvalues of A_{k+1} and $E_1, \ldots, E_s$ be the corresponding spectral subspaces of A_{k+1}, that is, restricted onto E_ℓ, the mapping $x \mapsto A_{k+1} x$ acts as multiplication by λ_ℓ. By Fact D.10, the subspaces E_ℓ are invariant subspaces for the matrices $A_1, \ldots, A_k$. Then, by the inductive hypothesis, we deduce that for every $\ell \leq s$ there exists an orthonormal basis in E_ℓ whose basis vectors are the eigenvectors of every one of $A_1, \ldots, A_k$. In addition, these vectors, like all vectors from E_ℓ, are eigenvectors of A_{k+1}. So, the union over $\ell \leq s$ of vectors from these orthonormal bases forms an orthonormal basis in the entire space (recall that $E_1, \ldots, E_s$ are mutually orthogonal, and their sum is the entire $\mathbf{R}^n$), and in this basis all matrices $A_1, \ldots, A_{k+1}$ become diagonal. ∎

Fact D.10 *Let $A, B \in \mathbf{R}^{n \times n}$ be commuting and let λ be a real eigenvalue of A. Then, the spectral subspace $E = \{x \in \mathbf{R}^n : Ax = \lambda x\}$ of A corresponding to λ is an invariant subspace for B (i.e., $Be \in E$ for every $e \in E$).*

Proof For $e \in E$ we have $A(Be) = B(Ae) = B(\lambda e) = \lambda(Be)$, that is $Be \in E$. ∎

Fact D.11 *If A is an $n \times n$ matrix and L is an invariant subspace for A (i.e., L is a linear subspace such that $Ae \in L$ whenever $e \in L$), then the orthogonal complement $L^\perp$ of L is an invariant subspace for the matrix $A^\top$. In particular, if A is symmetric and L is an invariant subspace for A, then $L^\perp$ is an invariant subspace for A as well.*

Proof Suppose L is an invariant subspace for A. Recall that $x \in L^\perp$ if and only if $x^\top y = 0$ for all $y \in L$. Therefore, for all $x \in L^\perp$ we have

$$y \in L \implies Ay \in L \implies 0 = x^\top Ay = (A^\top x)^\top y,$$

where the first implication follows since L is an invariant subspace for A, and the second implication follows since $Ay \in L$ and $x \in L^\perp$. Then, from $(A^\top x)^\top y = 0$ for all $y \in L$ we deduce that $A^\top x \in L^\perp$. ∎

Fact D.19

(i) *Let E and F be finite-dimensional linear spaces. The induced norm $\|\cdot\|_{F,E}$ is indeed a norm on the space $\mathrm{Lin}(E, F)$ of linear mappings from E to F.*

(ii) *Let E, F, G be finite-dimensional linear spaces equipped with the norms $\|\cdot\|_E, \|\cdot\|_F, \|\cdot\|_G$ respectively. Let $y = Ax : E \to F$ and $z = By : F \to G$ be linear mappings. Then,*

$$\|BA\|_{G,E} \leq \|B\|_{G,F} \|A\|_{F,E}.$$

Proof (i): The homogeneity and positivity of $\|\cdot\|_{F,E}$ are evident. To prove the triangle inequality, consider $A, B \in \mathrm{Lin}(E, F)$. Then, for any x we have

$$\|(A + B)x\|_F = \|Ax + Bx\|_F \leq \|Ax\|_F + \|Bx\|_F \leq \left(\|A\|_{F,E} + \|B\|_{F,E}\right)\|x\|_E.$$

This implies $\|(A + B)\|_{F,E} \leq \|A\|_{F,E} + \|B\|_{F,E}$ as desired.

(ii): Indeed, for every $x \in E$ we have $\|BAx\|_G \leq \|B\|_{G,F} \|Ax\|_F \leq \|B\|_{G,F} \|A\|_{F,E} \|x\|_E$. ∎

Fact D.20 *Let $\|\cdot\|$ be the spectral norm on $\mathbf{R}^{m \times n}$.*

(i) *For any $A \in \mathbf{R}^{m \times n}$, we have*

$$\|A\| = \max_{x,y} \left\{ y^\top A x : \|x\|_2 \leq 1, \|y\|_2 \leq 1 \right\},$$

hence also $\|A\| = \|A^\top\|$.

(ii) *For any $A \in \mathbf{S}^n$, we have $\|A\| = \max\{|\lambda_{\max}(A)|, |\lambda_{\min}(A)|\}$. Moreover, for any $A \in \mathbf{R}^{m \times n}$ we have*

$$\|A\|^2 = \|A^\top A\| = \lambda_{\max}(A^\top A) = \lambda_{\max}(AA^\top) = \|AA^\top\| = \|A^\top\|^2.$$

Proof (i): Indeed, $\|Ax\|_2 = \max_y \{y^\top Ax : \|y\|_2 \leq 1\}$ by Cauchy's inequality (Theorem B.1), implying that $\max_x \{\|Ax\|_2 : \|x\|_2 \leq 1\} = \max_{x,y} \{y^\top Ax : \|x\|_2 \leq 1, \|y\|_2 \leq 1\}$.

(ii): Indeed, let $A = U \operatorname{Diag}\{\lambda\} U^\top$ be the eigenvalue decomposition of the symmetric matrix A. Hence, $y^\top Ax = (U^\top y)^\top \operatorname{Diag}\{\lambda\}(U^\top x)$ with orthogonal U. Thus,

$$\max_{x,y} \{y^\top Ax : \|x\|_2 \leq 1, \|y\|_2 \leq 1\}$$
$$= \max_{u,v} \{v^\top \operatorname{Diag}\{\lambda\} u : \|u\|_2 \leq 1, \|v\|_2 \leq 1\}$$
$$= \max_{u,v} \left\{ \sum_i \lambda_i u_i v_i : \|u\|_2 \leq 1, \|v\|_2 \leq 1 \right\}$$
$$\leq (\max_i |\lambda_i|) \max_{u,v} \left\{ \sum_i |u_i v_i| : \|u\|_2 \leq 1, \|v\|_2 \leq 1 \right\}$$
$$\leq \max_i |\lambda_i|,$$

where the last inequality is due to Cauchy's inequality. Note also that $y^\top Ax = \max_i |\lambda_i|$ when x is the unit norm eigenvector of A corresponding to the maximum-magnitude eigenvalue, λ_{i_*}, of A, and $y = \operatorname{sign}(\lambda_{i_*})x$. Thus, when A is symmetric, we do have $\|A\| = \max_i |\lambda_i(A)|$.

For an arbitrary matrix A we have
$$\|A\|^2 = \max_{x:\|x\|_2 \leq 1} \|Ax\|_2^2 = \max_{x:\|x\|_2 \leq 1} x^\top (A^\top A)x = \lambda_{\max}(A^\top A),$$

where the last equality is given by the variational characterization of the largest eigenvalue of the symmetric positive semidefinite matrix $A^\top A$ (Theorem D.12). Moreover, $\|A^\top A\| = \lambda_{\max}(A^\top A)$ due to the already proved "symmetric" part of (ii). Thus, $\|A\| = \sqrt{\|A^\top A\|} = \sqrt{\lambda_{\max}(A^\top A)}$, Recalling that $\|A\| = \|A^\top\|$, we complete the verification of (ii). ∎

Fact D.21 *The vector-valued function $A \mapsto \lambda(A) : \mathbf{S}^n \to \mathbf{R}^n$ is Lipschitz continuous; specifically, denoting by $\|\cdot\|$ the spectral norm, for all $k \leq n$ we have*
$$|\lambda_k(A) - \lambda_k(A')| \leq \|A - A'\|, \quad \forall (A, A' \in \mathbf{S}^n).$$

Proof Indeed, by the VCE, denoting by $\mathcal{E}_k$ the family of all subspaces of codimension $k - 1$ in $\mathbf{R}^n$, we have
$$\lambda_k(A) = \min_{E \in \mathcal{E}_k} \max_{e \in E : \|e\|_2 = 1} e^\top A e,$$

and the right-hand side is clearly a Lipschitz continuous, with constant 1, w.r.t. $\|\cdot\|$, function of A.[1] ∎

---

[1] The justification of "clearly" is based upon two nearly evident facts (check them!): (1) whenever $e \in \mathbf{R}^m, h \in \mathbf{R}^n$, the function $e^\top B h$ is a Lipschitz continuous, with constant $\|h\|_2 \|e\|_2$ w.r.t. $\|\cdot\|$, function of $B \in \mathbf{R}^{m \times n}$, and (2) let $Q \subseteq \mathbf{R}^k$, $\|\cdot\|$ be a norm on $\mathbf{R}^k$, and $\{f_\alpha(\cdot) : \alpha \in \mathcal{A} \neq \varnothing\}$ be a family of real-valued functions on Q that are Lipschitz continuous, with common constant L with respect to $\|\cdot\|$. If the function $f(x) = \sup_{\alpha \in \mathcal{A}} f_\alpha(x)$ is finite at some point of Q, it is finite everywhere on Q and is Lipschitz continuous, with constant L, w.r.t. $\|\cdot\|$. It should be added that in the latter statement, Q can be replaced with an abstract metric space (a nonempty set equipped with distance

Fact D.23 *Let A be a positive semidefinite $m \times m$ matrix, and B be a positive semidefinite matrix such that $B^2 = A$. Then, $B = A^{1/2}$.*

Proof When $B^2 = A$, the matrix B clearly commutes with A, implying by Fact D.9 that the matrices admit simultaneous diagonalization, i.e.,

$$A = U \operatorname{Diag}\{\lambda_1, \ldots, \lambda_m\} U^\top, \quad B = U \operatorname{Diag}\{\upsilon_1, \ldots, \upsilon_m\} U^\top$$

where U is an orthogonal matrix. Thus, the relation $A = B^2$ reads

$$\begin{aligned}
U \operatorname{Diag}\{\lambda_1, \ldots, \lambda_m\} U^\top &= (U \operatorname{Diag}\{\upsilon_1, \ldots, \upsilon_m\} U^\top)(U \operatorname{Diag}\{\upsilon_1, \ldots, \upsilon_m\} U^\top) \\
&= U \operatorname{Diag}\{\upsilon_1, \ldots, \upsilon_m\} \underbrace{(U^\top U)}_{=I_m} \operatorname{Diag}\{\upsilon_1, \ldots, \upsilon_m\} U^\top \\
&= U \operatorname{Diag}\{\upsilon_1^2, \ldots, \upsilon_m^2\} U^\top,
\end{aligned}$$

and so $\lambda_i = \upsilon_i^2$ for all i. Moreover, B is positive semidefinite, implying that $\upsilon_i \geq 0$. The bottom line is that $\upsilon_i = \lambda_i^{1/2}$ for all i, that is, $B = A^{1/2}$. ∎

Fact D.24 *Let $f(x)$ be a continuously differentiable real-valued function on an interval (a, b), (where $-\infty \leq a < b \leq +\infty$) of the real axis. The function $\phi(X) = \operatorname{Tr}(f(X))$ is continuously differentiable on the open set $\mathcal{D}$ of $\mathbf{S}^m$ composed of all matrices with spectrum from (a, b), and*

$$\left.\frac{d}{dt}\right|_{t=0} \phi(X + tH) = \operatorname{Tr}(f'(X)H) \quad \forall X \in \mathcal{D}, H \in \mathbf{S}^m.$$

In other words, $\phi(\cdot)$ is continuously differentiable on the open set $\mathcal{D}$ and its gradient, taken w.r.t. the Frobenius Euclidean structure on $\mathbf{S}^m$, is $\nabla \phi(X) = f'(X)$.

Hint: In order to prove Fact D.24, it makes sense to verify its validity for algebraic polynomials and then use the fact that a continuously differentiable real-valued function on an interval (a, b) of the real axis is the limit, in the sense of uniform convergence along with the first-order derivative on compact subsets of (a, b), of a sequence of polynomials.

Proof Following the Hint, let us start with the case when $f(x) = \sum_{i=0}^{k} a_i x^i$ is an algebraic polynomial. In this case the matrix-valued function $f(X) = a_0 I_m + a_1 X + \cdots + a_k X^k : \mathbf{S}^m \to \mathbf{S}^m$ is clearly infinitely many times differentiable, and for all $X, H \in \mathbf{S}^m$ we have

$$\left.\frac{d}{dt}\right|_{t=0} f(X + tH) = \sum_{i=1}^{k} a_i \Big(\sum_{j=1}^{i} X^{j-1} H X^{i-j}\Big),$$

and thus

$$\begin{aligned}
\left.\frac{d}{dt}\right|_{t=0} \phi(X + tH) &= \left.\frac{d}{dt}\right|_{t=0} \operatorname{Tr}(f(X + tH)) \\
&= \operatorname{Tr}\left(\left.\frac{d}{dt}\right|_{t=0} f(X + tH)\right) \qquad [\text{as } \operatorname{Tr}(Z) \text{ is a linear function of } Z]
\end{aligned}$$

$d(\cdot, \cdot)$, possessing properties 1–3, see the discussion on Euclidean distance between a pair of points in $\mathbf{R}^n$ on page 369, between points of Q, with Lipschitz continuity, with constant L w.r.t. d, of the function $f : Q \to \mathbf{R}$ meaning that $|f(x) - f(x')| \leq L d(X, x')$ for all $x, x' \in Q$).

$$= \mathrm{Tr}\left(\sum_{i=1}^{k} a_i \Big(\sum_{j=1}^{i} X^{j-1} H X^{i-j}\Big)\right)$$

$$= \sum_{i=1}^{k} a_i \sum_{j=1}^{i} \mathrm{Tr}\big(X^{j-1} H X^{i-j}\big)$$

$$= \sum_{i=1}^{k} a_i \sum_{j=1}^{i} \mathrm{Tr}(H X^{i-1}) \qquad \text{[by Fact D.1]}$$

$$= \mathrm{Tr}\left(H \Big(\sum_{i=1}^{k} a_i i X^{i-1}\Big)\right)$$

$$= \mathrm{Tr}(H f'(X)).$$

Next, Corollary D.22 implies that $\mathcal{D}$ is an open set. Now let f be a continuously differentiable real-valued function on the interval $(a,b) \subseteq \mathbf{R}$, and $f_t(x)$, $t = 1, 2, \ldots$, be polynomials such that $f_t(\cdot)$ and $f'_t(\cdot)$ converge as $t \to \infty$, uniformly on every segment $[a', b'] \subset (a, b)$, to $f(\cdot)$ and $f'(\cdot)$, respectively; it is well known that such a sequence exists. Taking into account that the spectral norm (i.e., the maximum of the magnitudes of the eigenvalues) of $g(X)$ is upper bounded by the uniform norm of $g(\cdot)$ on the spectrum of the matrix X, i.e., $\sigma(X)$, we immediately conclude that as $t \to \infty$ the functions $\phi_t(x) := \mathrm{Tr}(f_t(X))$ converge to $\phi(X)$ uniformly on closed and bounded subsets of $\mathcal{D}$, and the gradients $\nabla \phi_t(X) = f'_t(X)$ converge to $f'(X)$ uniformly on closed and bounded subsets of $\mathcal{D}$. By the standard results of Analysis, it follows that $\phi(\cdot)$ is continuously differentiable on $\mathcal{D}$ and its gradient is given by $\nabla \phi(X) = f'(X)$. ∎

Fact D.31 *For any $A, B \in \mathbf{S}^n$ such that $A \succeq B$, and for any $V \in \mathbf{R}^{n \times m}$, we always have*

$$V^\top A V \succeq V^\top B V.$$

Proof Indeed, we should prove that if $A - B \succeq 0$, then also $V^\top (A - B) V \succeq 0$. This is immediate as the quadratic form $y^\top (V^\top (A - B) V) y = (Vy)^\top (A - B)(Vy)$ of y is nonnegative along with the quadratic form $x^\top (A - B) x$ of x. ∎

D.4 Exercises

Exercise D.1 1. Find the dimension of $\mathbf{R}^{m \times n}$ and point out a basis in this space.
2. Build an orthonormal basis in $\mathbf{S}^m$.

Exercise D.2 In the space $\mathbf{R}^{m \times m}$ of square $m \times m$ matrices, there are two interesting subsets: the set $\mathbf{S}^m$ of *symmetric* matrices $\{A : A = A^\top\}$ and the set $\mathbf{J}^m$ of *skew-symmetric* matrices $\{A = [A_{ij}] : A_{ij} = -A_{ji}, \forall i, j\}$.

1. Verify that both $\mathbf{S}^m$ and $\mathbf{J}^m$ are linear subspaces of $\mathbf{R}^{m \times m}$.
2. Find the dimension of $\mathbf{S}^m$ and point out a basis in $\mathbf{S}^m$.
3. Find the dimension of $\mathbf{J}^m$ and point out a basis in $\mathbf{J}^m$.
4. What is the sum of $\mathbf{S}^m$ and $\mathbf{J}^m$? What is the intersection of $\mathbf{S}^m$ and $\mathbf{J}^m$?

Exercise D.3 Is the "three-factor" extension of Fact D.1 valid, at least in the case of square matrices X, Y, Z of the same size? That is, for square matrices X, Y, Z of the same size, is it always true that $\text{Tr}(XYZ) = \text{Tr}(YXZ)$?

Exercise D.4 Given $P \in \mathbf{S}^p$, $Q \in \mathbf{R}^{r \times p}$, and $R \in \mathbf{S}^r$, consider the matrices

$$A = \begin{bmatrix} P & Q^\top \\ Q & R \end{bmatrix}, \quad B = \begin{bmatrix} P & -Q^\top \\ -Q & R \end{bmatrix}, \quad C = \begin{bmatrix} R & Q \\ Q^\top & P \end{bmatrix}, \quad D = \begin{bmatrix} R & -Q \\ -Q^\top & P \end{bmatrix}.$$

Prove that $\lambda(A) = \lambda(B) = \lambda(C) = \lambda(D)$. Thus, the matrices A, B, C, D simultaneously are/are not positive semidefinite. As a consequence, the Schur Complement Lemma says that when $R \succ 0$, we have $A \succeq 0$ if and only if $P - Q^\top R^{-1} Q \succeq 0$; since $A \succeq 0$ if and only if $C \succeq 0$, we see that the same lemma says that when $P \succ 0$, we have $A \succeq 0$ if and only if $R - Q P^{-1} Q^\top \succeq 0$.

Exercise D.5 Let $\mathbf{S}^n_{++} := \text{int}\, \mathbf{S}^n_+ = \{X \in \mathbf{S}^n : X \succ 0\}$, and consider $X, Y \in \mathbf{S}^n_{++}$. Then, $X \preceq Y$ holds if and only if $X^{-1} \succeq Y^{-1}$ (the $\succeq$-antimonotonicity of X^{-1}, $X \in \mathbf{S}^n_{++}$). Is it true that from $0 \prec X \preceq Y$ it always follows that $X^{-2} \succeq Y^{-2}$?

Exercise D.6 Let $A, B \in \mathbf{S}^n$ be such that $0 \preceq A \preceq B$. For each one of the following, either prove the statement or produce a counterexample:

1. $A^2 \preceq B^2$;
2. $0 \preceq A^{1/2} \preceq B^{1/2}$.

Exercise D.7 A matrix $A \in \mathbf{S}^n$ is called *diagonally dominant* if it satisfies the relation

$$a_{ii} \geq \sum_{j \neq i} |a_{ij}|, \quad i = 1, \ldots, n.$$

Prove that every diagonally dominant matrix A is positive semidefinite.

Exercise D.8 Prove the following matrix analogy of the scalar inequality $ab \leq \frac{a^2 + b^2}{2}$ for $a, b \in \mathbf{R}$:

$$AB^\top + BA^\top \preceq AA^\top + BB^\top, \quad \forall A, B \in \mathbf{R}^{m \times n}.$$

Exercise D.9 1. Let I_k denote the $k \times k$ identity matrix, and let A be an $m \times n$ matrix. Prove that the following three properties are equivalent to each other:
- $A^\top A \preceq I_n$;
- $AA^\top \preceq I_m$;
- $\begin{bmatrix} I_m & A \\ A^\top & I_n \end{bmatrix} \succeq 0.$

2. Let $A_1, \ldots, A_k$ be $n \times n$ matrices such that

$$A_1^\top A_1 + \cdots + A_k^\top A_k \preceq I_n.$$

For each of the following, either prove the statement or produce a counterexample:
- $A_1 A_1^\top + \cdots + A_k A_k^\top \preceq I_n$;
- $\begin{bmatrix} A_1 A_1^\top & A_1 A_2^\top & \cdots & A_1 A_k^\top \\ A_2 A_1^\top & A_2 A_2^\top & \cdots & A_2 A_k^\top \\ \vdots & \vdots & \ddots & \vdots \\ A_k A_1^\top & A_k A_2^\top & \cdots & A_k A_k^\top \end{bmatrix} \preceq I_{kn}.$

References

[Axl15] S. Axler, *Linear algebra done right, 3rd edition*, Undergraduate Texts in Mathematics, Springer, 2015.

[BNO03] D. Bertsekas, A. Nedic, and A. Özdağlar, *Convex analysis and optimization*, Athena Scientific, 2003.

[BTN] A. Ben-Tal and A. Nemirovski, *Lectures on modern convex optimization: analysis, algorithms, and engineering applications*, SIAM 2001 and https://www.isye.gatech.edu/~nemirovs/LMCOLN2024Spring.pdf 2023.

[BV04] S. Boyd and L. Vandenberghe, *Convex optimization*, Cambridge University Press, 2004.

[Edw12] C. H. Edwards, *Advanced calculus of several variables*, Courier Corporation, 2012.

[Gel89] I. M. Gelfand, *Lectures on linear algebra*, Dover Books on Mathematics, Dover Publications, 1989.

[HUL93] J.-B. Hiriart-Urruty and C. Lemarechal, *Convex analysis and minimization algorithms, I: Fundamentals, II: Advanced theory and bundle methods*, Springer, 1993.

[IT79] A. D. Ioffe and V. M. Tikhomirov, *Theory of extremal problems*, Nauka, 1974 (in Russian). English translation: Studies in Mathematics and its Applications, v. 6, North-Holland, 1979.

[Nem24] A. Nemirovski, *Introduction to linear optimization*, World Scientific, 2024.

[Nes18] Yu. Nesterov, *Lectures on convex optimization, 2nd edition*, Springer, 2018.

[Pas22] D. S. Passman, *Lectures on linear algebra*, World Scientific, 2022.

[Roc70] R. T. Rockafellar, *Convex analysis*, Princeton University Press, 1970.

[Rud13] W. Rudin, *Principles of mathematical analysis, 3rd edition*, McGraw Hill India, 2013.

[Str06] G. Strang, *Linear algebra and its applications*, Belmont, CA: Thomson, Brooks/Cole, 2006.

Index

Aff($\cdot$), affine hull, 356
Argmin, 175
a.k.a., also known as, xviii
affine
  basis, 360
  combination, *see* combination, affine
  dependence/independence, 358
  dimension
    of affine subspace, 360
    of set, 28
  hull, 356
    structure of, 358
  plane, 364
  span, 356
  subspace, 356
    representation by linear equations, 361
affinely
  independent set, 358
  spanning set, 358
annulator, *see* orthogonal complement

bd, boundary, 15
rbd, relative boundary, 17
ball
  Euclidean, 5, 23
  in norm, 5
  unit, 5
base
  of cone, 106
basic orth, *see* standard basic orth, 345
basis
  in $\mathbf{R}^n$, 346
  in linear subspace, 346
  orthonormal, 353
    existence of, 354
boundary, 15
  relative, 17

Cone($\cdot$), conic hull, 11
ConeT($\cdot$), conic transform, 20
Conv($\cdot$), convex hull, 8
cl, closure, 14
$\overline{\text{ConeT}}(\cdot)$, closed conic transform, 23
characteristic polynomial, 403
closure, 14

combination
  affine, 357
  conic, *see* conic combination
  convex, *see* convex combination
  linear, *see* linear combination
combination, linear, *see* linear combination
complementary slackness
  in convex programming, 265
  in linear programming, 58
cone, 9
  definition of, 9
  dual, 99, 101
  ice cream, 10
  Lorentz, 10, 239, 248
  normal, 177
  pointed, 101, 105
  polyhedral cone, 10, 127
    extreme rays of, 125
  positive semidefinite, 10, 414
    self-duality of, 415
  radial, 176
  recessive, 95
  regular, 239
  second-order, *see* cone, Lorentz
  self-dual, 102
cone-convexity, 239, 277
cone-monotonicity, 278
conic
  combination, 10
  duality, 255
    theorem, 255
  hull, 11
  program, 255
  sets, 9
Conic Duality Theorem, 256
conic programming, 255
conic transform, 20
  closed, 23
constraints
  active, 235
  of optimization program, 52, 235
convex
  combination, 7
  hull, 8

problem/program, 236
    in cone-constrained form, 239, 252
    sets, *see* sets, convex
    definition of, 3
Convex Programming Duality Theorem, 249
    in cone-constrained form, 253
convex programming optimality conditions
    in Karush–Kuhn–Tucker form, 267
    in saddle point form, 264
Convex Theorem of the Alternative, 239
    in cone-constrained form, 243
convexity
    criterion for smooth functions, 165
    of function, 153
    of function, strict, 153
    of set, 3
    strict, sufficient condition for, 166
coordinates, 347
    barycentric, 361

Det($\cdot$), determinant, 348
Dg$\{X\}$, vector of diagonal entries of square matrix X, xviii
Diag$\{X_1, \ldots, X_K\}$, block-diagonal matrix with diagonal blocks $X_1, \ldots, X_K$, xviii
Diag$\{x\}$, Diagonal matrix with entries of vector x on the diagonal, xviii
Dom, domain (of a function), 157
dim, dimension, 28
derivatives, 385
    calculus of, 391
    computation of, 392
    directional, 387
    directionsl of convex function, 199
    existence, 391
    of higher order, 394
    representation of, 389
determinant, 348
differential, *see* derivatives
dimension, 346
    of a set, 28
dimension formula, 347
dual of optimization program, *see* Lagrange dual

eigenvalue decomposition
    of a symmetric matrix, 403
    simultaneous, 405
eigenvalues, 403
    variational description of, 405
    vector of, 404
        Lipschitz continuity, 409
eigenvectors, 403
ellipsoid, 7
Euclidean
    norm, 367
    structure, 351
Euclidean distance, 369

extreme
    direction, 106
    ray, 106
extreme points, 91
    of polyhedral sets, 118

Farkas' Lemma
    Homogeneous, 45
    Inhomogeneous, 50
feasibility of optimization problem
    essentially strict, 238
    strict, 238
Fourier–Motzkin elimination, 40
function, affine minorant of, 187
functions
    affine, 153
    characteristic, a.k.a. indicator, 157
    cone-convex, 239, 277
        calculus of, 277
    continuous, 375
        basic properties of, 377
        calculus of, 376
    convex
        below boundedness of, 171
        calculus of, 159
        closures of, 190
        convexity of sublevel sets of, 156
        definition of, 153
        differential criteria of convexity, 161
        domains of, 156
        elementary properties of, 155
        Fenchel conjugates of, 203
        Karush–Kuhn–Tucker optimality, conditions for minimizers of, 178
        Legendre transforms of, 203
        Lipschitz continuity of, 171
        lower semicontinuous, 184
        maxima and minima of, 175
        proper, 184
        subdifferential of, 193
        subgradients of, 193
        values outside the domains of, 156
    differentiable, 385–408
    domains of, 379
    epigraphs of, 380
    hypograph, 381
    Lipschitz continuity of, 171
    lower semicontinuous, 185, 380
    lsc, a.k.a. lower semicontinuous, 185, 380
    Minkowski, 224
    of eigenvalues of symmetric matrices, 212
    of symmetric matrices, 410
    permutation symmetric, 167, 212
    positive homogeneity of, 155
    proper, 379
    regular, 186
    spectral, 212
    subadditive, 155

sublevel sets of, 156
superlevel sets of, 305
support, 223
taking values in extended real axis, 379
uniformly continuous, 377
upper semicontinuous, 381
usc, a.k.a. upper semicontinuous, 381

General Theorem of the Alternative, 49
gradient, 389
Gram–Schmidt orthogonalization, 354
GTA, *see* General Theorem of the Alternative

Hessian, 398
hull
   affine, *see* affine hull
   conic, *see* conic hull
   convex, see convex hull8, *see* convex hull
hypercube, 6
hyperoctahedron, 5
hyperplane, 364
   supporting, 90

Im A, image $\{y : \exists x : y = Ax\}$ of matrix A, 350
int, interior, 14
iff, if and only if, xviii
   rint, relative interior, 16
image space of a matrtix, *see* Im
inequality
   Cauchy, 368
   gradient, 169
   Hölder, 208
   Jensen, 155
   moment, 209
   trace, 216
   triangle, 367
   Young, 207
inner product, 351
   Frobenius, 355, 402
   representation of linear forms, 352
interior, 14
   relative, 16

Ker$A$, kernel $\{h : Ah = 0\}$ of matrix A, 350
Karush–Kuhn–Tucker
   optimality condition for cone-constrained problem, 269
   optimality condition for inequality-constrained problem, 269
   point of cone-constrained problem, 269
   point of inequality-constrained problem, 269
kernel of a matrix, *see* Ker
KKT, *see* Karush–Kuhn–Tucker

Lin($\cdot$), linear span, 345
$\mathbf{L}^m$, Lorentz cone, 10

Lagrange
   dual, 250
      in cone-constrained form, 252
   duality, *see* Lagrange dual
   function, 249
      for problem in cone-constrained form, 252
      saddle points of, 251
   multipliers, 251
Legendre transform, 203
lemma
   $\mathcal{S}$-Lemma, 257
   $\mathcal{S}$-Lemma, Inhomogeneous, 259
   on Schur Complement, 418, 424
   Dubovitski–Milutin, 102, 103
   Minimax, 301
limit
   inferior, 371
   superior, 371
line, 363
linear
   combination, 7, 344
   dependence/independence, 345
   form, 351
   mapping, 347
      representation by matrices, 348
   subspace, 344–345
      basis of, 346
      dimension of, 346
      orthogonal complement of, 352
      representation by homogeneous linear equations, 362
linear combination, 344
linear programming (LP)
   Duality Theorem, 56
   necessary and sufficient optimality conditions, 58
   problem/program, 53
   program, 129
      bounded/unbounded, 129
      dual of, 54
      feasible/infeasible, 129
      solvability of, 130
      solvablc/unsolvable, 129
linear span, 345
linear subspace, parallel to affine subspace, 356
linearly dependent/independent set, 345
   spanning set, 347
local optimality, 284

majorization, 132
Majorization Principle, 133
mappings, *see* functions
mathematical programming program (MP), *see* optimization program
matrices
   doubly stochastic, 124
   Frobenius inner product of, 402
   negative (semi)definite, 414
   nonsingular, 349

orthogonal, 404
    positive (semi)definite, 407
        square roots of, 414
    spaces of, 402
    spectral norm of, 410
    symmetric, 403
        eigenvalue decomposition of, 403
        functions of, 410
minimizer
    global, 175
    local, 175

$\mathcal{N}(\mu, \Theta)$, 148
nonnegative orthant, 10
norm
    conjugate, 210
    convexity of, 155
    dual, 210
    induced, 409
    properties of, 367
    spectral, 409
nullspace of a matrix, see Ker

objective, of optimization program, 52, 235
optimal value, 52, 235
    in parametric convex problem, 269
optimality conditions
    in Karush–Kuhn–Tucker form, 267
    in linear programming, 58
    in saddle point form, 264
optimality, local, 284
optimization problem, linear, 53
optimization program, 52, 235
    below bounded, 52, 236
    constraints of, 52, 235
    convex, 236
    domain of, 52, 235
    feasible, 52, 235
    feasible set of, 52, 235
    feasible solutions of, 52, 235
    objective of, 52, 235
    optimal solutions of, 52, 235, 236
    optimal value of, 52, 235
    solvable, 52, 236
orthogonal
    complement, 352
    projection, 352
orths, standard, see standard basic orths

Persp($\cdot$), perspective transform, 20
    of a function, 160
pointed cone, 105
polar, of convex set, 108
polyhedral
    cone, 10
    representation, 38
    set, 4

polytope, 9
program, semidefinite, 323
proper
    convex function, 157

Rec($\cdot$), see recessive cone
$\mathbf{R}^m_+$, nonnegative orthant, 10
$\mathbf{R}^{m \times n}$, space of $m \times n$ real matrices, 402
$\mathbf{R}^n$, space of real column vectors of dimension n, 343
rank($\cdot$), rank, 349
recessive cone, 95
recessive direction, 95
regular cone, 239
regular solution, 284
representation, polyhedral, 38

$\mathbf{S}^m$, space of $m \times m$ real symmetric matrices, 402
$\mathbf{S}^m_+$, see cone, positive semidefinite, 414
saddle points, 251, 297
    existence of, 300
    game theory interpretation of, 297
    structure of the set of, 299
Schur Complement Lemma, see lemma on Schur Complement
self-dual cone, 102
semicontinuity
    lower and upper, 380
    upper, 381
separation
    strong, 84
    theorem, 85
sets
    closed, 371
    closure, 373
    compact, 373, 374
    conic, see conic sets
    convex, 3
        calculus of, 11
        topological properties of, 17
    interior, 372
    open, 371
    polyhedral, see polyhedral sets
        calculus of, 42
        extreme points of, 118
        structure of, 126
similarity transformation, 403
simplex, 9
Slater condition, , 238
    in cone-constrained form, 242
    relaxed, in cone-constrained form, 242
solution, regular, 284
span, linear, see linear span
standard basic orths, 345
subdifferential, 193
subgradient, 193

subspace, spectral, of matrix, 405
symmetry principle, 179

Taylor expansion, 400
Theorem
   Carathéodory, 28
      in conic form, 31
   Duality Theorem in LP, 55
   Eigenvalue Decomposition, 403
   Eigenvalue Interlacement, 408
   Hahn–Banach, 200
   Helly, 32
   Helly, II, 36
   Kirchberger, 71
   Krein–Milman, 92
      in conic form, 107
   of the Alternative, Convex, 239
      in cone-constrained form, *see* Convex Theorem of the Alternative in cone-constrained form
   of the Alternative, General, *see* General Theorem of the Alternative
   on convex programming duality, *see* Convex Programming Duality Theorem
   on linear programming duality, *see* Linear Programming Duality Theorem
   Radon, 31
   Separation, 85
   Shapley–Folkman, 70
   Sion–Kakutani, 300, 304, 305
trace inequality, 215
transform
   closed conic, 27
   conic, *see* conic transform
   perspective, *see* perspective transform

vertex, 91

w.l.o.g., without loss of generality, xviii
w.r.t., with respect to, xviii
weak duality
   in convex cone-constrained programming, 253
   in convex programming, 249
   in LP, 57

Printed in the United States
by Baker & Taylor Publisher Services